Recent Progress in Science and Technology

Recent Progress in Science and Technology

Edited by **Amber Hall**

Willford Press

New York

Published by Willford Press,
118-35 Queens Blvd., Suite 400,
Forest Hills, NY 11375, USA
www.willfordpress.com

Recent Progress in Science and Technology
Edited by Amber Hall

International Standard Book Number: 978-1-68285-308-5 (Hardback)

Printed in China.

Contents

Preface

Advances in science and technology have revolutionized the human existence in the past three decades. This book provides a comprehensive overview of the progress happening all around the world in the diverse fields of science and technology. It particularly delves into the discipline of materials science and discusses its industrial applications. A number of latest researches exploring various interdisciplinary scientific and technological advances have been included in this book.

This book is a result of research of several months to collate the most relevant data in the field.

When I was approached with the idea of this book and the proposal to edit it, I was overwhelmed. It gave me an opportunity to reach out to all those who share a common interest with me in this field. I had 3 main parameters for editing this text:

1. Accuracy – The data and information provided in this book should be up-to-date and valuable to the readers.
2. Structure – The data must be presented in a structured format for easy understanding and better grasping of the readers.
3. Universal Approach – This book not only targets students but also experts and innovators in the field, thus my aim was to present topics which are of use to all.

Thus, it took me a couple of months to finish the editing of this book.

I would like to make a special mention of my publisher who considered me worthy of this opportunity and also supported me throughout the editing process. I would also like to thank the editing team at the back-end who extended their help whenever required.

Editor

ACCELERATED METHODS FOR ESTIMATING THE DURABILITY OF PLAIN BEARINGS

Myron Czerniec[1]

[1] Institute of Technological Systems of Information, Lublin University of Technology, Nadbystrzycka 36, 20-612 Lublin, Poland, e-mail: wm.itsi@pollub.pl

ABSTRACT

The paper presents methods for determining the durability of slide bearings. The developed methods enhance the calculation process by even 100000 times, compared to the accurate solution obtained with the generalized cumulative model of wear. The paper determines the accuracy of results for estimating the durability of bearings depending on the size of blocks of constant conditions of contact interaction between the shaft with small out-of-roundedness and the bush with a circular contour. The paper gives an approximate dependence for determining accurate durability using either a more accurate or an additional method.

Keywords: enhanced numerical methods, durability evaluation, plain bearings, accuracy deviations.

INTRODUCTION

Plain bearings find numerous applications in machine design as well as in various devices that facilitate human life. The literature on the subject provides simplified methods [1–14] for estimating the durability of these bearings where it is assumed that the shaft and bush have an ideal (i.e. circular) section of contours. However, in reality these components of slide bearings are produced with out-of-roundness deviations (ovality, trilobing, tetralobing) that have – as demonstrated in [15, 16] – an effect on the durability of bearings.

The kinematics of wear of the shaft with out-of-roundedness can be investigated using either the cumulative model of wear [15–17] or the generalized cumulative model of wear [18] for the interval-discrete interaction between bearing components. The cumulative model of wear allows estimating the degree of wear and durability of a bearing for single-area contact of the components, while the generalized cumulative model of wear is applied to examine mixed contact.

The above models entail dividing the contour of the shaft with out-of-roundedness into discretization intervals $\Delta\alpha_2$, where both initial contact pressures and contact area are assumed to be constant. In a full rotation of the shaft with out-of-roundedness, there is j-th interaction with the bush ($j = 360°/\Delta\alpha_2$).

Bearing components undergo wear during rotation, hence the initial contact parameters change and their wear is cumulative.

With application of the interval-discrete pattern of tribocontact interaction, the time of computations is considerably long, particularly at low values of the discretization interval $\Delta\alpha_2$ of shaft contour that ensure the most accurate numerical solution of the problem. An increase in $\Delta\alpha_2$ leads to a proportional decrease in the estimated durability of a bearing. Nonetheless, the computation time still remains considerably long. In order to decrease effectively the time of computations, the express method [19] with an interval-range-block calculation pattern has been developed. The method is based on the assumption that parameters of contact interaction for particular intervals of discretization remain constant during a certain number of shaft rotations (block size). Such an approach ensures a shorter time of computations, proportionally to the size of a block with constant conditions of interaction between the bearing components. The present paper presents the results of this problem as solved by the express method using various calculation methods.

CALCULATION METHODS

The slide bearing (Fig. 1) carries the load N, while shaft 2 rotates with the angular velocity ω_2. The bearing has a radial clearance of $\varepsilon = R_1 - R_2$, where $2R_1$ denotes the nominal diameter of the bush and $2R_2$ denotes the diameter of the shaft. Under the action of load, the bearing components contact each other in the $2R_2\alpha_0$ area where contact pressures $p(\alpha)$ are generated. The bearing is made of different materials, whose properties and resistance to wear differ.

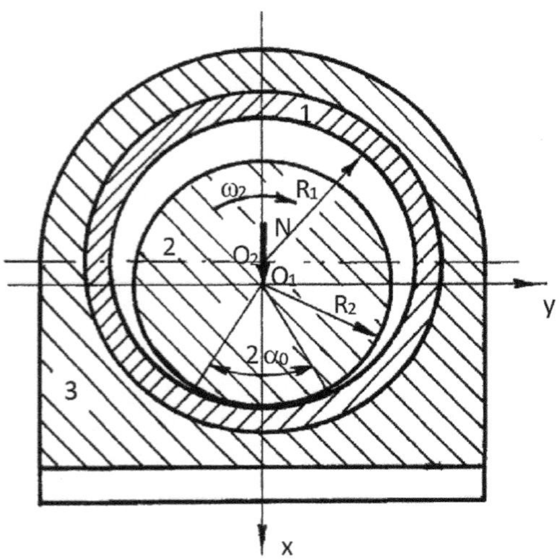

Fig. 1. Schematic diagram of a slide bearing

Due to the fact that the components of bearing exhibit low ovality, they produce either single-area (Fig. 2a) or double-area (Fig. 2b) contact. During a rotation of the shaft (α_2>0), two types of interactions are possible: full single-area and mixed (single – double – single-area). The interaction between bearing components with the contour ovality L_1 and L_2 is described by the following contact parameters:

- single area contact (Fig. 2a): $2\alpha_{0\delta}(\alpha_2)$ – variable contact angle, contact pressures $p(\alpha_2, \delta)$, maximum contact pressures $p(0, \delta)$, contact area $W = 2R_2\alpha_{0\delta}$;
- mixed contact (Fig. 2b): 2γ – symmetric contact angle or asymmetric contact angles $2\gamma_1(\alpha_2)$, $2\gamma_2(\alpha_2)$; symmetric contact pressures $p(\gamma, \delta)$ and asymmetric contact pressures $p(\alpha_2, \delta)$; symmetric contact area $W_{1,2} = 2R_2\gamma$ and asymmetric contact areas $W_1 = 2R_2\gamma_1$, $W_2 = 2R_2\gamma_2$.

The contact angle 2λ of bearing components is determined by the method discussed in the study [20]. In the case of the symmetric double-area contact (Fig. 2b), the forces $N_1 = N_2 = N/(2\cos\lambda)$.

However, for the asymmetric contact $N_1 \neq N_2$, $\lambda_1 \neq \lambda_2$, $2\lambda_1 \neq 2\lambda_2$, $p(\lambda_1, \delta) \neq p(\lambda_2, \delta)$ [17]. The ovality of contours of the bearing components is described by the deviations $\delta_1 = R_1 - R_1'$, $\delta_2 = R_2' - R_2$ (Fig. 2).

a)

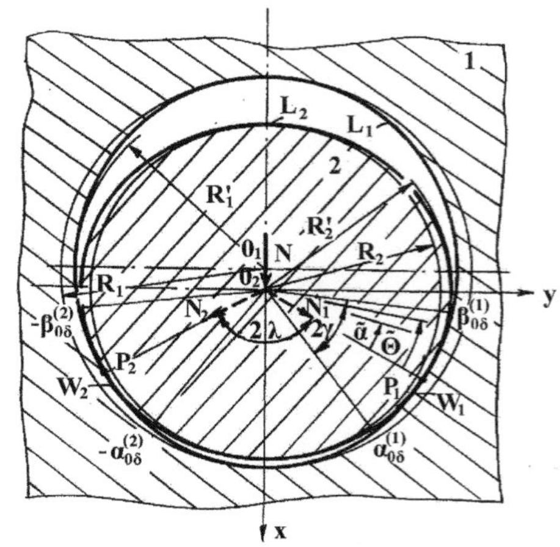

b)

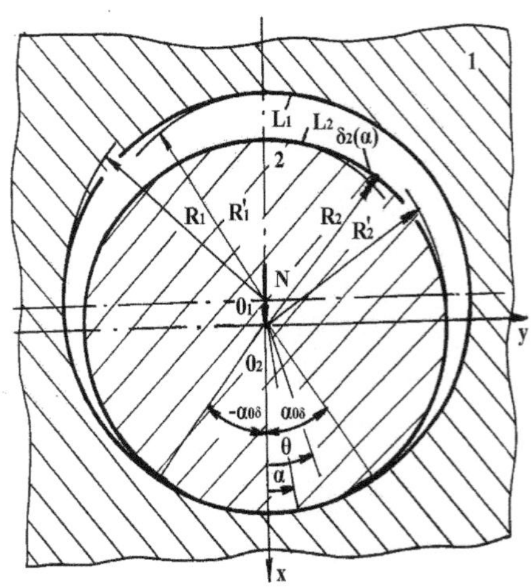

Fig. 2. Single-area (a) and double-area (b) contact of shaft and bush

Using the generalized cumulative model of wear [18], a numerical calculation was performed for the problem of full single-area contact and mixed contact of the bearing components, where the following parameters were applied: N = 0.1 MN; R_2 = 50 mm; v = 0.0628 m/sec – sliding velocity; f = 0,04 – sliding friction factor at boundary friction; ε = 0.21; 0.41 mm; δ_1 = 0 mm; δ_2 = 0; 0,05; 0,1; 0,15; 0,2; 0,3; 0,4 mm; $\delta_1 + \delta_2 \leq \varepsilon$; n_2 = 12 rpm – number of rotations of the shaft; h_{1*} = 0,3 mm – maximum wear of the bush; the shaft is made of steel (hardened + tempered), while the bush is made of bronze CuSn5Zn5Pb5; $\Delta\alpha_2 = 10^0$ – the discretization interval of the shaft contour; $B = 72000, 7200, 720, 12, 1$ rotations – block sizes.

RESULTS

The results of solving the problem are listed in:

- Tables 1, 2 – the accurate method based on the interval-block pattern ($B = 1$ rotation);
- Tables 3, 4 (numerator) – the approximate method based on the interval-block pattern ($B = 12, 720, 7200, 72\ 000$ rotations);
- Tables 3, 4 (denominator) – the more accurate method based on the results of the approximate solution.

An increase in radial clearance leads to decreased durability of the bearing.

Here, the actual wear h_1 of the bush is always somewhat higher than the maximum wear h_{1*} (Tables 3, 4).

The bearing durability h_{1*} was estimated more accurately based on its approximate durability n_2 and the corresponding wear $h_1 > h_{1*}$ of the bush in the following way: $n_{2*} = n_2(h_{1*}/h_1)$. Tables 5 and 6 give the deviations Δ_B of the solutions.

Table 1. Bearing durability n_{2*} as calculated by the accurate method ($\varepsilon = 0.21$ mm)

δ_2, mm	$B = 1$ rot	
	n_{2*}, rot	h_{1*}, mm
0.2	1 722 413	0.3
0.15	1 663 970	0.3
0.1	1 651 362	0.3
0.05	1 413 160	0.3
0.0	1 332 336	0.3

Table 2. Bearing durability n_{2*} as calculated by the accurate method ($\varepsilon = 0.41$ mm)

δ_2, mm	$B = 1$ rot	
	n_{2*}, rot	h_{1*}, mm
0.4	1 239 410	0.3
0.3	1 195 707	0.3
0.2	1 226 752	0.3
0.1	1 047 977	0.3
0	982 501	0.3

Table 3. Bearing durability n_{2*} calculated by the approximate method ($\varepsilon = 0.21$ mm)

δ_2, mm	$B = 72\ 000$ rot		
	n_2/n_{2*}	h_1/h_{1*}	Δ, %
0.2	1 656 000	0.30098	0.33
	1 650 535	0.3	
0.15	1 656 000	0.31155	3.85
	1 592 244	0.3	
0.1	1 584 000	0.30110	0.37
	1 578 139	0.3	
0.05	1 368 000	0.30570	1.9
	1 342 008	0.3	
0	1 296 000	0.31040	3.47
	1 251 029	0.3	

δ_2, mm	$B = 720$ rot		
	n_2/n_{2*}	h_1/h_{1*}	Δ, %
0.2	1 722 240	0.3001	0.03
	1 721 723	0.3	
0.15	1 663 920	0.30012	0.04
	1 663 255	0.3	
0.1	1 650 960	0.3001	0.03
	1 650 415	0.3	
0.05	1 412 410	0.3000	0.00
	1 412 410	0.3	
0	1 332 000	0.3001	0.03
	1 331 560	0.3	

δ_2, mm	$B = 7200$ rot		
	n_2/n_{2*}	h_1/h_{1*}	Δ, %
0.2	1 720 800	0.30098	0.33
	1 715 121	0.3	
0.15	1 663 200	0.30116	0.39
	1 691 205	0.3	
0.1	1 648 800	0.30110	0.37
	1 642 700	0.3	
0.05	1 411 200	0.30110	0.37
	1 405 979	0.3	
0	1 324 800	0.30048	0.16
	1 322 680	0.3	

δ_2, mm	$B = 12$ rot		
	n_2/n_{2*}	h_1/h_{1*}	Δ, %
0.2	1 722 396	0.3000	0.00
	1 722 396	0.3	
0.15	1 663 944	0.3000	0.00
	1 663 944	0.3	
0.1	1 651 348	0.3000	0.00
	1 651 348	0.3	
0.05	1 413 156	0.3000	0.00
	1 413 156	0.3	
0	1 332 320	0.3000	0.00

Table 4. Bearing durability n_{2*} calculated by the approximate method (ε = 0.41 mm)

δ_2, mm	B = 72 000 rot		
	n_2/n_{2*}	h_1/h_{1*}	Δ, %
0.4	1 224 000	0.313 110	4.37
	1 170 499	0.3	
0.3	1 152 000	0.306 900	2.31
	1 125 399	0.3	
0.2	1 224 000	0.317 040	5.68
	1 154 477	0.3	
0.1	1 008 000	0.309 107	3.04
	977 400	0.3	
0	936 000	0.307 786	2.60
	911 708	0.3	

δ_2, mm	B = 7200 rot		
	n_2/n_{2*}	h_1/h_{1*}	Δ, %
0.4	1 238 400	0.300 932	0.31
	1 234 561	0.3	
0.3	1 195 200	0.301 104	0.37
	1 190 778	0.3	
0.2	1 224 000	0.301 188	0.40
	1 219 153	0.3	
0.1	1 044 000	0.300 923	0.31
	1 040 788	0.3	
0	979 200	0.301 191	0.40
	975 313	0.3	

δ_2, mm	B = 720 rot		
	n_2/n_{2*}	h_1/h_{1*}	Δ, %
0.4	1 239 120	0.300 096	0.03
	1 238 724	0.3	
0.3	1 195 200	0.300 041	0.01
	1 195 037	0.3	
0.2	1 226 160	0.300 132	0.04
	1 225 746	0.3	
0.1	1 047 600	0.300 098	0.03
	1 047 258	0.3	
0	982 080	0.300 092	0.03
	981 779	0.3	

δ_2, mm	B = 12 rot		
	n_2/n_{2*}	h_1/h_{1*}	Δ, %
0.4	1 239 408	0.300 001	0.00
	1 239 403	0.3	
0.3	1 195 700	0.300 000	0.00
	1 195 697	0.3	
0.2	1 226 611	0.300 003	0.00
	1 226 000	0.3	
0.1	1 047 970	0.300 002	0.00
	1 047 963	0.3	
0	982 500	0.300 003	0.00
	982 489	0.3	

The analysis of the obtained results demonstrates that an increase in block size leads to higher deviations Δ_B. At radial clearance ε = 0.21 mm (Table 5), for B = 720 rotations, the estimated durability decreases by 0.04–0.058% compared to the accurate durability (B = 1 rot); for B = 7200 rotations it decreases by 0.42–0.730%; for B = 72000 rotations it decreases by 4.17–6.1% depending on shaft ovality value. In the case of shafts with a circular section (δ_2 = 0), the deviations will be the highest. If radial clearances increase, the deviation Δ_B increases as well, as shown in Table 6 for ε = 0.41 mm. It is assumed that block B = 7200 rotations will be optimum for the discretization interval $\Delta\alpha_2 = 10^0$;

Table 7 list an additive method based on the adjusted interval-block pattern (B = 72000 → 7200 → 720 → 12 → 1 rot), using the formula:

$$n_{2*} = x_1 B_{72000} + x_2 B_{7200} + x_3 B_{720} + x_4 B_{12} + x_5 B_1 , \quad (1)$$

where: x_1 is the maximum number of basic blocks B_p, for which $h_1 < h_{1*}$;

x_2, x_3, x_4 are the respectively maximum number of successive block sizes which will satisfy a given condition;

x_5 is the number of blocks B = 1 rotation that ensure obtaining accurate value of wear h_{1*}.

The results of investigating the effect of the size of blocks of constant conditions of contact on the durability of a bearing with shaft ovality demonstrate that the deviations from the accurate result caused by blocks with a size of up to 10000 rotations do not exceed 1%, which is also the number of times by which the computation time is reduced.

For practical reasons, the additive method based on the adjusted interval – block pattern is most rational to apply to solve the problem. Apart from a much simpler and faster procedure of numerical calculation, the accuracy of results obtained is similar to the results obtained with a more accurate method (Tables 5, 6). Analyzing the durability estimation results (Tables 5–7), it is found that the durability $n_{2*}^{(D)}$ of the accurate solution can be determined based on the durability $n_{2*}^{(Dk)}$ or $n_{2*}^{(A)}$ in the following way:

Table 5. Deviations of the approximate durability with regard to the accurate one (ε = 0.21 mm)

δ_2, mm	B, rot	n_{2*}, rot	Δ_B, %
	1	1 722 413	0.000
	12	1 722 396	0.001
0.2	720	1 721 723	0.040
	7200	1 715 121	0.423
	72 000	1 650 535	4.173
	1	1 663 970	0.000
	12	1 663 944	0.001
0.15	720	1 663 255	0.032
	7200	1 656 714	0.436
	72 000	1 592 244	4.311
	1	1 651 362	0.000
	12	1 651 348	0.001
0.1	720	1 650 415	0.057
	7200	1 642 700	0.525
	72 000	1 578 139	4.434
	1	1 413 160	0.000
	12	1 413 156	0.001
0.05	720	1 412 410	0.053
	7200	1 405 979	0.508
	72 000	1 342 008	5.035
	1	1 332 336	0.000
	12	1 332 320	0.001
0	720	1 331 560	0.058
	7200	1 322 680	0.725
	72 000	1 251 029	6.103

Table 6. Deviations of the approximate durability with regard to the accurate one (ε = 0.41 mm).

$_2$, mm	B, rot	n_{2*}, rot	Δ_B, %
	1	1 239 410	0.000
	12	1 239 396	0.001
0.4	720	1 238 729	0.055
	7200	1 234 561	0.398
	72 000	1 170 499	5.56
	1	1 195 707	0.000
	12	1 195 697	0.001
0.3	720	1 195 037	0.056
	7200	1 190 778	0.41
	72 000	1 125 399	5.88
	1	1 226 752	0.000
	12	1 226 700	0.004
0.2	720	1 225 746	0.082
	7200	1 219 153	0.619
	72 000	1 154 477	5.891
	1	1 047 977	0.000
	12	1 047 963	0.001
0.1	720	1 047 258	0.069
	7200	1 040 788	0.681
	72 000	977 400	6.731
	1	982 501	0.000
	12	982 500	0.000
0	720	981 779	0.074
	7200	975 313	0.732
	72 000	911 708	7.205

Table 7. Bearing durability for different sizes of basic blocks (ε = 0.41 mm)

δ_2, mm	B = 1 rot	B_p = 7200 rot		B_p = 72 000 rot	
	n_{2*}, rot	n_{2*}, rot	Δ_Σ,%	n_{2*}, rot	Δ_Σ, %
0.4	1 239 410	1 232 469	0.56	1 170 375	5.57
0.3	1 195 707	1 189 011	0.56	1 125 160	5.90
0.2	1 226 752	1 219 141	0.62	1 154 341	5.90
0.1	1 047 977	1 040 733	0.69	975 973	6.87
0	982 501	975 301	0.73	910 501	7.33

Note: Δ_Σ – deviations of the additive durability from the accurate durability (B = 1 rotation).

$$n_{2*}^{(D)} \approx n_{2*}^{(Dk)} + B \text{ or } n_{2*}^{(D)} \approx n_{2*}^{(A)} + B_p, \qquad (2)$$

where: (Dk) and (A) denote, respectively, the more accurate and additive methods for solving the problem.

Tables 8 and 9 list durability differences $\Delta n_2 = n_{2*}^{(D)} - n_{2*}^{(Dk)}, n_{2*}^{(A)}$ which confirm the formula (2).

CONCLUSIONS

1. Values of deviations of durability determined by the more accurate and additive methods were compared to those obtained by the accurate method.
2. It is found that the computation time decreases directly proportional to the size of blocks of constant conditions of interaction.

Table 8. Differences $\Delta n_2 = n_{2*}^{(D)} - n_{2*}^{(A)}$ for the durability D and Dk ($\varepsilon = 0.41$ mm)

δ_2, mm	B, rot	n_{2*}, rot	Δn_2, rot
0.4	1	1 239 410	–
	12	1 239 396	14
	720	1 238 729	681
	7200	1 234 561	4849
	72 000	1 170 499	68 911
0.3	1	1 195 707	–
	12	1 195 697	10
	720	1 195 037	670
	7200	1 190 778	4929
	72 000	1 125 399	70 308
0.2	1	1 226 752	–
	12	1 226 700	52
	720	1 225 746	1006
	7200	1 219 153	7599
	72 000	1 154 477	72 275
0.1	1	1 047 977	–
	12	1 047 963	14
	720	1 047 258	719
	7200	1 040 788	7189
	72 000	977 400	70 577
0	1	982 501	–
	12	982 500	1
	720	981 779	722
	7200	975 313	7188
	72 000	911 708	70 793

3. We have got an approximate dependence for estimating accurate durability of a bearing based on the more accurate and additive method.
4. It is found that the additive method is a simple and effective way of solving the problem of estimating the durability of a slide bearing.

REFERENCES

1. Andrejkiv A.E., Czerniec M.V.: Ocenka kontaktnogo vzaimodiejstvija truszczichsia dietaliej maszyn. Naukova Dumka, Kijev, 1991.
2. Goriaczeva I.G., Dobyczin N.M.: Kontaktnyje zadaczi v tribologii. Maszinostrojenije, Moskva, 1988.
3. Kovalenko E.V.: K rasczetu iznaszivanija soprjażenija vał – vtułka. Mechanika Tverdoho Tieła, 6, 1982, 66–72.
4. Kuźmenko A.G.: Metody rozrachunkiv na znoszuvannja ta nadijnist. TUP, Chmelnyckyj, 2002.
5. Kragielskij I.V., Dobyczin N.M., Kombalov V.S.: Osnovy rasczetov na trenie i iznos. Maszinostroenie, 1977, pp. 526.
6. Tepłyj M.I.: Opredielienije kontaktnych paramietrov i iznosa v cilindriczeskich oporach skolżenija. Trenije i Iznos, 6, 1987, 895–902.
7. Tepłyj M.I.: Kontaktnyje zadaczi dla oblastej s kruhowymi granicami. Lwow, Wyszcza Szkoła, Wyd. LGU, 1983, pp. 176.
8. Usov P.P.: Vnutrennij kontakt cilindriczeskich tieł blizkich radiusov pri iznaszivanii ich poverchnostiej. Trenije i Iznos, 3, 1985, 404–414.
9. Zwierzycki W.: Prognozowanie niezawodności zużywających się elementów maszyn. Radom, Wyd. ITE, 1998.
10. Sorokatyj R.V.: Uzahalnennia metodu tryboelementiv dla modeljuvannia procesiv znoszuvannia v pidszypnykach kovzannia. Problemy Trybologii, 2, 2007, 36–45.
11. Sorokatyj R.V.: Reszenije iznosokontaktnych zadacz metodom triboelementov v sredie koniecznoelementnogo pakieta ANSYS. Problemy Trybologii, 3, 2007, 9–17.
12. Kolmogorov V.L., Kharlamov V.V., Kurilov A.M., Pavlishko S.V.: Friction and wear model for a heavily loaded sliding pair. Part II. Application to an unlubricated journal bearing. Wear, 197, 1996, 9–16.

Table 9. Differences $\Delta n_2 = n_{2*}^{(D)} - n_{2*}^{(A)}$ for the durability D and A ($\varepsilon = 0.41$ mm)

δ_2, mm	B = 1 rot	B_p = 7200 rot		B_p = 72 000 rot	
	n_{2*}, rot	n_{2*}, rot	Δn_2, rot	n_{2*}, rot	Δn_2, rot
0.4	1 239 410	1 232 469	6941	1 170 375	69 035
0.3	1 195 707	1 189 011	6696	1 125 160	70 547
0.2	1 226 752	1 219 141	7611	1 154 341	72 411
0.1	1 047 977	1 040 733	7244	975 973	72 004
0	982 501	975 301	7200	910 501	72 000

13. Rezaei A., Ost W., Paepegem W.V., Degrieck J., Debaets P.: Experimental study and numerical simulation of the large-scale testing of polymeric composite journal bearings: two-dimensional modeling and validation. Wear, 270, 2011, 431–438.

14. Rezaei A., Paepegem W. V., Debaets P., Ost W., Degrieck J.: Adaptive finite element simulation of wear evolution in radial sliding bearings. Wear, 296, 2012. 660–671.

15. Czerniec M.V., Liebiedieva N.M.: Ocinka kinetyky znoszuvannia trybosystem kovzannia pry najavnosti ovalnosti konturiv ich elementiv za kumuliacijnoju modellu. Problemy Trybologii, 4, 2005, 114–120.

16. Chernets M.V., Andreikiv O.E., Liebiedieva N.M., Zhydyk V.B.: A model for evaluation of wear and durability of plain bearing with small non-circularity of its contours. Materials Science, 2, 2009, 279–290.

17. Czerniec M., Żydyk W., Czerniec J.: Symulacja zużywania łożyska ślizgowego z niekołowością konturów. Cz. 1. Uogólniony model liniowy zużywania. W: Projektowanie i sterowanie procesami. Lublin, Wyd. Politechniki Lubelskiej, 2013, 76–89.

18. Czerniec M., Żydyk W., Czerniec J.: Symulacja zużywania łożyska ślizgowego z niekołowością konturów. Cz. 2. Uogólniony model kumulacyjny zużywania. W ks.: Projektowanie i sterowanie procesami. Lublin, Wyd. Politechniki Lubelskiej, 2013, 90–104.

19. Czerniec M.V.: Addytyvnyj metod ocinky dovhovicznosti pidszypnyka kovzannia. Problemy Trybologii, 4, 2013, 20–24.

20. Chernets M.V.: A Contact Problem for a Cylindrical joint with Technological Faceting of the Contours of its Parts. Materials Science, 6, 2009, 859–868.

2

ASSESSMENT OF SURFACE QUALITY FOR CHOSEN MILLING STRATEGIES WHEN PRODUCING RELIEF SURFACES

Jan Varga[1], Jozef Stahovec[1], Jozef Beno[1], Marek Vrabeľ[1]

[1] Department of Technologies and Materials, Faculty of Engineering, Technical University in Košice, Mäsiarska 74, 040 01 Košice, Slovak Republic, e-mail: jan.varga@tuke.sk; jozef.stahovec@tuke.sk; jozef.beno@tuke.sk; marek.vrabel@tuke.sk

ABSTRACT

The paper deals with design and modeling of the relief surfaces that are produced in milling. Modeled and real surface quality is presented for the chosen fragments of the relief surfaces. Fragmentation of the relief surfaces has been made by the surface sampling. Milling strategies are compared with regard to surface formation. Surface quality was checked with regard to applied cutting conditions.

Keywords: relief surface, fragmentation of the surface, milling strategies.

INTRODUCTION

The subject of this paper is the design and manufacturing of parts with complex shaped surfaces. Relief surface is widely used in the design of complex product with molds and dies features. These surfaces are often produced by 3 and 5-axis computer numerical control machine tools using ball-end milling cutter. Mansour [1] deal with automatically generate part programs for machining sculptured surfaces by using CAD/CAM packages. For milling relief surface it used the ball milling. Siller [2] predicts the cycle time of high-speed milling for sculptured surfaces with high feed rates. The experiments and predictions were focused on representative surfaces of dies and molds, whose geometric complexity and complexity distribution were, modified parametrically (a set of representative sculptured surfaces with spherical caps and a forging die surface). The prediction of cycle time under high feed rate cutting conditions studied by Monreal [3] and Kim [4] for CNC programs of roughing operations with a relatively large proportion of long tool paths and zigzag patterns. In these studies, the proposed methods for cycle time prediction are based on characterization of machine tool acceleration and principles of constant acceleration motion. Feng [5] presents a new approach for the determination of efficient tool paths in the machining of sculptured surfaces using 3-axis ball-end milling. The objective is to keep the scallop height constant across the machined surface such that redundant tool paths are minimized. The effectiveness of the present approach is demonstrated through the machining of a typical sculptured surface. The surface texture depends on the optimal set of parameters, so it is needed for the circumspect design of the process [6].

EXPERIMENTAL DETAILS

Milling tests were performed on 3 axes CNC milling machine tool Emco Mill 155 equipped with digital control system Heidenhain TNC 426. Milling experiments were conducted by using cutting coolant. In the experiments end mill and ball nose solid carbide cutting tools were used. Cutting tools parameters and cutting conditions are shown in Table 1. Cutting conditions were chosen based on the recommendations of the tools manufacturer and the possibilities of the machine. Aluminum alloy with dimensions 145×140×30 mm was used as a workpiece material. 3D model of part was designed in CAD software Solid Works

2012 and CAM software SolidCAM was usedto manufacture this part . Part model and machined part is shown in Figure 1.

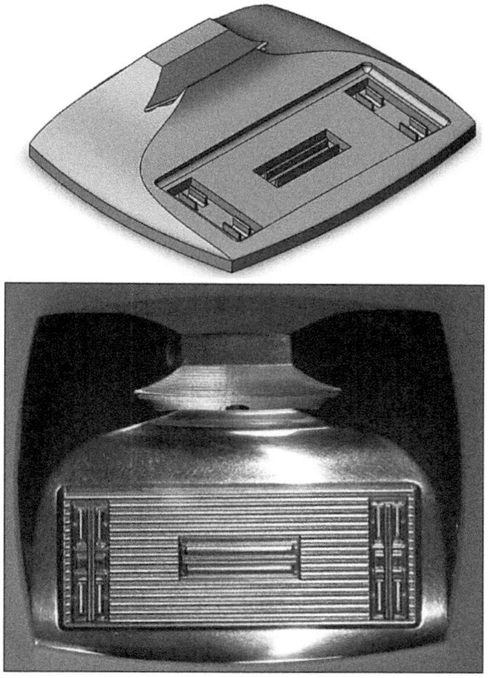

Fig. 1. Part model and final part

SOLID MODEL WITH RELIEF SURFACE

There are 5 types of representation of complex surfaces in CAD systems. Different types of these representations are shown in Figure 2.

Analytical representation

The basic advantage of analytical representation is its accuracy. Descriptions of the objects represented using spline frontier model consist of points, curves and spline surfaces. Such models contain full information to describe the model. This model representation is used for very precise geometric modeling in CAD/CAM systems.

Wireframe model

Wireframe model consists of points, which are connected to curves. In this case two points are connected with line. The model created in this representation has many limitations and disadvantages, because there are missing data about the volume between lines.

Planar model

This kind of model works with surfaces in three-dimensional Cartesian coordinate system. They are used in the areas of aerospace, shipbuilding and automotive industries, where the processing of complex surfaces placed great demands. The procedure consists in creating of vertices, edges, and finally defines the area [7].

Volume model

Model is defined by their borders (e.g. walls, edges, lines, curves and points).

Voxel model

Voxel models are similar to bitmap graphics. When this model is used, the object is decomposed into elementary volume units – cubes. The basic unit of description is a Voxel, which is a volume unit representing the value of the cell in three-dimensional Cartesian coordinate system.

SURFACE QUALITY AND USED MILLING STRATEGY

By the sequence operations of technological process framework more machining strategies were designed, which divides each operation into multiple sections. Using multiple strategies it is possible to precisely control of tool paths when machining complex surfaces, that model includes.

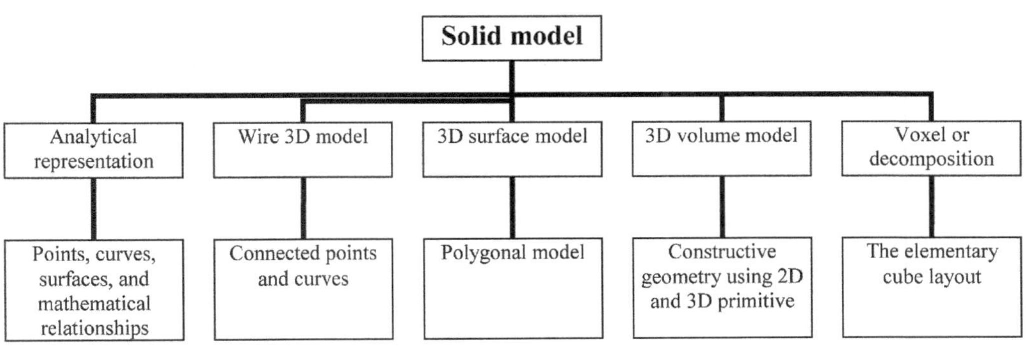

Fig. 2. Representation of solid model

Roughing milling operation was divided into two stages, basic shape roughing and roughing pockets respectively. 3D Milling strategy was used in roughing operation and end mill with a diameter 12 mm was employed in machining. Corresponding cutting conditions are shown in Table. 1 Depth of cut a_p was set at 1–3 mm and width of cut a_e was set to 4.8 mm. Start-up angle of the tool into the material was 5°. Individual tool paths after roughing and machined surface after this operation are shown in Figure 3 [8].

Semi-finishing operation was divided into milling residual material from previous operations and milling head contour parts. 3D milling and contouring strategy was chosen. In this operation end mill with diameter 6 mm was used. The depth of cut a_p was in range 1–2 mm and the width of cut a_e was set to 1 mm. Tool paths after semi-finishing shows Figure 4 and machined surface after the semi-finishing operation are shown in Figure 4. By finishing external surface was used ball nose mill, diameter Ø 8 mm. Was used HSS strategy-morph between two boundary curves, i.e. strategy between the two guide curves. Tool paths after finishing external surface and real manufactured part are shown in Figure 5.

Finishing operation of the internal surfaces, which was divided into 19 sections, as this is the completion of functional bearing surfaces. Were used HSS-morph between two boundary curves (between the two guide curves), HSS Parallel to curves (parallel under the curve) and 3D Milling strategy. Was used ball nose mill, diameter 2 mm. Tool paths in finishing the internal surfaces Figure 6 shows and the subsequently surface after the operation shows Figure 6. The largest share of time had operations in which ball nose mill, diameter 2 mm was used [8].

Table 1. Tool parameters and cutting conditions for milling

Tool	Diameter D [mm]	Number of flutes	Feed XY [mm/min]	Feed Z [mm/min]	Spin rate S [RPM]	Depth of cut a_p [mm]	Width of cut a_e [mm]
End mill	6	2	200	70	5000	1–2	1
End mill	12	2	400–485	150	3750–5000	1–3	4,8
Ball nose mill	2	2	60	20	5000	0,3–0,75	0,2–0,8
Ball nose mill	4	2	120	40	5000	0,5	1,2
Ball nose mill	8	2	250	80	5000	0,5	0,4

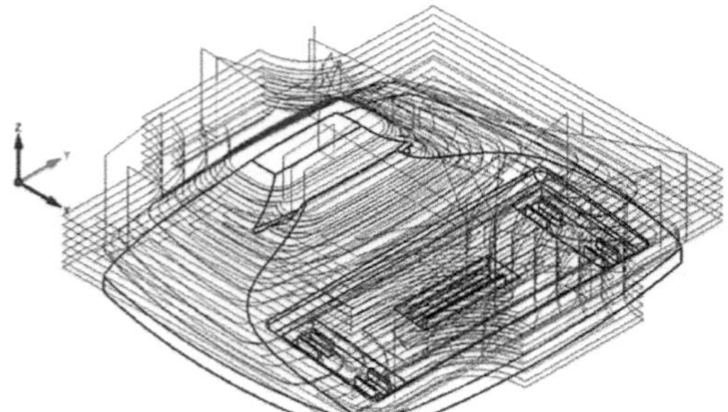
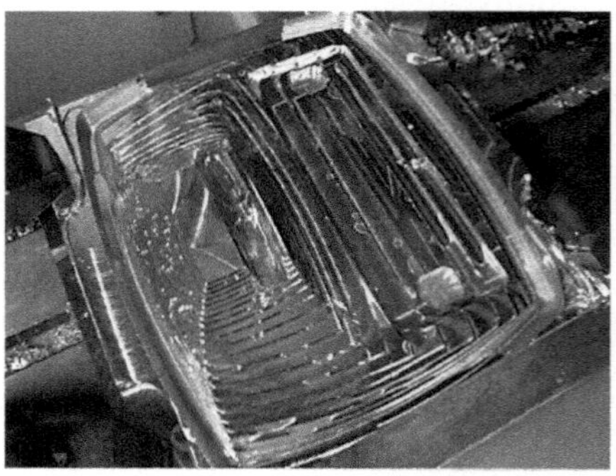

Fig. 3. Roughing

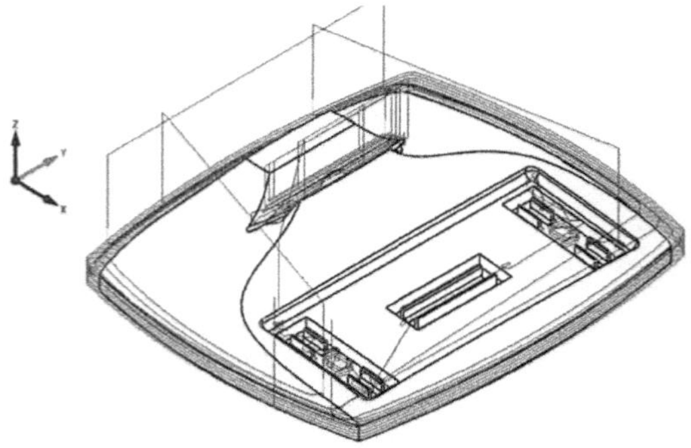

Fig. 4. Semi-finishing

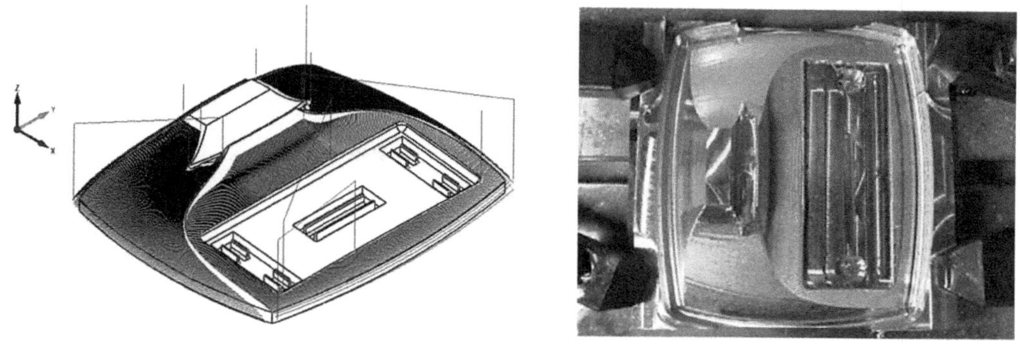

Fig. 5. Finishing external surface

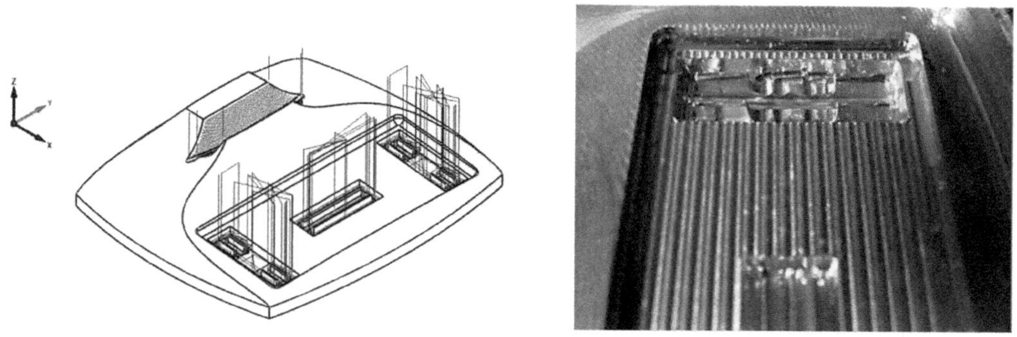

Fig. 6. Finishing internal surface

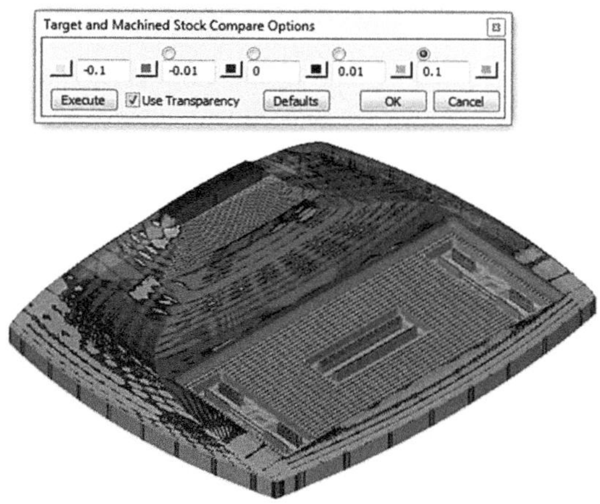

Fig. 7. Target and machined stock compare options

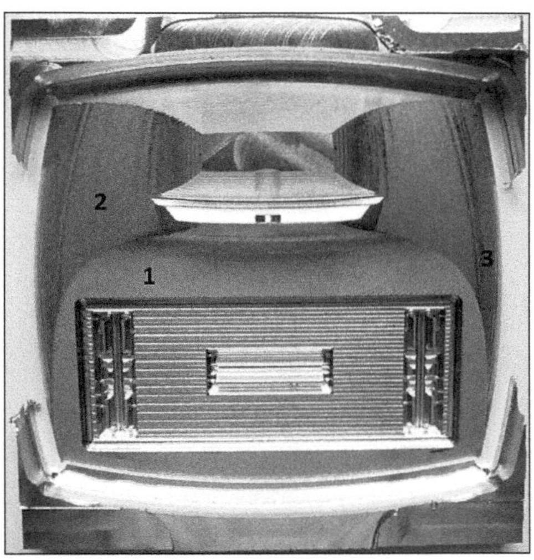

Fig. 8. Fragmentation relief surface

The estimated time of proposed operations was 8 h 10 min and the resulting quality of machined relief surface shown in Figure 7. Estimated deviation of machined surfaces has been –0.01 mm and in some places up to –0.1 mm. CAM system offers possibility to compare deviations between target and machined model and proof achieved surface quality after machining.

FRAGMENTATION OF THE RELIEF SURFACE

The tool marks are clearly visible in Figure 8 and fragments with sampling are shown in Figure 9. Ball nose mill creates on machined surface peaks and valleys – scallop height *Rz*. Acceptable surface roughness value *Rz* for that particular part was set to 5 μm. Figure 9 illustrates different chosen fragments of parts on machined surface. The machined relief surfaces were finished by grinding and hand polishing.

CONCLUSION

The presented study was performed concerning the relief surface and its aim was to determine

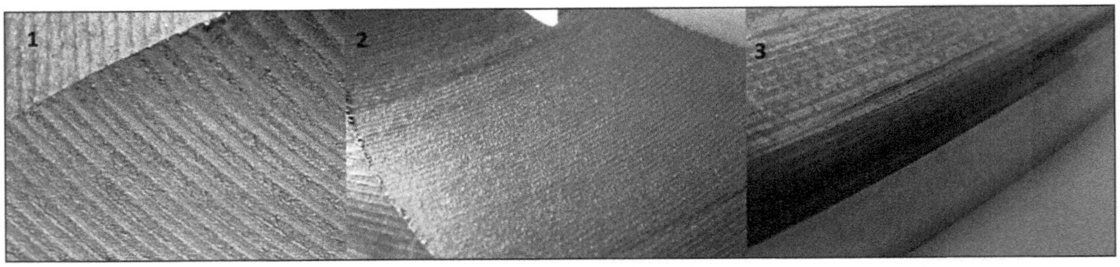

Fig. 9. Fragment with sampling

appropriate procedure to select tool path strategy when machining parts consisting of relief surfaces. The resulting procedure was chosen from several alternatives while the criterion of selecting was the resultant surface quality, production time and milling strategies, as well. Such a sort of analysis was performed as a proper use of the available CAM system. Machined surface of the workpiece produced bears characteristic marks of the ball nose milling cutters. Their locations and height of the resultant cusps were changed according to the inclination of the tested fragment area. Surface quality was checked according to the applied cutting conditions. Generally, the areas with highest inclination of the surface fragments was found out to achieve desired scallop height, therefore, these areas require a smaller amount to apply hand finishing operations.

Acknowledgements

This work was supported by the project VEGA1/0500/ 12 Research on Quality Improvement when Milling Formed Surface by Cutters with Advanced Coatings, granted by Scientific grant agency of Ministry of Education, Science, Research and Sport of Slovakia and Slovak Academy of Science.

REFERENCES

1. Mansour S.: Automatic generation of part programs for milling sculptured surfaces. Journal of Materials Processing Technology 127, 2002, 31-39.
2. Siller H., Rodriguez C.A., Ahuett H.: Cycle time prediction in high-speed milling operations for sculptured surface finishing. Journal of Materials Processing Technology 174, 2006, 355-362.
3. Monreal M., Rodriguez C.A.: Influence of tool path strategy on the cycle time of high-speed milling. Computer-Aided Design 35, 2003, 395-401.
4. Kim Bo H., Choi Byoung K.: Machining efficiency comparison direction-parallel tool path with contour-parallel tool path. Computer-Aided Design 34, 2002, 89-95.
5. Feng H.-Y., Huiwen L.: Constant scallop-height tool path generation for three-axis sculptured surface machining. Computer-Aided Design 34, 2002, 647-654.
6. Balázs M., Beňo J., Ižol P., Maňková I.: Surface quality of sculpture surface in case of 3D milling. [online]. Dostupné na internete: <http://www.uni--obuda.hu/users/mikob/2011itc.pdf>.
7. Spisak E., Fabian M.: Strojárske technológie s CAx podporou. 1. vydanie. Košice: Technická univerzita v Košiciach, Strojnícka fakulta, 2010, p. 397.
8. Svihura P.: Návrh a overenie výroby propagačného predmetu využitím CAD/CAM systémov, TU, SjF, diplomová práca, 2013, p.104.

3

EFFECT OF BRUSHING PARAMETERS UPON EDGE STATES AFTER WIRE BRUSHING OF AZ91 HP MAGNESIUM ALLOY

Jakub Matuszak[1], Kazimierz Zaleski[1]

[1] Department of Production Engineering, Mechanical Engineering Faculty, Lublin University of Technology 36 Nadbystrzycka Str., 20-618 Lublin, Poland, e-mail: j.matuszak@pollub.pl; k.zaleski@pollub.pl

ABSTRACT
When the tool leaves the part during machining operations, burrs can form as a result of material plastic flow. Wire brushing is one of burr removing methods. Low cost of brushes which can be mounted in a milling center gives the perspective of developing this method. The article was aimed at presenting the results of brushing process upon edge radius after wire brushing of AZ91HP magnesium alloy. Three kinds of brushes were tested with different types of fibers and the diameter of a single fiber. Brass and steel brushes were used.

Keywords: edge state, wire brushing, magnesium alloy.

INTRODUCTION

Brushing tools have been used for surface finishing operations for many years. The most popular applications of brushing are cleaning and deburring as well as removal of old varnish coats and corrosion. Brushes can be classified according to the shape and type of fiber which generally can be divided into metallic, non-metallic and natural fiber. Besides simple cleaning operations brushing can be used for more sophisticated applications, such as generating specific surface topography, as well as desirable properties of surface layer. Surface roughness can be controlled after brushing and depends on cutting parameters and type of brush [11]. Surface roughness decrease when the fiber tips of the brush cut or round up at the tops of the highest peaks of the profile after wire brushing. This effect may be desirable when it is important to increase material ratio which influences the wear behavior of parts in frictional contact [15]. On the other hand, in many situations it is desirable to develop surface topography, especially before applying protective or decorative coatings, during adhesive bonding technology and soldering process. Increased developed interfacial area ratio (Sd_r) contributes to increased adhesion. Brushing process can increase Sd_r parameter. Wire brushing affects the properties of surface layer.

The results presented in the article [5] show the ability to generate compressive stresses in the surface layer after wire brushing. Transforming the tensile residual stress into the compressive meant that fatigue cracks were shorter by 20–40 microns, compared to baseline (150–200 microns).This treatment led to an increase in fatigue strength by tens of percent.

The authors of articles [8, 9] applied wire brushing under varying conditions to surfaces of AZ31 magnesium alloy sheets for the purpose of grain refinement. Ultrafine grains were obtained in all wire-brushed sheets. The mean grain size depended mainly on the feed speed. The 0.2% proof stress and tensile strength slightly increased after the wire-brushing. The wire-brushed sheet shows high corrosion resistance for the salt spray test, which was approximately 4 times higher than that of a starting sheet. The results confirm the concept that wire brushing can improve mechanical properties and corrosion resistance of AZ31 magnesium alloy sheet.

The article [14] presents results of microhardness study after wire brushing of 6061-T6 aluminum alloy. In order to improve the understanding of the process of material removal, scanning electron microscopy was used.

The authors of the article [10] presented results of surface layer properties after wire brushing of 3103 and 7075 aluminium alloys. Significant increase of microhardness was observed for low values of feed rates and high values of rotational speed after wire brushing of 3103 and 7075 aluminium alloys.

Milling is one of the most commonly used processes in industry for machining parts of precise sizes and shapes. Unfortunately, when the tool leaves the part during machining operations, burrs can form as a result of material plastic flow. Currently there are many different definitions and classifications of burrs depending on the manufacturing process, shape, mechanism of burr formation and the type of treatment [2, 6, 12].

In many cases, there is a need to change the edge states after milling, because edges with burrs can sometimes cause difficulties in assembly operations, where high accuracy is needed. On the other hand, if burrs do not have to be removed from a workpiece for accuracy reasons, they can be very sharp and dangerous for workers and consumers, because they can cause injuries and scratches. According to ISO 13715 edge state is a geometrical shape and the size of the edge. External edge of the part can be sharp, with undercut or with burr [7].

State of an edge can be changed with the use of a variety of methods. The most common ones include hand deburring, grinding, milling. There is a group of methods using abrasive media such as vibratory finishing, abrasive flow machining, and magnetic abrasive finishing. Burrs can be removed by using chemical, electrochemical or thermal treatment. These methods are widely described in [1, 3, 4, 13].

Wire brushing is one of burr removing methods. Low cost of tools which can be mounted in a milling center gives the perspective of developing this method.

The article was aimed at presenting the results of brushing process upon edge radius after wire brushing of AZ91HP magnesium alloy. Three kinds of brushes were tested with different types of fibers and the diameter of a single fiber. Brass and steel brushes were used.

MATERIALS AND EXPERIMENTAL CONDITIONS

The workpiece material used during the experiment was AZ91HP magnesium alloy, commonly used in aerospace industry and many other branches of industry, due to low density and good mechanical properties such as castability and machinability. Cubicoid samples with dimensions of 15x10x50 mm were used. The chemical composition and physical properties are presented in Table 1.

Experiments were carried out on a vertical milling center FV 580a. Before brushing the samples were milled with the use of fixed parameters.

The experiment was carried out with the use of three different circular brushes with a diameter of 120 mm, brush-B1: brass filament with fiber diameter 0.2 mm, brush B2: steel filament with fiber diameter 0.2 mm, brush B3: steel filament with fiber diameter 0.3 mm. The main difference between these brushes is that the brass brash is very flexible. The use of fibers with a bigger diameter results in greater rigidness. Dry brushing was carried out. The brushing parameters are presented in Table 2. For each sample two passes were carried out, first pass with the right hand spindle rotation and second return pass with the left hand spindle rotation. Two passes provide the stability radius of the rounded edge. Specimen orientation and brushing kinematics are presented in Figure 1.

To represent the edge outline after brushing replica method was used with Plastiform LK-AD two-component mass. Polymerization started after mixing the two components. Its characteristics

Table 1. Chemical composition and physical properties of AZ91HP magnesium alloy

Chemical composition (%)	Cu	0.016	Physical properties	Rm (MPa)
	Mn	0.17		20–240
	Mg	89.61		
	Zn	0.72		$Rp_{0.2}$ (MPa)
	Si	0.03		150–170
	Fe	0.002		
	Al	9.45		

Table 2. Brushing parameters used during the experiments

No.	Feed rate, v_f (mm/min)	Cutting speed, v_c (m/min)	Feed-in, a (mm)
1	370	422	3
2		844	
3		1266	
4		1689	
5	140	1266	
6	370		
7	1000		
8	3700		

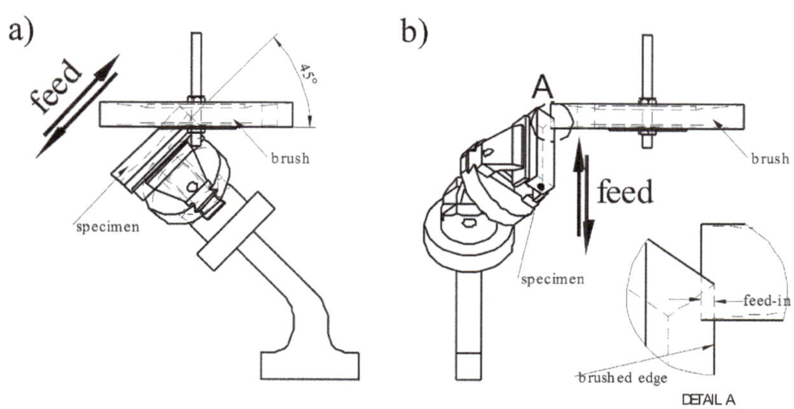

Fig. 1. Specimen orientation during brushing on a vertical machining center FV 580a: a) front view, b) side view

ensure moulding with high level of detail accuracy and faithful replicas. Cross sections of replicas were analyzed under a microscope. Scaled images were imported into AutoCAD software. An example replica specimen is shown in Figure 2. Circle radius inscribed in a rounded edge determines the edge radius after brushing. The experiment was repeated five times. The radius was measured six times at each edge.

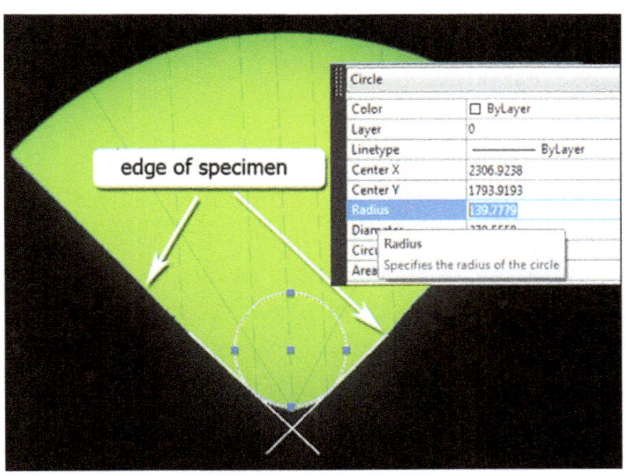

Fig. 2. Radius determining method

RESULTS

Milling process led to the burr formation on the specimen edges. The average value of burr height of AZ91HP magnesium alloy was 43µm. In order to remove burrs, brushing process with different parameters was used. Table 3 presents an example a replica cross section views of rounded edges for B2-brush and corresponding views of real edges after wire brushing. Figure 3 presents edge states after brushing with different parameters for brass brush. B1-brush with very flexible brass fibber is unable to remove burrs for most of the used parameters. Only at the highest brushing speed when the kinetic energy is greatest, burrs were removed, but the edge radius was very small. Another case in which the B1-brush provide effective burrs removal was for feed rate v_f = 140 mm/min. In this case the contact time of brush with specimen is long enough to remove burrs. However, from an economic point of view, it is important to reduce the time of deburring process.

B2 and B3 brushes provide an effective deburring process for the whole range of technolog-

Table 3. Replica cross sections and real views of rounded edges after brushing

v_f = 140 mm/min	v_f = 370 mm/min	v_f = 1000 mm/min	v_f = 3700 mm/min
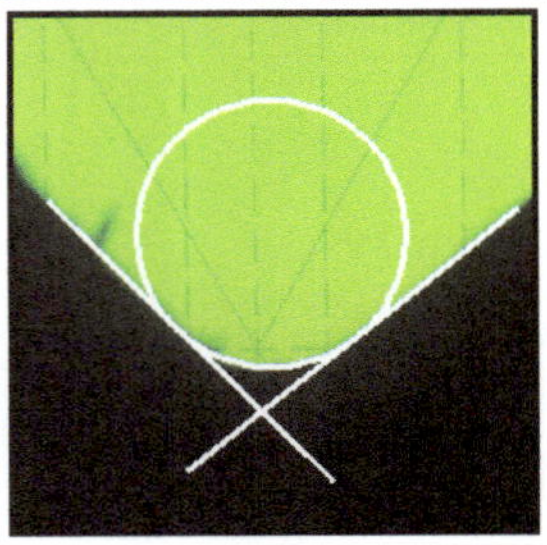	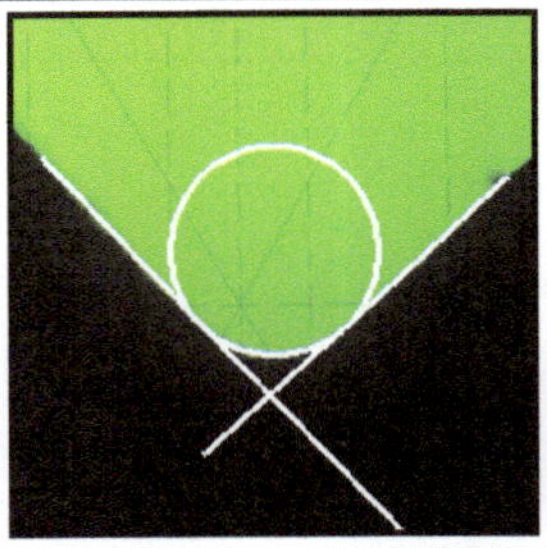	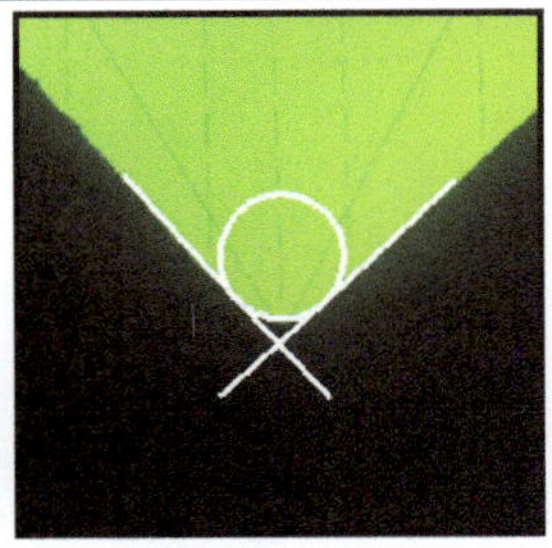	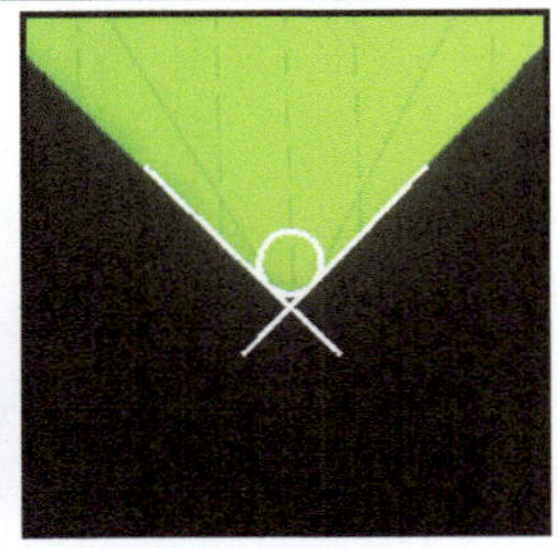
Real view of brushed edges			
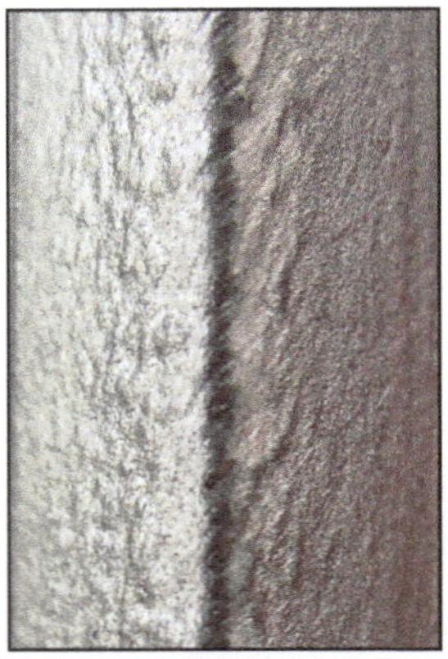			

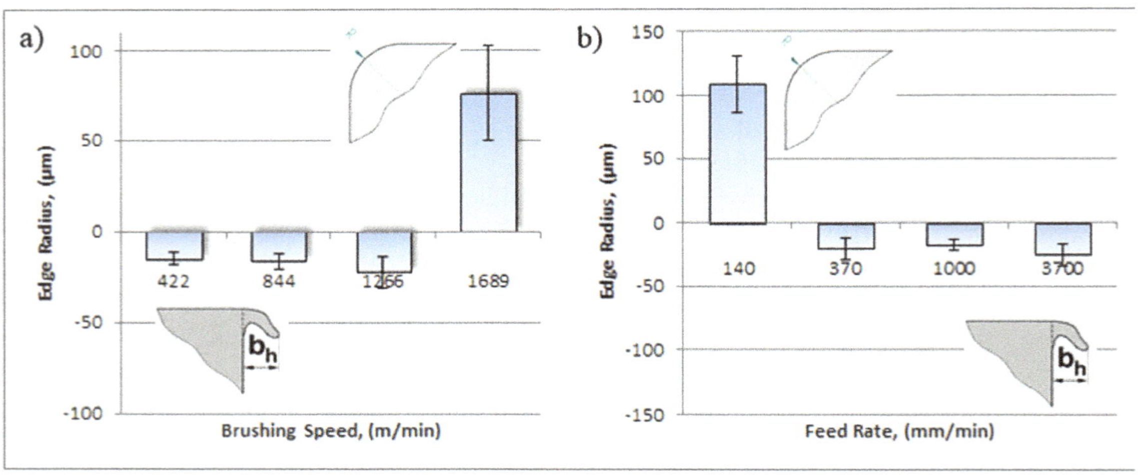

Fig. 3. Influence of brushing parameters upon edge radius for B1 brush: a) variable brushing speed, b) variable feed rate

ical parameters. When brushing speed increases, the edge radius increases as well. Edge radius changes ranging from 120 µm to 600 µm were observed for B2-brush, what is shown in Figure 4.

The highest values of edges radius compared to other brushes were observed for B3-brush. Tool rigidity caused a strong impact on the area at the edge. The largest edge radius was almost 700 µm for feed rate v_f = 140 mm/min.

Strong plastic deformation for the highest brushing speed and the lowest feed rate translate into a higher standard deviation.

CONCLUSIONS

In this study a simple and inexpensive method for deburring process was presented. The effect of brushing parameters upon edge radius of AZ91HP magnesium alloy was investigated. Additionally, three types of brushes were tested for the impact of the fiber material, fiber diameter, on edge radius. The following conclusions summarize the results of the performed experiments:

- tool stiffness is important in deburring process,

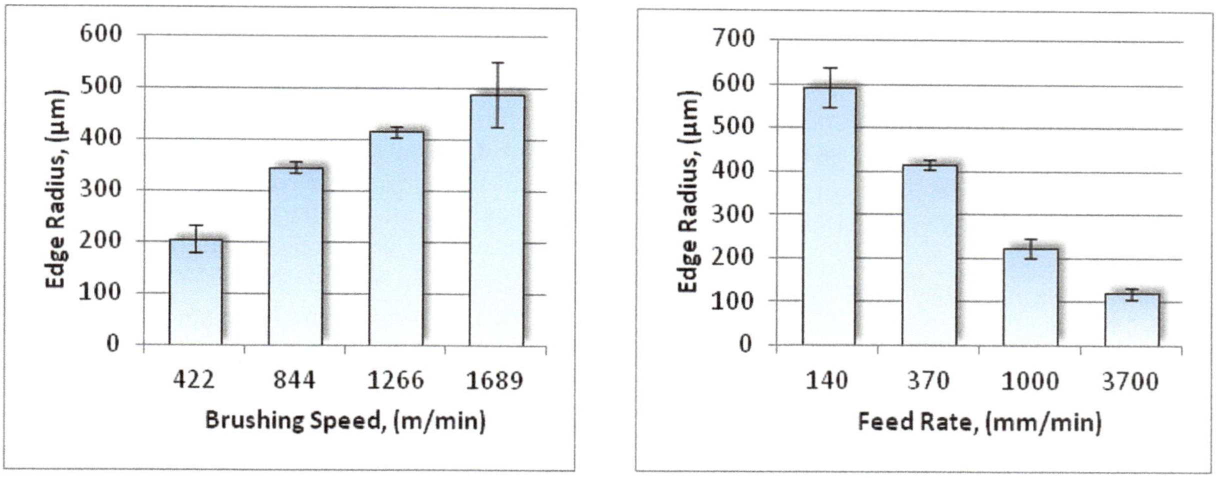

Fig. 4. Influence of brushing parameters upon edge radius for B2 brush: a) variable brushing speed, b) variable feed rate

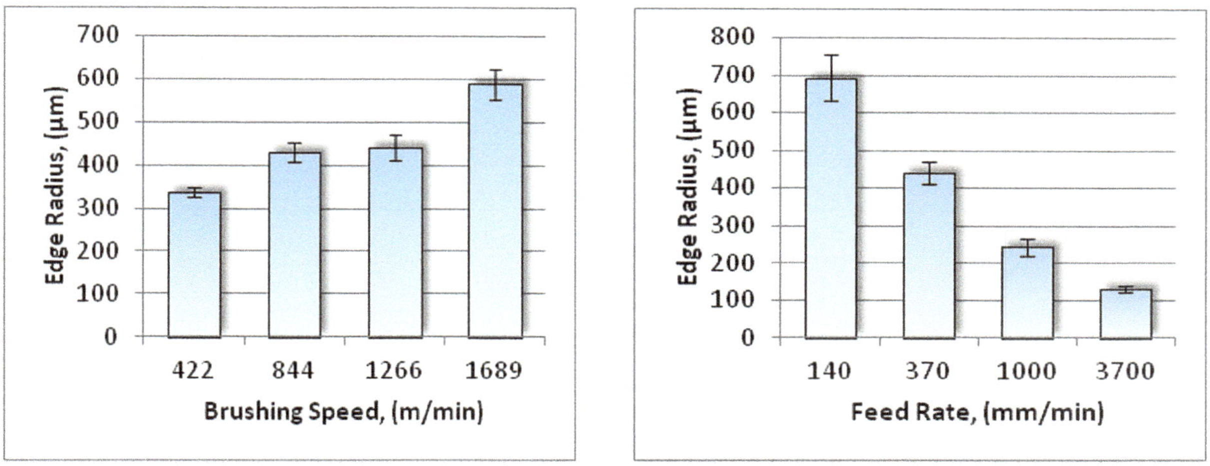

Fig. 5. Influence of brushing parameters upon edge radius for B3 brush: a) variable brushing speed, b) variable feed rate

- the state of the edge can be precisely controlled by the brushing parameters,
- with the increase of brushing speed edge radius increases,
- the edge radius increases with the decrease of feed rate, which is due to a longer contact time with the brushed edge.

During brushing process, a significant increase in temperature was observed. Future research will focus on the analysis of the surface layer after deburring by wire brushing.

Acknowledgement

Financial support of Structural Funds in the Operational Programme - Innovative Economy (IE OP) financed from the European Regional Development Fund – Project "Modern material technologies in aerospace industry", Nr POIG.01.01.02-00-015/08-00 is gratefully acknowledged.

REFERENCES

1. Aurich J.C., Dornfeld D., Arrazola P.J., Franke V., Leitz L., Min S. Burrs—Analysis, control and removal. CIRP Annals - Manufacturing Technology, 2009, 58(2): 519-542.
2. Chern G.-L. Experimental observation and analysis of burr formation mechanisms in face milling of aluminum alloys. International Journal of Machine Tools and Manufacture, 2006, 46(12-13): 1517-1525.
3. Cichosz P., Kuzinovski M. Metody wykonywania fazek i gratowania krawędzi. Cz. 1. Mechanik, 2011, 7: 353-359.
4. Cichosz P., Kuzinovski M. Metody wykonywania fazek i gratowania krawędzi. Cz. 2. Mechanik, 2011, 8-9: 674-681.
5. Fredj N.B., Nasr M.B., Rhouma A.B., Braham C., Sidhom H. Fatigue Life Improvements of the AISI 304 Stainless Steel Ground Surfaces by Wire

Brushing. Journal of Materials Engineering and Performance, 2004, 13(5): 564-574.

6. Gillespie L.K., Blotter P.T. The Formation and Propierties of Machining Burrs. Journal of Engineering for Industry, 1976, 98: 66-74.
7. ISO, ISO13715:2000(E), in Technical drawings - Edges of undefined shape - Vocabulary and indications, 2000.
8. Kitahara H., Yada T., Hashiguchi F., Tsushida M., Ando S. Mg alloy sheets with a nanocrystalline surface layer fabricated by wire-brushing. Surface and Coatings Technology, 2012.
9. Kitahara H., Yada T., Tsushida M., Ando S. Microstructure and Evaluation of Wire-brushed Mg Sheets. Procedia Engineering, 2011, 10: 2737-2742.
10. Matuszak J., Zaleski K. Badania właściwości warstwy wierzchniej stopu aluminium po obróbce szczotkowaniem. [In:] Współczesne problemy w technologii obróbki przez nagniatanie. Ed. Przybylski W., 2011 (T.3): 307-318.
11. Matuszak J., Zaleski K. Effect of cutting fluids upon surface quality after wire brushing of 7075 aluminum alloys. [In:] Innovative manufacturing technology, Ed. P. Rusek, 2012: 203-214.
12. Matuszak J., Zaleski K. Effect of milling parameters upon burr formation during AZ91 HP Magnesium Alloy Face Milling. [In:] New materials and it technologies in production engineering. Ed. J. Lipski, A. Świć, LTN Lublin 2011: 31-42.
13. Oczoś K.E. Doskonalenie procesów obróbki ściernej. Cz. IV. Gładzenie, dogładzanie i obróbka luźnym ścierniwem. Mechanik, 2010, 11: 783-793.
14. Overholser R.W., Stango R.J., Fournelle R.A. Morphology of metal surface generated by nylon/abrasive filament brush. International Journal of Machine Tools and Manufacture, 2003, 43(2): 193-202.
15. Schmähling J., Hamprecht F.A. Generalizing the Abbott–Firestone curve by two new surface descriptors. Wear, 2007, 262(11-12): 1360-1371.

4

ANALYSIS OF THE SURFACE PROFILE AND ITS MATERIAL SHARE DURING THE GRINDING INCONEL 718 ALLOY

Martin Novák[1], Nataša Náprstková[1], Jerzy Józwik[2]

[1] Faculty of Production Technology and Management, Jan Evangelista Purkyně University in Ústí nad Labem, Pasteurova 1, 400 96 Ústí nad Labem, Czech Republic, e-mail: novak@fvtm.ujep.cz; naprstkova@fvtm.ujep.cz

[2] Faculty of Mechanical Engineering, Lublin University of Technology, 36 Nadbystrzycka Str., 20-618 Lublin, Poland, e-mail: j.jozwik@pollub.pl

ABSTRACT
Grinding is still an important method for surface finishing. At FPTM JEPU research, which deals with this issue is conducted. Experiments are carried out with grinding various materials under different conditions and then selected components of the surface integrity are evaluated. They include roughness *Ra*, *Rm* and *Rz*, Material ratio curve (Abbott Firestone curve) and also the obtained roundness. This article deals with grinding nickel Inconel 718 alloy, when selected cutting grinding conditions were used and subsequently the surface profile and the material ratio curve were measured and evaluated.

Keywords: grinding, Abbott Firestone curve, Inconel.

INTRODUCTION

The importance of processing nickel alloys has been currently increasing, particularly due to their good properties, such as high *Rm* and their use for relatively high temperatures. The experiment deals with grinding a specific nickel alloy marked as Inconel 718 (NiCr1NbMo).

It is a super-alloy having a high content of nickel and it is used at extreme temperatures and in strongly oxidizing environments. This alloy was originally developed for the construction of space shuttles, but today it is also used in civilian and military sphere, for example in the aerospace industry, where it is used in the manufacture of jet engines, in the automotive industry for exhaust systems, valves, etc., in the chemical industry for pipes producing, heat exchangers, etc. This material is difficult to machine with interesting properties, which has a great potential for the future, which is why it was chosen for this experiment [1, 2].

Inconel is a family of austenite nickel-chromium-based superalloys. Table 1 presents a chemical compound of this alloy, according to the materials producer [3].

One of the parameters used to describe the surface roughness and profile is called Material ratio curve (Abbott Firestone curve). This curve

Table 1. Limiting Chemical composition of Inconel 718 [3]

INCONEL 718				
Element (% by mass)				
C	Si	Mn	P	S
0.08 max	0.35 max	0.35 max	0.015 max	0.015 max
Cr	Mo	Ni	Al	B
17.0–21.0	2.8–3.3	50.0–55.0	0.2–0.8	0.006 max
Co	Cu	Ti	Fe	Nb+Ta
1.0 max	0.3 max	0.65–1.15	balance	4.75–5.50

shows the relationship between the depressions (places without material) and protrusions (places with the material). Whether they are depressions or protrusions it will be determined by the fact if the inequalities take positive or negative values from the profile core. For each finishing technology, Abbott Firestone curve is characteristic and it has particular relevance for assessing loaded functional surfaces, solving problems of friction, wear and lubrication. The value of upper part profile *Rpk* is the profile part which wears rapidly, for example it is used for assessing the nature and speed of the running-in sliding or rolling bearings. The lower part *Rvk* is again important for lubrication of the contact surfaces [4, 5, 6, 7].

The paper describes experiments and their results, which were conducted at FPTM JEPU with grinding Inconel 718.

EXPERIMENT

As mentioned above, the experiments were carried out for the material designated as INCONEL 718 (NiCr19NbMo). It is a hardenable alloy based on Ni-Cr containing a large quantity of iron, niobium, molybdenum, together with minor amounts of aluminum and titanium. It combines the corrosion resistance and high creep rupture strength to 700 °C with excellent weldability, including resistance to cracking. It is used in the aerospace industry, for gas turbines, in rocket engines, in the extraction of oil and gas and in the nuclear industry. In the experiment the alloy was subjected to spectral analysis, for verifying the expected chemical composition (Table 2) which was confirmed (see Table 1) [1, 3].

For the realization of the experiment a grinding wheel with grinding material SiC (silicon carbide) was chosen. As a material of this abrasive wheel green silicon carbide is applied. For the experiment semisynthetic process liquid was used. It was a 5% aqueous solution.

Grinding of all samples was carried out on a cylindrical grinder BU-16. The total of 12 samples were grounded and measured. Then a combination of cutting conditions were selected (see Table 3 and 4). The experiment utilized two methods of grinding, longitudinal and grooving [1, 2].

MEASUREMENTS AND ANALYSIS

As already mentioned above, profile and material ratio of the surface was evaluated on the machined surfaces, which are some of the parameters for the assessment of the surface quality after machining. The measurements of surface profile and material ratio were carried out on a device Hommel Tester T1000 according to DIN EN ISO 4287.

As mentioned above, the surface profile provides information on the segmentation of the surface, which is caused by many factors acting during machining, e.g. selected cutting conditions, trembling, rigidity of system, edge wear etc. Material ratio curve provides us with information about the structure of surface profile, therefore, the ratio of the core profile for dips and protrusions. The curve course can be used to evaluate the expected behavior of the component (running in the drive shaft, wear the loading area etc.). If the curve is decreasing slowly, it is a solid profile with a small volume of wells, and therefore, the surface is with very

Table 2. The chemical composition of the sample from Inconel 718 alloy according to the spectral analysis

INCONEL 718				
Element (% by mass)				
C	Si	Mn	P	S
0.015	0.01	0.06	0.002	0.0002
Cr	Mo	Ni	Al	B
17.96	2.99	53.55	0.53	0.002
Co	Cu	Pb	Sn	Ti
0.06	<0.1	0.0001	0.0009	0.98
Bi	Ca	Mg	Se	Fe
<0.00003	0.0005	0.0006	0.0001	rest
Nb+Ta				
4.965				

Table 3. Cutting conditions for longitudinal grinding

Longitudinal grinding, wheel SiC			
Sample No.	Cutting speed v_c [m·s^{-1}]	Feed f [m·min^{-1}]	Workpiece speed v_w [m·min^{-1}]
1	35	0.02	15
2		0.04	
3	40	0.02	
4		0.04	
5	45	0.02	
6		0.04	

Table 4. Cutting conditions for grooving grinding

Grooving grinding, wheel SiC			
Sample No.	Cutting speed v_c [m·s^{-1}]	Infeed v_f [mm·min^{-1}]	Workpiece speed v_w [m·min^{-1}]
7	35	0.13	15
8		0.17	
9		0.26	
10	40	0.13	
11		0.17	
12		0.26	

good antiwear properties. Steeply declining curve by contrast, notes strong cracks surface and thus the unfavorable surface properties and poor prognosis. [4, 5, 9, 11]

The first surfaces were obtained by the longitudinal grinding. Figures 1 to 6 present the profiles of samples from 1 to 6 ground according to the cutting conditions (see Table 3).

The largest waviness is showed in sample 1. Also, sample 5 shows in comparison with other samples worse quality, dips and protrusions are not so crowded as in sample 1 but in turn, they are considerably deeper. In comparison with all samples machined by longitudinal grinding, the best quality is showed in sample 2 (v_c= 35 m·s^{-1}, f = 0.04 m·min^{-1}) and 6 (v_c=45 m·s^{-1}, f= 0.04 m·min^{-1}).

If we look then at material ratio of individual samples (Fig. 7–12) the above mentioned facts are to some extent observable here too. Here, however, the worst material ratio curve is showed in sample 5 (v_c = 45 m·s^{-1}, f= 0.02 m·min^{-1}), although waviness was worst for sample 1, so its material ratio curve belongs to those advantageous (this is due to the amount and ratio of dips and protrusions). In comparison with all samples machined by longitudinal grinding, advantageous material ratio curve is showed in sample 3 (v_c = 40 m·s^{-1}, f= 0.02 m·min^{-1}) and sample 6 again (v_c = 45 m·s^{-1}, f= 0.04 m·min^{-1}).

Furthermore, the evaluated surfaces were obtained by grooving grinding, as mentioned above, there were used two cutting speeds v_c for three infeeds v_f. Generally, it can be stated that both, profiles and material ratio curves had worst values in comparison with the longitudinal grinding, as shown in Figures 13 to 24. Within grooving grinding sample 8 fared preferably

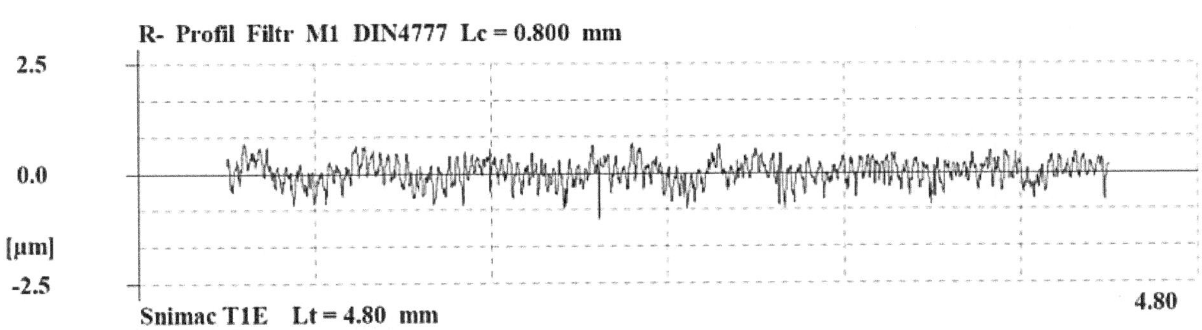

Fig. 1. Surface profile, longitudinal grinding, v_c = 35 m·s^{-1}, f= 0.02 m·min^{-1}, sample 1

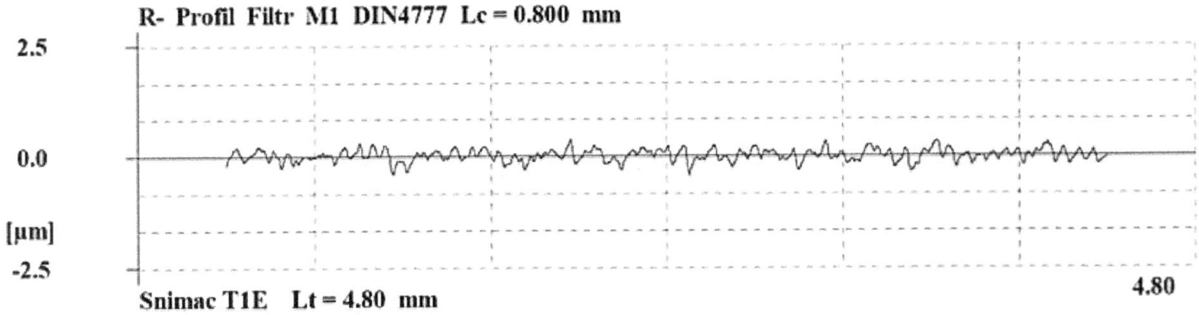

Fig. 2. Surface profile, longitudinal grinding, v_c = 35 m·s^{-1}, f = 0.04 m·min^{-1}, sample 2

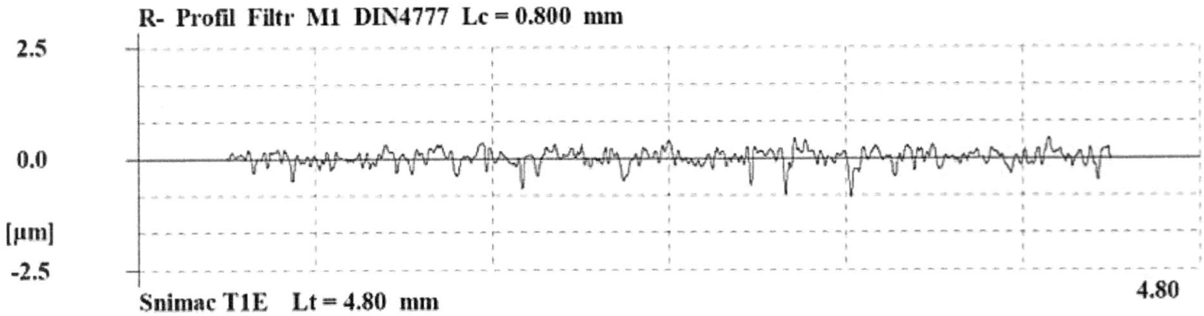

Fig. 3. Surface profile, longitudinal grinding, v_c = 40 m·s^{-1}, f = 0.02 m·min^{-1}, sample 3

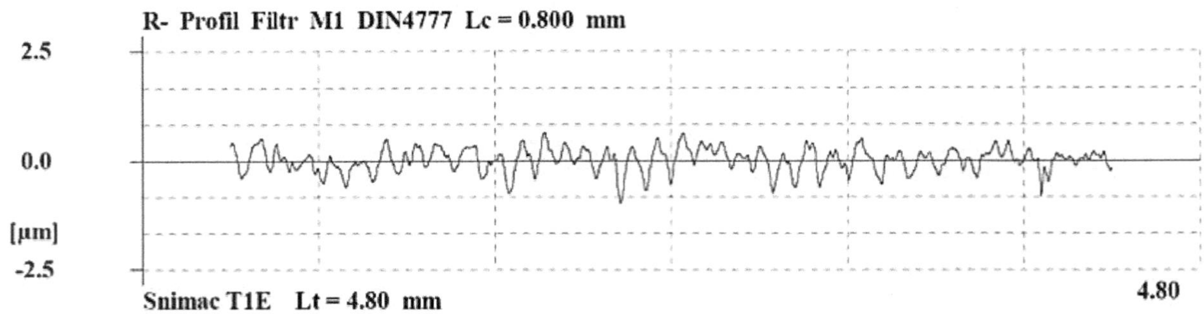

Fig. 4. Surface profile, longitudinal grinding, v_c = 40 m·s^{-1}, f = 0.04 m·min^{-1}, sample 4

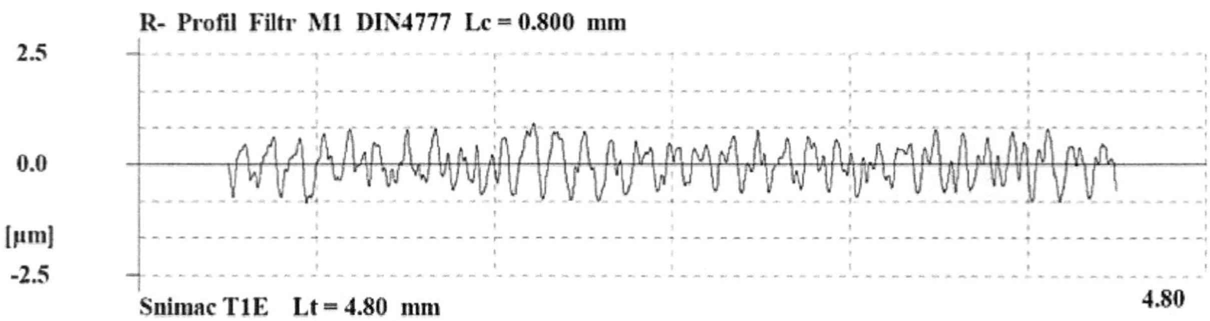

Fig. 5. Surface profile, longitudinal grinding, v_c = 45 m·s^{-1}, f = 0.02 m·min^{-1}, sample 5

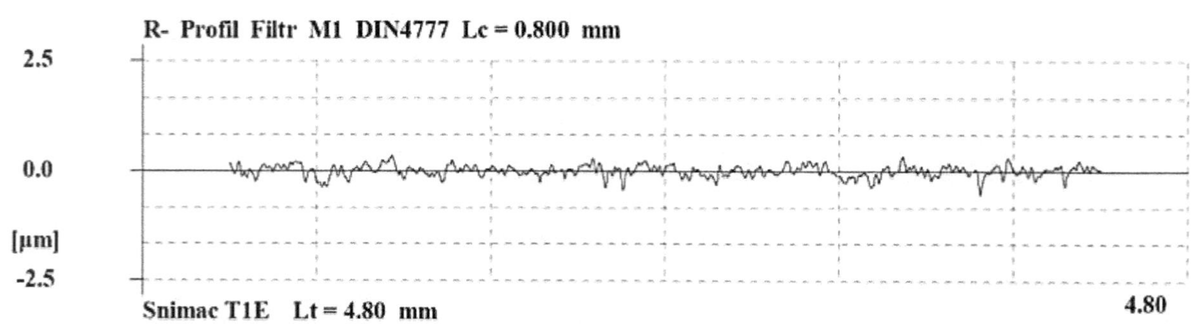

Fig. 6. Surface profile, longitudinal grinding, v_c = 45 m·s^{-1}, f = 0.04 m·min^{-1}, sample 6

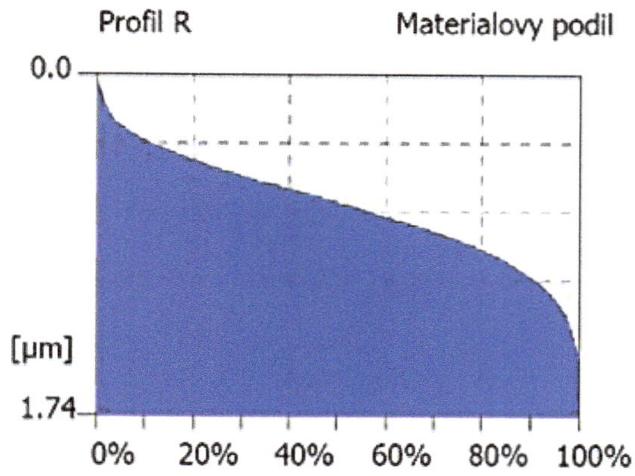

Fig. 7. Material ratio curve, longitudinal grinding, v_c=35 m·s^{-1}, f=0.02 m·min^{-1}, sample 1

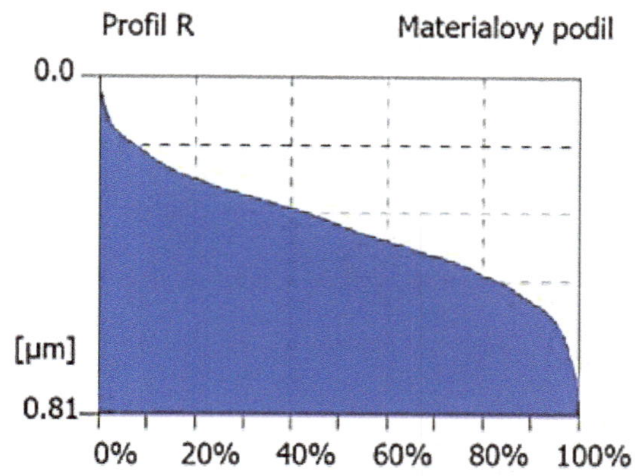

Fig. 8. Material ratio curve, longitudinal grinding, v_c=35 m·s^{-1}, f=0.04 m·min^{-1}, sample 2

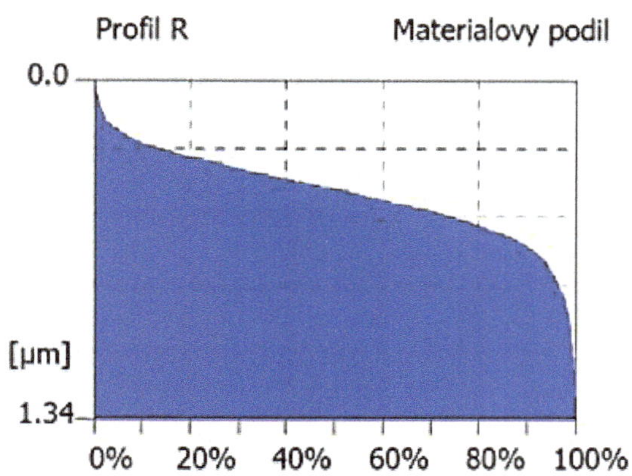

Fig. 9. Material ratio curve, longitudinal grinding, v_c=40 m·s^{-1}, f=0.02 m·min^{-1}, sample 3

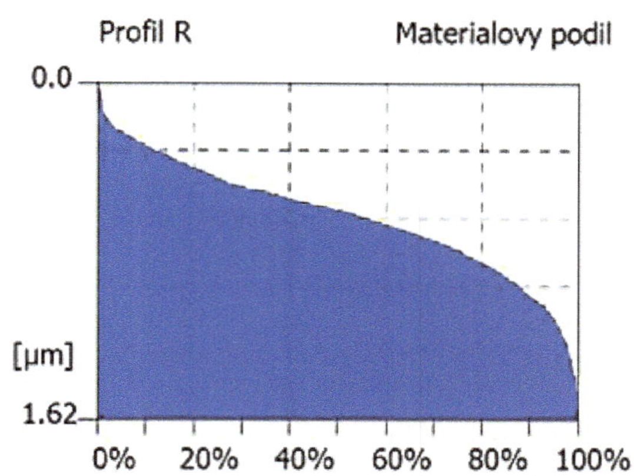

Fig. 10. Material ratio curve, longitudinal grinding, v_c=40 m·s^{-1}, f=0.04 m·min^{-1}, sample 4

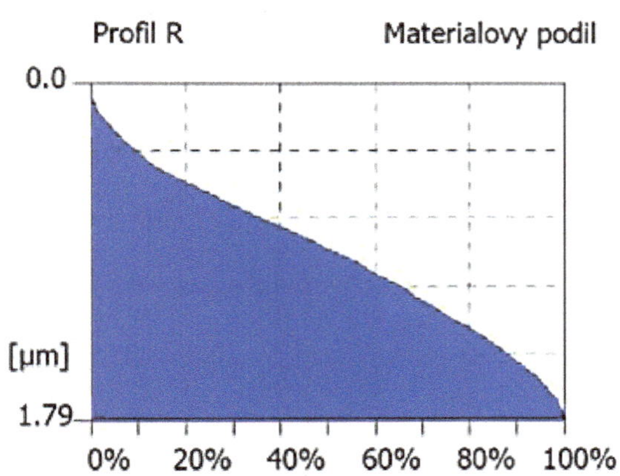

Fig. 11. Material ratio curve, longitudinal grinding, v_c=45 m·s^{-1}, f=0.02 m·min^{-1}, sample 5

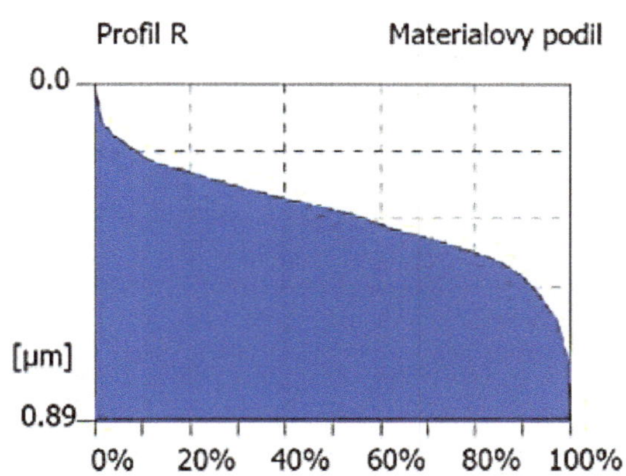

Fig. 12. Material ratio curve, longitudinal grinding, v_c=45 m·s^{-1}, f=0.04 m·min^{-1}, sample 6

(v_c = 35 m·s^{-1} and v_f = 0.17 mm·min^{-1}) in terms of the profile curve and in terms of the material ratio curve it is sample 9 (v_c = 35 m·s^{-1} and v_f = 0.26 mm·min^{-1}). The worst results in the context of the material ratio curve are showed in sample 11 (v_c = 40 m·s^{-1} and v_f = 0.17 mm·min^{-1}).

CONCLUSION

Machining of nickel alloys is currently a popular technology due to the expansion of nickel alloys use in the industry, and because they are difficult to machine materials, due to their hardness, this subject is necessary to pay attention to. At FPTM

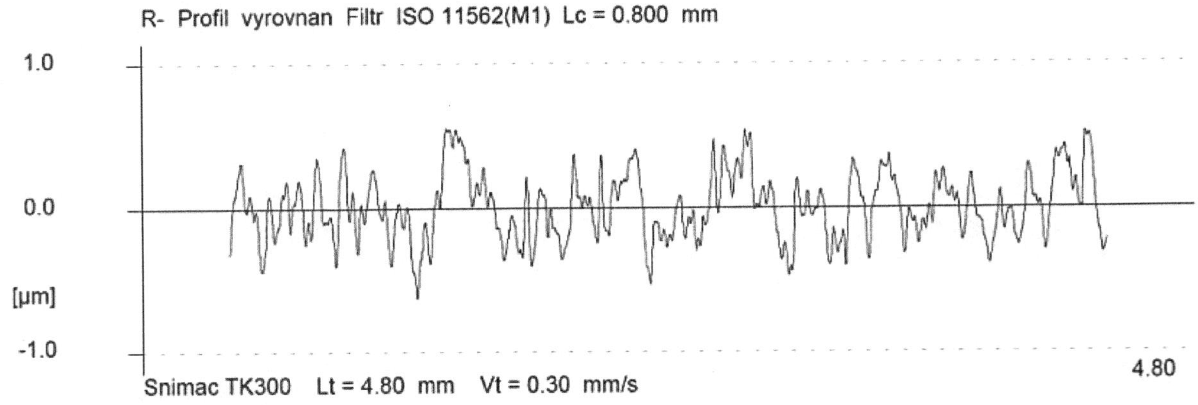

Fig. 13. Surface profile, grooving grinding, $v_c = 35\ \mathrm{m \cdot s^{-1}}$, $v_f = 0.13\ \mathrm{mm \cdot min^{-1}}$, sample 7

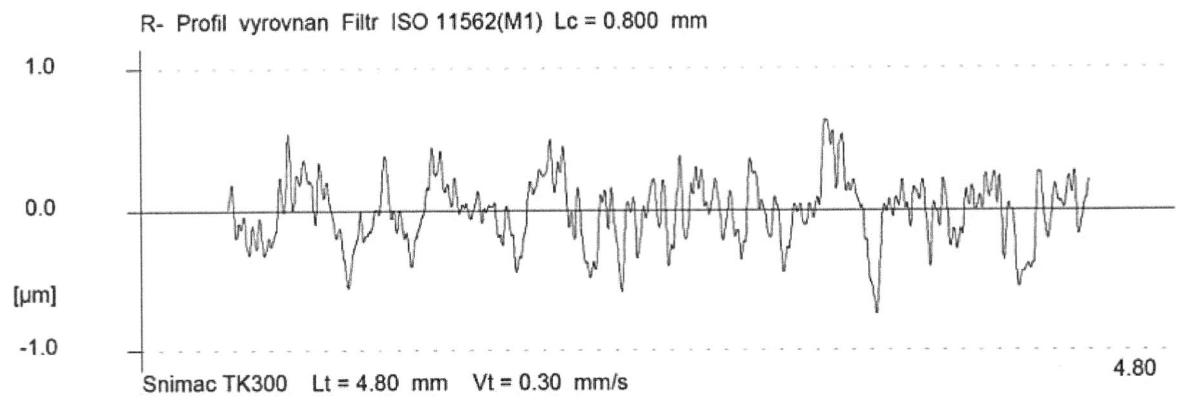

Fig. 14. Surface profile, grooving grinding, $v_c = 35\ \mathrm{m \cdot s^{-1}}$, $v_f = 0.17\ \mathrm{mm \cdot min^{-1}}$, sample 8

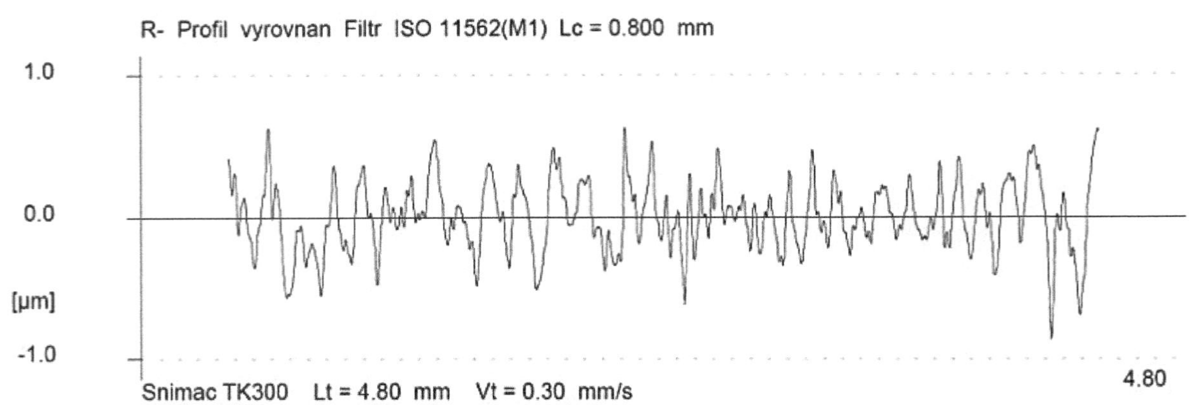

Fig. 15. Surface profile, grooving grinding, $v_c = 35\ \mathrm{m \cdot s^{-1}}$, $v_f = 0.26\ \mathrm{mm \cdot min^{-1}}$, sample 9

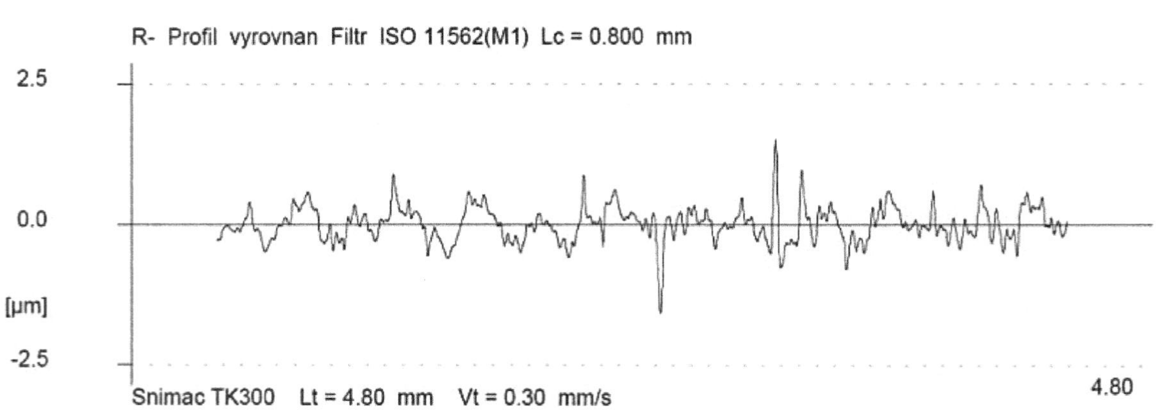

Fig. 16. Surface profile, grooving grinding, $v_c = 40\ \mathrm{m \cdot s^{-1}}$, $v_f = 0.13\ \mathrm{mm \cdot min^{-1}}$, sample 10

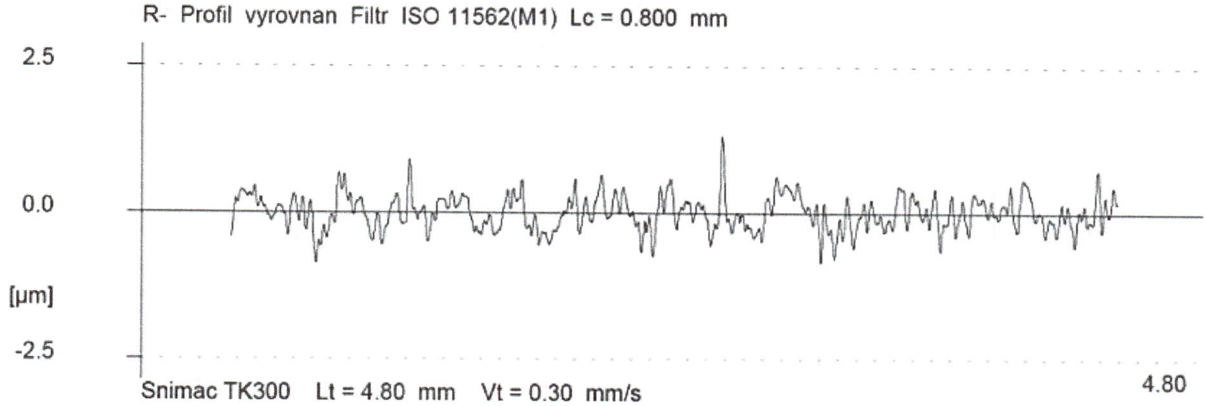

Fig. 17. Surface profile, grooving grinding, v_c = 40 m·s^{-1}, v_f = 0.17 mm·min^{-1}, sample 11

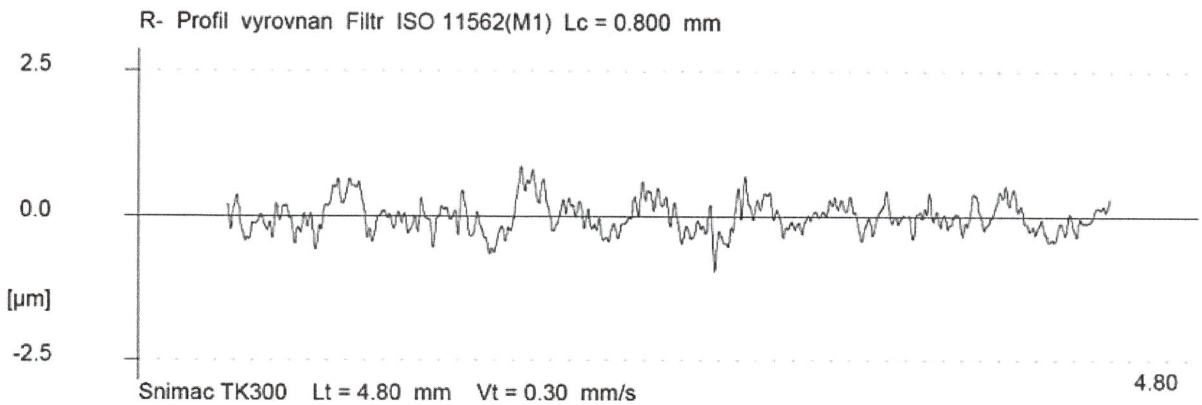

Fig. 18. Surface profile, grooving grinding, v_c = 40 m·s^{-1}, v_f = 0.26 mm·min^{-1}, sample 12

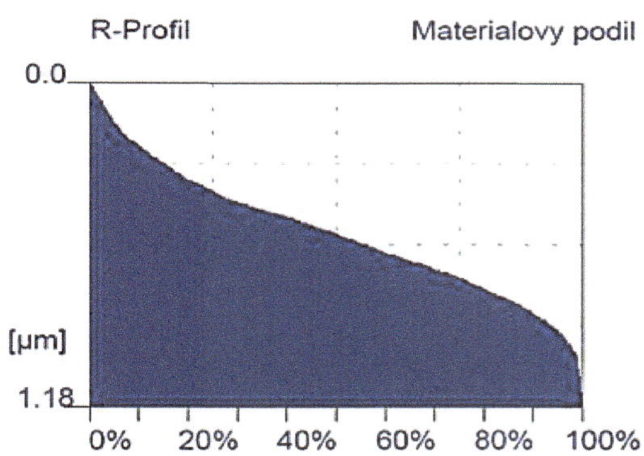

Fig. 19. Material ratio curve, grooving grinding, v_c= 35 m·s^{-1}, v_f=0.13 mm·min^{-1}, sample 7

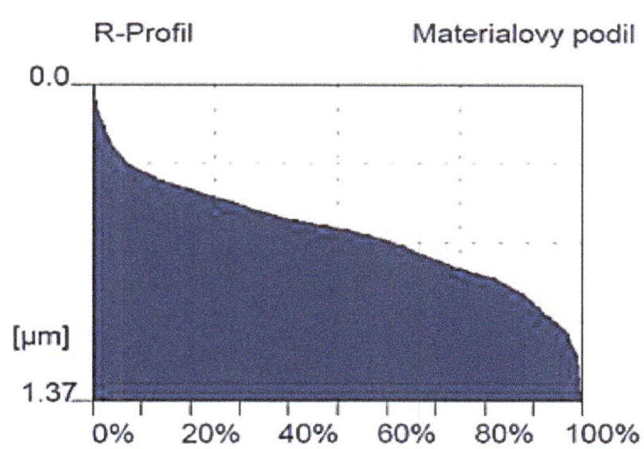

Fig. 20. Material ratio curve, grooving grinding, v_c=35 m·s^{-1}, v_f=0.17 mm·min^{-1}, sample 8

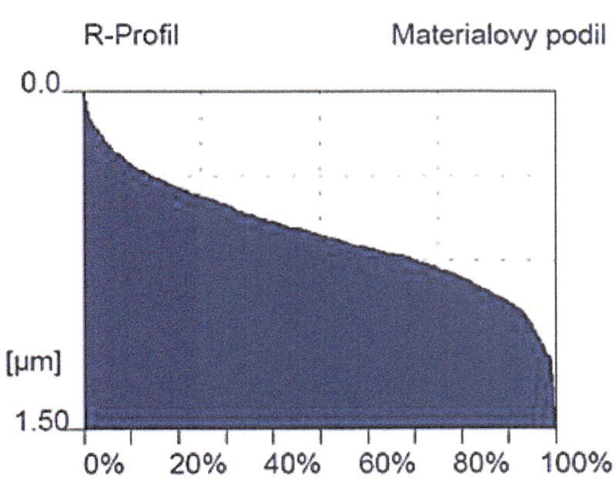

Fig. 21. Material ratio curve, grooving grinding, v_c=35 m·s^{-1}, v_f=0.26 mm·min^{-1}, sample 9

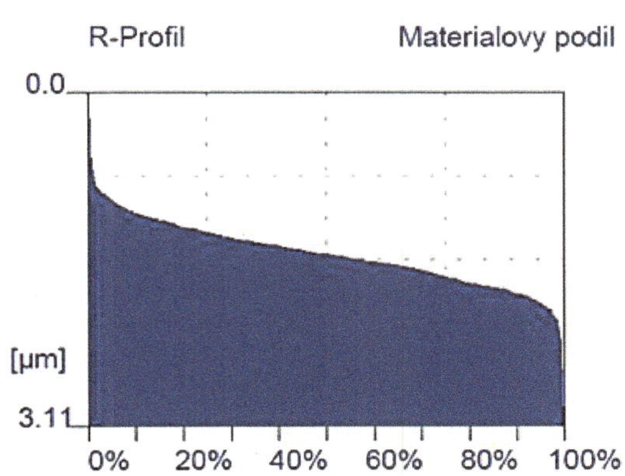

Fig. 22. Material ratio curve, grooving grinding, v_c=40 m·s^{-1}, v_f=0.13 mm·min^{-1}, sample 10

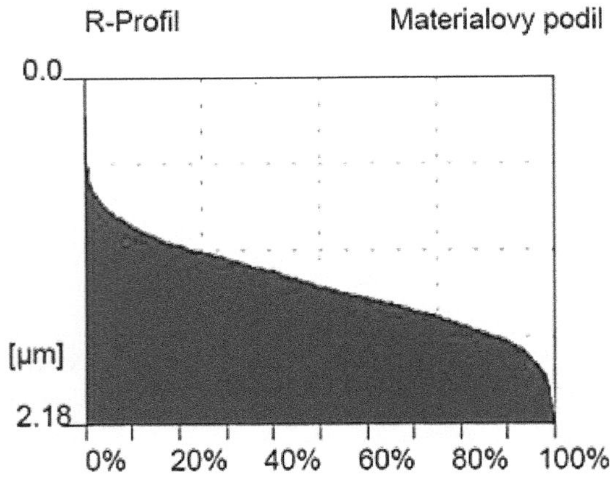

Fig. 23. Material ratio curve, grooving grinding, v_c=40 m·s^{-1}, v_f=0.17 mm·min^{-1}, sample 11

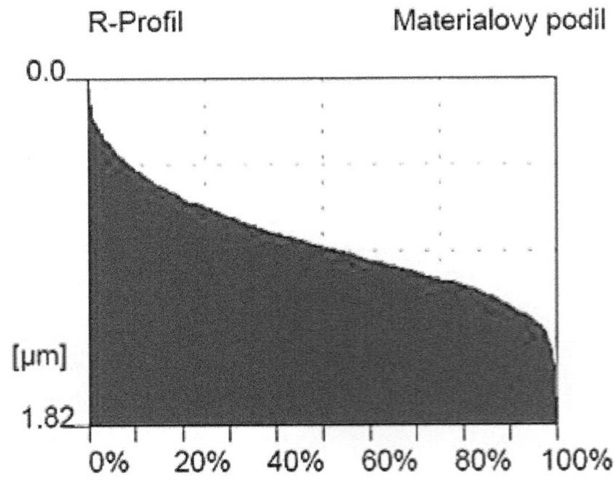

Fig. 24. Material ratio curve, grooving grinding, v_c=40 m·s^{-1}, v_f=0.26 mm·min^{-1}, sample 12

JEPU there are experiments carried out in this area and one of their components introduces this contribution. Experiments were carried out with grinding specific Inconel 718 alloy. The samples were ground longitudinally and by grooving grinding.

Generally, practically all samples appeared ground by longitudinal grinding and preferably for grinding conditions v_c = 45 m·s^{-1}, f = 0.04 m·min^{-1} (sample 6) were better for all cutting conditions because the profile has a relatively flat pits and protrusions, nor on the profile was any waviness visible. Furthermore, the material ration curve decreases relatively slowly in this sample, which again means a solid profile with a small volume of pits and protrusions. It is therefore stated in the experiment that in the context of performed measurements the best profile and the material ratio curve was acquired for cutting conditions for longitudinal grinding with cutting conditions v_c = 45 m·s^{-1}, f = 0.04 m·min^{-1}.

The described experiments are part of the extensive research carried out at FPTM JEPU in Ústí nad Labem.

Acknowledgements

Above mentioned results were created by means of the project in J.E. Purkyně University in Usti nad Labem "Materials and Human Resources for Environment".

REFERENCES

1. Bauer R. Obrábění těžkoobrobitelných materiálů. 2008. [Available at: www http://www.mmspektrum.com/clanek/obrabeni-tezkoobrobitelnych-materialu.html – cit. 2014-3-13].
2. Byrne P. Turning milling and grinding processes. Arnold, London 1996.
3. INCONEL® alloy 718 (UNS N07718), 2007. [Available at: www http://www.hpalloy.com/alloys/descriptions/INCONEL718.html – cit. 2014-3-13].
4. Novák M. Surface quality of hardened steels after grinding. Manufacturing Technology, 11(11), UJEP, Ústí nad Labem, 2011, 55–59.
5. Novak M. Surfaces with high precision of roughness after grinding. Manufacturing Technology, 12(13), 2012, 66–70.
6. Valíček J., Rusnák J., Müller M., Hrabě P., Kadnar M., Hloch S., Kušnerová M. Geometrické aspekty drsnosti povrchu klasických a netradičních technologií. Jemná Mechanika a Optika, 53(9), 2008, 249–253.
7. Kocman K. Optimalizace dokončovacích operaci výrobních procesů, Strojírenská Technologie, 17(3), 2012, 164–169.
8. Jusko O. Vývoj a inovace brousicích nástrojů. Strojírenská Technologie, 15(1), 2010, 17–22.
9. Novák M. Studium jakosti broušeného povrchu kalených ocelí, část I. – drsnost povrchu. Strojírenská Technologie, 16(6), 2011, 26–33.
10. Holešovský F., Novák M. Grinding and its influence to ground surface durability. Proceedings of International Conference on Advances in Materials and Processing Technologies, Paris. Amer INST Physics, Melville, NY, USA, 2010.
11. Kouřil K., Cep R., Janásek A., Kříž A., Stančeková D. Surface integrity at reaming operation by MT3 head. Manufacturing Technology, 14(2), 2014, 193–199.

5

MEASUREMENTS OF A STEEL CHARGE EMISSIVITY UNDER STRONG IRRADIANCE CONDITIONS

Agnieszka Benduch[1], Rafał Wyczółkowski[1]

[1] Department of Industrial Furnaces and Environmental Protection, Technical University of Czestochowa, Al. Armii Krajowej 19, 42-200 Częstochowa, Poland, e-mail: abenduch@wip.pcz.pl

ABSTRACT

Steel bars are manufactured in the rolling process, whereby they are characterized by strain hardening and poor plastic properties. In many application cases such properties are improper, therefore, additional heat treatment is required. Crucial influence on the products quality after heat treatment has an appropriate selection of process parameters. In many modern technologies of heat treatment the charge of porous structure is subjected to the heating process. Proper control of heat treatment parameters of bundles of rods requires knowledge on their thermal properties. However, it also requires accurate identification of complex heat transfer processes occurring in the porous material. Such analysis, with respect to bundles of bars, provide a response of qualitative nature of the heat exchange area of these charges. The article describes the emissivity measurements of samples of the steel charge using a thermal imaging camera.

Keywords: porous charge, thermal radiation, thermal resistance, long elements bundles, effective thermal conductivity, radiative thermal conductivity.

INTRODUCTION

For several steel products, one of the stages of the manufacturing process is heat treatment. This is related to the need to heat the workpiece up to a temperature whose value is determined by the position of the critical points in the equilibrium diagram of Fe-Fe3C. In practice, the recommended values of the heat treatment temperature are selected from the standards [1].Another important factor is the heat treatment time. This is due to the fact that the transformation processes of the material's microstructure which occur during heating are diffusive in character. Too short a heating time does not assure the transformation to be completed, whereas too long heating can, in turn, lead to a grain growth, excessive phase coagulation, oxidation and decarburization increase and, due to the energy consumption of heating, it is not economical. For these reasons, proper selection of the heat treatment parameters has a decisive impact on the quality of products.

The charge warm-up time can be theoretically based on the solutions of the Fourier-Kirchhoff equation [2, 3]. Obtaining the correct solution requires the determination of the net heat flux absorbed by the heated material. This value can be calculated with the use of the method of brightness radiation balance [4]. For this reason, the analysis of the radiative heat transfer in the system is a key issue for the calculations of steel charge heating. Obtaining the correct measurement results requires precise knowledge of the emissivity of heated steel. This value can be read from the tabular data reported in the literature [5–7]. However, this information should be considered as merely indicative. This is due to the fact that the emissivity of steel depends on a number of factors, which can take very different values. The most reliable source of knowledge on this parameter are measurements performed on samples made of the material of interest.

The article describes the emissivity measurements of samples taken from a batch of steel performed with the use of a thermal imaging camera. The measurements were carried out in the temperature range of 200–700 °C.

DEFINITION AND TYPES OF EMISSIVITY

Emissivity is a fundamental property of radiant bodies. Generally, it determines the derogation relating to the ability of thermal radiation emission of real bodies from a black body for the same temperature [5]. This formulation can be presented in a form of the following relation:

$$\varepsilon = \frac{\dot{e}}{\dot{e}_C} \tag{1}$$

where: $\dot{e}$ – stream of the radiation emission of a grey body;
$\dot{e}_C$ – stream of the radiation emission of a black body.

According to the radiative heat transfer theory there are several types of emission. In general, emissivity can be divided into the total and the monochromatic emissivity, as well as the directional and the half space emissivity. This allows to distinguish four basic types of this quantity, which are [4, 5]:

- directional monochromatic emissivity – $\varepsilon_{\lambda,\varphi}$;
- half-space monochromatic emissivity – ε_{λ};
- directional total emissivity – ε_{φ};
- half-space total emissivity – ε.

The value of emissivity is affected by many factors. The total emissivity depends on temperature; moreover, the monochrome emissivity varies with the wavelength. Both types of emission depend on the viewing angle φ of the emitting surface. At the same time, emissivity is an integral feature of the material itself, depending on its internal structure, the nature of the substrate and the working conditions of the emitting material. Determining the impact of various parameters on the emissivity value poses serious difficulties and it is not possible theoretically. This can be done only empirically by means of examining the emissivity as a function of various parameters, while maintaining the stability of the other ones. Emissivity measurements can be performed directly or indirectly [6]. Direct methods are based on measuring the radiant size of the tested body and the reference body, as well as determining – mostly based on their ratio – the values of ε. These methods are versatile and can be applied to any measurement temperatures and to all types of materials. They are usually used to measure the total directional emissivity or the monochromatic emissivity. Indirect methods usually allow the determination of the half space emissivity and they rely on the measurement of the optical or thermal quantities associated with the heat exchange process. On the basis of the measured quantities, the searched for emissivity value is determined from the energy balance equations. We can distinguish between several indirect methods [5, 6]:

- Calorimeter – involving an indirect measurement of the amount of energy given up radiatively by the body whose emissivity is being measured;
- Total reflection – involving the measurement of reflectivity, based on which the radiation absorptivity A is determined, the value of which is equal to that of emissivity. This method is most often used to determine the emissivity at room temperature,
- Measurement of the physical quantities, such as resistivity and optical constants of polarized radiation, based on which the emissivity is determined directly or indirectly, by way of integrating, in a specific wavelength range l and the previously calculated monochromatic emissivity.

RESEARCH METHODOLOGY

The emissivity measurements within the frames of the research were performed by a direct-type method. In general, the method consists of a simultaneous measurement of the temperature of the examined body by means of a thermocouple and a pyrometer [6]. During the tests, the value of emissivity in the pyrometer should be adjusted in such a away so that the temperature shown by the device can correspond to the value of the contact measurement. If this condition is fulfilled, we can assume that the emissivity set in the pyrometer is equal to the emissivity of the examined object. In the case of tests conducted at higher temperatures, the presented method has a certain flaw, which is the fact that, under such conditions, the measurement result is affected by the radiation stream emitted by the environment which reaches the examined surface. In the theory of radiative heat exchange, this stream is called irradiance. The environment of the examined object is understood as all the surfaces emitting heat radiation towards this object. The higher the ambient temperature, i.e. the temperature of these surfaces, the higher the irradiance. This phenomenon causes the fact that, during the tests, the radiation related to the emission of the examined object itself is the only one which does not reach the pyrometer's lens. The radiation reflected by that surface may be a significant part. Pyrometers, due to their simple measurement model, does not allow for a correction of this phenomenon. And so, the emissivity value obtained with their use, in direct measurements, can be burdened with a hard-to-estimate error.

The effect of irradiance in emissivity tests can be eliminated by way of using a thermovision camera instead of a pyrometer. That is because thermovision cameras, in comparison to pyrometers, characterize in a much more complex mathematical measurement model. The principle of those models, which results from the situation occurring during a thermovision measurement, is illustrated in Figure 1 [7]. As can be seen, the measurement result is affected by the following objects being the sources of radiation: the environment (1), the examined body (2) and the atmosphere (3). The symbols visible in the discussed figure denote the following: e – the emissivity of the examined object, t – the atmosphere transmittance, W_{refl}, W_{obj}, W_{atm} – the power of particular radiation sources which reaches the camera, T_{refl}, T_{obj}, T_{atm} – the temperatures of those sources.

And so, three heat radiation streams reach the camera lens during the measurements. They are: the stream emitted by the examined object, the stream emitted by the environment (the irradiance stream) and reflected from the examined object and the stream emitted by the atmosphere. Those streams are also attenuated by the medium present in the way of the measurement. The total infrared radiation power recorded by the camera is expressed by the following equation:

$$W_{tot} = \varepsilon \cdot \tau \cdot W_{obj} + (1-\varepsilon) \cdot \tau \cdot W_{ref} + (1-\tau) \cdot W_{atm} \quad (2)$$

Relation (2) can be treated as a simplified mathematical model of the thermovision measurement [7]. It should be noted that the stream of the radiation reflected by the examined object and the stream of the atmosphere radiation, to which powers W_{refl} and W_{atm} correspond, are factors which disturb the measurement. However, modern cameras perform an automatic compensation of those factors. This is realized by way of introducing the following data into the camera menu: the emissivity of the examined object, the environment temperature, the distance from the examined object and the temperature and humidity of the atmosphere. From this we can infer that, in the direct measurement of the emissivity with the use of a camera, we can make a compensation of a negative effect of the radiation. This is realized by way of setting the correct value of the ambient temperature in the camera's menu. A correctly adjusted value of this parameter eliminates the effect of the irradiance on the measurement result. This possibility was used during the described examinations.

RESEARCH IMPLEMENTATION

The emissivity investigations were performed for three low-carbon steel samples. Two of them were made of bars, 30 and 40 mm in diameter. They characterized in a raw state of the surface. The third sample was made of a square profile, 40 mm in diameter. The samples were marked

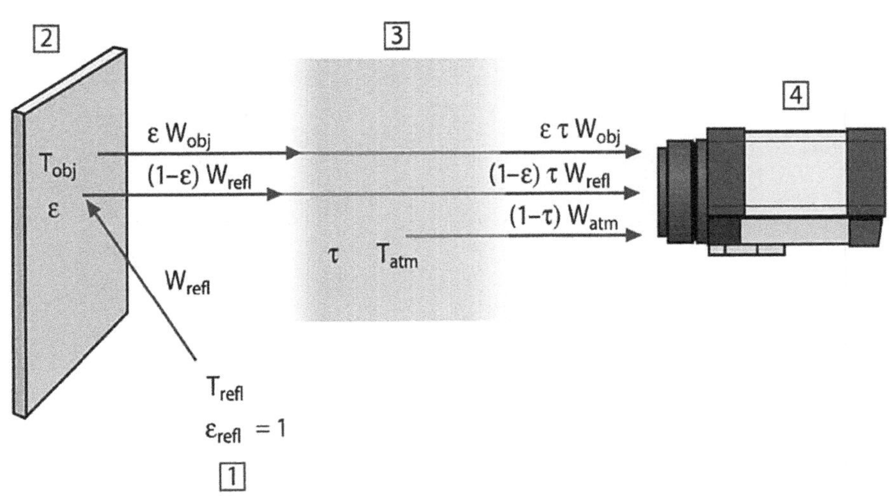

Fig. 1. Explanation of the principle of the mathematical model of a thermovision measurement

as follows: bar 30, bar 40 and the section. All the samples were about 200 mm long. Thermocouples type K were adjusted to each sample by means of heat-resisting silicon. They were shell thermocouples, 0.5 mm in diameter. Those detectors were used for a contact measurement of the temperature of the examined samples. An image of samples: bar 30 and bar 40 with the adjusted thermocouples is shown in Figure 2.

Fig. 2. Image of samples: bar 30 and bar 40 with shell thermocouples adjusted to their surface

An electric chamber furnace was used to heat the samples. The furnace chamber had the following diameters: 1000×300×250 mm (length, width, height). The examined samples were situated in such a way so that the surface with the adjusted thermocouples would be directed towards the charging hole of the furnace. The way of the sample arrangement in the furnace is shown in Figure 3a.

The test methodology required observation of the samples by means of a thermovision camera. For this reason, the charging hole of the furnace was covered by a ceramic needled cloth board with a hole cut out in the middle of it (Fig. 3b). The purpose of this procedure was to limit the heat loss from the chamber. At the same time, this assured stability of the thermal conditions in the furnace area during the tests.

The experiment itself consisted in heating the samples together with the furnace up to 700 °C. After reaching that value, the heating system of the furnace was activated. In consequence, in the further stage, the furnace slowly cooled down together with the samples. During that process, at the specified time intervals, thermogram registration of the examined samples was performed. Each thermogram corresponded to a specific temperature of the furnace. This quantity was measured by means of a thermocouple being part of the equipment of the regulation system of the furnace. At the moment of the thermogram registration, also registered were the temperatures measured by the thermocouples adjusted to the surface of the samples. For the assumed test range, a total of 10 thermograms were registered. Two exemplary thermograms are presented in Figure 4.

The further test methodology consisted in an analysis of the obtained thermograms. To that end, a special software was used, which was part of the equipment of the camera used for the tests. The thermogram analysis was connected with introducing the proper values of the ambient temperature T_{refl} and the emissivity e. The value of temperature T_{refl} was selected to be the temperature value of the furnace corresponding to a given thermogram. Next, the emissivity of given samples was set in such a way so that the sample temperature shown by the camera would correspond to the temperature shown by the thermocouple. After this condition was fulfilled, it was assumed that the set value of emissivity corresponded to the emissivity of the examined sample.

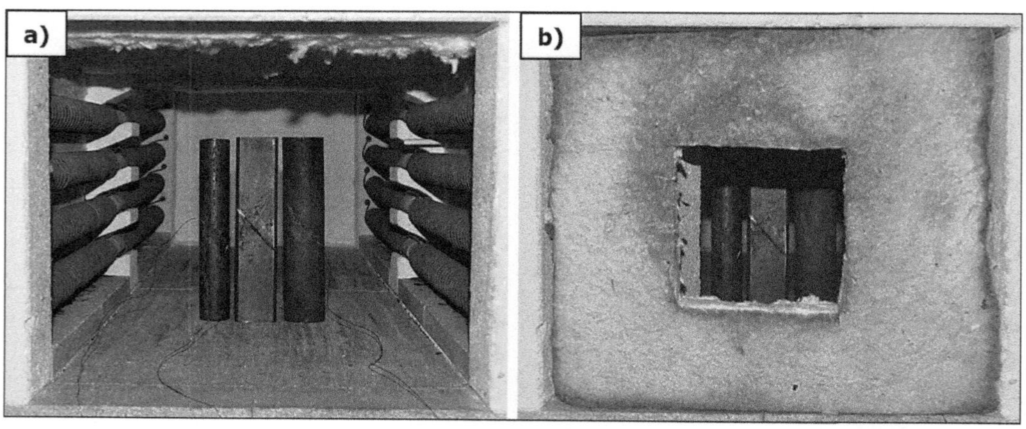

Fig. 3. a) Image of the examined sample in the furnace chamber, b) furnace charging hole covered by a ceramic needled cloth board

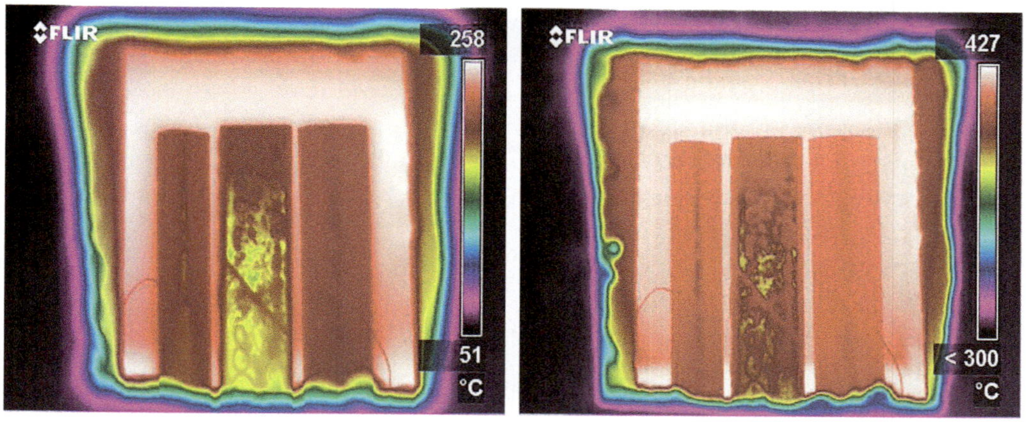

Fig. 4. Exemplary thermograms registered during the tests

The emissivity values of the samples in the function of temperature obtained in this way are presented in Figure 5. For all the samples, emissivity in the function of temperature increases almost linearly. For the bars, the obtained results were very similar. In the considered temperature range, the measured values of parameter *e* for those samples varied from 0.67 to 0.78. Slightly higher values were obtained for the sample made from the profile. The emissivity of that charge, together with the temperature, changed within the range of 0.83–0.93. This effect can be easily explained. The sample made from the profile, due to its low mass, heated up rapidly. At the same time, this resulted in a rapid oxidation of surface. The scale being the effect of this process, in comparison to the initial surface, has a higher emissivity.

The results of particular samples from Figure 5 were approximated by the least square method with the use of linear functions. The following relations were obtained in this way:

sample bar 30 $\varepsilon = 0{,}0002 \cdot t + 0{,}643$ (3)

sample bar 40 $\varepsilon = 0{,}0002 \cdot t + 0{,}635$ (4)

sample section $\varepsilon = 0{,}0002 \cdot t + 0{,}798$ (5)

It is worth mentioning that all the samples characterize in the same dynamics of emissivity increase. This is proved by the similar values of the slopes in equations (3) – (5).

The importance of the effect of irradiance on the values of emissivity determined by the presented method is shown in a diagram in Figure 6, where we can see the values of parameter *e* obtained for samples: bar 30 and bar 40, in the case when, in the analysis of the thermograms, the ambient temperature value was assumed as 25 °C (297 K). Figure 4 does not include the results for the profile. For this element, with an underrated value of the ambi-

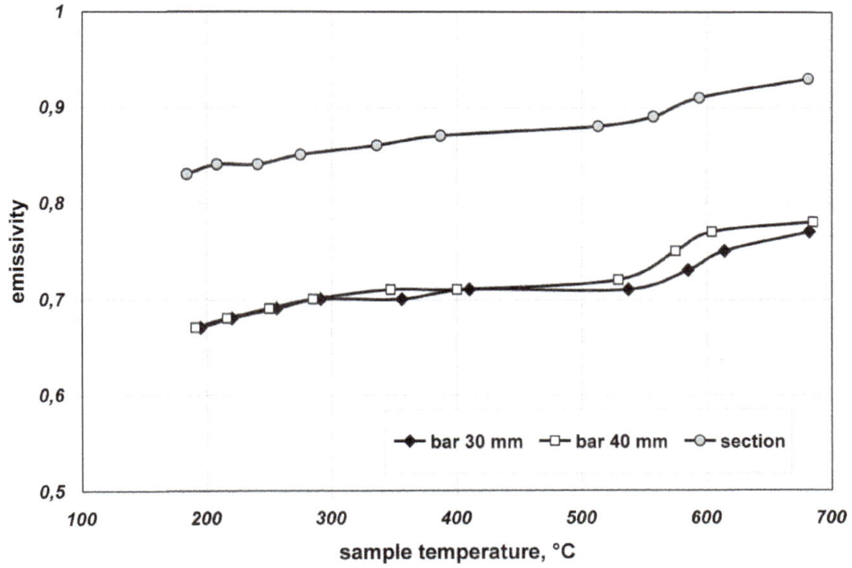

Fig. 5. Emissivity of the examined samples in the function of temperature

ent temperature, it was impossible to adjust the emissivity to the range of up to 1.0. As can be seen, with an underrated value of temperature T_{refl}, which means almost no irradiance of the samples, we obtained an overrated value of emissivity. The excessive values, as compared to the correct results, increase together with the temperature and equal about 0.1–0.2. Precisely, the differences between the results in Figure 3 and 4 can be expressed by means of a relative difference of emissivity. This parameter was defined as follows:

$$\Delta\varepsilon = \frac{\varepsilon_{25} - \varepsilon}{\varepsilon} \cdot 100\% \qquad (6)$$

where: ε_{25} – emissivity determined for the ambient temperature of 25 °C,
ε – emissivity determined for the ambient temperature equal to the temperature of the furnace.

The changes of parameter $\Delta\varepsilon$ depending on the sample temperature are shown in Figure 7. The data included there show clearly how important is the selection of the proper ambient temperature in the determination of emissivity. For the initial temperature range, the result discrepancy equals about 15%. However, for higher temperatures, parameter $\Delta\varepsilon$ increases and reaches the values of 24% and 27% at 700 °C.

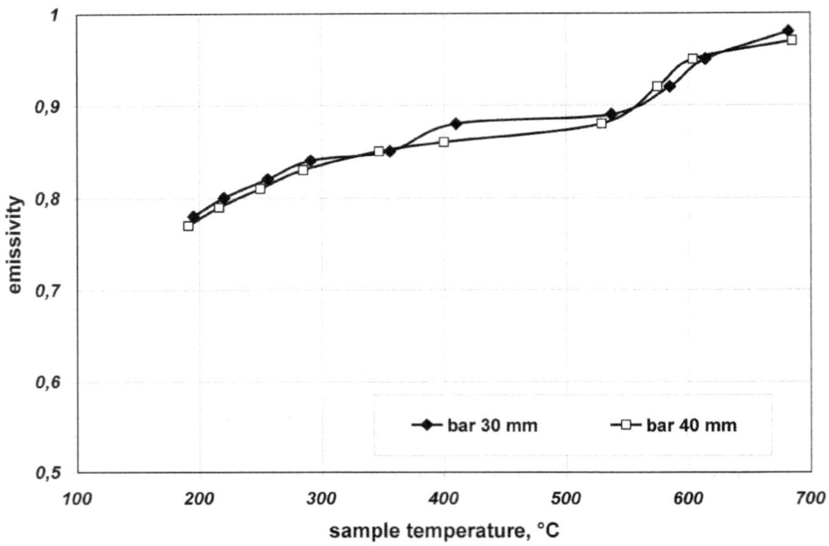

Fig. 6. Emissivity of the examined samples in the function of temperature determined for the ambient temperature of 25 °C

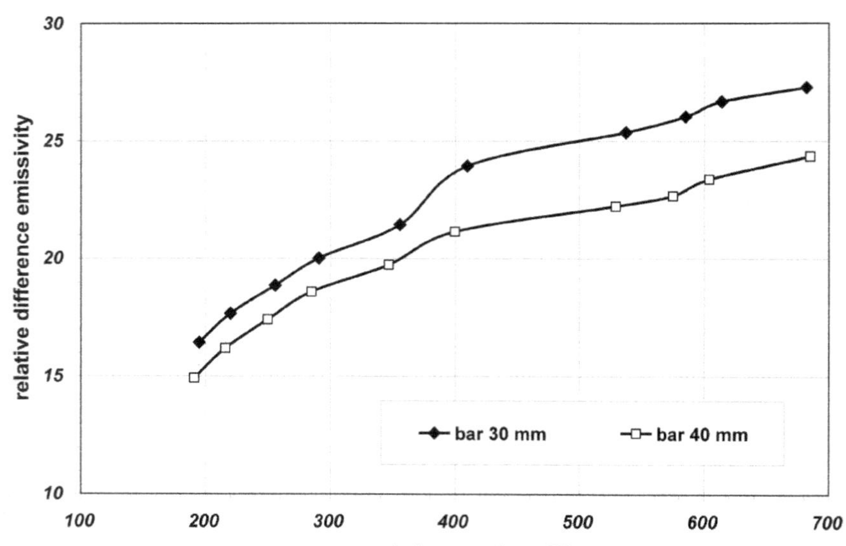

Fig. 7. Relative difference of emissivity for samples: bar 30 and bar 40 in the function of temperature

CONCLUSION

The results of emissivity measurements for steel charge samples were performed by way of a direct method consisting in a simultaneous measurement of the temperature of the examined charges with the use of thermocouples and a thermovision camera. The samples were heated in an electric chamber furnace in the temperature range of 200–700 °C. This way of heating caused a strong irradiance of the sample surfaces. The application of a thermovision camera allowed for a correction of this disturbance phenomenon. It was performed in the analysis of the thermograms by way of introducing an appropriate value of the ambient temperature. The value of this parameter is crucial for the quality of the obtained emissivity measurement results.

REFERENCES

1. Rudnicki Z.: Radiation heat flow in the industrial furnaces. Wydawnictwo Politechniki Śląskiej, Gliwice 1998 (in Polish).
2. Kostowski E.: Thermal radiation. Wydawnictwo Politechniki Śląskiej, Gliwice 2009 (in Polish).
3. Cengel Y.A.: Heat transfer – a practical approach. Second Edition, Mc Graw Hill, 2002.
4. Wiśniewski S., Wiśniewski T.S.: Heat transfer. Wydawnictwo WNT, Warszawa 2012 (in Polish).
5. Lienhard J.H.I.V., Lienhard J.H.V.: A Heat Transfer Textbook. Third Edition. Phlogiston Press, Cambridge Massachusetts.
6. Burakowski T.: Emissivity of heating resistor alloys. BOINTE IMP, Warszawa 1976 (in Polish).
7. Minkina W.: Thermovision measurements – instruments and methods. Wydawnictwo Politechniki Częstochowskiej, Częstochowa 2004 (in Polish).

6

INVESTIGATION AND ASSESSMENT OF OCCUPATIONAL RISK ON THE METAL CUTTING MACHINE TOOL STAND

Jerzy Józwik[1], **Paweł Pietras**[2]

[1] Department of Production Engineering, Mechanical Engineering Faculty, Lublin University of Technology 36 Nadbystrzycka Str., 20-816 Lublin, Poland, e-mail: j.jozwik@pollub.pl

[2] Graduate student of Mechanical Engineering Faculty, Lublin University of Technology, Poland

ABSTRACT

The paper analyzes occupational risk at selected metal cutting machine tool stands (turning, milling, drilling and grinding stands) and presents a set of recommendations formulated to minimize hazards at the above mentioned stands. The assessment is based on the analysis that allowed for selecting the best method of occupational risk assessment, measurements, observations, interviews and conclusions drawn from the obtained results. Occupational risk assessment forms were devised, what allowed for selecting appropriate action to increase the safety of selected machinery operators. The guidelines were assessed in terms of their practical implementation and expected utilitarian results.

Keywords: occupational risk, metal cutting machine tools, acoustic hazards, vibrations, occupational risk assessment forms.

INTRODUCTION

Metal machining is the most common manufacturing method that the machine industry development is based on. The observed trends of increasing efficiency of metal machining entail increasing machinery operating speeds, which, in turn, increases the risk of accidents at such stands. The most common hazards include: all kinds of mechanical injuries, exposure to high levels of noise, body exposure to vibrations, risk of electric shock, high temperature of chips, chemical substances present in cutting tool lubricants and their toxic fumes. These hazards pose a certain risk for operators of machine tool stands. In order to be reduced, occupational risk should first be assessed, as only then the sources of main hazards can be identified and eliminated. Under point 7 of §2 of the general health and safety regulations and the norm PN-N-18002:2000, occupational risk is defined as: „the probability of undesired events connected with the performed work causing damage, particularly adverse health effects of employees, due to occupational hazards resulting from work environment or manner of work" [13].

Occupational risk is comprised of many factors; under §39 of the general health and safety regulations, they are divided into 8 categories: injuries, diseases, occupational diseases, employee health, harmful and dangerous factors, employee qualifications, state of safety and health at work, and many others [1–4].

The concept of occupational risk pertains to all economic activities. Apart from casualties, it also involves material losses. Yet, given that under labor law the human being is considered the most important, it was therefore decided that occupational risk should always concern human resources [5-24].

OBLIGATION OF OCCUPATIONAL RISK ASSESSMENT

Occupational risk assessment is one of the obligations of every employer, irrespective of type and level of harmful, dangerous and noxious factors in workplace. This obligation pertains to all work positions and is regulated by relevant laws [19].

Under article 226 of the Labor Code, every employer is obliged to assess and record occupational risk involved in work performed, to inform employees about occupational risk and safety measures protecting from hazards as well as to administer essential preventive measures reducing the risk.

Under §39 of the general health and safety regulations issued by the Minister of Labor and Social Policy, the employer is obliged to conduct and record occupational risk assessment. The employer is obliged to take records of the administered essential preventive measures [18].

The above requirements result from the European Union directives. Under article 6 of the framework directive 89/391/EWG, employers should prevent occupational risk by minimizing hazards to safety and health of their employees. The employer is required to follow the rules specified in the above mentioned directive which include: hazard prevention, assessing hazards that cannot be eliminated, eliminating sources of hazards, implementing new technological solutions, replacing dangerous operations with operations that pose less danger or no danger at all, proper training and instruction of the staff, proper preventive policy and its coherent development, giving priority to collective preventive measures over individual preventive measures [18].

According to the latest amendments to the Labor Code, that is, the revised version of article 235, effective from July 3, 2009, the employer is obliged to determine causes of occupational diseases as well as the character and extent of hazards causing these diseases. The definition of occupational disease was transferred to the Labor Code from the law on social security in respect of accidents at work and occupational diseases [6].

Under article 227 of the Labor Code, employers are obliged to counteract occupational diseases. To this aim, they should administer measures that prevent occupational diseases and other diseases involved in the performed work. These measures include: keeping in proper operational condition, both devices that limit or eliminate health-harmful factors in work environment and devices that are used to measure these factors, conducting at their own expense the examination and measurement of factors harmful to health, to record and store the examination results, as well as to make them accessible to employees [19].

Even though legal regulations do not unequivocally name the person responsible for conducting occupational risk assessment, it should be conducted by competent persons (a company employee, external consultant, etc.). The norm PN-N-18002:2000 recommends that the following data be used in assessing occupational risk: information about stand localization and tasks performed there, information about employees operating the stand, particularly about those to whom special criteria are applied such as pregnant women, young and inexperienced employees or the disabled. Occupational risk assessment involves conducting relevant analyses (technical analysis of machinery and devices used, instructions for stand operators, etc.), measurements of harmful or dangerous factors that occur at the stand (allowable concentration values and other values determined for a given stand according to the legal regulations and norms [13]), visual inspections and staff interviews (gathering information about work measures used, preventive measures, materials and performed operations as well as the manner and time of performing these operations by the employees). The analysis of immeasurable values should be performed, most of all, by close observation of work environment, manner of performing work at a given stand and outside it, as well as by conducting invaluable staff interviews [13].

Those who conduct occupational risk assessment should possess appropriate knowledge and information about the following: identified hazards, risk types and how they are posed; materials, machines and technologies used at work; wok procedures and organization; staff exposure to materials used; type, probability, frequency and duration of hazard exposure; effects of exposure to certain harmful factors; norms and legal requirements concerning risk types at a given workplace. Such persons need to know and understand the rules and the idea of occupational risk assessment; additionally, it is recommended that they should be capable of inventing corrective or/and safety measures and of assessing their effectiveness [6].

When conducting risk assessment, relevant norms and legal regulations concerning the stand being analyzed should be observed. What should also be taken into account is the data about hazards and their sources detected on a given stand earlier, effects, accidents, occupational diseases, failures that have taken place, etc. [13].

Risk level assessment connected with the work performed at a given stand should be conduced whenever new stands are introduced or the existing stands are to be changed. Occupational risk assessment is also necessary when technological process or work organization is to be modified. Also, such analysis should be conducted for preventive reasons in order to avoid hazards resulting form the worsening of machinery operational condition, staff rotation or other factors that went undetected in previous inspections [13].

PURPOSES OF OCCUPATIONAL RISK ASSESSMENT

The main purposes of conducting occupational risk assessment include:

- identifying hazards which occur on the tested stands,
- determining the probability of their occurrence,
- determining the effect of the identified hazards on employees and work environment,
- selecting the optimal stand equipment, materials, safety measures and methods of work organization, and, having identified the hazards, even selecting adequate employees to operate such stands, [16]
- establishing priorities concerning the reduction or total elimination of hazards [16],
- constant improvement of safety and health at work [16],
- demonstrating to both employees and control bodies (e.g. National Labor Inspection agents) that the risk is known and appropriate measures to eliminate it have been taken.

Occupational risk assessment begins with determining machine function, together with the production profile of this machine, its efficiency, materials used, spatial limits and application type, planned cycle life, functions and modes of operation, predicted machine malfunctions and probable failures, employees responsible for operating the machine, and involuntary machine operation or its reasonably predictable misuse. Examples of involuntary machine operation or its reasonably predictable misuse include: machine control loss, reflex action in the case of machine malfunction, failure or damage, improper behavior resulting from concentration loss or inattention, operating the machine in an irresponsible manner, haste, etc. When assessing occupational risk, hazards taken into consideration should include both the ones that may occur at the stage of determining machine functions (design, production) and, most importantly, those that may occur at all stages of machine life cycle (Table 1).

Having identified the hazards, risk assessment for every hazardous situation being considered (for all hazards) should be conducted. The risk depends on such factors as the level of damage caused by a given hazard (*L* – low, *M* – medium, *H* – high) and the probability of damage occurrence (Figure 1), resulting from exposing an employee or employees to the hazard, a hazardous event and technical and human capabilities of preventing or limiting the damage.

Occupational risk assessment can be conducted using various tools, such as risk tables, risk charts, numerical methods, and the like. When assessing occupational risk based on risk calculation results, it needs to be established if safety measures are necessary to be taken and when the required risk reduction is to be obtained. Risk assessment results need to be recorded. The risk assessment records should include the procedure applied and results obtained, as well as such data as machine tech-

Table 1. Hazard identification when assessing occupational risk

Hazards which have to be considered by producers …	….and in all stages of machine life cycle
• mechanical hazards	• transport, assembly and mounting
• electrical hazards	• start-up
• thermal hazards	• repairs
• hazards caused by noise	• verification of units and elements
• hazards caused by vibrations	• renovation
• hazards caused by radiation	• regeneration of units and elements
• hazards caused by materials and substances	• normal operation and failure elimination
• hazards caused by lack of ergonomic solutions	• maintenance and cleaning
• hazards resulting from slipping, tripping or falling	• disassembly
• hazards connected with machine working environment	• utilisation
• hazards resulting from the above hazards combined	

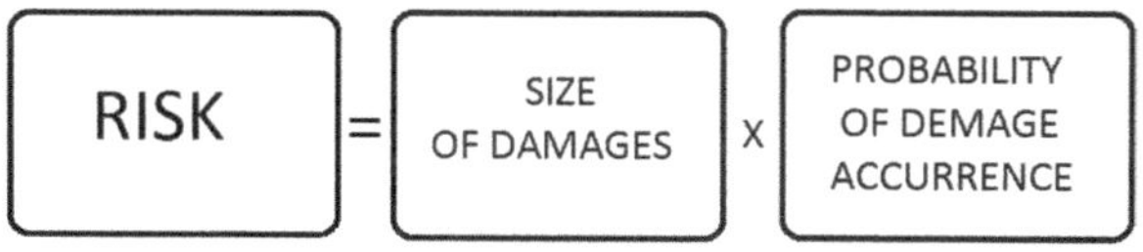

Fig. 1. Risk calculation and assessment

nical details, assumptions made, all identified hazards and dangerous situations, events, data used and their sources, description of safety measures applied, description of risk-reduction aimed that are feasible using the safety measures applied, residual risk connected with a given machine, documents prepared in the course of risk assessment.

METHODS OF OCCUPATIONAL RISK ASSESSMENT

There are numerous methods of occupational risk assessment. Given the accuracy of risk assessment and type of data, the methods can be divided into two categories: qualitative and quantative [2, 16–22].

The qualitative methods are employed if there is no access to statistical databases providing such information as number and type of accidents, dangerous events, occupational diseases, time of exposure to harmful factors, and the like [2]. If there is no access to such information, one of the methods of event occurrence should be employed. The risk (R) can be expressed as a function (2).

$$R = f(P, S) \tag{1}$$

where: S – the outcomes caused by the event,
P – the probability of outcome occurrence,
R – the assessed occupational risk.

This value can be determined in a descriptive manner (linguistically, qualitatively) or by means of numbers (quantitatively). The outcome probability employed in the qualitative methods is not mathematically defined, but it is rather a measurement which states that outcomes of different events are assessed at different levels. The risk is expressed as a combination of event frequency (probability) and consequences of their outcomes. The qualitative methods include: Preliminary Hazard Analysis (PHA), Risk Score, the method of five steps to risk assessment, risk graph and risk matrices according to the norm PN-N-18002 [2].

EXPERIMENTAL TESTS AND THEIR RESULTS

Operating metal working machine stands always poses numerous hazards, the applied safety measures notwithstanding. A vast number of such hazards occur over time and they are impossible to avoid. This results, among others, from changes in work environment, machinery wear, failures, and ignoring the industrial safety regulations on part of employees [2]. The Polish Norm PN-80/Z-08052 – Work protection. Dangerous and harmful factors in the work process describes hazards that should be taken into consideration when assessing occupational risk, yet not all of the hazards listed therein pertain to machining stands. Out of these hazards, 7 most important ones have been selected for the sake of the present paper. They include: exposure to high noise level, vibrations (measurable factors), risk of falling down on the same level, injures caused by movable machine parts, the risk of being thumped or cut by protruding machine elements, the risk of burns and electric shock [8]. The fundamental safety measures against the above hazards involve proper machinery arrangement and using shields protecting against injuries caused by chips and cooling agent splashing. Each stand should be equipped with a valid industrial safety manual. As for the measurable factors, periodic inspections of their maximum allowable concentration (MAC) and maximum allowable intensity (MAI) should be conducted. The pathway distance between the machines should be of at least 0.75 m, and there should be 2 m^2 of free floor space per employee. The room should be provided with proper lighting and ventilation [2].

The study was undertaken to assess occupational risk involved in operating metal cutting machine tools such as: turning lathes – S-32, Cu-401 and C-11; milling machines – FWF-32, FYA -32 and FNB-26, drilling machines: WS-15 and WKA-25; as well as grinding machines: SPC-20 and SWA-10. The study involved measuring noise and vibrations generated by the metal cutting machine tools as well as visual inspection necessary to assess immeasurable hazards. The noise measurement was done in the manner specified by the norm PN-N-01307:1994. Requirements for conducting measurements-namely, the microphone was at a height of 1.2 m over the floor and its distance from the machine was the same as the distance of the operator from the machine. The noise measurements were done using a Sonopan P-01 meter. Another thing used was a correction filter type A, which is applied to measure maximum intensity level of sound A. Such filter best reflects the noise and frequencies the human ear is exposed to. The maximum allowable sound level is 85 dB, standardized by legal regulations and norms [11].

In order to measure vibration level, a vibration meter, SEQUOIA Triaxal Acceleration Computer, was used. It was connected to a signal converter and computer monitoring the vibrations. The sensor was mounted on the machine housings so as to best examine the vibrations that an operator is exposed to. The vibration level was examined in the Y axis because vibrations in this direction were the highest and affected the machine operators most. The obtained results were copied to a spreadsheet and – based on the data— relevant charts were made as well as the standard deviation σ and the maximum allowable value of vibration acceleration in the direction of the Y axis (1) were determined [16].

$$a_{hv,30min,dop} = \max\{\sqrt{a_{y1}^2}, \sqrt{a_{y2}^2}, \sqrt{a_{y3}^2}, \ldots, \sqrt{a_{yn}^2}\} \quad (1)$$

It was assumed that vibrations affect the human body via the upper limbs (local vibrations) and the time of their action is shorter than 30 minutes. In accordance with such data, the acceptable value of short-term exposure to vibrations was taken as $a_{hv,30min,dop} = 11.2\ m/s^2$.

COMPARISON OF THE TESTED MACHINE GROUPS

Table 2 and Figure 2 offer a comparison of results of measured noise level emitted on the tested metal cutting machine tool stands. Such comparison allowed for distinguishing two groups of machines: one including machines which pose lowest hazard for their operators' health and one which comprises machines posing the highest hazard for their operators' health.

Table 3 and Figure 3 compare the obtained standard deviation σ and peak value a_{hvmax} for vibration acceleration measurements in the direction of the Y axis for the examined metal cutting machine tool stands. The acceptable acceleration value was taken as $a_{hv,30min,dop} = 11.2\ m/s^2$.

In occupational risk assessment, the most significant parameter caused by local vibrations, that is, the ones affecting human body via the upper limbs, is the vibration acceleration peak value a_{hvmax}; the acceptable value for this vibration type is 12.8 m/s^2. From Table 3 it can be inferred that as many as five out of the examined machine tool stands do not satisfy the above mentioned condition. Out of all the machines, it is the milling machine FYA-32 that poses a higher hazard in this respect, while the drilling machine WS-15 is the safest.

Table 2. Comparison of minimum, maximum and mean noise intensity levels on all the tested stands EA

Noise intensity level [db]	Turning lathes		Milling machines		Grinding machines		Drilling machines	
	idle run	operation test	idle run	operation test	idle run	operation test	idle run	operation test
Mean	80.3	87.9	85.33	86.11	91.00	96.00	88.25	85.50
Min	68	82	82	83	90	96	87	82
Max	87	93	92	88	92	96	89	88

Table 3. Comparison of vibration parameters of the tested machine groups, $a_{hv,30min,dop} = 11.2\ m/s^2$

Group	Symbol	Minimum standard deviation σ [m/s²]	Maximum standard deviation σ [m/s²]	Mean standard deviation $\sigma_{śr}$ [m/s²]	Peak value a_{hvmax} [m/s²]
Turning lathes	S-32	1.27	7.79	5.29	28.1
	C-11	0.25	3.47	1.64	9.86
	CU401	0.51	7.28	3.23	22.3
Milling machines	FYA-32	0.25	13.3	3.53	40.2
	FWF-32	0.16	2.29	0.79	10.1
	FNB-26	1.44	9.91	4.98	19.36
Drilling machines	WKA-25	0.17	4.06	1.27	10.1
	WS-15	0.17	0.18	0.18	4.3
Grinding machines	SWA-10	0.08	1.44	0.76	5.21
	SPC-20	0.25	0.59	0.42	14.06

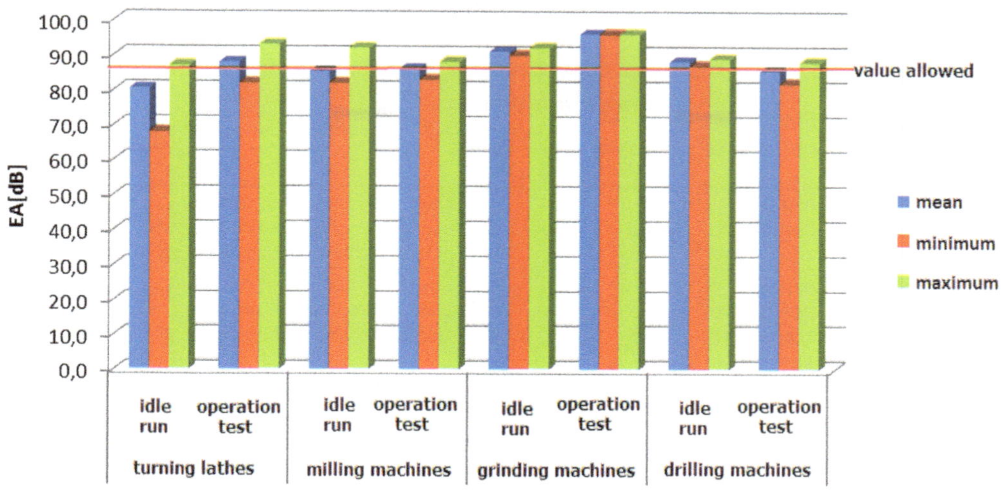

Fig. 2. Comparison of noise levels emitted by machines on tested metal cutting machine tool stands, EA_{dop}= 85dB – marked in the chart by a horizontal line

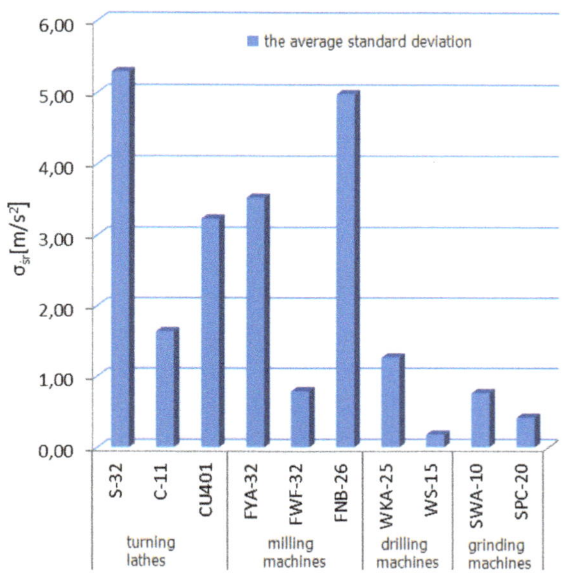

Fig. 3. Comparison of mean standard deviations of vibration acceleration in the direction of the *Y* axis of the tested stands

From a technological point of view, the standard deviation σ is a more important parameter describing vibrations. The bar chart in Figure 3 illustrates a comparison of mean standard deviations, which allows for selecting each time the least technically efficient machines, that is, the machines which have the highest mean standard deviation $\sigma_{śr}$.

OCCUPATIONAL RISK ASSESSMENT FORM FOR THE TESTED STANDS

In order to devise occupational risk assessment forms for the examined stands, it was first necessary to measure noise level and vibration acceleration, conduct inspection and carry out interviews with employees.

The method selected to conduct occupational risk assessment complied with the norm PN-N-18002/2000 due to its transparency and the possibility of assessing both measurable and immeasurable values. This method is best known and most often described in the literature, and, as a result, most often employed by all kinds of companies, including those dealing with metal machining. The forms contain modifications and recommended preventive measures and means of reducing the risk involved in operating the examined stands. In addition to that, occupational risk after the application of the suggested recommendations was also assessed. The devised occupational risk forms are shown as tables. Table 4 is an example of the risk assessment form devised for a turning lathe S-32. Similar forms were made for other tested machines.

CONCLUSIONS

The analysis of the available works and the conducted research allowed for assessing occupational risk on the selected metal cutting machine tool stands. Based on the conducted experimental tests, the following have been found:

- the fastest, most transparent and, at the same time, simplest method of occupational risk assessment is the method which complies with the norm PN-N-18002:2000.
- in the examined machine groups, the safest machine tools include milling and drilling stands,

Table 4. Occupational risk assessment form for a metal cutting machine tool stand (occupational risk assessment form devised for a turning lathe S-32, risk assessment levels: *L* – low, *M* – medium, *H* – high)

Hazards	Source of hazard	Potential hazard effects	Before assessment			Recommendations	After implementing recommendations		
			probabilty (P)	risk (T)	level (L)		probabilty (P)	risk (T)	level (L)
Noise	Machines	Hearing damage, neuroses	M	L	L	Keep the machine in proper operating condition, provide employees with hearing protectors.	L	L	L
Vibrations	Machines	Nervous system damage, cardiovascular system damage, vibration white finger	M	M	M	Keep the machine in proper operating condition, provide vibration damping devices.	M	L	L
Falling on the same level	Slippery, uneven ground	Contusions, fractures, internal injuries	M	L	M	Provide proper footwear. Keep the workplace in order; wipe up any spilt liquids immediately.	M	L	L
Injuries caused by movable machine parts	Cutting tools, power units	Personal injury, death	H	H	H	Use shields and proper tools, wear protective clothing (without any loose elements) when operating machinery.	M	M	M
Thumps, cuts	Improperly mounted object, chips	Personal injury, death	H	H	H	Mount the workpiece properly. Do not remove chips by hand, use shields and safety goggles.	M	M	M
Burns	Cooling liquid, chips	Burn wounds	M	H	H	Equip the stand with shields against hot chips.	L	L	L
Electric shock	Electrical breakdowns	Death	H	M	H	Perform periodic inspections of electrical installations	M	L	L

- the operation of turning lathes and grinding machines involves a higher risk of accidents and failures; the safest stands are: the drilling machine WKA-25, the milling machines FNB-26 and FWF-32; while the highest hazard is posed by the turning lathe S-32 and grinding machine SPC-20,
- the most common problem identified in occupational risk assessment is lack of protection against chips and fast rotating elements breaking off a machine or tool, which can cause grave injuries, cuts, or even death,
- emitted noise level should be taken into account – it is recommended that personal protectors be used as in the case of all the machines, except for the turning lathe S-32, the noise emission level is very close to the threshold limit value of 85dB or even exceeds it, especially in the case of grinding stands SPC 20 and SWA 10, where – at operation – this value is exceeded by 11dB.
- the group of machines in the worst technical condition includes: turning lathes – S32, whose mean standard deviation σ = 5,29 m/s^2, and peak value a_{hvmax} = 28,1 m/s^2, CU401, whose mean standard deviation σ = 3,23 m/s^2, and peak value a_{hvmax} = 22,3 m/s^2, and the milling machines – FYA-32 (mean standard deviation σ = 3,53 m/s^2; a_{hvmax} = 40,2 m/s^2) and FNB-26 (mean standard deviation σ = 4,98 m/s^2; a_{hvmax} = 19,36 m/s^2). In the case of these stands, the rigidity of the machine tool-chuck-workpiece-tool system should be enhanced and vibration damping tools should be applied as well,
- the group of machines in the best technical condition includes: the turning lathe C11, the milling machine FWF-32, the drilling machines WKA-25 and WS-15, as well as the grinding machines SWA-10 and SPC-20, where the acceptable peak value a_{hvmax} was not exceeded, and the values of standard deviation σ are very low ($<$ 1.5 m/s^2),
- except for the drilling machines, all the other machines generate higher noise in operation tests than in idle running,
- there is no constant relationship between the rotational speed and emitted noise level EA because for some machines, such as the turning lathe CU401 or the milling machine FYA-32, an increase in the speed n leads to increased noise emission, while for other machines, for example the turning lathe C-11, the drilling machines WKA-25 and WS-15, it resulted in decreased emitted noise levels.
- carrying out the necessary maintenance repairs to improve the technical condition of

the examined machines will both ensure the safety to their operators and it will enhance the quality of machining conducted on these stands.

REFERENCES

1. Booth R.T. Risk Assessment Workbook. Aston University, Birmingham 1989.
2. Bryła R. Ocena ryzyka zawodowego na stanowisku obróbki skrawaniem metali. Stal - Metale i Nowe Technologie, 5-6, 2007.
3. Karczewski J.T., Rączkowski B. Ocena ryzyka zawodowego na stanowisku – tokarz, frezer. ODDK, Gdańsk 2000.
4. Kinney G.F., Wiruth A.D. Practical Risk Analyses for Safety Management. Naval Qweapons Centre, China Lake, 1976.
5. Lind N. Tolerable risk. International Conference on Safety, Risk and Reliability Trends in Engineering. IABSE, Malta 2001, Conference Report, ETH-Hönggerberg, Zürich 2001.
6. Norma EN 1050:1996. Safety of machinery – Principles for risk assessment.
7. Norma PN IEC 300-3-9:1999. Zarządzanie niezawodnością. Przewodnik zastosowań. Analiza ryzyka w systemach technicznych.
8. Norma PN-80/Z-08052. Ochrona pracy. Niebezpieczne i szkodliwe czynniki występujące w procesie pracy. Klasyfikacja.
9. Norma PN-91/N-01353. Dopuszczalne wartości przyspieszenia drgań oddziałujących na organizm człowieka przez kończyny górne i metody oceny narażenia.
10. Norma PN-EN ISO 14121-1:2008. Maszyny. Bezpieczeństwo. Zasady oceny ryzyka.
11. Norma PN-N-01307:1994. Hałas. Dopuszczalne wartości hałasu w środowisku pracy. Wymagania dotyczące wykonywania pomiarów.
12. Norma PN-N-18001:2004. System zarządzania bezpieczeństwem i higiena pracy. Wymagania.
13. Norma PN-N-18002:2000. System zarządzania bezpieczeństwem i higiena pracy. Ogólne wytyczne do oceny ryzyka zawodowego.
14. Norma PN-N-18002:2000. Systemy zarządzania bezpieczeństwem i higieną pracy. Ogólne wytyczne do oceny ryzyka zawodowego.
15. Projekt normy PN IEC 60300-3-9:2004. Zarządzanie niezawodnością. Przewodnik zastosowań. Analiza ryzyka w systemach technicznych
16. Romanowska-Słomka I., Słomka A. Zarządzanie Ryzykiem Zawodowym. Tarnobus Sp. z o.o. Tarnobrzeg 2001.
17. Romanowska-Słomka I. Ryzyko zawodowe – procedury, metody, zagrożenia, Wrocław 2006.
18. Smoliński D. Ocena ryzyka zawodowego na stanowisku pracy – poradnik, Wrocław 2001.
19. Stec D. Zasady BHP w praktyce, Kraków 2009.
20. Szczelina K. Ogólne wytyczne do analizy ryzyka zawodowego, Kraków 2009.
21. Tokarz W. Zarządzanie bezpieczeństwem pracy. Ocena ryzyka zawodowego. Zał. 4F. Wyd. FORUM, Poznań 2009.
22. Werner K. Ocena ryzyka zawodowego na stanowisku tokarza. Przyjaciel Przy Pracy, 6, 2009: 22-23.
23. www.gdansk.oip.pl/64.html (15.04.2013).
24. www.poznan.oip.pl/cutenews//data/doc/ocena_ryzyka.pdf (14.04.2013).

7

EFFECT OF WATER ABSORPTION ON THE MECHANICAL PROPERTIES OF FLAX FIBER REINFORCED EPOXY COMPOSITES

Umit Huner[1]

[1] Faculty of Engineering, Department of Mechanical Engineering, Trakya University, Edirne 22030, Turkey, e-mail: umithuner@trakya.edu.tr

ABSTRACT

Flax fiber reinforced epoxy composites were subjected to water immersion tests in order to study the effects of water absorption on the mechanical properties. Epoxy composites specimens containing 0, 1, 5 and 10% fiber weight were prepared. Water absorption tests were conducted by immersing specimens in a de-ionized water bath at 25 °C and 90 °C for different time durations. The tensile and flexural properties of water immersed specimens subjected to both aging conditions were evaluated and compared alongside dry composite specimens. The percentage of moisture uptake increased as the fiber volume fraction increased due to the high cellulose content. The tensile and flexural properties of reinforced epoxy specimens were found to decrease with increase in percentage moisture uptake. Moisture induced degradation of composite samples was significant at elevated temperature.

Keyword: natural fiber, mechanical properties, epoxy composites.

INTRODUCTION

Various lignocellulosic fillers (Table 1) can be used (flax, hemp, kenaf, sisal, jute) whose physical properties such as density, tensile strength, modulus of elasticity, etc. are related to the internal structure and chemical composition of fibers [1].

In recent years, the use of flax fibers as reinforcement in composites has gained popularity due to an increasing requirement for developing sustainable materials. Flax fibers are cost effective and offer specific mechanical properties comparable to those of synthetic fillers. Composites made of flax fibers with thermoplastic, thermoset, and biodegradable matrices have exhibited good mechanical properties [2]. Disadvantages of these materials are high moisture absorption capacity. It has been long known that epoxy and epoxy composites easily absorb water when exposed to humid environments. This reduces the stable lifetime of the material. Experiments show that the water in epoxy is present in two states [2]; firstly, distributed water molecules between the polymer chain and condensed water in fractures and cavities. Lignocellulose natural fibers are highly hydrophilic due to the presence of an OH group. The incorporation of these fibers into the matrix increases water uptake of the composites by forming a hydrogen bond between water hydroxyl groups of cellulose, hemicellulose, and lignin [3]. The hydrophilic character of natural fibers leads to low compatibility with hydrophobic polymer matrices and also to a poor dimensional stability as water uptake swells fibers [3].

The alternative of natural fibers as a reinforcing material in polymer composite materials is still attracting much interest at both the academic and industrial level. Currently, various researchers and material scientists all over the world are focusing their attention on effective utilization of these natural fibers in polymer composites.

Anter et al. [4] have reported the effect of water absorption on hardness properties for Epoxy reinforced with glass fibers in their research article. They used the Epoxy resin as matrix re-

Table 1. A comparison between the physical properties of selected natural fibers [5–8]

Fibre	Density (g·cm^{-3})	Tensile strength (MPa)	Young's modulus (GPa)	Elongation at break (%)
Flax	1.5	345–1.500	27.6	2.7–3.2
Jute	1.3–1.49	393–800	13–26.5	1.16–1.5
Sisal	1.45	468–700	9.4–22	3–7
Hemp	1.7	690	70	1.6

inforced by [0–90° Woven Roving and Random] with volume fraction 25%. The shore D hardness of all samples investigated before and after immersion in water at room temperature. Results of the work show that the value of hardness done at room temperature decreases with increasing the time of immersion in water.

Chandramohan [5] et al studied the effect of dry and wet conditions on tensile and hardness properties of bio-epoxy composites. In their research, natural fibers like Sisal (Agave sisalana), Banana (Musa sepientum) & Roselle (Hibiscus sabdariffa), Sisal and banana (hybrid), Roselle and banana (hybrid) and Roselle and sisal (hybrid) are fabricated with bio epoxy resin using molding method. They studied flexural rigidity and hardness of Sisal and banana (hybrid), Roselle and banana (hybrid and Roselle and sisal (hybrid) composite at dry and wet conditions.

Dhakal et al. [6] studied the effect of water absorption on mechanical properties of hemp fiber reinforced of unsaturated polyester matrix composites. The composites specimens containing 0, 0.10, 0.15, 0.21 and 0.26 fiber volume fraction were prepared. Water absorption tests were conducted by immersing specimens in a de-ionized water bath at 25 °C and 100 °C for different time durations. The percentage of moisture uptake increased as the fiber volume fraction increased due to the high cellulose content. They found that the tensile and flexural properties of composite specimens decrease with increase in percentage moisture uptake.

The knowledge of water and water vapor sorption in composites and in polymer matrices is considered to be of the most importance [3]. Therefore, it is important to study water absorption behavior in detail in order to evaluate not only the consequences that the water absorbed may have, but also how this water uptake can be minimized in some way.

The objective of this work was to compare the influence of both fiber reinforcement and water uptake on mechanical properties of flax fiber reinforcement epoxy composites and characteristics of the water absorption.

MATERIALS AND METHOD

Materials

The matrix material used in this study was based on a commercially available epoxy, Trade Name "MGS LR 160 (Hexion)" supplied by Dost Kimya Ltd. The matrix was mixed with curing catalyst, MGS LH 160 (Epıkure) at a concentration of 0.40 w/w of the matrix for curing.

The reinforcement was a natural flax fiber (Fig. 1) bundle obtained by a biological retting process in which some waxes and pectin were removed. These fibers were kindly supplied by Vanes Ltd. For applications as composites reinforcement, fibers have to be separated individually to be homogeneously distributed.

The fibers prior to use were washed thoroughly with distilled water because fibers generally contain some impurities that may affect the bonding between the fibers and the polymer matrix. And then fibers dried in a vacuum oven (Ecocell 55 comport) at 130±2 °C for 24 h.

Specimen preparation

All specimens in this study were manufactured by hand layup technique. The mould that was used in this work for casting process is made

Fig. 1. Flax bundle

of PMMA (Plexiglas) with dimension of (200, 200, 10) mm and it has cavities which shape of tensile and flexural test samples dimensions. The mould surface was allowed to dry after cleaning it with a thinner solution. After drying, the surface was coated with anti-adhesive agent. The polymeric material is prepared by mixing the epoxy resin with the hardener in (100:40) ratio at room temperature, the mixing was very slow, and using glass rod for (5 min) until it becomes homogenous. The flax bundle is added to the epoxy in different weight fractions of (1, 5, 10)% and continuously mixing until it becomes homogenous, the mixing is completed after (2 min), then the mixture is poured into the mould. Figure 2 shows composite samples.

Water absorption tests

Water absorption studies were performed following the ASTM D 570-98 method (Memmert water bath). Water absorption of the composites was determined after 2 h and 24 h immersion in distilled water at 25 °C and 90 °C (Fig. 3a). Five specimens of each formulation were dried in an oven (Fig. 3b) for 24 h at 103±2 °C.

The dried specimens were weighed with a precision of 0.001 g (Sartorius ED 224S model) and were placed in distilled water. At the end of the immersion periods, the specimens were removed from the distilled water, the surface water was wiped off using blotting paper, and wet weight values were determined. Water absorption percent was calculated using the following formula:

$$M\,(\%) = (m_t - m_0)/m_0 \times 100, \tag{1}$$

where: m_0 and m_t denote the oven-dry weight and weight after time t, respectively.

Mechanical properties

Tensile and flexural properties are the most commonly investigated mechanical properties of natural filler or synthetic filler reinforced plastic composites. In this study, both, tensile and flexural tests were performed using an Instron Univer-

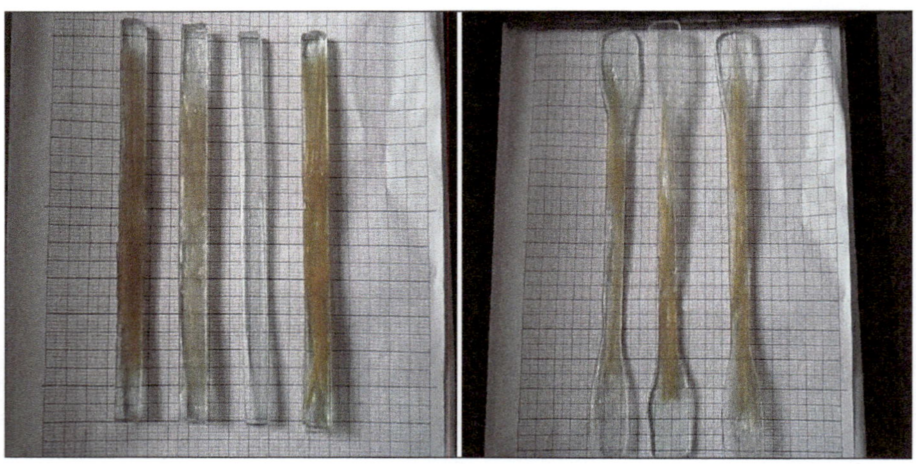

Fig. 2. Flexural and tensile test specimens

a) b)

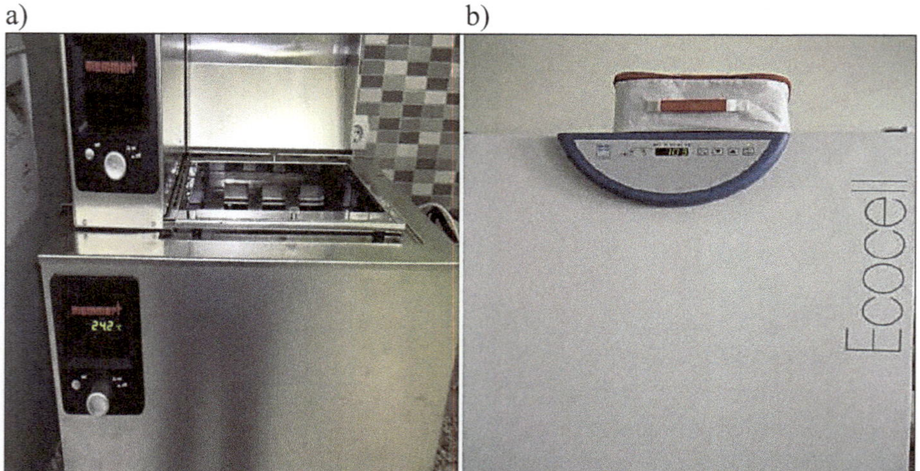

Fig. 3. a – water bath (left), b – drying oven (right)

sal Testing Machine Model 8501, equipped with a 500 kg load cell, strain-gauge extensometer (Instron, model 2620) after conditioning at 23±2 °C according to ISO 527 standard and ASTM D790, respectively. The cross-head speed used for the type IA tensile specimens was 5 mm/min. For the Flexural test (three point bending) a specimen with nominal dimensions of 80×10×4 mm^3, a span of 32 mm and a cross-head of 1 mm/min were used.

RESULTS AND DISCUSSION

Water absorption test results

The effect of fiber loading on the water absorption flax fiber reinforced composites with increase in immersion time is shown in Figure 4. It is evident from the figure that the rate of moisture absorption increases with increase in fiber weight. The effect of fiber loading is concerned composites with 10 wt% flax fiber loading shows higher water absorption rate as compared to 5 wt% and 1 wt% fibers loading. The reason may be due to the flax fibers contain abundant polar hydroxide groups, which result in a high moisture absorption level of natural fiber reinforced polymer matrix composites.

Another effect of increasing water absorption rate of composites is the nature of flax fiber, voids inside the composites lead to the formation of micro channels. After a time immersion in water, the presence of these micro voids and cracks in the composite surface results in the movement of water molecules to the material by capillary action [7]

Figure 4b shows that the absorption equilibrium was reached much faster at higher temperature (12 h) and the amount of sorbed water was lower than that observed at 25 °C. According to the Le Chatalier's principle [3], the absorption process should be exothermal.

To have durable flax fibre reinforced composites, some significant studies have been conducted. Improving the poor environmental and dimensional stability of lignocellulosic materials is good to modify the tensile properties of flax fibres [8]. In the study by Stamboulis et al. [8], the environmental behavior of flax mat reinforced composites is investigated by monitoring the moisture absorption and swelling, and measuring the residual mechanical properties of the composites at different moisture levels. It confirmed that the moisture absorption and swelling of the Duralin treated flax composites is approximately 30% lower than that of the composites based on untreated flax fibres. Dhakal [6] found that hemp fiber reinforced polyester corresponding to 0, 0.15, 0.21 and 0.26 fibre volume fractions, respectively, immersed at room temperature for 888 h is 0.879, 5.63, 8.16 and 10.97%, respectively. Water absorption value increased with increasing fiber loading. Similar results have been obtained in this study. This phenomenon can be explained by considering the water uptake characteristics of flax fibre. When the composite is exposed to moisture, the hydrophilic flax fibers swell. As a result of fibre swelling, micro cracking of the brittle thermosetting resin (like epoxy) occurs [6].

Tensile and flexural test results

The percentage of water absorption of the composites was evaluated as the difference between the initial and equilibrium absorption

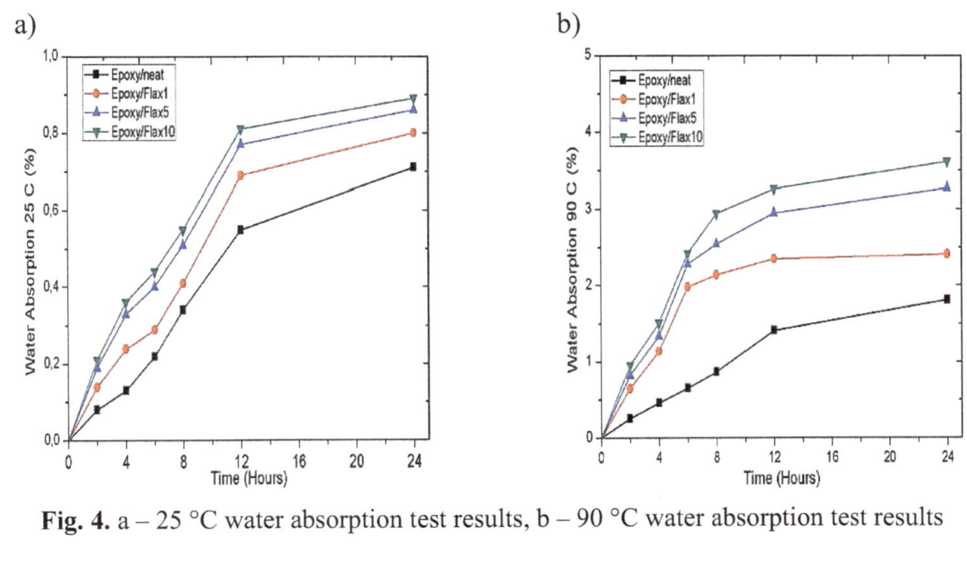

Fig. 4. a – 25 °C water absorption test results, b – 90 °C water absorption test results

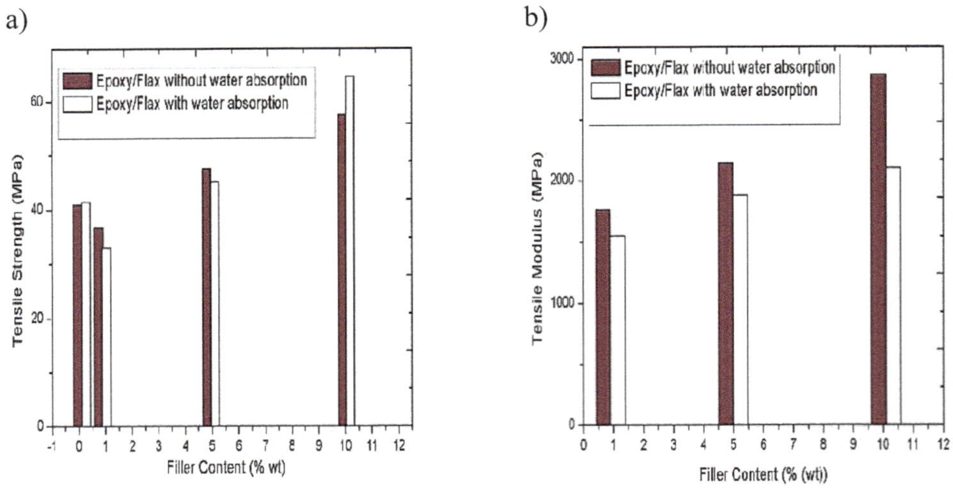

Fig. 5. a – tensile strength results with/without immersed water, b – tensile modulus results with/without immersed water 90 °C

weights of the specimen. The influence of fiber loading on the tensile strength and tensile modulus of the composites is shown in Figures 5a and 5b respectively.

Figure 5 shows that tensile strength of all specimens decreased with increasing water absorption rate. These results are because of water penetration inside polymers decreasing the connection between fiber and polymer material. Composite materials reinforced with flax fibers had many channels and capillary tube which allowed for water molecules to penetrate inside the materials and acting along the interface between epoxy and flax fiber causing swells in the samples. Then the bonds between resin and fibers will break. So the strength of the composite material will decrease. For the 10% flax fiber reinforcing the ultimate tensile stress of wet samples is higher than that for dry samples. This could be due to the fact that high amounts of water causes swelling of the fibres, which could fill the gaps between the fibre and the polymer–matrix, and eventually could lead to an increase in the mechanical properties of the composites. Dhakal [6] et al has reported similar results in their research for hemp/polyester composites. The tensile stress drops by 38% in their study. Theoretically, high tensile strength of a natural fibre reinforced composite could be achieved by increasing the amount of the fibre used. Singleton et al. [9] investigated the effect of fibre volume fraction on the mechanical properties of flax mat/recycled HDPE composites by film stacking and compression moulding. It was observed that the tensile strength and modulus increased with an increase in fibre volume fraction (0%, 10%, 18%, 20% and 30%).

Figure 5 shows the tensile properties of composites. All of the composites show higher tensile modulus than the neat epoxy resin. In particular, the tensile modulus strongly increases using as reinforcement 10% wt of flax fibers. In particular, the 1%, 5% and 10% wt flax fiber composites show tensile modulus about 5%, 21% and 71% higher than epoxy resin, respectively. Dhakal [6] et al. have reported the reduction in tensile modulus for 3, 4 and 5 layer hemp reinforced specimens compared to dry specimens is 61%, 97% and 87%, respectively.

The simple tension and flexure modulus (Fig. 6) measurements can hence differ significantly when the material is inhomogeneous and anisotropic. A flexural test is highly influenced by the properties of the specimen closest to the top and bottom surfaces, whereas a simple tension test reflects the average property through the thickness [10]. In a flexural test, the total bending deflection is a combination of a bending deflection and a shear deflection as shown in Eq.:

$$\delta = Pl^3/48\ EI + Pl/4G. \qquad (2)$$

The flexural stress (Fig. 6a) drops incrementally as the fibre volume fraction increases hence increased moisture uptake percentage. The decrease in flexural properties after water immersion can be related to the weak fibre-matrix interface due to water absorption.

The flexural modulus, however, is not adversely affected by moisture absorption. The increase in flexural modulus is more pronounced with higher fibre content specimens, hence higher moisture content. It would be intuitive to assume that the effect of fibre reinforcement to be less

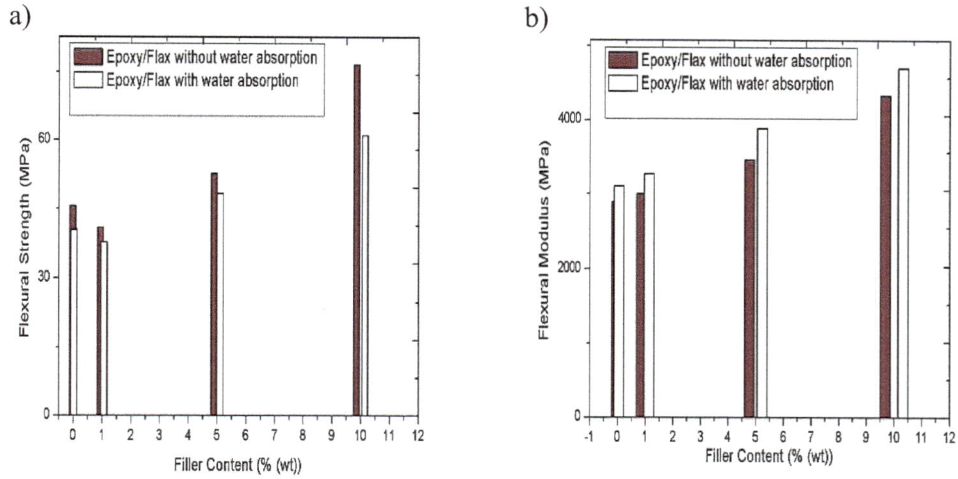

Fig. 6. a – flexural strength test results with/without immersed water, b – flexural modulus with/without immersed water 90 °C

critical for the flexural failure stress than in tensile failure mode. This is because the flexural samples fail in combination of compression, shear and tension mode [6].

CONCLUSION

Flax fibres are cost-effective materials have specific mechanical properties which have a potential to replace synthetic fibres as reinforcement in composite. Their main disadvantage is the variability in their properties. Environmental effects (e.g. high relative humidity) will degrade the tensile properties of flax fibres. A suitable chemical treatment (e.g. Silane, plasma) can increase the tensile strength and strain of the flax fibres.

REFERENCES

1. Baley C., Busnel F., Grohens Y., Sire O., Influence of chemical treatments on surface properties and adhesion of flax fibre-polyester resin. Composites. Part A. 37. 2006, 1626–1637.
2. Yan L., Chouw N., Jayaraman K., Flax fibre and its composites – a review. Composites, Part B, 56, 2014, 296–317.
3. Turmanova S., Dimitrova A., Vlaev L., Comparison of water absorption and mechanical behaviors of polypropylene composites filled with rice husks ash. Polymer-Plastics Technology and Engineering, 47, 2008, 809–818.
4. Faik H., Hind A., Hasan S., Effect of water absorption on hardness property for epoxy reinforced by glass fibers. Journal of University of Anbar for Pure Science, 5(3), 2011.
5. Chandramohan D., Marımuthu K., Tensile and hardness tests on natural fiber reinforced polymer composite material. IJAEST, 6, 2011, 97–104.
6. Dhakal H.N., Zhang Z.Y., Richardson M.O.W., Effect of water absorption on the mechanical properties of hemp fiber reinforced unsaturated polyester composites. Composites Science and Technology, 2006.
7. Narendar R., Priya Dasan K., Effect of chemical treatment on the mechanical and water absorption properties of coir pith/nylon/epoxy sandwich composites. International Journal of Polymer Anal. Charact., 18, 2013, 369–376.
8. Stamboulis A, Baillie C.A., Peijs T., Effects of environmental conditions on mechanical and physical properties of flax fibres, Composites, Part A, 32(8), 2001, 1105–1115.
9. Singleton A.C.N., Baillie C.A., Beaumont P.W.R., Peijs T., On the mechanical properties, deformation and fracture of a natural fibre/recycled polymer composite. Composites, Part B, 34(6), 2003, 519–526.
10. Wambua P., Ivens J., Verpoest I.. Natural fibres: can they replace glass in fibre reinforced plastics? Composites Science and Technology, 63, 2003, 1259–1264.

8

EFFECT OF VIBRATION SHOT PEENING PARAMETERS UPON SHAPES OF BEARING CURVES OF ALLOY STEEL SURFACE

Kazimierz Zaleski[1], **Agnieszka Skoczylas**[1]

[1] Lublin University of Technology, 36 Nadbystrzycka Str., 20-618 Lublin, Poland, e-mail: k.zaleski@pollub.pl; a.skoczylas@pollub.pl

ABSTRACT

The article presents the results of studies on the effect of technological parameters of vibration shot peening upon the shape of bearing curves and the parameters of that curve. The topography of surface formed in result of vibration shot peening and pre-machining, which was grinding, were analyzed. The material used in the studies was alloy steel 30HGSA. The shot peening process was performed on an appliance equipped with a kinematic vibrator. Shot peening time t, vibrator oscillations (vibrations) frequency v, as well as vibrator oscillations amplitude A were changed within: t = 1–15 min, v = 17,5–35 Hz, A = 10,3–17,3 mm. For measuring surface roughness and determining the bearing curve the appliance T8000 RC120-400 by Hommel-Etamic was used together with the software. In result of the machining performed surface roughness decreased, and the parameters of bearing curve improved.

Keywords: vibration shot peening, bearing curve, surface roughness.

INTRODUCTION

Vibration shot peening is one of the variants of dynamic burnishing, characterized by striking effect of shot peening elements, most often in a form of balls, upon the treated object. The formation of the surface of geometrical structure during shot peening is a complex structure, including "knocking out" hollows, leveling micro-elevations formed in earlier phases of shot peening and accruing hardening of the surface layer in result of condensation traces of blows [12].

Roughness of treated objects surfaces can be characterized by means of numerous parameters [16]. The most frequently applied roughness parameters (Ra, Rz) do not allow for a full assessment of operating properties of treated surfaces. Important information about tribological properties of a given surface is provided by bearing curve, also called the Abbott-Firestone curve [1, 11, 15].

The (mating) surfaces of a tribological pair should be divided by a layer of a smearing liquid. The hydrodynamic (buoyant force) of that liquid depends upon the geometrical structure of the (mating) surfaces [3]. The bearing curve of the surface whose shape depends upon the method of machining that surface, affects the thickness of the smearing film [10].

A number of authors present the results of studying various treatments of machining methods affecting the shape of bearing curve of the machined surface. The effect of the technological parameters of turning treatment upon the shape of Abbott-Firestone curve was presented in the paper [9], and the influence of the kind of abradant upon the shape of that curve of the ground surface was shown in this paper as well as in [8]. One of the methods of finishing treatment, providing for obtaining the surface with high load capacity is honing. That is why in many works the effect of honing upon the indices of Abbott-Firestone curve is analyzed [2, 4, 14].

Studies of the effect of burnishing upon load capacity of the machined surface. In this paper [7] the results of comparative studies of indices of the load capacity share of nodular cast iron after grinding and after roller burnishing. Shot peening allows forming hollows on the surfaces working in the conditions of seizing, which are grease pockets. It was found that the application of grease pockets influences the decrease of the wear of frictionally mating elements, as well as the increase of friction travel to seizing friction pair [5, 6].

Previous studies of surface quality after vibration shot peening comprised the effect of vibration shot peening parameters upon surface roughness, expressed by the parameter *Ra* [13]. The objective of the currently presented studies was to determine the effect of vibration shot peening parameters upon the shape bearing curve of the machined surface.

RESEARCH METHODOLOGY

In the experimental studies we used rectangular prism samples of dimensions: 4×15×100 mm, made of alloy steel 30HGSA. After initial, forming treatment the samples underwent toughening to the hardness of 36–38 HRC, and then they were ground with the application of small feed and intense cooling.

The vibration burnishing was performed on a special appliance, equipped with a kinematic vibrator (Figure 1), giving the possibility of changing the vibration amplitude and frequency. The appliance consists of the working chamber (4), situated on the upper plate of the vibrator (1). Samples (2) are attached to the bottom of the chamber. These are covered by loose balls (3), which, as a result of vibrations caused by the vibrator, collide with the treated surface, causing the effect of shot peening.

Vibration shot peening was accomplished for the following technological parameters: vibrator's vibration amplitude $A = (10.3–17.3)$ mm, frequency of the vibrator's vibrations $v = (17.5–35)$ Hz and shot peening time $t = (1–15)$ min. The diameter of shot penning balls was 9.4 mm. We know from the previous studies [13] that these parameters significantly affect the state of surface layer.

Measurements of surface roughness, topography and bearing curve parameters were performed with the use of Hommel-Etamic T8000 RC120-400 appliance and the software which is compatible with the equipment.

As the parameters of Abbott-Firestone curve the following were determined: *Rk* – depth of roughness body, that is a part of the profile, excluding the distinguishable elevations and deep hollows, *Rpk* – reduced height of elevations, *Rvk* – reduced depth of the hollow and the material ratio at depth *c* roughness profile, *Rmr* on the level of reference $c = 1$ µm [15, 16]. In order to determine the material share of roughness profile the zero level was assumed and on the level of 5% below the level containing extreme elevations, for which the reference level was determined. The current practice is to determine the bearing length coefficient in this way [11]. This allows for eliminating the effect of extreme elevations upon the result of measuring the relative material ratio at depth c roughness profile. Figure 2 shows the graphic interpretation of bearing curve for the roughness profile.

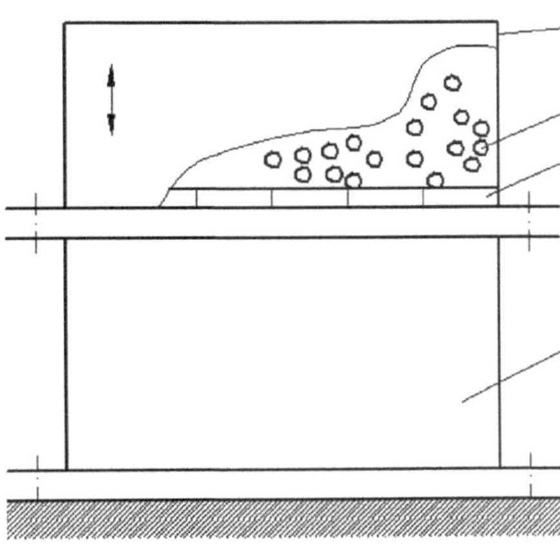

Fig. 1. Schema of a stand for vibration shot peening: 1 – vibrator, 2 – specimens, 3 – peening ball, 4 – chamber

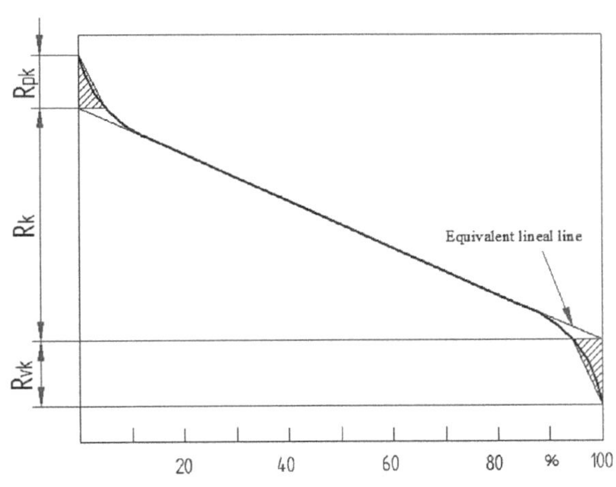

Fig. 2. Parameters bearing curve (based on [15])

STUDY RESULTS

The accomplishment of these studies allowed for presenting the effect of vibration shot peening parameters upon the parameters of bearing curve for roughness profile and upon the shape of Abbott-Firestone curve.

Figure 3 shows the bearing curve and topography for the surface after grinding. The surface after grinding is characterized by a unidirectional arrangement of micro-irregularities in uniform (equal) spacing. Deep scratches and sharp elevations are visible. The geometrical structure formed after grinding should be qualified as unidirectional. The Abbott-Firestone curve is a curve with a large angle of inclination, representing a surface with sharp vertexes. The profile of such a surface with small bearing is called a curve with a "thin center" [11].

The study results demonstrate that the technological parameters of vibration shot peening affect surface roughness and the parameters of Abbott-Firestone curve.

The graph presented in figure 4 shows that the change of shot peening time affects the parameters of bearing curve. Increasing burnishing time within the range from 1 to 5 minutes causes the decrease of surface roughness and the parameters: *Rk*, *Rvk*, *Rpk*, what can be justified by the growing number of strikes falling on a surface unit. When the time is longer, a slight increase of bearing curve parameters takes place with reference to the minimal values achieved for $t = 5$ min.

In the aspect of the two surfaces mating with each other, an important parameter is the material ratio at depth *c* *Rmr*, determined in the accomplished studies on the reference level $c = 1$ μm. According to the supposition, the increase of burnishing time causes the increase of *Rmr* value (Figure 5). It should be explained by the increased number of strike traces on the machined surface, which contributes to achieving a less rough surface.

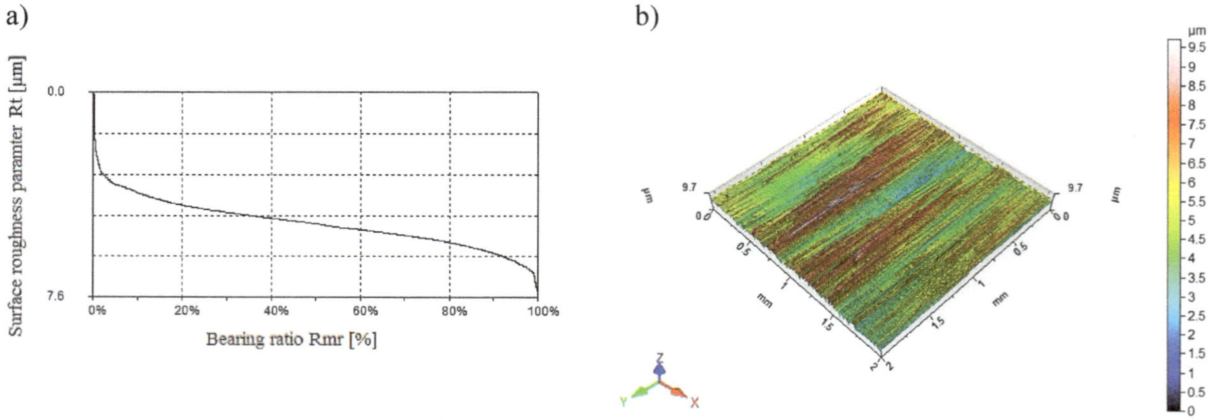

Rt = 5.28 μm (± 0.5868); Rk = 2.03 μm (±0.1881)
Rpk = 0.79 μm (±0.0774); Rvk = 1.02 μm (±0.1303)

Sa = 0.994 μm, St = 9.71 μm, Sp = 4.49 μm,
Sv = 5.22 μm

Fig. 3. Bearing curve, surface topography and surface roughness parameters 2D and 3D after grinding

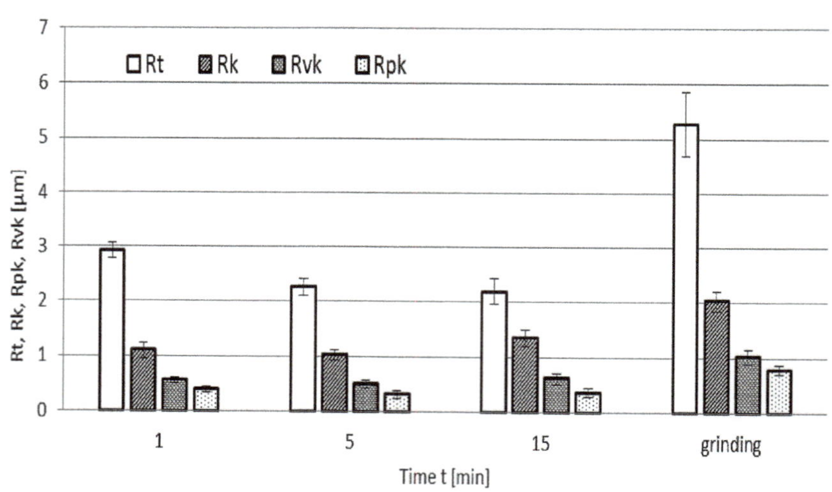

Fig. 4. Effect of vibration shot peening time onto surface roughness parameter and bearing ratio curve parameters

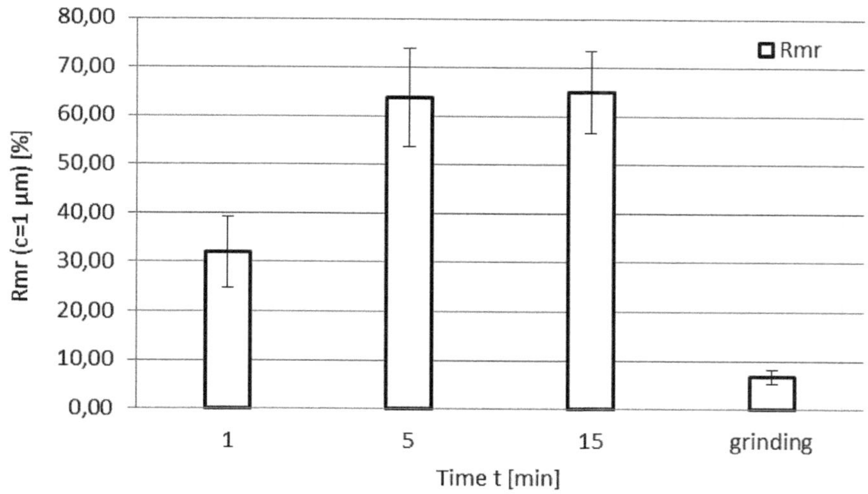

Fig. 5. Effect of vibration shot peening time onto material ratio at depth *c Rmr*

Figure 6 presents the effect of vibrator vibrations frequency upon surface roughness and the parameters: *Rk*, *Rvk*, *Rpk*. The increase of vibrator vibrations frequency is accompanied by increased impact energy and the number of impact falling on a surface unit, which is translated into higher values of material ratio at depth *c* in the roughness profile. The most advantageous machining effects were achieved for the frequency v = 23.5 Hz. For the frequency v = 17.5 Hz the energy of impact does not achieve large values, which causes a smaller change in the surface roughness level Rt. In the case when v = 35 Hz "knocking out" of larger hollows in the machined surface takes place, which affects surface roughness and the parameters of Abbott-Firestone curve. The effect of vibrator vibrations frequency upon the material ratio at depth *c* is shown in Figure 7. At the beginning there is a visible increase of Rmr parameter, which later decreases for the maximal frequency.

Considering the effect of vibrator vibrations amplitude upon the parameters of bearing curve (Figures 8 and 9) an increase of the examined parameters can be noticed, together with the increased amplitude. That is connected with the increased energy of balls colliding with the shot peened samples, the consequence of which is increased surface roughness. That may lead to a deteriorating resistance of that surface to abrasive wear (attrition). The optimal conditions, allowing for obtaining advantageous conditions of bearing curve parameters, as well as low roughness, were obtained for the amplitude A = 14 mm. The application of low amplitude values, the consequence of which is lower impact energy, causes incomplete levelling of micro- irregularities after machining preceding shot peening, a too high value of the amplitude, in turn, causes deformation of the machied surface.

Having performed the studies one can notice that for all sets of shot peening parameters the

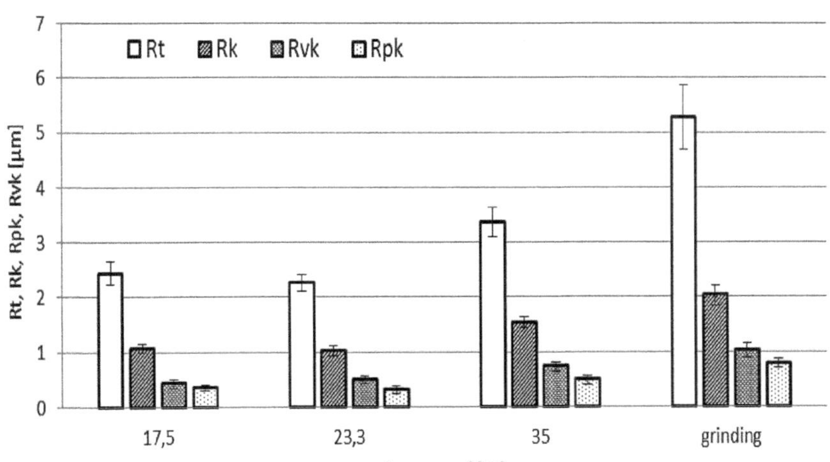

Fig. 6. Effect of vibrator vibrations frequency onto surface roughness parameter and bearing curve parameters

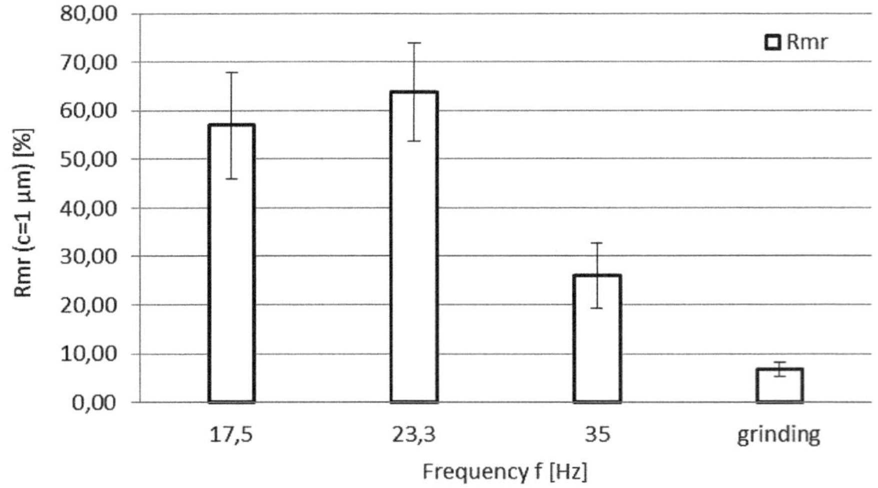

Fig. 7. Effect of vibrator vibrations frequency onto material ratio at depth *c Rmr*

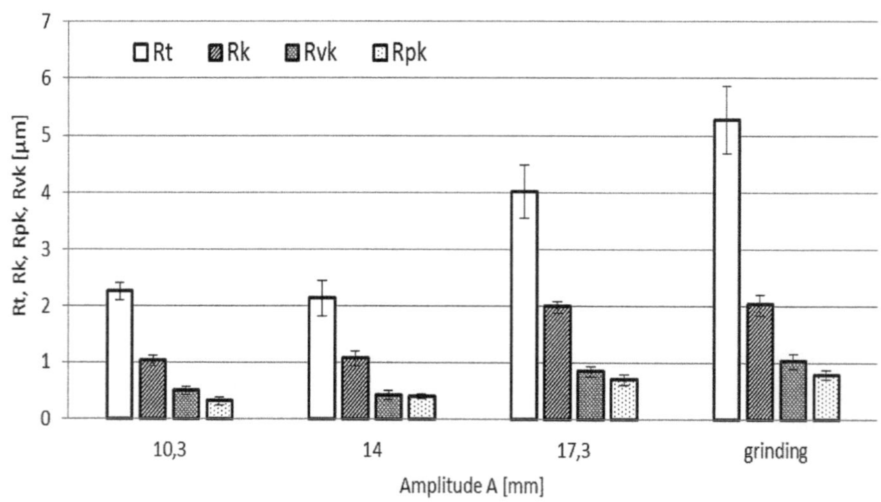

Fig. 8. Effect of vibrator vibrations amplitude onto surface roughness parameter and bearing curve parameters

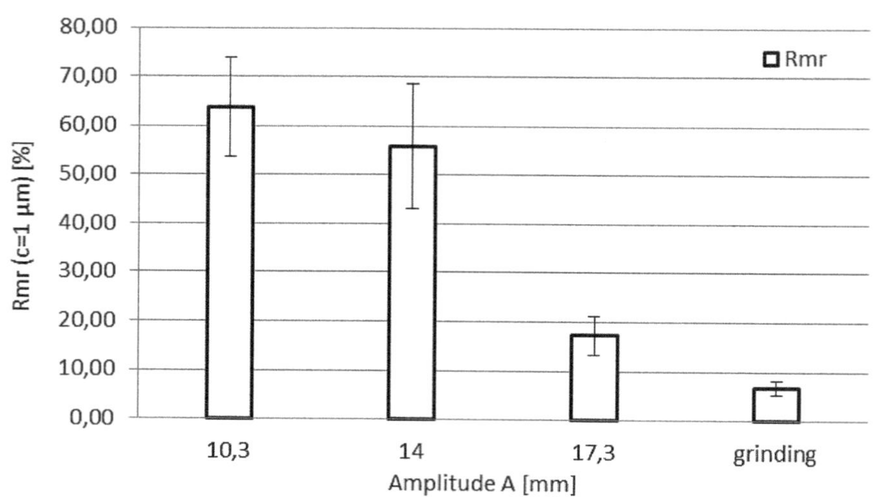

Fig. 9. Effect of vibrator vibrations amplitude onto material ratio at depth *c Rmr*

a)

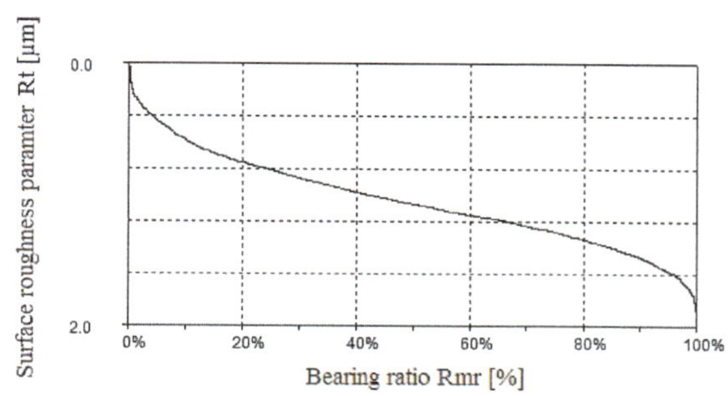

b)

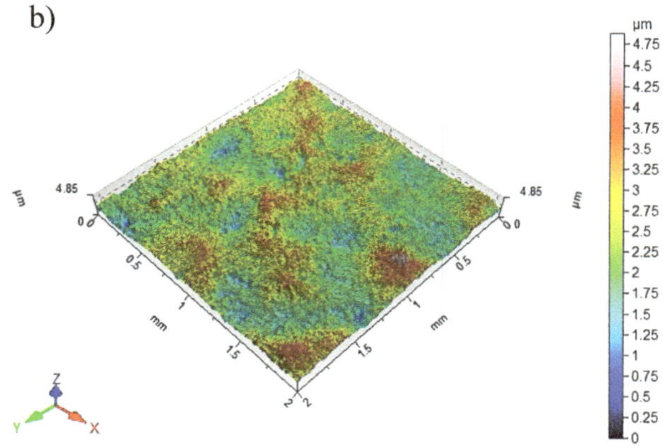

Rt = 2.27 µm (± 0.1524); Rk = 1.02 µm (±0.0939)
Rpk = 0.31 µm (±0.0662); Rvk = 0.49 µm (±0.0679)

Sa = 0.419 µm, St = 4.88 µm, Sp = 2.52 µm,
Sv = 2.36 µm

Fig. 10. Bearing curve, surface topography and surface roughness parameters 2D and 3D after vibratory shot peening (machining parameters d = 9.4 mm, t = 5 min, v = 23.3 Hz, A = 14 mm)

obtained values of roughness and bearing curve were lower with reference to surface after grinding. The *Rpk* parameter, maintained during surface lapping, after shot peening achieved from 11% to 61% smaller values than after grinding. This proves high abrasion resistance of surfaces machined by vibration shot peening. In the case of Rvk parameter smaller values were achieved: from 17% to 58% in relation to the ground surface. That may mean that the surfaces treated in this way have smaller oil maintenance capacity.

The geometrical structure of the surface after vibration shot peening (Figure 10) is disordered. The total roughness profile is visibly decreased, and there are characteristic hollows, formed in result of the balls striking against the sample surfaces. The bearing curve of roughness profile is of degressive-progressive nature. It has flattened vertexes of micro-irregularities, which probably means increased abrasion resistance of the treated surface.

CONCLUSIONS

In this paper we examined the effect of technological parameters of vibration shot peening upon the shape of bearing curve, as well as the parameters related to Abbott-Firestone curve. The following results sum up the conducted studies.

In the effect of vibration shot peening the *Rpk* parameter decreased from 11% to 61% with reference to values after grinding. That proves the increase of abrasion resistance. For *Rvk* parameter the change ranged from 17% to 58%, which meant that the surfaces machined in this way have lower oil maintenance capacity. The material ratio at depth c of surface roughness profile is from 2.5 to 9.5 times greater after vibration shot peening than after grinding. This allows for increasing the surface mating time in frictional pair.

The profile bearing curve after shot peening assumes degressive-progressive character and has a small angle of inclination. Technological parameters of vibration shot peening affect the parameters of bearing curve.

REFERENCES

1. Adamczak S., Normalizacja pomiarów struktury geometrycznej powierzchni. Cz. 8. Mechanik, 8-9, 2005, 690–692.
2. Corral I.B., Calvet J.V., Salcedo M.C., Use of roughness probability parameters to quantify the material removed in plateau-honing. International Journal of Machine Tools & Manufacture, 50, 2010, 621–629.
3. Czarnecki H., Analiza teoretyczna wpływu stereometrii powierzchni na działanie pary tribologicznej. Tribologia, 4, 2005, 19–31.
4. Deepak L.K., Rammamoorthy B., Surface topography characterization of Automotive cylinder liner surfaces using fractal methods. Applied Surface Science, 280, 2013, 332–342.
5. Gałda L., Możliwości kształtowania regularnej geometrii czopów łożysk ślizgowych w procesie obróbki nagniataniem udarowym. Tribologia, 3, 2004, 111–119.
6. Koszela W., Gałda L., Dzierwa A., Pawlus P., Sęp J., Wpływ kieszeni smarowych na zatarcie pary ciernej stal-brąz. Tribologia, 3, 2009, 51–58.
7. Laber S., Wpływ stanu warstwy wierzchniej na właściwości tribologiczne żeliwa sferoidalnego

ferrytycznego po nagniataniu. Tribologia, 1, 2010, 51–59.

8. Niżankowski C., Matras A., Wpływ gatunku ścierniwa na dokładność wałków stalowych po szlifowaniu ściernicowym. Inżynieria Maszyn, 15 (3), 2010, 55–68.

9. Petropoulos G.P., Torrance A.A., Pandazaras C.N., Abbott curves characteristics of turned surfaces. International Journal of Machine Tools & Manufacture, 43, 2003, 237–243.

10. Sahlin F., Almqvist A., Larrson R., Glavatskih S., Rough surface flow factors in full film lubrication based on a homogenization technique. Tribology International, 40, 2007, 1025–1034.

11. Wieczorowski M., Cellary A., Chajda J., Przewodnik po pomiarach nierówności powierzchni czyli o chropowatości i nie tylko. Zakład Poligraficzno-Wydawniczy M-Druk, Poznań 2003.

12. Zaleski K., Kształtowanie wybranych właściwości warstwy wierzchniej elementów metalowych w procesie dynamicznego nagniatania rozproszonego. Wydawnictwo Politechniki Lubelskiej, Lublin 2008.

13. Zaleski K., Modelowanie chropowatości powierzchni i stopnia umocnienia warstwy wierzchniej stali 30HGSA po nagniataniu wibracyjnym. Mechanik, 3, 2007, 194–198.

14. Żebrowski H., Kowalski M., Pomiary chropowatości i stereometryczny opis struktury powierzchni gładzonych. Prace Naukowe Instytutu Technologii Maszyn i Automatyzacji Politechniki Wrocławskiej, 74, 1999, 227–234.

15. PN-EN ISO 13565-2. Struktura geometryczna powierzchni: metoda profilowa; powierzchnie o warstwowych właściwościach funkcjonalnych.

16. PN-EN ISO 4287. Struktura geometryczna powierzchni: metoda profilowa.

9

MECHANISM OF ORIGIN OF STRUCTURAL VIBRATIONS IN CONICAL ROLLER BEARINGS

V.I. Marchuk[1], L.M. Ravenets[1], S.V. Marchuk[1], Antoni Świć[2]

[1] Lutsk National Technical University, 75 Lvivska Str., Lutsk 43018, Ukraine, e-mail: info@kptm.lutsk.ua

[2] Institute of Technological Information Systems, Lublin University of Technology, Nadbystrzycka 36, 20-618 Lublin, Poland, e-mail: a.swic@pollub.pl

ABSTRACT

This paper investigates the mechanism of origin of structural, structural and technological defects of rollers. The technique for integrated indicator of vibration working surfaces of the rings to determine the level of life of the finished part to the operation as part of the bearing and predict the vibroacoustic characteristics of rolling bearings. It was established that technological defects cause low-frequency and high-frequency vibrations. The question about the extent to which it is necessary to strengthen the tolerances on the parameters of bearings on which vibration level is determined not errors bearing parts and their structural properties. Calculated values of the amplitudes vibroacceleration due to the rigidity of the bearing vibrations are so small that in some cases adopted precision calculations turned enough to detect such vibrations. Thus, when tested on the vibro-acoustic installations structural vibration does not play an important role.

Keywords: vibroactivity, geometric undulation, structural vibration microtopography, design and technological factors.

INTRODUCTION

One of the main tasks that domestic bearing industry faces nowadays is to reduce the vibration and noise of rolling bearings. The main direction of solving this problem is to improve the technology of bearings manufacture: improving geometric accuracy of rings and rolling bodies (balls and rollers), reducing the roughness of the working surface, using isotropic to elastic properties of the starting materials, ensuring highly refined lubricants and parts and assembly cleanliness, proper conditions of installation and operation. Obviously, the implementation of the above mentioned measures is expensive and must be economically justified. Details of the modern high-precision and low-noise bearings must meet high requirements. To implement them in terms of mass and, moreover, serial production is extremely difficult and very expensive. Regarding this fact the question arises to what extent it is necessary to strengthen the tolerances on the parameters of bearings and where lower limit by which vibration level is determined not by errors of bearing parts and their structural properties.

It is known that even in cases when the bearing can be considered ideal in terms of the absence of errors in geometry and isotropy of parts' material, the presence of indissoluble elastohydrodynamic lubricant film between rings and rolling bodies completely clean from contamination of the lubricant, the bearing nevertheless vibrates. This vibration is common not only to perfect but also to ordinary bearings. The vibration is due to their structure and is called structural vibration, mainly due to two reasons [5, 8, 11].

The first reason is the bending deformation of the bearing rings forces acting on them from rolling bodies. Rings of bearing are folded under contact loads from rolling bodies, taking a form of a polyhedron that rotates with them. Such bending deformations are transferred to other parts of machines or distributed in the environment (for example air) in a form of acoustic waves, i.e. noise. The frequency of this vibration is the frequency of rolling bodies passing through the ring and multiple frequencies: qzf_c, where $q = 1, 2, ...$; z – number of rolling bodies; f_c – rotational speed of the separator. Most machines and devices of practical importance may have primarily bending deformation of the outer bearing rings [3, 5, 9, 11].

The second reason of the structural vibration is oscillation of stiffness of the bearing during rotation as a result of the radial component of the load. As the rolling bodies move to the line of the radial load, the stiffness of the bearing varies periodically with frequencies multiple to frequency of balls passing on non-rotating ring. Considering the most common scheme of rotating of inner and non-rotating outer rings we observe that as a result of fluctuations of stiffness relative displacement of the bearing rings occurs on the same frequencies as in the first case [5, 6].

It should be noted that despite the obviousness of these mentioned causes of structural vibration of bearings in scientific and technical literature there are no systematic quantitative evaluations of its significance. It is clear that the level of this vibration is relatively small because of the high rigidity of bearing rings, small values of the ratio of radial to axial load under conditions specific to the operation of most sizes of bearings. In bearings of low precision this vibration is not noticeable. At the same time, with increasing of bearings precision, structural vibration plays a much more visible role, and therefore, it requires scientific and reasonable evaluation of its significance [2, 4].

The assessment of the significance of structural vibration of bearings can be done in two ways. Firstly, comparing the calculated values with the values of the amplitudes, which are recorded practically. Second, comparing calculated values of the amplitudes of the structural vibration and the vibration caused by errors of bearings. In both cases the amplitudes of the harmonics vibration at frequencies qzf_c must be compared. There are no difficulties with the first way of assessment the causes if the problem of structural vibration of bearing is solved and amplitudes of vibratory acceleration of outer rings during practical measurements are known. The second method requires some explanation.

The theory of no ideal ball bearings with geometric errors of the working surfaces of the cones and balls allows to link errors spectrum with the spectrum of vibration [5, 11]. Since the structural vibration occurs only at frequencies qzf_c. To compare with the spectrum of vibrations of non-ideal ball bearing it should be highlighted that these harmonic components. The theory shows that the vibration at frequencies qzf_c is caused only by the geometry errors of the outer ring (as summing isotropy of the elastic properties of the material) and the harmonics of errors with numbers $\lambda = qz \pm 1$. Average value of amplitude of vibratory displacement of out erring as a rigid body in this case is [3]:

$$a_{f=qzf_c} = \sqrt{a^2_{\lambda=q\ -1} + a^2_{\lambda=q\ +1}}\ ,$$

$$A_{f=qzf_c} = \sqrt{a^2_{\lambda=q\ -1} + a^2_{\lambda=q\ +1}} \qquad (1)$$

first harmonic of vibratory displacement that is the most important (at a frequency of passing rolling bodies in the outer ring zf_c) is specified by harmonic geometry errors of the outer ring with numbers $\lambda = z + 1$ and $\lambda = z - 1$ that regardless of phase emerge in (1) mean-square.

Thus, to assess the importance of structural vibration of roller bearing with another method we should calculate the amplitude of the first harmonic of the structural vibration of the two considered sources and compare them directly with amplitudes of harmonic waviness of the outer ring of numbers $\lambda = z + 1$ and $\lambda = z - 1$.

We solve the problem of structural vibration of roller bearing, specified by two above-mentioned reasons based on that decision and calculate the amplitude of vibratory accelerations and vibratory displacements of surfaces of the outer rings, the most commonly used in machinery and tool engineering of roller bearings with their characteristic operating conditions at frequencies qzf_c and make comparison of calculation results with actual according to the accuracy of the rings and vibration levels for the same bearings. Experimental studies of the vibration level were conducted on automated measuring complex DVK [4].

VIBRATION FROM CHANGES OF STIFFNESS OF ROLLER BEARING

To solve the problem of structural vibration of roller bearing, caused by variability of its rigidity during rotation under combined loads we should solve the problem of static of perfect bearing, which involves determining the position of the inner ring as to the outer under the applied load. A characteristic feature of a roller bearing as no rigid deformation system is static uncertainty, and therefore, the direct analytical solution of the direct problem of statics in general case is impossible [1, 11]. Thus, in the calculation of the roller bearing initially the inverse problem is solved: the reaction vector is determined $\overline{R} = [F_x, F_y, F_z, m_y, m_z]^T$ by responses to the vector of relative displacement of the rings $\overline{\Delta} = [x, y, z, Q_y, Q_z]^T$ then by approximate analytical methods [11] or numerically [1] the direct problem of statics is solved.

Nonlinearity of dependence $\overline{R} = \overline{R}(\overline{\Delta})$ makes it impossible to exact the solution of the direct problem of static roller bearings under conditions of combined loading. During the solution of the inverse problem the bearing is divided into z mutually independent contact groups, where z is a number of rollers, each of which consists of a roller and two rings. For each roller on the relative position of the rings contact force is calculated if defined model of the contact group, and then by summing over all the rollers the vector of responses is calculated.

We use a generally accepted Hertz model of contact group roller [10] – the ring that connects the contact force Q with the contact deformation ω with following dependence:

$$Q(\omega) = K_H \omega^{1,5},$$

where: K_H – Hertz coefficient [11].

The placement of j-th roller as to the fixed point of the outer ring is defined by angle $\varphi_j = \gamma + 2\pi(j-1)/n$, where φ_j – angle of the separator bending on the outer ring. We introduce the denotations and also basic rule of signs [10]. In this case, the deformation of the j-th contact group ω_j is determined by the vector relatively to displacement of rollers in the following way:

$$\omega(\overline{\Delta}, \varphi_j) = \begin{cases} \left\| L(\varphi_j)\overline{\Delta} - \overline{l}' \right\|, f \ \left\| L(\varphi_j \overline{\Delta} - \overline{l}') \right\| \succ 0 \\ 0, f \ \left\| L(\varphi_j \overline{\Delta} - \overline{l}') \right\| \le 0 \end{cases},$$

where:

$$L(\varphi_j) = \begin{bmatrix} 1 & 0 & 0 & -R_B\cos\varphi_j & -R_B\sin\varphi_j \\ 0 & \sin\varphi_j & \cos\varphi_j & 0 & 0 \end{bmatrix}$$

$$\overline{l}' = l' \begin{bmatrix} \sin\tau_0 \\ \cos\tau_0 \end{bmatrix}$$

l' – initial distance between the inner and outer rings is defined in the radial cutting plane passing through the axis of the j- th and the center of the inner ring.

$$\cos\tau = (R_b + r_b) - (R_z + r_z)/l',$$

where:

$$l' = r_z + r_{\hat{a}} - d_m;$$

φ – contact angle;

r_z, r_b – in accordance the initial radius of the outer and inner rings;

R_z, R_b – finite radii of the outer and inner rings;

d_m – diameter of the roller.

Working contact of the angle of j-th roller of rings is calculated by the formula:

$$\tau(\overline{\Delta}, \varphi_j) = arctg\left(\frac{L^1(\varphi_j)\overline{\Delta}}{L^2(\varphi_j)\overline{\Delta}}\right),$$

where: L^1– the first and L^2 – second row of the matrix $L(\varphi)$.

The component of the vector load on roller bearing $\overline{F}$ is caused by the contact forces from the j-th roller,

$$\overline{F}\varphi_j = [F_x, F_y, F_z, m_y, m_z]^T \ \varphi_j = \overline{E}(\tau)Q(\omega) \quad (2)$$

where:

$$\overline{E}(\tau) = [-\sin\tau, 0, -\cos\tau, R_B\sin\tau, R_B\cos\tau, -r_B]^T$$

Let us denote by $T(\varphi_j)$ matrix that relates the coordinates of the vector $\overline{F}\varphi_j$ in the basis OXZ with coordinates of the same vector in the basis $\varphi_j X\varphi_j Y\varphi_j\varphi$:

$$T(\varphi_j) = \begin{bmatrix} 1 & 0 & 0 & 0 & 0 \\ 0 & \cos\varphi_j & \sin\varphi_j & 0 & 0 \\ 0 & -\sin\varphi_j & \cos\varphi_j & 0 & 0 \\ 0 & 0 & 0 & \cos\varphi_j & \sin\varphi_j \\ 0 & 0 & 0 & -\sin\varphi_j & \cos\varphi_j \end{bmatrix}$$

Then the solution of the inverse problem of static roller bearing has the form:

$$\overline{R}(\overline{\Delta}) = \sum_{\varphi_j} T(\varphi_j)\overline{F}(\varphi_j) = \sum_{\varphi_j} T(\varphi_j)\overline{E}(\tau)Q(\omega) \quad (3)$$

Direct static problem, which involves determining the vector of relative displacement of rings $\overline{\Delta}$ at angle of separator bending and vector of load $\overline{P} = [P_x, P_y, P_z, G_y, G_z]^T$ that should be considered as a problem of determining the root $\overline{\Delta}$ of the equation $\overline{R}(\overline{\Delta}) = \overline{P}$. Numerical solution of this problem is convenient to carry out using the method of Newton-Raphson. To this end, for this purpose at each step of iterations as to the known rules of differentiation tangent stiffness matrix of the form is calculated:

$$\frac{\partial R}{\partial \overline{\Delta}} = \sum_{\varphi_j} T(\varphi_j)\left\{\frac{\partial \overline{E}_1}{\partial \tau}(\tau)\frac{\partial \tau}{\partial \overline{\Delta}}(\varphi_j, \overline{\Delta})Q(\omega) + \right. \quad (4)$$
$$\left. + \overline{E}(\tau)\frac{\partial Q}{\partial \omega}(\omega)\frac{\partial \omega}{\partial \overline{\Delta}}(\varphi_j, \overline{\Delta})\right\}$$

The calculations terminates when $\left\|\overline{R}(\overline{\Delta}) - \overline{P}\right\|$ becomes less than 0.01 (in the units of H, mm). The implementation of this algorithm for solution of the problem of static on the computer proved its high efficiency – the number of iterations required by the proper choice of the first approximation does not exceed 7. As calculations of many types of roller bearings have shown significant value is only the first harmonic vibration in the frequency spectrum zf_c, harmonics of higher or small amplitude and practice do not matter.

The comparison of the calculated values of the vibratory acceleration amplitudes of roller bearings due to bending vibrations of the outer ring in the axial load characteristic for testing of bearings for driven units [4] with the values established by standards for low-noise vibration of roller bearings shows that the calculated value is at 11–23 dB below established standards for the relevant frequency range. Calculated values of the vibratory acceleration amplitudes due to the rigidity of the bearing vibrations are so small that in some cases adopted precision calculations were not enough to detect such vibrations. Thus, when tested on the vibroacoustic installations structural vibration does not play an important role.

To assess the importance of structural vibration of roller bearing value of its amplitude should be compared with the tolerances for amplitude of harmonic undulation of bearings components. At present, the general tolerances for the undulation of tracks of rolling outer rings, for example 2-class accuracy, reaches 0,1–0,2 m, depending on the size of the rings. Thus, the structural vibration is significant if its amplitude is larger than 0,1–0,2 microns. As it follows from [11], parameters that define the structural vibration during measurement of vibration on the vibro-acoustic setup always at least one order less than specified undulation tolerances, and therefore, structural vibration in this case does not play any role. The amplitudes of the structural vibration of roller bearings caused by bending deformations of the outer ring under severe axial tension are equal or approximately equal to the tolerances for undulation of rolling ring tracks of bearing accuracy class 2. Structural vibration amplitudes caused by fluctuating rigidity under severe tension and a very high ratio of radial stress to axial in most cases have much lower tolerance for waviness.

The performed analysis shows that the vibration of roller bearings under the characteristic conditions of control and exploitation conditions is due not to the design (structure) peculiarities of the bearing but technological errors, lubrication conditions and other factors. The level of structural vibrations of modern precision roller bearings does not exceed the level of vibration caused by technological reasons, including errors of macro geometry of tracks of bearing rings, and the undulation of working surfaces.

The most important is the structural vibration of roller bearing due to bending vibrations of the outer rings as a result of their loading by rollers that move. Oscillation amplitude of surfaces of outer rings in these cases can be quite large; such fluctuations may have a significant source of acoustic noise.

The vibration of the bearing due to fluctuations in its rigidity during rotation under combined loads in real bearings is very small and is the source of vibrations in machines and devices that can be ignored [5, 10, 11].

CONCLUSIONS

The mechanism of origin of constructural, structure and process defects of roller bearings is investigated. The relationship between the defect and performance indicators is determined on the example of changes of vibro-acoustic characteristics of the bearing. This fact served as the basis for development of a strategy of quality of components prediction and performance indicators of bearing on the stage of the design process. To diagnose common causes of vibrations, general spectrum of fluctuations of bearing is advisable to divide into three bands – low frequency oscillations (frequency from 50 to 1500 Hz), medium (1500–3000 Hz) and high frequency vibration (1500–10 000 Hz). It was established that technological defects cause low-frequency and high-frequency vibrations. The proposed technique according to integrated indicator of vibration of working surfaces of the rings determines the level of life of the finished part for the exploitation as a part of the bearing and predicts the vibro-acoustic characteristics of roller bearing.

REFERENCES

1. Avdieiev A.M. at al. Methods of Struggle with Vibration of Rolling Bearings. Scientific Engineering Digest "Bearing Industry", issue 1, NII Navtoprom, 1984, 32–38 [in Russian].
2. Balmot V.B. Theory and Methods to Improve Accuracy and Reduce Vibration and Ball Bearing Assemblies. Extended abstract of Doctor's thesis. 1987, 36 p. [in Russian].
3. Danylchenko Yu.M., Kuznetsov Yu.M. Precision Spindle Assemblies for Rolling-contact Bearings (theory and practice). Ternopil, Kyiv, Ekonomichna Dumka, 2003, 344 p. [in Ukrainian].
4. Marchuk V.I. Modeling of Dynamical System of Centreless Grinding of Working Surfaces of Roller Bearings. Scientific Notes "Ingeneering Mechanics", issue 14, 2004, 187–198 [in Ukrainian].
5. Rahulskis K.M., Yurauskas A.Yu. Vibrations of Bearings. L.: Mashinostroieniie, Leningrad Department, 1985, 119 p. [in Russian].
6. Świć A.: Elastyczne systemy produkcyjne. Technologiczno-organizacyjne aspekty projektowania i eksploatacji. Wydawnictwo Politechniki Lubelskiej, Lublin 1998.
7. Świć A. Gola A.: Elements of Design of Production Systems. Methodology of Machine Tool Selection in Casing-Class FMS, Management and Production Engineering Review. Vol. 1 (2), 2010, 73–81.
8. Świć A, Taranenko W, Szabelski J. Modelling dynamic systems of low-rigid shaft grinding. Eksploatacja i Niezawodnosc – Maintenance and Reliability 2 (50), 2011, 13–24.
9. Świć A. Technologia obróbki wałów o małej sztywności. Wyd. Politechniki Lubelskiej, Lublin 2009.
10. Dimentberh F.M. and Kolesnikov K.S. (Eds.) Vibrationin Technics: Directory. Fluctuations of Machines, Constructions and its Elements, 1980, p. 544 [in Russian].
11. Zhuravliov V.F., Balmot V.B. Mechanics of Ball Bearings Gyroscopes. Machinebuilding, 1986, p. 248 [in Russian].

10

INVESTIGATION OF THE INFLUENCE OF MOLD ROTATIONAL SPEED ON THE CAST WALL THICKNESS IN THE ROTATIONAL MOLDING PROCESS

Tomasz Jachowicz[1], Janusz W. Sikora[1]

[1] Department of Polymer Processing, Mechanical Engineering Faculty, Lublin University of Technology, 36 Nadbystrzycka Str., 20-618 Lublin, Poland e-mail: t.jachowicz@pollub.pl; janusz.sikora@pollub.pl

ABSTRACT

This paper presents the rotational molding process. The general principles of this polymer processing technology have been described. The main applications have been introduced and leading advantages and typical disadvantages of rotational molding process have been discussed. Based on the conducted experimental tests, the influence of changing one selected technological parameter, which characterized rotational molding process, on selected geometrical features of the polymer cast has been determined. Rotational mold's speed around axes was changed and a thickness of cast walls has been measured. Laboratory test stand, processing properties of polymer, also test program and experimental test methodology have been described.

Keywords: rotational molding process, mold rotational speed, polymer cast.

INTRODUCTION

The technological process of polymer molding can be divided into normal and rotational molding as well as slush molding and casting [3, 6]. In the rotational molding process, material in a liquid or plasticized state is placed in a mold form and subjected to centrifugal action or, additionally, to external pressure. In rotational molding, the mold is rotated around two or more axes at the same time, sometimes with an auxiliary circular motion [2, 7]. In centrifugal casting, which is a type of rotational molding, the mold is rotated around one axis only [4, 14].

Rotational molding is predominantly used to produce tanks, tubes, collar tubes, containers, lids, cases, canoes, and other axially symmetric parts, especially those which cannot be produced using other methods of polymer processing due to their exceptionally large external dimensions [3, 7]. Other from the advantages of the rotational molding technology, in contrast to different methods of production of hollow elements such as blow molding [17], there is the profitable influence of the processing shrink on removing the cast from the casting mold. The material applied to rotational molding has to have the shape of the powder or microgranules, what it gets the special methods of extruding the most often [16] using special extruder machines with the diverse construction [15].

The rotational molding process consists of a correct relation between rotational speeds around the axes, as it fundamentally affects both material flow direction and its even distribution in the mold [2, 6, 7]. Adequate technological parameters for the rotational molding process, such as an appropriate ratio of rotational speed around the main axis and auxiliary one, is fundamental to produce casts with the required wall thickness. The adequate parameter selection also entails good mixing, which ensures appropriate polymer homogenization in the mold; in effect, the polymer mechanical strength is adequate and its other properties are satisfactory. Hence, the cast meets the expected functional properties and high quality [4, 6]. Most often, the speed ratio is determined in manufacturing conditions for particular cast types only by means of experimental tests and is a business secret.

The available literature offers descriptions of studies in the rotational molding process dealing with melt polymer motion in the mold [1, 8], heat exchange between mold and polymer [18], and the behavior of various polymers in this process [5, 10, 19]. A development of modern calculation methods in recent years has led to developing programs which allow for the simulation of phenomena occurring during the rotational molding process [11, 12, 13, 20], mainly phenomena such as polymer heating [18], flow [1] and cooling [11], thanks to which the production preparation period is less time consuming and errors can be eliminated at the stage of developing cast design (examples of such special software are Rotolog and RotoSim as well as COMSOL, a multi-purpose computer program which allows for modeling and simulating various physical phenomena, including some aspects of rotational molding).

The present paper describes the experimental tests undertaken to examine the influence of rotational speeds of two axes of the mold on wall thickness of casts produced in the rotational molding process.

EXPERIMENTAL

Test stand

To determine the influence of the ratio between main axis speed and auxiliary axis speed of the mold in rotational molding on wall thickness of produced casts, experimental tests were performed using a laboratory machine for rotational molding available in the Department of Polymer Processing at Lublin University of Technology. The cast shape resembled a rectangular cube, and some of its edges were rounded (to facilitate the removal of the cast).

The machine control system allows for gearless speed change of both the main and auxiliary axes. The main axis speed can be adjusted within the range from 0 to 33 rpm, whereas the auxiliary axis speed is changed within the range from 100 to 0% of the main axis speed. This means that at 100%, the auxiliary axis makes as many rotations as the main axis, while at 0% the mold rotation around the auxiliary axis is stopped, as a result of which it is possible to realize the centrifugal molding process (in which the mold rotates around the main axis only) instead of rotational molding. Changing the rotational speeds of the main and auxiliary axes is done independently, which therefore makes it possible to adjust the reciprocal ratio of the main axis speed to the auxiliary axis speed. Such a possibility is vital for adequate selection of both rotational speeds, which will ensure that casts with adjustable wall thicknesses can be produced. Figure 1 shows this laboratory machine for rotational molding.

The rotational molding machine consists of the following systems:

- a rotational system,
- a heating system,
- a base equipped with a casing and a fan for cooling the mold,
- a control and regulation system.

Fig. 1. Laboratory machine for rotational molding: 1 – heating system, 2 – rotational system, 3 – mold, 4 – control and regulation system, 5 – base, 6 – fan

The rotational system of the machine consists of the following elements: a main axis rotation unit, an auxiliary axis rotation unit, a mold, a support structure, a main axis rotation power unit, and an auxiliary axis rotation power unit. The support structure of the machine is in a form of a spatial frame made of aluminum angles. On the frame, the power units of arm and mold rotation as well as bearing casings are mounted. The road wheels make it possible for the rotational system to move along the rails in the machine base so that the mold can be placed in the heating chamber [9].

The heating system consists of a heating chamber and temperature sensors that transmit signals to the control and regulation system. The chamber consists of a chamber jacket, heaters, insulation, a chamber plate and a chamber door.

The rotational and heating systems as well as the rotational system casing and the fan for cooling the mold are mounted on the base.

The control and regulation system ensures that the rotational speed of the mold around the main and auxiliary axes can be changed; the system also allows for adjusting and stabilizing temperature inside the heating chamber (using resistance heaters and thermostats) [9].

Technical data of the rotational molding machine are given in Table 1.

Table 1. Technical data of rotational molding machine [20]

Technical parameter	Value
Mold maximum dimensions, mm	200 x 200 x 200
Main axis maximum speed, rpm	33
Auxiliary axis maximum speed, rpm	33
Rotation unit weight, kg	80
Heating unit with base weight, kg	300
Machine total weight, kg	400
Mold form dimensions, mm	180 x180 x 180
Mold rotation power, kW	0.25
Arm rotation power, kW	0.25
Heating chamber power, kW	12

Tools used

The mold shown in Figure 2 is made of steel plates. Their surface, which makes the internal walls of the mold, is polished to facilitate removal of the molded cast. The mold is closed from the top and bottom by means of

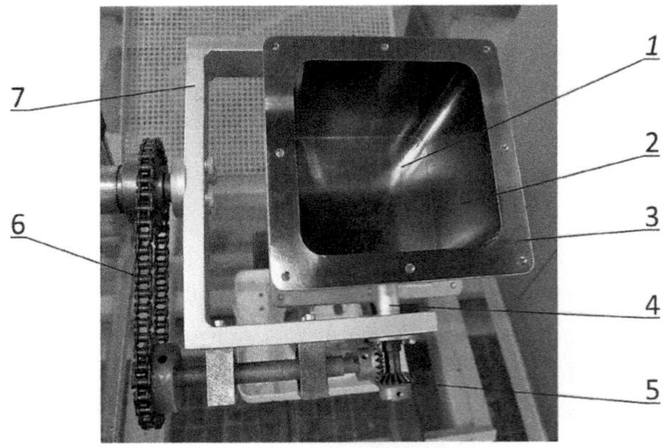

Fig. 2. Auxiliary axis rotation unit and mold for rotational molding: 1 – bottom head of mold, 2 – mold plate, 3 – collar for fastening upper head, 4 – journal bearing, 5 – bevel gear, 6 – chain transmission, 7 – arm

heads which are screwed to the collars. There are journals welded to the two mold plates; the journals are at the same time mounted to the bearings located in the arm. On the longer journals there is a bevel gear which communicates rotational motion [9].

Polymer used in the experiment

The polymer used in the experiment was LUPOLEN 3621 M RM, a medium-density polyethylene PE-MD in a powder form, produced by LyondellBasell Industries. This polyethylene kind is used particularly in the rotational molding process due to its good flow rate to density ratio, as well as its good tensile impact strength and thermal stability. The melt flow rate of this polymer is 7.5 g/10 min (at a temperature of 190 °C and load of 2.16 kg) and its density is 935.5 kg/m^3 [9]. Its Vicat softening temperature is 113 °C, and the recommended processing temperature range is 180–210 °C. Its mechanical properties are as follows: tensile modulus – 700 MPa, tensile stress at yield – 17 MPa, tensile strain at yield – 10%, tensile impact strength at a temperature of –30 °C – 104 kJ/m^2, tensile impact strength at a temperature of 23 °C – 213kJ/m^2 [9].

Experimental test program

In the experimental tests, auxiliary axis speed was considered a variable factor. Given the capabilities of the control and regulation system of the rotational molding machine, it was assumed that auxiliary axis speed n_{sa} was 100, 80 and 60% respectively of the main axis speed n_{pa}. Table 2 shows values of coefficient R_{oR}, or the ratio of main axis speed n_{pa} to auxiliary axis speed n_{sa}, calculated with the formula [6]:

$$R_{oR} = \frac{n_{pa}}{n_{sa}} \tag{1}$$

Table 2. Values of n_{pa}, n_{sa} and R_{oR}

No.	Axis speed		Coefficient
	main axis speed n_{pa}, rpm	auxiliary axis speed n_{sa}, rpm	R_{oR}
1	30	30	1
2		24	1.25
3		18	1.67

The factor examined directly was the thickness of wall b of the produced cast. In order to describe changes in thickness of particular walls of the cast depending on the ratio between main axis speed n_{pa} and auxiliary axis speed n_{sa}, the coefficient of cast wall thickness, W_{bs}, was determined; it was calculated as the ratio of wall thickness in a given measuring point of the cross section to average wall thickness in a given section. If the value of W_{bs} is greater than 1, it means that the cast walls are thicker, especially in the corners and on the edges of the rectangular cast, whereas W_{bs} lower than 1 means that the cast walls are thinner, what can be particularly noticed in their geometric center.

The constant factors included such technological parameters of rotational molding as: main axis speed of 30 rpm and polymer weight of 300 g. The dwell time of the mold in the heating chamber was 20 minutes, while the cooling time of the mold was 25 minutes. The temperature inside the heating chamber was 230 °C and cooling air flow rate was 300 m^3/h.

The disturbing factors included changes in ambient temperature, polymer humidity, air relative humidity and voltage fluctuations in the mains supply. It has been observed that these disturbing factors have no significant effect on the experiment results.

Experimental test methodology

In order to produce casts, 30 g batches of LUPOLEN 3621 MRM were prepared using a laboratory scales. After checking if the bottom head was installed accurately, the mold was filled with the prepared polymer. Next, the upper head of the mold was installed and tight fixed. After starting the rotational motion around the main and auxiliary axes and setting appropriate speed values of n_{pa} and n_{sa}, the rotating mold was placed in the heating chamber of the rotational molding machine heated to a temperature of 230 °C. After 20 minutes, the mold was removed from the chamber and then cooled for 25 minutes. The rotating mold was cooled in an airflow generated by a fan. After cooling the mold and stopping its rotational motion, the upper and bottom heads of the mold were unscrewed and the produced cast was hand removed.

The process of rotational molding was performed three times; each time the mold speed around the auxiliary axis was decreased according to the adopted experimental test program. Figure 3 shows one of the produced casts.

The wall thickness of the produced casts was measured using a slide caliper equipped with a digital measuring device. In order to make the measurements, the casts were cut in three reciprocally perpendicular planes S1, S2 and S3 shown in Figure 4. The cast walls were denoted in the following way: C denote the walls adjacent to the mold walls connected to the journal bearings, W denote the walls perpendicular to the main axis of rotation, while P denote the walls which are adjacent to the upper and bottom heads closing the mold. The measuring points in which the thickness of the sides of the cast was measured were disposed in even spaces, measuring approx. 10 mm.

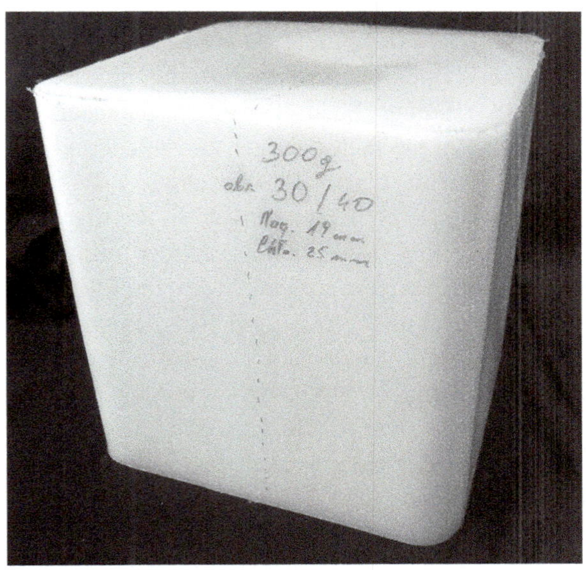

Fig. 3. Cast produced in rotational molding

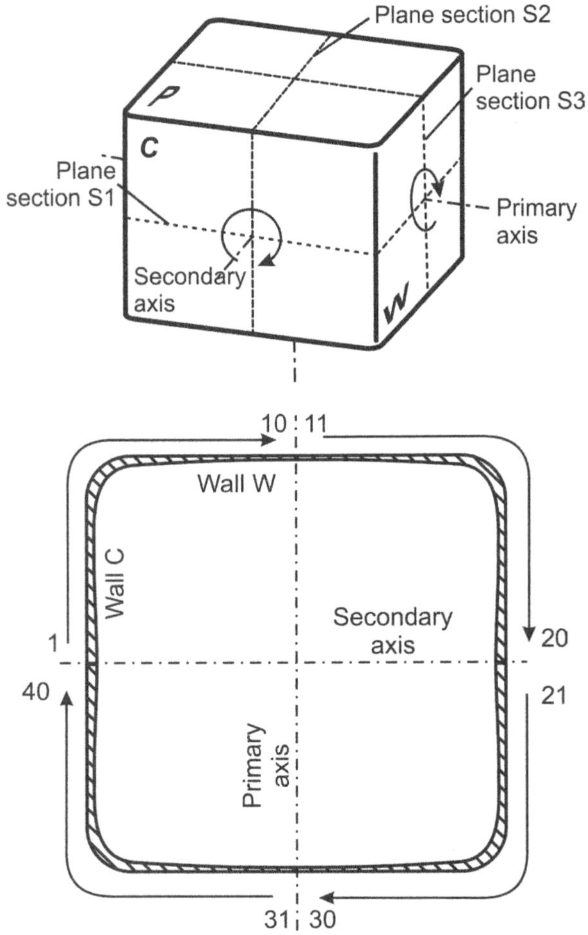

Fig. 4. a) positions of plane sections S1, S2 and S3 relative to main and auxiliary axes of rotation, b) example of position of measurement points at plane section S1

EXPERIMENT RESULTS

Figures 5, 6 and 7 present changes in average wall thickness depending on a quarter of the cross section shape obtained after cutting the cast by planes S1, S2 and S3. Increasing R_{oR} caused a clear change in the melt polymer movement and the forming of walls *C* and *W* of the cast. At R_{oR} = 1, in the geometric centre of walls *C* in the thermal centre resulting from the mass concentration of the journal bearings, average wall thickness was 2.81 mm and as R_{oR} was increasing, average wall thickness was decreasing to 2.46 mm and 2.36 mm respectively, which corresponds to a decrease in wall thickness by 12% and 16%. An increase in R_{oR} led to a decrease in cast wall thickness in the rounded corners (in plane section S1) from 3.36 mm at R_{oR} = 1 to 2.66 mm at R_{oR} = 1.25 and 2.11 mm at R_{oR} = 1.67, which corresponds to a decrease in the rounded corner thickness by 20 and 37%. The highest value of wall thickness, over 5 mm, was measured in the acute corners (of plane sections S2 and S3), which is shown in Figures 6 and 7. In the plane perpendicular to the main axis of rotation, the wall thickness on the cast edges changed from 5.32 mm to 4.60 mm, together with an increase in R_{oR}, whereas in the plane perpendicular to the auxiliary axis of rotation, a change in the value of R_{oR} had no significant effect on changing the wall thickness on the cast edge – it changed from 5.30 mm to 5.18 mm together with an increase in R_{oR} (Fig. 7).

Figure 8 shows the effect of speed ratio R_{oR} values on average thickness of walls *C*, *P* and *W*, determined in the geometric centre of these walls. The highest value of average thickness can be noted in the case of walls *C*, which resulted from an unfavorable distribution of the mold material mass, that is – the journal bearings connected to the relevant mold walls. An increase in R_{oR} led to a proportional decrease in average wall thickness, the most considerable decrease was observed in the case of walls *C* (from 2.68 mm to 2.27 mm, which corresponds to 15%). In the case of *P* walls, a decrease in thickness was smaller, by 12%, from 0.92 mm to 0.81 mm, yet it occurred already at R_{oR} = 1.25. Average thickness of walls *W*, determined in their geometric centre (hence on the main axis of rotation) decreased the least, by 10% (from 0.87 mm to 0.78 mm), and it did not occur until R_{oR} was of 1.67.

In order to describe the extent of changes in the thickness of particular walls, depending on the ratio between the main axis speeds n_{pa} and auxiliary axis speeds n_{sa}, wall thickness coefficient W_{bs} was calculated. In the areas where cast walls were thicker, W_{bs} was greater than 1, which could especially be observed in the corners and on the edges of the cast, whereas when W_{bs} was smaller than 1, it meant that the cast walls had a decreased thickness, which was especially noticeable in the central area of the cast walls. Figure 9 illustrates changes in the values of coefficient W_{bs} depending on the ratio of main axis speed to auxiliary axis speed, determined in plane section S1. The higher the difference between the mold speed around the main axis and its rotational speed around the auxiliary axis, the more evident the change in cast wall thickness, especially on the walls parallel to the mold axis of rotation. At R_{oR} = 1, the changes in W_{bs} ranged approx. from 0.60 to 1.3, while at the maximum difference between main axis speed n_{pa} and auxiliary axis speed n_{sa}, at

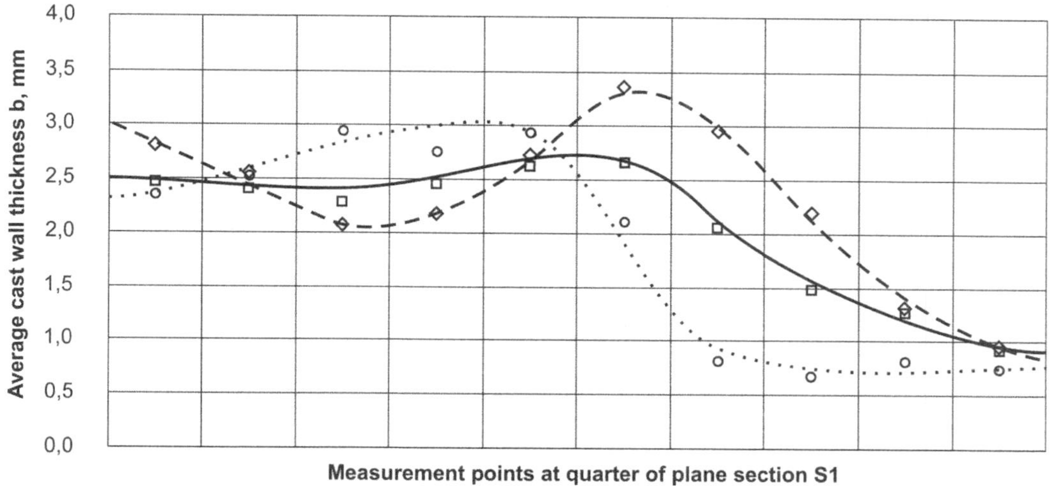

Fig. 5. Changes in average cast wall thickness *b* in plane section S1: $R_{oR} = 1$ (full line), $R_{oR} = 1.25$ (dashed line), $R_{oR} = 1.67$ (dotted line)

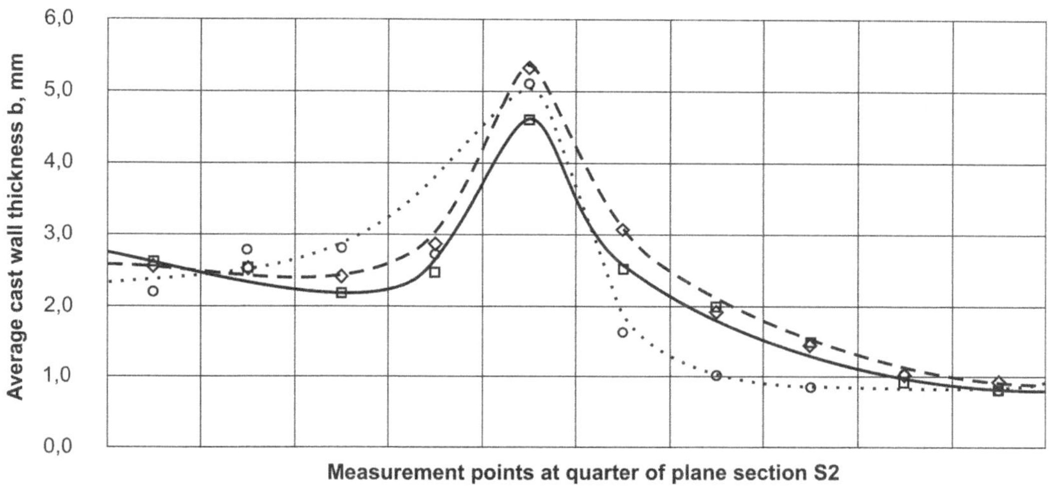

Fig. 6. Changes in average wall thickness *b* in plane section S2: $R_{oR} = 1$ (full line), $R_{oR} = 1.25$ (dashed line), $R_{oR} = 1.67$ (dotted line)

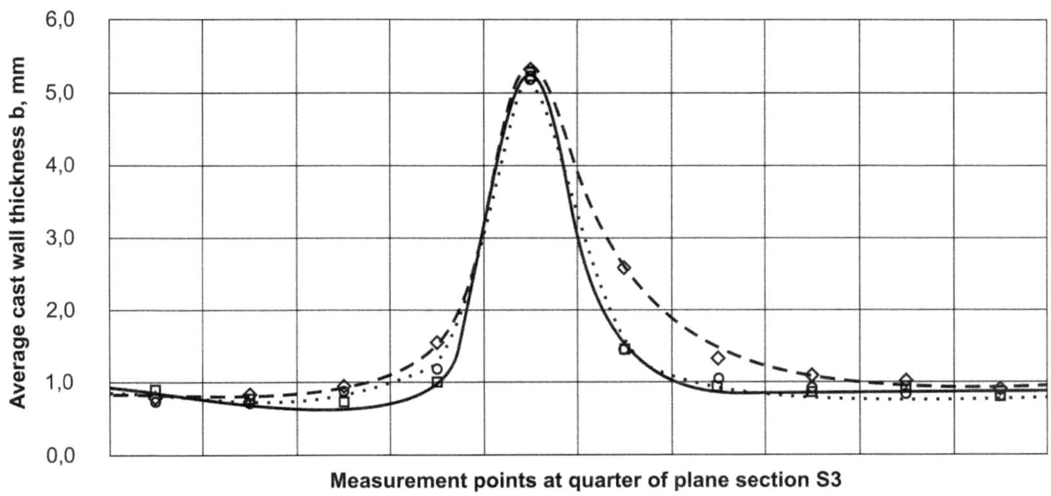

Fig. 7. Changes in average wall thickness *b* in plane section S3: $R_{oR} = 1$ (full line), $R_{oR} = 1.25$ (dashed line), $R_{oR} = 1.67$ (dotted line)

a)

b)

c)

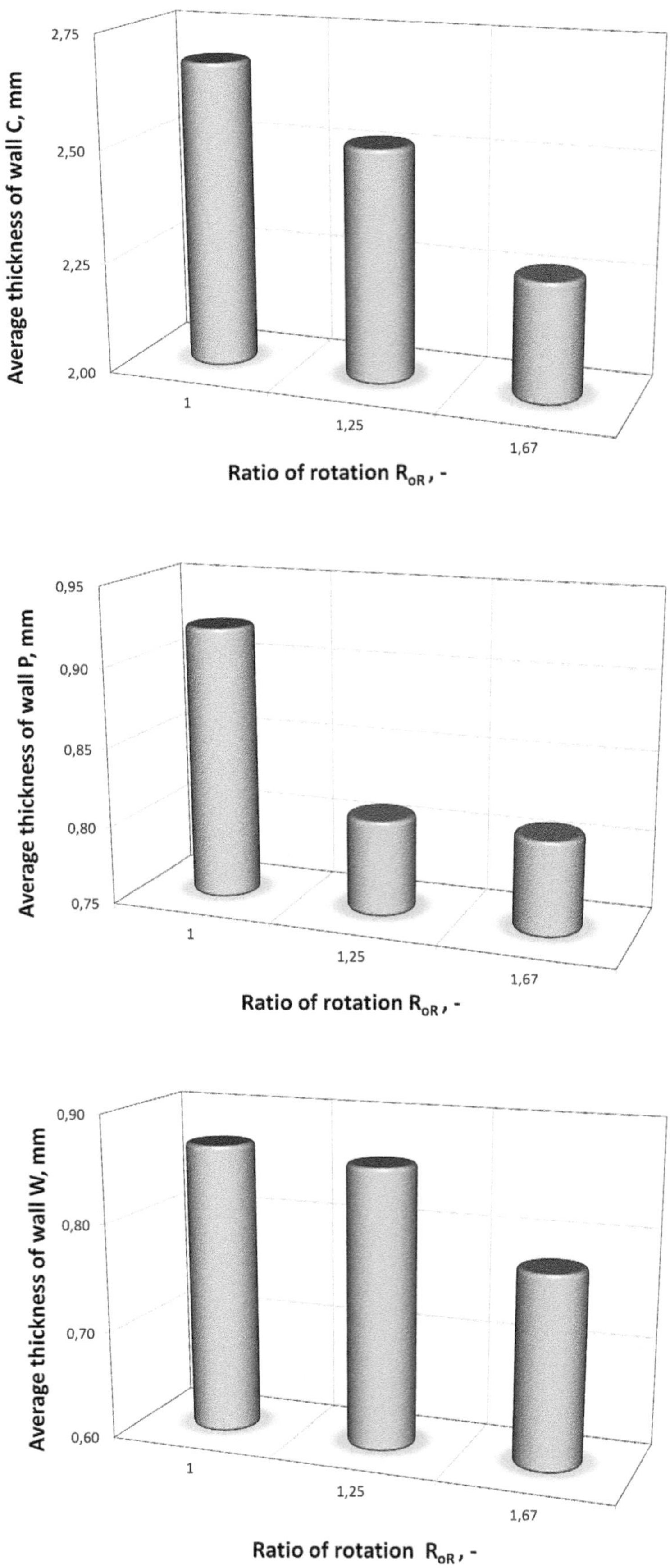

Fig. 8. Dependence of average wall thickness determined in geometric centre of cast on speed ratio R_{oR}: a) wall *C*, b) wall *P*, c) wall *W*

$R_{oR} = 1.67$, differences in the value of W_{bs} ranged from approx. 0.20 up to 1.6.

The results obtained in the experimental tests undertaken to investigate the influence of the changing speed ratio R_{oR} on wall thickness of casts produced by rotational molding, allowed for evaluating the unevenness of wall thickness of the produced casts. An increase in the value of R_{oR} meant a clear difference in wall thicknesses, which resulted in decreased thickness in their geometric centre, while the walls were thicker in the corners and on the edges. Owing to this, the casts in a form of rectangular cubes with varying rigidity were produced; the more rigid the casts were, the greater amount of polymer moved towards the cast edges, yet the wall strength was relatively small, which results in worse functional properties of these casts. Decreasing the auxiliary axis speed led to a change in the movement of the plasticized polymer along the walls parallel to this axis, while a relative increase in the main axis speed caused that the melt polymer moved away from walls W which were the least thick. Moreover, it was observed that walls *C* were clearly thicker in their geometric centre, which is the area where, in terms of the design, the mold was connected to the journal bearings. This metal mass concentration whose thickness was higher than on walls *W* and *P* leads to the occurrence of a thermal centre that keeps the polymer in the plasticized state in this area longer and leads to uneven cast cooling. As a result, particles of the melt polymer can adhere to this increased temperature area more easily, creating in this way a thicker layer. Compared to the other walls, wall *C* is nearly twice as thick at $R_{oR} = 1$. The unfavorable wall thickness distribution in plane section S3 was corrected by setting the ratio R_{oR} to a value of 1.25, which is illustrated by the dotted line in Figure 9. However, it should be noticed that such a solution is not optimum – the wall thickness is partially uniform near the journal bearings where the thermal centres occur, yet the thickness of the other walls is decreased because the melt polymer is affected by the centrifugal forces of varying values, which results from a difference between the main axis speed and auxiliary axis speed.

CONCLUSIONS

Based on the conducted experimental tests undertaken to investigate the influence of the speed ratio R_{oR} of two axes of rotation on wall thickness of casts in form of rectangular cubes produced in the rotational molding process, the following has been observed:

- an increase in R_{oR} leads to a decrease in average thickness of all cast walls in their geometric centre, which proves that the polymer mass moves towards the corners of the mold form; the largest decrease in average wall thickness has been observed on walls *C* perpendicular to the auxiliary axis of rotation,
- changing values of R_{oR} results in the smallest differences in wall thickness in plane section S3 parallel to the main axis of rotation and perpendicular to the auxiliary axis of rotation; the curves illustrating changes in wall thickness in particular measuring points in these cutting planes are similar,

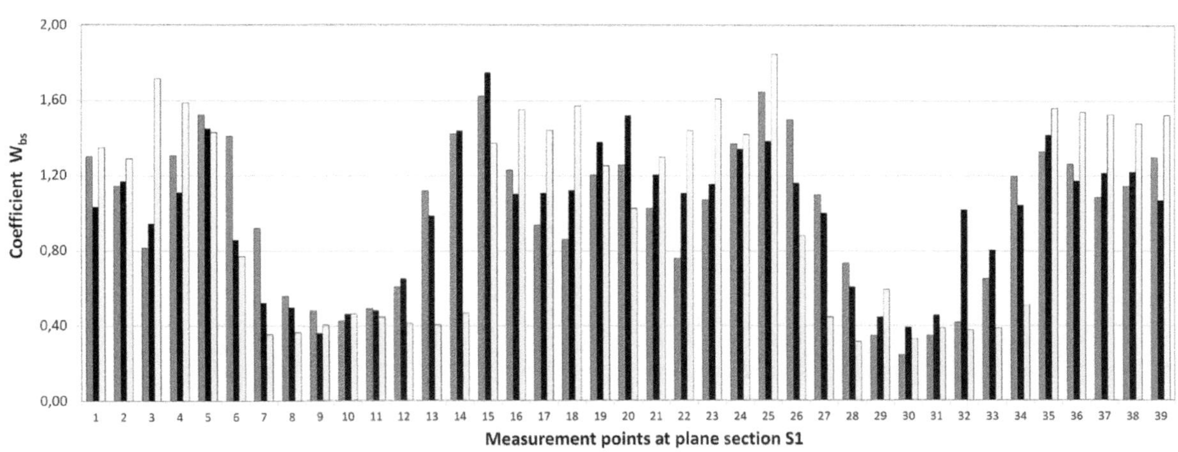

Fig. 9. Changes in values of W_{bs} depending on speed ratio R_{oR} determined in plane section S1: $R_{oR} = 1$ (grey bar), $R_{oR} = 1.25$ (black bar), $R_{oR} = 1.67$ (white bar)

- changing values of R_{oR} result in clear differences in wall thickness in plane section S1 that overlaps with the main and auxiliary axes of rotation of the mold,
- increasing values of R_{oR} in plane section S1 causes a gradual increase in wall thickness in the direction from the auxiliary axis of rotation to the cast corner (of wall C) and a decrease in wall thickness in the direction from the cast corner to the main axis of rotation (of wall W), and the most favorable change in wall thickness has been observed at $R_{oR} = 1.25$,
- differences of several percent in speeds of both axes entail a clear change in wall thickness which equals tens of percent. In the experiments it was found that decreasing the speed of one axis by 20% leads to a several percent difference between the thickness of the thickest wall and the narrowest one in the pointwise symmetric cast, which allows for formulating a speculation that it may have a negative effect on strength properties of such cast, for example its rigidity,
- as for the mold used to produce casts for the experimental tests, an unfavorable effect of extensive mold material mass concentration has been observed (in the area where the journal bearings were located and where the thermal centers occurred), which leads to unevenness of mold cooling and affects negatively wall thickness uniformity.

REFERENCES

1. Aissa A.A., Duchesne C., Rodrigue D.: Characterization of polymer powder motion in a spherical mold in biaxial rotation. Polymer Engineering & Science, 52, 2012. 953-963.
2. Anandha Rao M., Throne J. M.: Principles of rotational molding. Polymer Engineering and Science, 12, 4, 2004, 237-264.
3. Beall G., Throne J.: Hollow Plastic Parts: Design and Manufacture. Hanser Gardner Publications, USA, Cincinnati 2004.
4. Beall G.: Rotational Molding. Design, Materials, Tooling and Processing. Hanser Gardner Publications, USA, Cincinnati 1998.
5. Calò E., Massaro C., Terzi R.: Rotational Molding of Polyamide-6 Nanocomposites with Improved Flame Retardancy. International Polymer Processing, 27, 2012, 370-377.
6. Crawford R., Kearns M.: Practical Guide to Rotational Moulding. Rapra Technology Limited, UK, Shawbury 1998.
7. Crawford R., Throne J.: Rotational Moulding Technology. William Andrew Inc., USA, New York 2002.
8. Gogos G.: Bubble removal in rotational molding. Polymer Engineering & Science, 44, 3 2004, 88-394.
9. Information taken from the following websites: Clips Poly Engineering, Fixopan, Metalchem Toruń, Naroto, Persico, Plastigi, Rotoline, Rotomachinery Group, Roto Plastic International, Roto Plastics, R.S. Engineers, LyondellBasell Industries, and others.
10. Kulikov O., Hornung K., Wagner M.: Novel Processing Additives for Rotational Molding of Polyethylene. International Polymer Processing, 05, 2009, 452-462.
11. Lim K.K., Ianakiev A.: Modeling of rotational molding process: Multi-layer slip-flow model, phase-change, and warpage. Polymer Engineering & Science, 46, 7, 2006, 960-969.
12. Olson L.G., Crawford R., Kearns M., Geiger N.: Rotational molding of plastics: Comparison of simulation and experimental results for an axisymmetric mold. Polymer Engineering & Science, 40, 8, 2000, 1758-1764.
13. Olson L. G., Gogos G., Pasham V.: Axisymmetric finite element models for rotational molding. International Journal of Numerical Methods for Heat & Fluid Flow, 9, 5, 1999, 515-542.
14. Rusu G., Rusu M., Rusu E., Stolieru A., Teaca C. A.: Direct centrifugal molding of nylon 6-based product from ϵ-caprolaktam. Polymer - Plastics Technology and Engineering, 39, 2000, 233-247.
15. Sikora J.W.: Chapter 5. Screw Extrusion. In Advances in Polymer Processing: From macro to nano scales, Eds. Sabu T., Yang W. Oxford - Cambrigde - New Delhi, Wodhead Publishing Limited 2009.
16. Sikora J. W.: Review: Increasing the Efficiency of the Extrusion Process. Polymer Engineering and Science, 48, 9, 2008, 1678-1682.
17. Sikora R., Jachowicz T.: Wpływ czasu ochładzania na skurcz przetwórczy wytworów otrzymanych metodą wytłaczania z rozdmuchiwaniem. Polimery, 45, 2000, 713-719.
18. Sobotka V., Perot E., Maazouz A., Delaunay D.: Experimental Analysis of Heat Transfer in Rotational Molding Process. International Polymer Processing, 01, 2008, 93-102.
19. Szostak M.: Wpływ recyklatów PE na właściwości wyrobów odlewanych rotacyjnie. Inżynieria i Aparatura Chemiczna, 5, 47, 2008, 47-48.
20. Wang F., Yao W.G., Qiao C.D., Jia Y.X.: Finite element analysis of the physical gelation process of PVC plastisol during rotational molding. Acta Polymerica Sinica, 9, 2012, 1035-1041.

11

DIAGNOSTICS OF WORKPIECE SURFACE CONDITION BASED ON CUTTING TOOL VIBRATIONS DURING MACHINING

Jerzy Józwik[1], Dariusz Mika[2]

[1] Department of Production Engineering, Mechanical Engineering Faculty, Lublin University of Technology, Nadbystrzycka 36, 20-816 Lublin, Poland, e-mail: j.jozwik@pollub.pl

[2] Center for the Study of Engineering, Institute of Technical Sciences and Aviation, The State School of Higher Education in Chelm, Depułtycze Królewskie 55, 22-100 Chelm, Poland, e-mail: dmika@pwsz.chelm.pl

ABSTRACT

The paper presents functional relationships between surface geometry parameters, feedrate and vibrations level in the radial direction of the workpiece. Time characteristics of the acceleration of cutting tool vibration registered during C45 steel and stainless steel machining for separate axes (X, Y, Z) were presented as a function of feedrate *f*. During the tests surface geometric accuracy assessment was performed and 3D surface roughness parameters were determined. The *Sz* parameter was selected for the analysis, which was then collated with RMS vibration acceleration and feedrate *f*. The *Sz* parameter indirectly provides information on peak to valley height and is characterised by high generalising potential i.e. it is highly correlated to other surface and volume parameters of surface roughness. Test results presented in this paper may constitute a valuable source of information considering the influence of vibrations on geometric accuracy of elements for engineers designing technological processes.

Keywords: vibration, diagnostics, machining, turning, surface geometric structure, roughness parameters.

INTRODUCTION

Despite tremendous progress in the construction of both machine tools and cutting tools, effective methods of eliminating (minimising) vibration in machining are yet to be found [1, 3, 4, 7, 11, 12]. It particularly concerns the process of machining flexible objects and those susceptible to heat (i.e. thin-walled parts and long, slender rollers) [4–6, 8, 9, 15] as well as machining with the use of highly flexible tools (mainly of high overhang) [1, 10]. The analysis of literature [4, 5, 12, 15] shows that defects caused by elastic strains are of high importance, as far as the accuracy of machining is concerned, and they constitute from 20 to 80% of total machining error. While machined, low-stiffness parts generate vibration, their machining, however, cannot be avoided [5, 8, 15]. A significant group of parts used in the aviation industry can serve as an example of such objects. Irrespective of increasing stiffness of machine tools and modern materials used for main bodies and other parts of machine tools (having high capacity for dampening vibration) as well as precise tools and holders, still, vibration has a considerable influence on constituting the dimensional and shape characteristics manufactured parts [2, 3]. The necessity to avoid vibration, especially self-induced one, usually enforces reduction of technological parameters of machining, which does not always seem the solution to a problem and is not economically viable [15]. In the process of machining, vibration has a number of sources, the principal of which are the force input, kinematic input and inertial input [3, 4, 7, 11, 12]. As a result of self-induced vibration while machining, almost periodic shape defects occur. This kind of vibration in the machine tool-clamp-object-tool system results from the oscillation of friction force and machining force values as well as forces responsible for transferring energy from propulsion into the machine tool-clamp-object-tool system. Vibration occurrence can be an outcome of multiple factors such as machining parameters, type of material and shape of machined object, the structure of a tool and its fastening as well as the steadiness of a machine tool itself. It has a negative influence on metal machining and seems particularly undesirable while finishing surfaces. What occurs as a result is high surface roughness, faster wear of a cutting blade and a high frequency noise which can be harmful to workers. Vibration occurs as a result of: inaccurate balancing of machine tool rotary parts or a revolving object being machined, intermittent character of the process of machining (i.e. consecutive use of a cutting blade while milling or a broach while broaching), kinematic transmission defects (i.e. graduation errors or eccentric performance of cutting blades in relation to the wheel rotation axis which trigger the appearance of periodic forces being transferred to machine tool bearings and guideways), external vibration (i.e. when another machine generating strong vibration works nearby). One more source of vibration can be a physical process of machining itself [1–12].

Vibration identification can be carried out with the use of contact sensors (*i.e.* accelerometers <acceleration sensors>), non-contact laser techniques (i.e. laser vibrometers, laser sensors) as well as vision methods facilitating movement analysis (i.e. using fast vision cameras and a PONTOS system belonging to GOM). Examples of such measurement systems and their practical application can often be found in scientific literature [13–16].

METHODOLOGY

Diagnostic assessment of surface geometry after machining medium carbon steel C45 and stainless steel based on tool vibration measured during the machining process was the subject of this study. Rollers d_w = 50 mm in diameter and l = 200 mm in length served as workpiece. The research was conducted in The Center for the Study of Engineering in The State School of Higher Education in Chelm with the use of a CNC CTX 310 eco machine. The tests were conducted for several machining parameters collated in Table 1 and the test bed setup is presented in Figure 1.

Table 1. Technological machining parameters

Material	Feed rate, *f* [mm/rev]				
Steel C45	0.1	0.2	0.3	0.4	0.5
	v_c = 250 m/min				
Stainless steel	0.1	0.2	0.3	0.4	0.5
	v_c = 280 m/min				

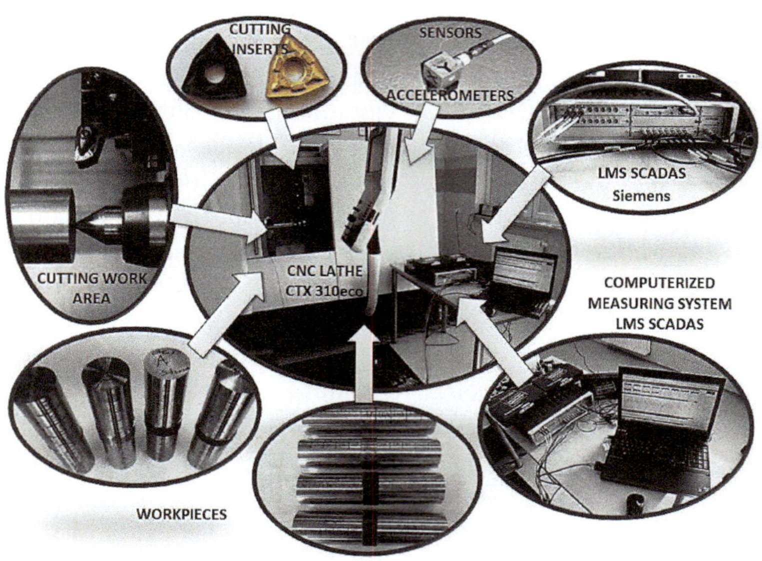

Fig. 1. Test stand; CTX 310 eco CNC, Coro Turn RC DWLNL 2020K08 rigid clamp, WNMG 08 04 08-PM 4225, WNMG 08 04 08-MR 2025 inserts for turning

Tools used for the study were turning tools with interchangeable inserts dedicated to materials used as samples in the tests. A WNMG 08 04 08-PM 4225 insert was used in the case of C45 steel and for the stainless steel a WNMG 08 04 08-MR 2025 insert was applied. The inserts were placed in a Coro Turn RC DWLNL 2020K08 rigid clamp. Siemens MS Scadas data acquisition hardware was used for recording vibration in the process. Piezo-electric tri-axis vibration sensor PCB Piezotronic was used in the research. Experimental tests were repeated four times for each set of machining parameters with the aim of assessing repeatability of measurements. The surface analysis was conducted using the *Sz* parameter for 3D surface roughness analysis, which is defined as 10-point surface height. The surface geometry of the machined workpiece is rather complex, therefore, its analysis requires advanced measurement tools. An example of 3D surface structure after machining is shown in Figure 2.

Additionally, the *Sz* parameter was included in the tests due to a distinct connection to the remaining parameters measured. The *Sz* parameter provides information on the height of irregularity of surface and is resistant to random peaks and valleys along the sample. The parameter also demonstrated a clear correlation with *Sa* and *Sq* parameters based on the character of surface ordinates distribution. *Sz* parameter seems to be characterised by great generalising properties (is strongly correlated with other surface and volumetric parameters). Alicona IF was used for the measurement of 3D surface roughness parameters.

RESULTS AND ANALYSIS

Vibration levels and their influence on stereometry during C45 steel machining

In the course of the conducted tests, time history of cutting tool vibration acceleration during machining in three perpendicular directions (X, Y, Z) defined in Figure 3 were registered. A coordinate system oriented in this manner applies to vibration measurement and results from the position of the sensor and the tool in the turret.

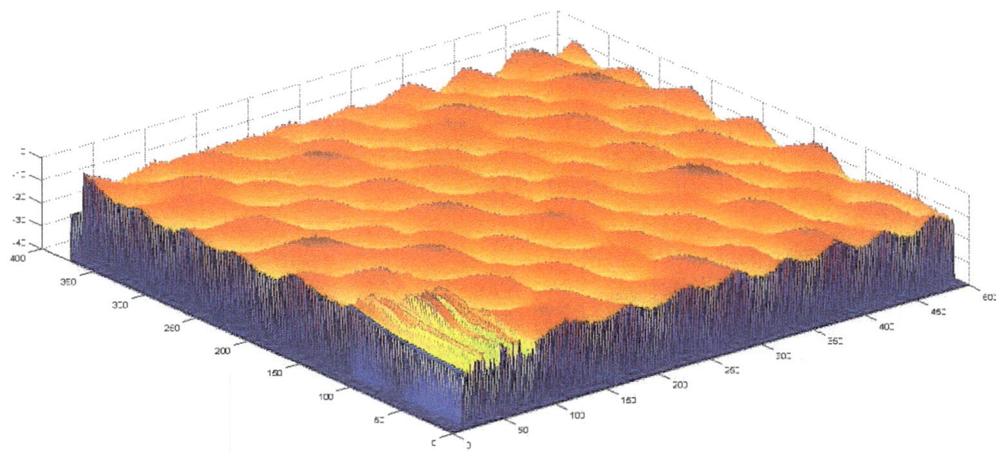

Fig. 2. An example of 3D surface structure after machining [17]

a) b)

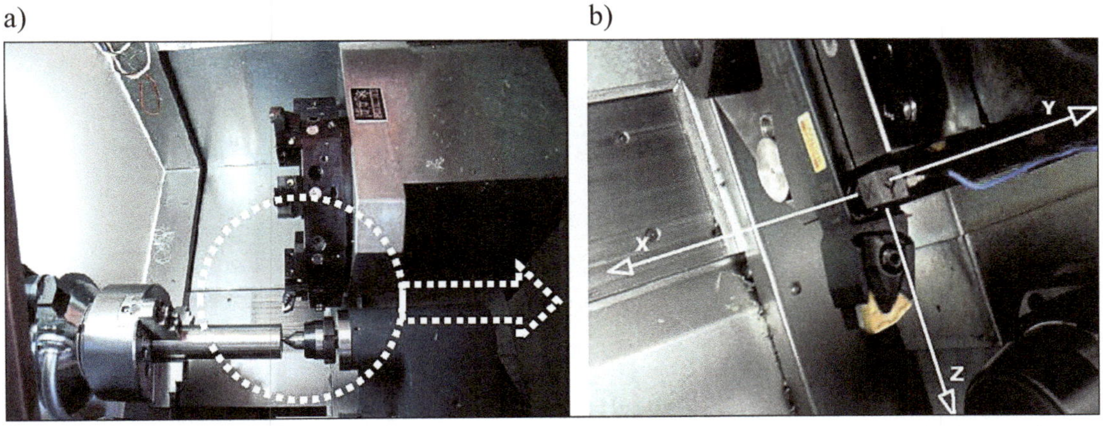

Fig. 3. Cutting stand: a) CNC working space, b) vibration directions measurement during turning

The system is independent of commonly applied numerically controlled axis positioning system of the analysed machine.

Figures 4–6 are a graphic representation of selected test results. Exceptionally interesting results were obtained in the Z axis direction (thrust force direction), where increasing values of the force should result in decreasing vibrations. Figure 4 shows time history of the cutting tool vibration acceleration during C45 steel machining in axes X, Y, Z defined in Figure 3. Based on the registered vibrations time history, it was found that feedrate demonstrates repeated, unequivocal influence on vibrations amplitude. Vibration increase has been observed both during machining high-quality constructional steel C45 (Figs. 4–5) and stainless steel (Figs. 8–9). Figure 5 presents results of cutting tool vibration time-frequency analysis in the form of spectrograms, for the X, Y and Z axes respectively. The spectrograms presented in the form of coloured maps display areas of accelerated vibrations of specific amplitudes at specific frequencies, derived from time history of vibration acceleration registered in tests. Figure 6 depicts distribution of energy density of the cutting tool vibration signal during C45 steel machining, with a characteristic peak of vibration energy concentration at approximately 3.7 kHz. The peak (Fig. 6) was detected for all measured axes and equals to maximum vibrations value 30 dB. The characteristic peaks at 3.7 kHz are a result of self-induced vibrations.

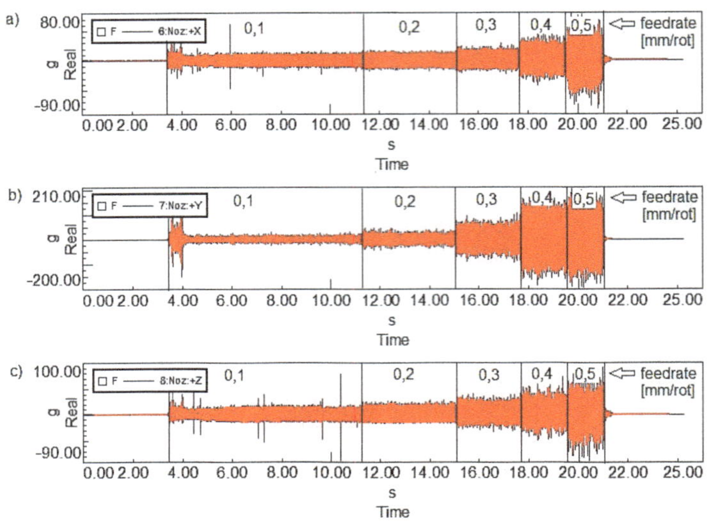

Fig. 4. Time history of cutting tool vibration during C45 steel machining for: a) X axis spectrogram, b) Y axis spectrogram, c) Z axis spectrogram

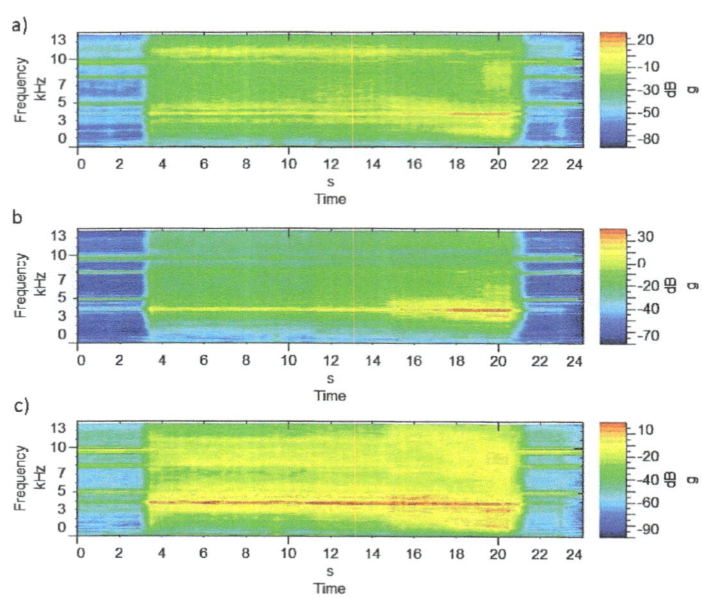

Fig. 5. Results of cutting tool vibration time-frequency analysis during machining C45 steel: a) X axis spectrogram, b) Y axis spectrogram, c) Z axis spectrogram

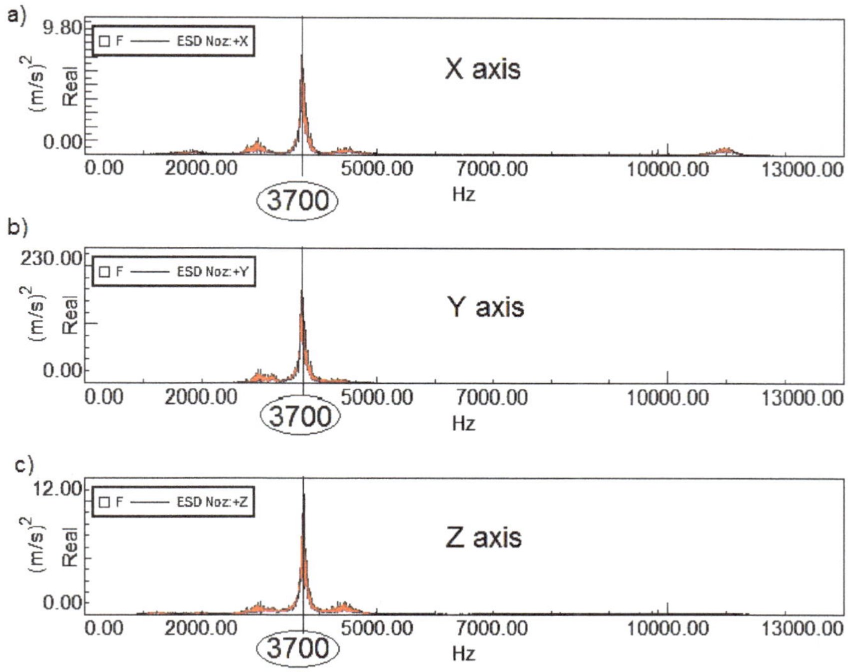

Fig. 6. Distribution of energy density of cutting tool vibration signal during C45 steel machining: a) for X axis, b) for Y axis, c) for Z axis

All conducted tests indicate that the Y axis, which is parallel to the cutting speed vector, is characterised by highest vibration values. This is caused by the highest value of the cutting force main component acting in that direction. The force function of the main component caused the highest value of the registered amplitude. In the case of C45 steel comparable vibration values were observed for both X (along the feedrate direction) and Z axis (thrust force direction). Vibrations in the Z axis (thrust force) direction seem interesting with regards to post machining surface condition. In the case of C45 steel vibrations in this axis are concentrated mainly at 3.7 kHz and are equal to 10 dB (with highest federate value f = 0.5 mm/rev). Table 2 collates the surface stereometry measurement results described with the 3D roughness parameter *Sz*. Vibrations were expressed using the RMS vibration acceleration expressed in *g* (grams). Table 2 indicates that in the case of C45 steel, surface roughness *Sz* increases as a function of feedrate from 14.98 mm (with f = 0.1 mm/rev) up to 37.81 mm (with f = 0.5 mm/rev) (Fig. 7a). An almost linear RMS vibration acceleration increase 14.98 *g* to 37.81 *g* implies an increase in surface roughness. It needs to be emphasised that increasing federate f values significantly influence roughness values.

Figure 7b implies various *Sz* surface roughness increase rates as a function of feedrate and RMS vibration acceleration. Test results have demonstrated a linear correlation of the analysed parameters as a function of feedrate f.

Table 2. Measurement results for specified machining parameters of C45 steel machining

Machining parameters	C45 steel				
Feedrate, f [mm/rev]	0.1	0.2	0.3	0.4	0.5
RMS vibration acceleration, [g]	5.3	7.6	9.2	9.6	11
Roughness, *Sz* [μm]	14.98	20.05	23.23	30.39	37.81
Cutting speed, v_c[m/min]	v_c = 250 m/min				
	Stainless steel				
Feedrate, f [mm/rev]	0.1	0.2	0.3	0.4	0.5
RMS vibration acceleration, [g]	3.4	3.6	5.8	44.1	92.3
Roughness, *Sz* [μm]	34.61	38.23	28.58	43.95	37.45
Cutting speed, v_c [m/min]	v_c = 280 m/min				

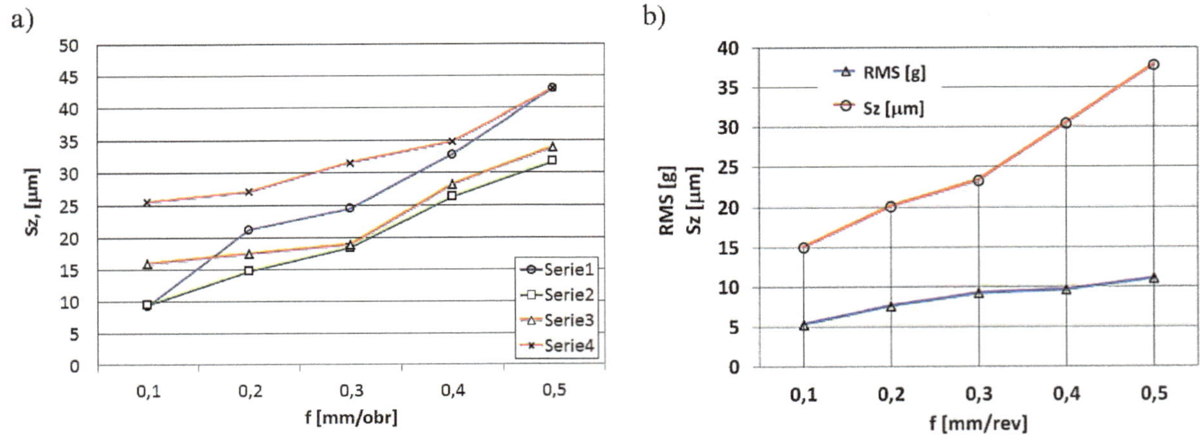

Fig. 7. *Sz* and RMS vibration acceleration parameters during C45 steel machining as a function of feedrate *f*: a) *Sz* value changes over 4 measurements, b) comparison of *Sz* and RSM vibration acceleration

Evaluation of vibration values and their influence on surface stereometry during stainless steel machining

Contrary to the results presented in literary sources, also stainless steel (similarly to C45) demonstrated growing changes in vibration acceleration amplitude as a function of feedrate, for all measurement axes defined in Figure 3, namely X, Y and Z (Figs. 8–9). Figure 8 shows time history of cutting tool vibration acceleration registered during stainless steel machining in three analysed axes X, Y and Z. During the tests it was discovered that vibrations in the direction (Z) of the thrust force prove to influence the surface geometric structure after machining to the largest extent.

Figure 9 presents results of time-frequency analysis of cutting tool vibration acceleration during stainless steel machining, whereas Figure 10 shows distribution of energy density of the cutting tool vibration acceleration signal. In the case of stainless steel the vibration acceleration amplitude in the Z axis direction was clearly higher compared with the X axis direction and comparable with the vibration acceleration across the Y axis. Based on the stainless steel tests, distinctly higher level of vibration was measured compared to high-quality steel C45 (Fig. 8). Maximum values were observed across the Y axis and equalled approximately 50 dB at 3.6 kHz and $f = 0.5$ mm/rev. The conducted time-frequency analysis of the cutting tool vibration signal during stainless steel machining revealed a characteristic peak at 3.6 kHz and for 10.6–10.7 kHz of the X axis.

Compared with C45, the peak found during stainless steel tests occurs at slightly lower fre-

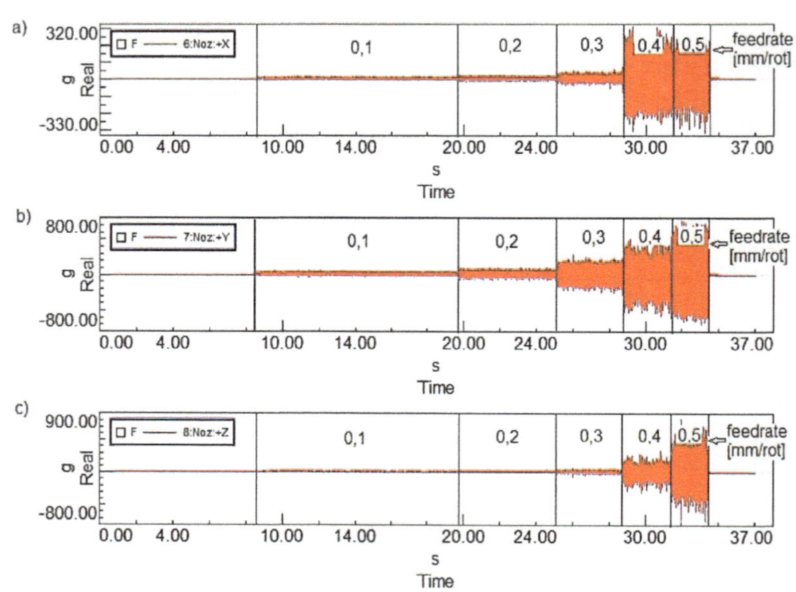

Fig. 8. Time history of cutting tool vibration acceleration registered during stainless steel machining for: a) X axis, b) Y axis, c) Z axis

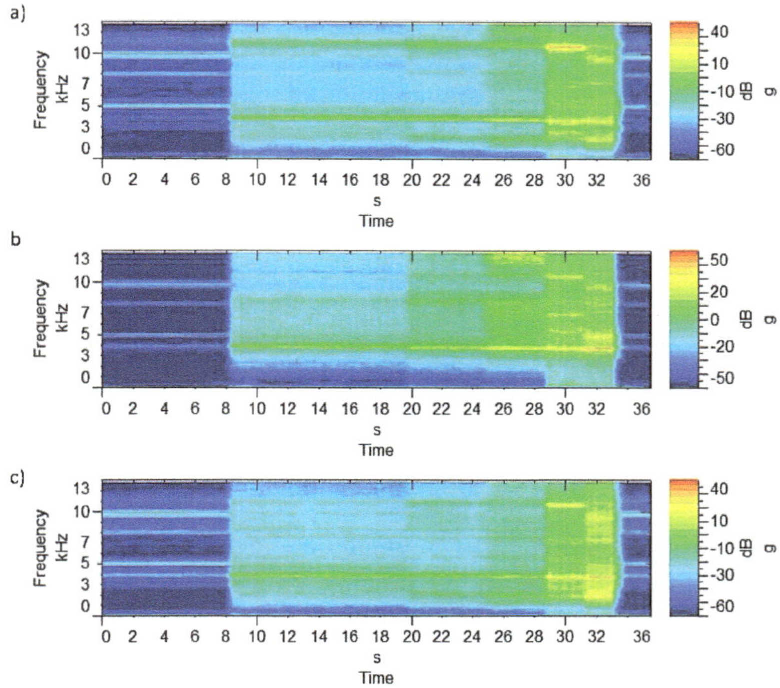

Fig. 9. Results of cutting tool vibration time-frequency a nalysis during machining stainless steel: a) X axis spectrogram, b) Y axis spectrogram, c) Z axis spectrogram

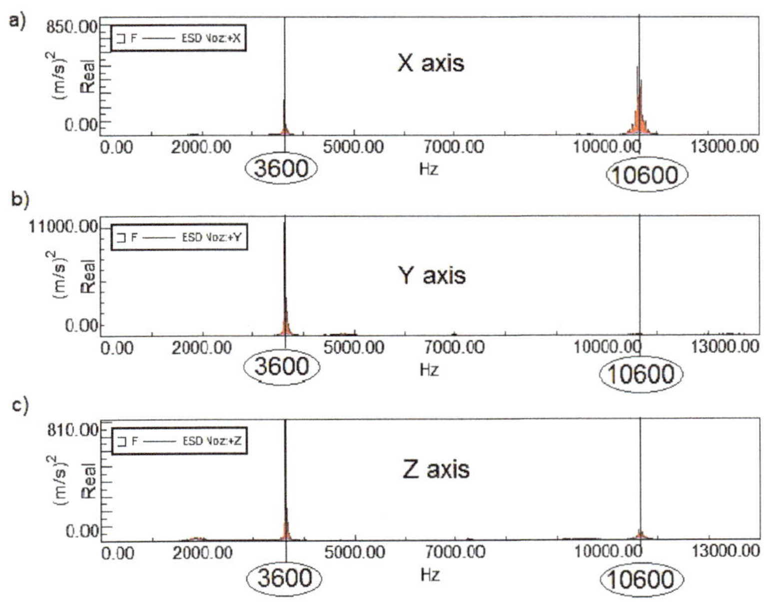

Fig. 10. Distribution of vibration signal density energy during stainless steel machining: a) for X axis, b) for Y axis, c) for Z axis

quency equal 3.6 kHz (presenting highest value for the Y axis at 50 dB). In the X axis (along the feedrate direction) an additional peak can be observed at approximately 10.6–10.7 kHz which equals 40 dB, also visible for the Z axis (the thrust force direction) mainly with f = 0.4 equal to approximately 20 dB. In the case of stainless steel, the main peak was observed at 3.5 kHz and is equal to 40 dB (f = 0.5) and 10.6–10.7 kHz amounting to 20 dB (f = 0.4). The observed peaks are most likely a result of self-induced vibration.

In the surface stereometry tests, the obtained surface roughness parameters presented considerably higher values than in the case of C45 steel (comparison of Fig. 7a and Fig. 11a). The *Sz* spatial roughness adopted maximum values amounting to 50–60 mm (Fig. 11a) in partial tests. Figure 11a indicates a great scatter of *Sz* values and a lack of a distinct tendency in each test run, which

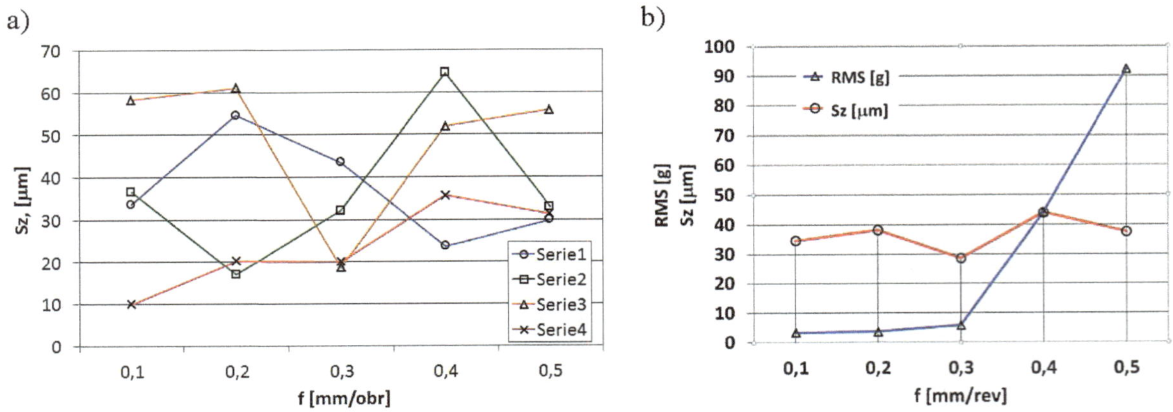

Fig. 11. *Sz* and RMS vibration acceleration parameters during stainless steel machining as function of feedrate *f*: a) *Sz* value changes over 4 measurements, b) comparison of *Sz* and RSM vibration acceleration

is completely different from the one observed for C45 steel (Fig. 7a). Lack of roughness change tendency is partly a result of slight changes in RMS cutting tool vibration acceleration within a range of feed values *f* (0.1–0.3 mm/rev). The vibration system is also relatively insensitive to changes in feed rate. Only highest values *f* = 0.4–0.5 mm/rev induced a sudden surge in RMS vibration acceleration values (Fig. 11b).

CONCLUSIONS

Based on conducted experimental tests a linear dependence of RMS vibration acceleration and *Sz* roughness of the C45 constructional carbon steel workpiece as a function of *f* feed can be observed. A changing intensity in roughness and RSM vibration acceleration increase as a function of *f* (different slopes of a curve depicting the discussed parameters – different slopes of a straight line) was found. Nevertheless, the obtained characteristics are unambiguous, linear and easily forecasted. Simultaneously, a strong scatter of results was observed for 3D structure of the workpiece surface *Sz* in stainless steel samples (as opposed to C45 steel). A lack of an increasing trend for RMS vibration acceleration as a function of *f* was identified for the *f* range amounting to *f* = 0.4–0.5 mm/rev and a significant increase in RMS vibration acceleration for higher *f* values (*f* = 0.4–0.5 mm/rev). The analysis of energy density distribution of the cutting tool vibrations signal during C45 steel machining and stainless steel demonstrated characteristic peaks at frequencies equal to 3.7 kHz (for C45 steel) and 3.6 kHz and 10.6–10.7 kHz (for stainless steel), which are a result of self-induced vibrations. Much higher values of vibration amplitude were revealed during stainless steel machining, which seems to result from material properties (mainly strength) as well as stainless steel machining resistance. This work presents graphic representation of the time-frequency analysis of cutting tool vibrations during C45 steel and stainless steel machining, which allow easy identification of amplitude and frequency vibration components dominant in the time history. Workpiece surface condition diagnostics based on cutting tool vibration is a relatively effective method when active machining process monitoring is concerned. However, it requires advanced measuring tools and programmes for the analysis of the obtained vibrations spectra and for estimating reliable parameters. Collecting and storing the data in a knowledge base may allow a quick and precise measurement of the surface state at the stage of production system design. Consequently, it would allow effective development of a technology which would minimise the number of defective units in production. The conducted tests prove a dependence of geometric characteristics of surface parameters on vibrations level. The conducted tests also indicate the need of further analysis of other factors influencing surface roughness, which need to be taken into consideration in modelling and developing diagnostic systems with the aim of separating their influence on surface geometry and vibration levels.

REFERENCES

1. Galewski M., Nadzorowanie drgań podczas frezowania szybkościowego smukłymi narzędziami z wykorzystaniem zmiennej prędkości obrotowej wrzeciona. Rozprawa doktorska, Gdańsk, 2007.

2. Józwik J., Kobyłka M., Badanie wpływu parametrów geometrycznych kieSzeni prostokątnej oraz warunków realizacji procesu skrawania na drgania podczas frezowania trochoidalnego. Postępy Nauki i Techniki, 8, 2011, 37–44.
3. Jemielniak K., Wypysiński R., Symulacja numeryczna drgań samowzbudnych – przegląd metod, możliwości i potencjalnych korzyści, Mechanik 8-9, 2013, 43–56.
4. Józwik J., Modelowanie ugięć sprężystych przedmiotów obrabianych w procesie skrawania toczeniem, Postępy Nauki i Techniki, 8, 2011, 183–191.
5. Józwik J., Lipski J., Błędy obróbki skrawaniem i ich prognozowanie z wykorzystaniem Sztucznych sieci neuronowych. Wyd. Politechnika Lubelska, Lublin 2014.
6. Józwik J., Filipiak P., Analysis of feedrate correction influence on corner radius errors of workpieces during milling, Journal of Machine Engineering, 9(1), 2009, 66–77.
7. Kęcik K., Rusinek R., Warminski J., Modelling of high-speed milling process with frictional effect, Journal of Muti-body Dynamics, Proceedings of the Institution of Mechanical Engineers Part K, 1(1), 2013, 3–11.
8. Kuczmaszewski J., Pieśko P., Wear of milling cutters resulting from high silicon aluminium alloy cast AISi21 CuNi machining. Maintenance and Reliability, 1(16), 2014, 37–41.
9. Kuczmaszewski J., Pieśko P., Włodarczyk M., Milling of thin-walled aviation elements – analysis of the model and verification tests, [In:] 6th International Conference – Supply on the Wings-Aerospace – the Global Innovation Driver, Frankfurt – Germany, 2011, 25–25.
10. Pieśko P., Zagórski I., Analiza dokładności obróbki frezami trzpieniowymi o zmiennej Sztywności tulei cienkościennych wykonanych z stopu AlMn1. [W:] A. Świć (red.) Innowacyjne procesy technologiczne, Lublin 2011, 33–46.
11. Rusinek R., Vibrations in cutting process of titanium alloy. Maintenance and Reliability, 3, 2010, 48–55.
12. Rusinek R. Warmiński J., Szabelski K., Drgania nieliniowe w procesie skrawania toczeniem. Monografia. IZT Sp. z.o., Lublin 2006.
13. Słodki B., Zębala W., Stanowisko do rejestracji obrazów szybkozmiennych w procesach skrawania. [W:] H. Latoś (red.) Obróbka skrawaniem zaawansowana technika. Wyd. Uczelniane Uniwersytetu Technologiczno-Przyrodniczego, Bydgoszcz 2009, 215–220.
14. Szydłowski M., Powałka B., Berczyński S., Stanowisko badawcze do wizyjnej inspekcji powierzchni obrobionej. Mechanik 8-9, 2013, 439–446.
15. Zębala W., Minimalizacja błędów obróbki przedmiotów cienkościennych. Inżynieria Maszyn, 15(3), 2010, 45–54.
16. Józwik J.: Analiza ruchu podczas obróbki frezarskiej przedmiotów cienkościennych z zastosowaniem wizyjnego systemu pomiarowego 3D. Mechanik, 8-9, 2014, 551–562.
17. Kacalak W., Szafraniec F., Kunc R. Remelska H.: Zastosowanie teorii fraktali do tworzenia i wizualizacji powierzchni o określonej topografii. [W:] Podstawy i technika obróbki ściernej, Łódź 2010.

THE INFLUENCE OF CURRANT MMA WELDING ON THE TENSILE STRENGHT OF JOINT

Piotr Penkała[1], Beata Płowaś[2]

[1] Institute of Technical Sciences and Aviation, The State School of Higher Education in Chełm, Pocztowa 54, 22-100 Chełm, Poland, e-mal: ppenkala@pwsz.chelm.pl

[2] Institute of Technical Sciences and Aviation, The State School of Higher Education in Chełm, Pocztowa 54, 22-100 Chełm, Poland, e-mal: bplowas@pwsz.chelm.pl

ABSTRACT

Welding with coated electrodes is still growing method of joining metals in construction machinery. If the electrodes are properly selected and current welding can weld materials in a wide range of geometric dimensions and species. With carefully selected ingredients one can introduce additional elements to the welded material and the joint during welding. An important issue in the electric welding is a welding current selection, depending on the thickness of the electrode and material thickness. The use of too low operating current cause instability and inadequate mechanical properties of the weld, resulting from insufficient melting of the material combined. Too high operating current results in the weld metal spraying and thermal overload of the electrode. For welding samples that were used to test the strength of the electrode used ESAB OK. 46.00 a diameter of 3.2 mm. Product of the company will provide a range of welding current ranging from 80 A to 150 A. The samples were welded currents amounting to 90 A, 115 A and 150 A. Due to problems with arc ignition and its instability is not made of welded samples current of 80 A.

Keywords: MMA welding, welding current, welded joint, tensile strength.

INTRODUCTION

The method MMA electrode coating is used, which consists of a metal core coated with compressed wrapper. Between the end of the electrode and the workpiece is produced electric arc. Arc ignition is a contact through the end of the electrode to the workpiece. The electrode melts and drops of molten metal electrodes are transferred through the arc to the molten weld metal welded to form a weld after solidification. Approaches welder electrode as its fusion to the workpiece so as to maintain a constant arc length, and at the same time moves the end of the melt, along the welding line. Melting electrode coating releases gases that protect the molten metal from the atmosphere and then solidifies and forms a puddle on the surface of the slag, which protects the solidifying weld metal from the influence of the environment. After laying a bead slag must be mechanically removed.

Electrode welding is used in all conditions and is, therefore, the most universal method throughout the welding industry. MMA method is a universal method because of the grade welded steel construction, position and place of welding. The main application is the welding of steel structures in shipbuilding and in most manufacturing industries, welding of pipelines, installation work on the construction sites, welding in the field and at altitudes and in areas with difficult access.

Welding design of machine elements is the most popular connection inseparable because they are cheap, simple and enable the achievement of significant savings in comparison with other design solutions. Welded joints replace items with large dimensions, casting or forging which is problematic for technological reasons. As compared to cast iron weight is obtained by using the structure of thinner walls, the materials of higher strength, less machining excess material.

WELDING PROCESS PARAMETERS BY MMA

Electrode welding, despite its huge popularity, is a method that requires considerable skills of the welder. Although manufacturers of power sources and additional materials through continuous improvement of its products, constantly striving to improve the ergonomics of welding, but the proper conduct of the welding process requires a combination of welder's experience and appropriately selected welding parameters. The basic parameters of manual welding with coated electrodes are [1–6]:

- the type and intensity of the welding current,
- arc voltage,
- arc length,
- welding speed,
- diameter of the electrode,
- slope of the electrode.

Welding current is a parameter that extent determines the amount of heat input to the weld and the speed of melting of the electrode to the greatest. In terms of the welding electrode, welding can be performed with:

- direct current (DC):
  - positive polarity (positive pole connected to the terminal electrodes, DC +),
  - negative polarity (negative pole connected to terminal electrodes, DC),
- alternating current (AC).

Type of welding current flows on the stability of arc, moving in the liquid metal droplets, the shape of the weld bead and the depth of submergence. Direct current arc provides a more stable and uniform distribution of the liquid metal in the arc, limiting the number of splashing even at low currents. Furthermore, with higher melting of the edges of the joined materials and tendency to shorten the circuit arc is lower. Some types of electrodes (mainly alkaline electrode designed for welding high-strength and high-alloy steels) require a very good arc stability, which can provide welding with positive polarity. In this configuration, the depth of fusion is increased, in comparison to the welding process with a negative polarity. AC welding is used rarely, but rather under "home" because of low cost of the device (welding transformer). Current value is selected depending on the type of material being welded, the thickness, the diameter of the electrode and the welding position. When welding current should be kept constant, stable value, which does not depend on the length of the electric arc. For this reason, it is recommended to steeply slope the characteristics static power sources, what ensures the maintenance of a constant current as a function of voltage (Figure 1). The steeper the current – voltage characteristics of the source the lower the change in current (I) when changing arc length (resulting in a change of the voltage U). For position welding a source of gently sloping characteristics may be used to adjust the length of the arc current.

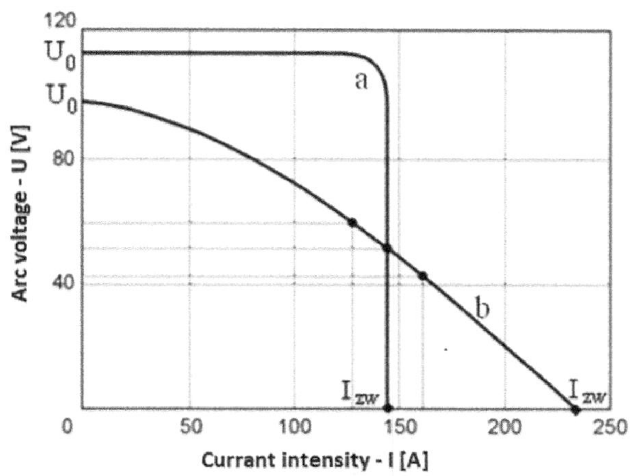

Fig. 1. Static current – voltage characteristics of sources for manual welding with coated electrodes: a – steeply sloping, b – gently sloping [1]

RESEARCH OF WELDED JOINTS

Research welds were carried out by destructive or non-destructive methods. In destructive tests a part of construction is destroyed in order to obtain a sample for testing. Destructive testing is used mainly in order to obtain information about the structure and mechanical properties of the joint. It is carried out on the samples whose dimensions and the research methodology are given in the relevant standards.

Non-destructive testing methods include those, which enclose the state of the object. They are made without prejudice to its consistency, or degradation of performance or functional characteristics. Non-destructive testing can be carried out at various stages of manufacture of welded joints: before, during and after welding [7].

The group of non-destructive tests enclose:

- visual,
- penetration,
- magnetic particle,
- radiographic,
- ultrasonic,
- tightness,
- eddy currents.

Destructive testing of welded joints are divided into:

- tensile test,
- hardness,
- impact,
- bending
- technological tests.

MECHANICAL PROPERTIES RESEARCH

The study of mechanical properties are an important group of tests that can be used for basic materials, additional and bonded joints. The study behaviour of the structure obtained on samples with specific dimensions, welded in the laboratory, may give different results than the whole structure under the actual load and the actual ambient conditions (temperature, humidity, pressure, etc.). It is also difficult to model the actual state of internal stress (welding) in the actual construction, resulting from subjecting its overlapping welding thermal cycles. The problem is also taken into account during the study of heterogeneity of a metallurgical bonded in the same joints, such as the heat – affected zone, face or ridge seam. Destructive tests can be used to determine the characteristics of the entire connection or a selected zone. The destructive methods for testing the mechanical properties are used to control:

- basic materials,
- additional materials,
- welded joints most commonly conducted on samples taken from the control joints made of the possible behaviour of the faithful implementation of a particular method of welding,
- test joints made during the development of new bonding technology, determine the welding conditions, the use of new materials, the fabrication of prototype,
- qualifying welders.

STATIC STRETCHING MATERIAL TESTING AND JOINTS

Static tensile test metal is included PN-EN 10002-1 and consists of subjecting the sample to a suitably shaped tensile force in the axial direction until the break. It is performed on samples of square, round, rectangular or annular depending on the shape and dimensions of metal, weld at ambient temperature of about (23 ± 5) °C. The basic tensile test is called static testing, although the load slowly increases at a predetermined speed. It is assumed, however, that corresponding to the specified deformation stresses appear immediately after operation of the load, that there is at any time able to balance stress of deformation. To a large extent this is true for elastic deformation, but in terms of plastic strain for many materials, such an assumption is inconsistent with reality. The standards provide for the limitation of the maximum speed of stretching. Maximal increase of stress in the elastic deformation range should not exceed 30 MPa/s. The growing burden should be slow and continuous for its maximum value.

Tensile testing is the basic and most common strain test, but one should keep in mind that the size of the characteristic obtained under tensile samples may not reflect the overall behaviour of the structure under load. For these reasons, some elements whose work load is mostly stretching, is subjected to a tensile test as a whole; for example: ropes, chains, wires, some of riveted joints or welded. Using tensile tests determined the following mechanical properties:

- tensile strength Rm [MPa],
- yield point Re [MPa],
- elongation at fracture A [%],
- sectional narrowing rates after fracture Z [%].

Static test of transverse stretching butt welded joints is the basic tensile test used in clinical properties of welded joints according to standard EN 895. It is performed to determine the tensile strength and the position of a breakthrough in the welded butt joints. The welded joint is heterogeneous in terms of ownership, and therefore does not specify the yield point and elongation of the sample, but only the tensile strength.

RESULTS OF STATIC STRETCHING THE SPECIMENS JOINTS BUTT- WELDING CURRANT OF 90 A

Based on the results in Table 1 and in accordance with Figure 2 and also observing the specimens revealed substantial differences in the values of strength and elongation. The differences in the results were due to the poor quality of the weld. Observing the joint after fracture noted a large amount of slag around the ridge and the absence of melting the parent material. Discrepancies welded joint were caused by too low a welding current. A low welding current specimens hampered during the arc, with the result that the above-mentioned inconsistency created.

Table 1. The results of the tensile test specimens welded at 90 A

Current of welding	No of specimen	Cross-sectional area $[mm^2]$	Force (max) [N]	R_m [MPa]
90 A	1	100	18 520	185.20
	2	100	28 204	282.04
	3	100	40 204	402.04
Expectation value:		100	28 976	289.76

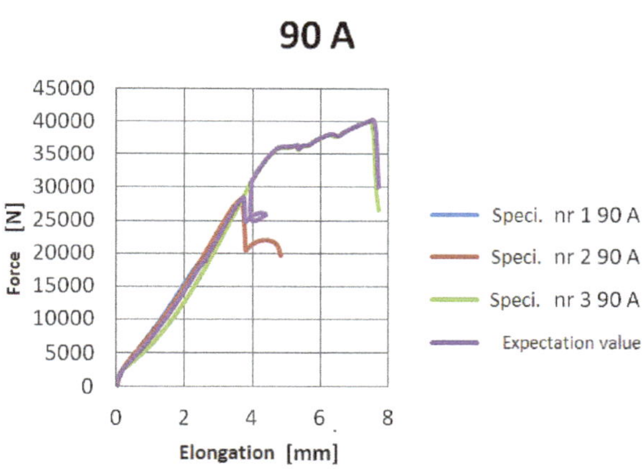

Fig. 2. Graph changes in the tensile test specimens welded current 90 A

RESULTS OF STATIC STRETCHING THE SPECIMENS JOINTS BUTT-WELDING CURRENT OF 115 A

Based on the results in Table 2 and Figure 3 one may conclude that a significant change of the force should be used to break the specimen. These values are considerably higher than in the previous study, that is welded to the current specimen 90 A. All three tested specimen were broken in the same place, i.e. at the site of the weld. After closer inspection of broken specimens we observed a small amount of slag and lack of fusion around the ridge. Such discrepancies could be due to problems with arc ignition in the surrounding ridge.

Table 2. The results of the tensile test specimens welded at 115 A

Current of welding	No of specimen	Cross-sectional area [mm]	Force (max) [N]	R_m [MPa]
115 A	1	100	34 333	343.33
	2	100	40 180	401.80
	3	100	44 684	446.84
Expectation value:		100	39 732.33	397.32

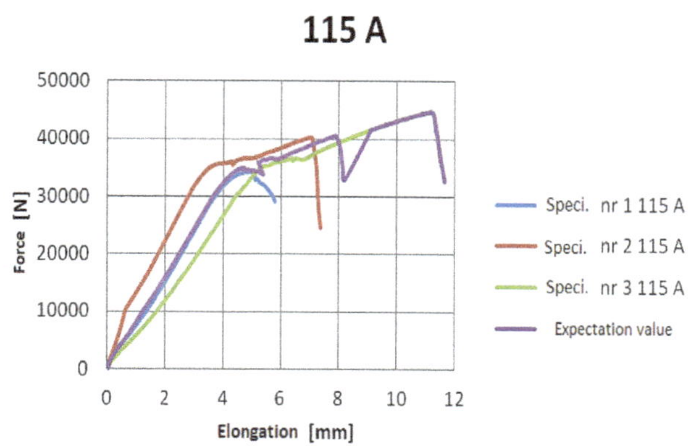

Fig. 3. Graph changes in the tensile specimens welded current 115 A

RESULTS OF STATIC STRETCHING THE SPECIMENS JOINTS BUTT-WELDING CURRANT OF 150 A

The data in table 3 and figure 4 show that a very significant amount of force is changed, which was to be used to break the specimen. These are significantly higher than previous attempts. The force required to break the three tested specimens is practically the same, the difference is just a few N, what is clearly showed in a graph in figure 4. In all three tested specimens there were local constrictions called the neck. The neck formed on both sides of the weld. Comparing the results it can be concluded that the strength of the weld made current 150 A is approximately equal to the strength of a solid material, as indicated by the results set out in table 4.

Table 3. The results of the tensile test specimens welded at 150 A

Current of welding	No of specimen	Cross-sectional area [mm]	Force (max) [N]	R_m [MPa]
150 A	1	100	50 579	505.79
	2	100	50 636	506.36
	3	100	50 565	505.65
Expectation value:		100	50 593.33	505.93

Table 4. The results of the tensile test specimens of solid material

Kind of specimen	No of specimen	Cross-sectional area [mm²]	Force (max) [N]	R_m [MPa]
Solid material	1	100	50 579	505.79
	2	100	50 614	506.14
	3	100	50 636	506.36
Expectation value:		100	50 610	506.10

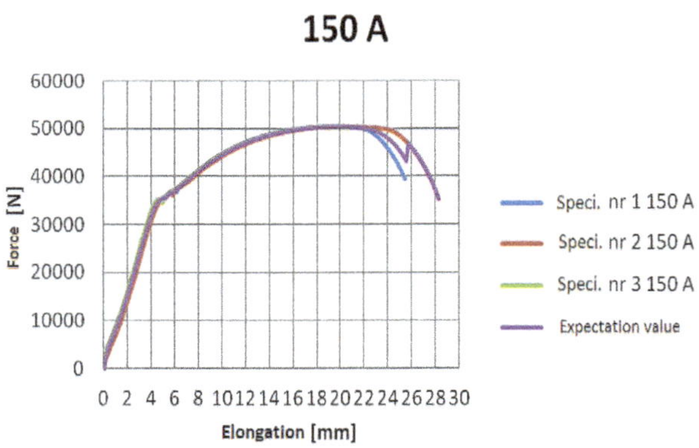

Fig. 4. Graph changes in the tensile specimens welded current 150 A

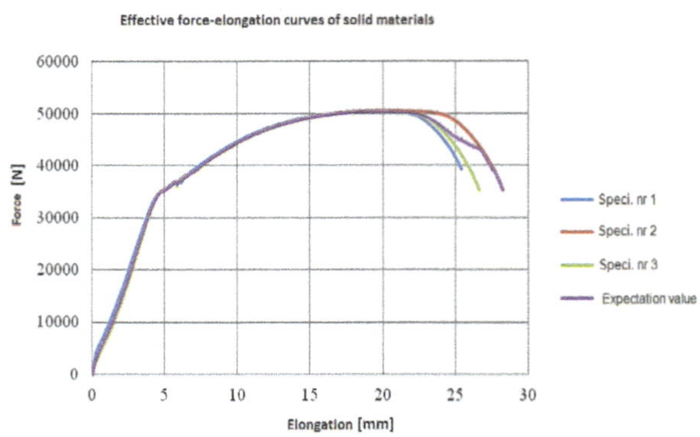

Fig. 5. Graph showing the relationship of strength and elongation for specimen of solid material

RESULTS OF STATIC STRETCHING THE SPECIMEN MATERIAL SOLID

Figure 5 and table 4 show linear increase in strain until a proportional limit. Then, after reaching the yield point the material enters the plastic state. Further increase of deformation tension caused non-linear growth until a noticeable narrowing of the neck called. Further stretching of the specimen caused its break under stress.

Table 5. Summary of the average results of the static tensile test

Kind of specimen	Cross-sectional area [mm²]	Force (max) [N]	R_m [MPa]
Welded – 90 A	100	28 976	289.76
Welded – 115 A	100	39 732.33	397.32
Welded – 150 A	100	50 593.33	505.93
Solid material	100	50 610	506.10

SUMMARY OF RESULTS OF STATIC TESTS AVERAGED STRETCHING WELDED JOINTS WITH SOLID MATERIAL

Based on the results in table 5 and 6 we have stated that the strength of the welded specimen of 150 A current is equal to a very rough estimate of the strength of a solid material. This situation is well illustrated in figure 6, where the lines charts of the specimens tested through the remaining parts thereof, among other things, a linear increase of strain, the achievement of the yield point, the transition into a state of plasticity and the transition into a state of deformation practically coincide.

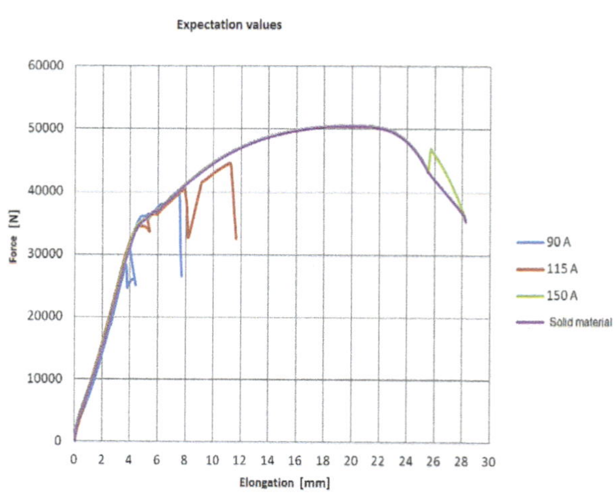

Fig. 6. Graph showing the dependence of strength and elongation for the results averaged

SUMMARY

Static tensile test confirms that the rupture of welded specimens a current of 90 A and 115 A, occurred at a much lower strength than welded current 150 A. After breaking the specimens welded current 90 A and 115 A a large amount of slag was observed in the root of the weld, and for sample welded current of 90 A as lack of fusion. In the case of welded specimens of current 150 A picked in the centre of the weld showed no abnormalities. The other two specimens were cut outside the weld, which indicates that the weld with properly chosen parameters has good strength properties.

Strength tests carried out showed that the welding current has a very significant impact on the strength and mechanical properties of welded joints. Non-conformities – welding slag around the ridge and the lack of penetration of the parent material were proved.

REFERENCES

1. Klimpel A.: Spawanie, zgrzewanie i cięcie metali. Technologie. WNT, Warszawa 1999.
2. Klimpel A.: Technologia spawania i cięcia metali. Wydawnictwo Politechniki Śląskiej, Gliwice 1997.
3. Pilarczyk J.: Poradnik Inżyniera. Spawalnictwo, Tom 2. WNT, Warszawa 2005.
4. Karpiński S., Moszumański J., Radwan-Wiatrowski K.: Laboratorium z podstaw spawalnictwa. Wyd. Politechniki Koszalińskiej, Koszalin 2001.
5. Ferenc K.: Spawalnictwo. WNT, Warszawa 2007.
6. Pilarczyk J.: Poradnik Inżyniera. Spawalnictwo. Tom 1, WNT, Warszawa 2003.
7. Ferenc K., Ferenc J.: Konstrukcje spawane. Połączenia (wydanie 3), WNT, Warszawa 2006.

13

STUDIES ON SELECTED PROPERTIES OF SURFACE LAYER OF C45 STEEL OBJECTS AFTER LASER CUTTING AND MILLING

Kazimierz Zaleski[1], Agnieszka Skoczylas[1]

[1] Department of Mechanical Engineering, Lublin University of Technology, 36 Nadbystrzycka Str., 20-618 Lublin, Poland, e-mail: k.zaleski@pollub.pl; a.skoczylas@pollub.pl

ABSTRACT

The article presents the results of studying the effects of technological parameters of milling upon surface roughness and microhardness of C45 steel objects after laser cutting. The metallographic structure formed as an effect of cutting by laser was also analyzed. The milling was performed on a FV-580a vertical machining centre. Depth of cut and feed per tooth were changed within the following range: $a_p = 0.09–0.18$ mm and $f_z = 0.02–0.17$ mm/tooth. To measure the surface roughness a Surtronic 3+ profile graphometer was used, whereas microhardness was measured with the use of a Leco LM 700AT microhardness tester. The surface roughness was significantly improved as a result of milling. The laser beam input and output zones were eliminated. Only a part of the layer hardened by laser cutting was removed while milling, in effect of which after milling the hardness of surface layer is much higher than hardness of the core.

Keywords: milling, laser cutting, surface roughness, microhardness.

INTRODUCTION

Laser cutting is a process, in which laser radiation (continuous or impulse) beam can cause melting of the material in the cutting crack, simultaneous melting and evaporation of the material or melting and/or combustion of the material [10].

The effect of coherent beam during cutting causes the formation of discrepancies on the surfaces and edges of the cut [19]. The occurrence of discrepancies hinders or prevents applying a given part in the subsequent stage of production. According to [19] discrepancies should mean irregularities or deviations from a specific shape or position of the shearing line. The following groups of discrepancies can be distinguished: discrepancies on cutting edges, discrepancies on surfaces, fractures, cinder and other discrepancies. The source of the resultant discrepancies is only the laser cutting process [19].

The assessment of the state of surfaces and edges is accomplished by means of the following indicators: surface roughness, perpendicularity tolerance (distance between two parallel straight lines, between which cutting profile is inscribed [18]), as well as that of cutting surface inclination, width of heat-affected zone (HAZ), kind of changes introduced by cutting, as well as mechanical and corrosive properties of the material [10, 18].

Surface roughness, formed as an effect of laser cutting depends upon striae, being formed on a cut surface as a result of laser beam fluctuation, is gas flow and hydrodynamic flow of the melted material [7]. On the treated surface two zones of differentiated roughness can be distinguished, which are separated from each other by the so-called boundary layer separation point (BLS) [6, 10]. The geometrical structure of the surface after laser cutting reveals parallel direction. The surfaces resulting from cutting should be classified as anisotropic random, less frequently as anisotropic periodical [17].

During laser cutting process, the beam, going deep into the material, gets deviated from perpendicularity. Depending on cutting conditions, deviation can equal from 0.3 mm to 0.75 mm, which should be considered at working out the subsequent technological operations [4].

The phenomena, which are unavoidable during cutting process are thermal changes within the area of cutting line, the formation of scale and flash on the opposite side to the position of laser [4]. On the cut through surface also shallow discontinuities and highly fragile fractures can be observed [1, 15]. The effect of thermal beam causes the formation of hardened layers in the surroundings of cutting crevice and cutting surface quenching. The hardened layers do have small thicknesses, but especially at the edges and quoins they are quite hard. The cutting edge surface is also oxidized [5, 13].

Laser cutting process is accompanied by high temperature gradient in the area of laser beam influence, which results in the formation of residual stress. The residual stress is deposed down to the depth of 2 mm from the cutting surface. Stress formation enhancing factor is also high cooling velocity. Crucial points in elements cut with a laser are areas near the cutting edge and corners [16].

In the effect of laser cutting the decrease of fatigue strength ranges from 40% to 65%, depending on the cut material, with reference to cutting with a stream of water and milling [11].

Currently one can meet technological solutions that reduce the formation of negative effects of laser cutting. They include correct selection of laser cutting parameters [2], covering the lower surface of cut sheet with a special chemical substance reducing the formation of metal overhang [4], as well as mechanical and manual removing of the formed scale and burrs [4]. Removing the metal overhang most often takes place with the use of belt grinders, manual files, wire brushes and by means of milling process dedicated for objects with larger dimensions and more complex shapes.

In connection with the occurrence of heat effect zone of a laser beam upon an object, the hardened layer milling process can be compared to quenched steel processing. The processing parameters selection depends upon the hardness of processed material.

In the paper [14] the effect of technological parameters of milling upon surface roughness was analyzed. It was found that cutting depth does not affect the roughness of the machined surface. It was also noticed that with the increase of cutting velocity v_c the decrease of surface roughness takes place. That is connected with mechanical vibrations, whose amplitude decreases with the increase of cutting velocity.

During machining of quenched materials quite true representation of the tooth edge shape occurs. At small feed values and slight disturbances „jaggedness" of cutting edges can have a much greater effect upon the machined surface roughness than the geometric-kinematical representation [8].

The object of the studies performed by the authors of this paper [3] was the assessment of the effect of milling parameters upon microhardness of stainless and unalloyed steel surface layer. The studies reveal that with the increase of cutting and feed velocity the surface layer microhardness increases. The depth of changes taking place equals about 120 μm and depends upon mechanical and thermal effects that take place.

Our previous studies [12] revealed that milling the surface after laser cutting is a complex process. The forces occurring during treatment of surface layer are larger than during milling the core material. What still remains to be analyzed is the kind of metalographic structure, microhardness and geometric structure of the surface (GSS) shaped by milling of the layer after laser cutting. For this purpose the studies of assessing the state of surface after laser cutting and then after finishing milling were undertaken. Knowledge of this problem is demanded. This results from the necessity of applying a given element in subsequent stages of production or assembly.

The purpose of milling was to remove negative effect of cutting process (striae on the cutting surface, error of contour shape and roundings of cutting edge). The cutting depth was selected in such a way as to leave a part of hardened layer, formed during laser cutting process.

METHODOLOGY OF RESEARCH

In the studies C45 steel samples were used (marking according to PN-93/H-84019) (Table 1). Samples of the dimensions 5×8×100 mm were made with the use of laser cutting system Amada 3000 W, with the application of standard parameters. Oxygen was used as a working gas.

Before the operation of cutting, a metal sheet was subjected to homogenizing annealing in the temperature of 860 °C within 30 minutes to reduce the state of residual stresses.

The milling process was conducted on vertical machining center FV-580a. Sandvik double – bit flycutters of the diameter of 20 mm with insert

Table 1. The chemical composition and strength properties of the C45 steel

Non-alloy steel C45								
C	Mn	Si	P	S	Cr	Ni	Mo	Fe
0.48	0.74	0.36	0.011	0.01	0.09	0.02	0.002	rest
Yield Point					R_e = 430 MPa			
Tensile strenght					R_m = 740 MPa			
Hardness					250 HB			

215880 APKT10 covered with TiN coat were used as tools. During milling cooling-lubricating liquid Mobil Cut was used. The examined samples were fixed in a special handle, which was put in a special vice so that the machined surface was, positioned parallel to the axis of machine tool spindle. The vice was fixed on the machine tool table.

Milling was performed for the following parameters: v_c = 102 m/mim; a_p = 0.09–0.18 mm, f_z = 0.02–0.17 mm/tooth. Figure 1 shows the kinematics of milling process and characteristic dimensions.

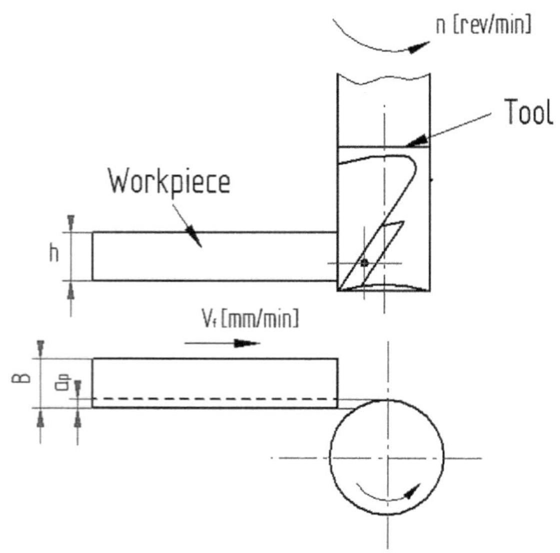

Fig. 1. Schematic illustration of milling process: *B* – sample width, *h* – sample height, a_p – cutting depth

Surface roughness measurements were performed with the use of laboratory prophilographometer Surtronic 3+ by Taylor Hobson, whereas topography and contour – with the use of T8000 RC120-400 appliance by Hommel-Etamic. The measurements were performed on the surface after laser cutting, approximately perpendicularly to the direction in which the radius acted. In connection with the occurrence of two characteristic zones, the measurements were performed in radius input and output zones. The same scheme was assumed for milled samples.

For the assessment of material structure the metalographic microscope Nikon MA 100 was used.

Microhardness measurements were performed on perpendicular microsections using Vickers's method at the bob weight equal to 100 g (HV 0.1), in accordance with the standard concerning hardness measurement with the use of Vickers's method. Microhardness tester LM700AT by Leco was used for these measurements For the obtained results of microhardness measurements the degree of strain hardening *e* from the relationship was determined:

$$e = 100 \cdot \frac{HV_{max} - HV_0}{HV_0} \ \% \tag{1}$$

where: HV_{max} – maximum microhardness for machining,

HVo – microhardness of material core (steel C45 before laser cutting).

Figure 2 shows the way of determining HV_0 and HV_{max}. The degree of strain hardening includes the effect of laser cutting and milling.

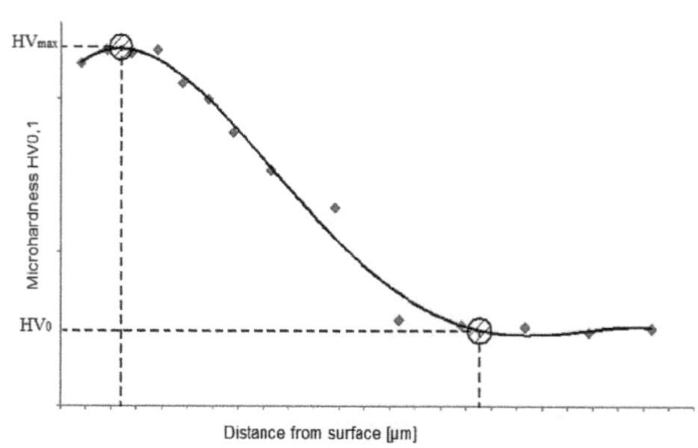

Fig. 2. The way of determining the microhardness distribution surface layer after laser cutting

STUDY RESULTS

Having finished the studies we analyzed the obtained results of surface roughness and surface layer microhardness measurements in the function of applied technological parameters of milling.

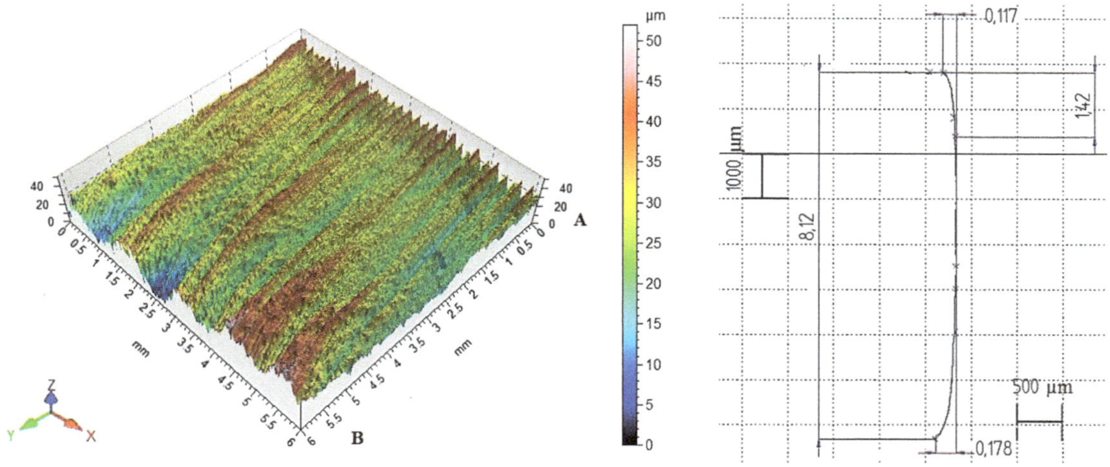

Sa = 4.63 µm, Sv = 27.5 µm, Sp = 24.5 µm, Sz = 52 µm

Fig. 3. Surface topography and contour after laser cutting

Figure 3 presents the topography of the surface and contour after laser cutting. There are visible zones of differentiated surface roughness. In the laser radius input zone (A) there occur uniform, rectilinear striae in small spaces between each other, whereas in the output zone (B), there are noticeable curvilinear striae, deviated from the intended track of laser beam relocation. Ra and Rz parameters for input zone equaled: Ra = 3.62 µm, Rz = 19.05 µm, for output zone: Ra = 4.21 µm, Rz = 21.11 µm.

In the case of contour a rounding of cutting edge is visible, as well as contour shape error, the maximum value of which is u_{max} = 0.178 mm (u – perpendicular dimension, describing the distance between two parallel straight lines, between which cutting profile is inscribed [18]).

C45 steel before laser cutting process had ferretic-pearlitic structure, of a band system. In the effect of laser cutting structural transformations take place. In the region of the cutting crack a heat affected zone (HAZ) was created, the width of which equals ca. 0.3 mm (Fig. 4a). In result of strong heating up of the material with a laser beam, and then cooling it down, quenching of the material took place in that zone. In HAZ martensite needles are visible, as well as bainite (near the cutting edge) (Fig. 4b), then ferrite+ martensite (in the middle of HFZ) and the structure of native material.

Figure 5 shows the distribution of C45 steel surface layer microhardness. In effect of laser cutting surface layer microhardness increases more than two times in the edge zone, then slowly decreases to achieve the value of core microhardness. The depth of changes taking place equals more than 300 µm.

In the graph showing the effect of machine cutting depth upon surface roughness (Fig. 6) a slight decrease of Ra parameter in the scope of

a) b)

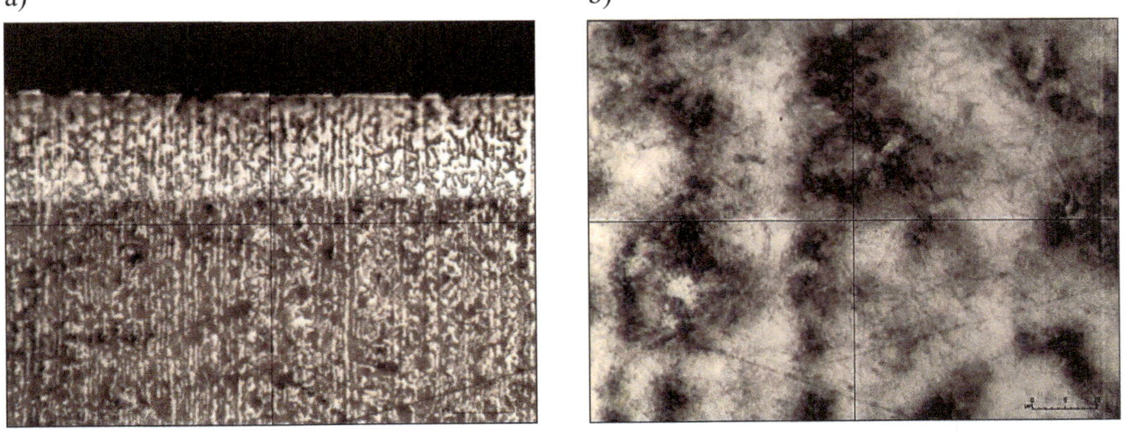

Fig. 4. Microstructure of C45 steel after laser cutting: a) edge-zone + core (magnification × 50) b) heat affected zone (magnification × 500)

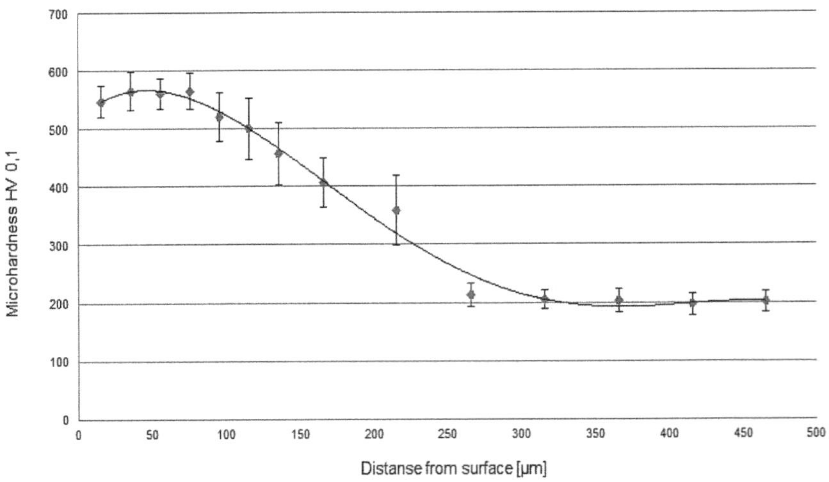

Fig. 5. Microhardness distribution of C45 surface layer after laser cutting

a_p = 0.09–0.12 mm, and for larger cutting depths the Ra parameter increases.

Greater value of Ra parameter for a_p = 0.09 mm is probably related to larger share of plastic and elastic strains of the surface layer occurring in the process of removing the thin layer of material. Roughness of surface after milling is a few times smaller than after laser cutting. It is difficult to indicate the input and output zones . The differences between zones are slight.

Figure 7 shows the effect of feed upon surface roughness. In accordance with our suppositions, with the increase of feed per tooth the Ra parameter slightly increases. Higher value of surface roughness for f_z = 0.02 mm/tooth may be caused by elastic return of unremoved part of the cut layer.

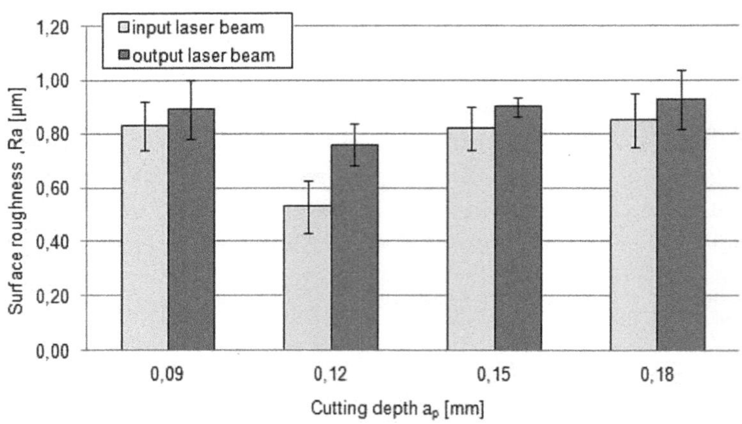

Fig. 6. Surface roughness parameters Ra as function cutting depth a_p

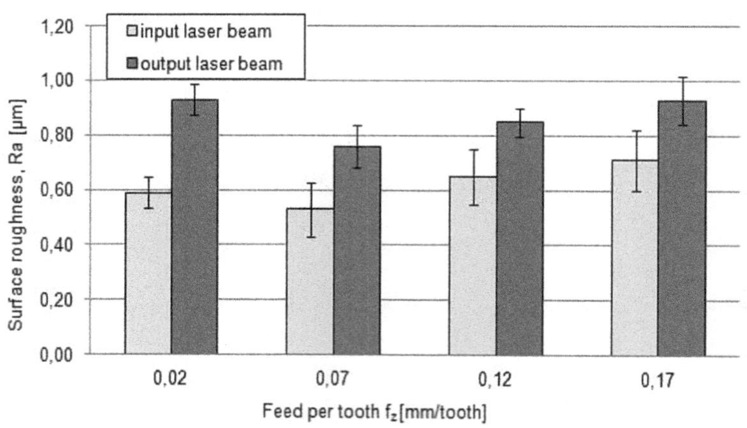

Fig. 7. Surface roughness parameters Ra as function feed per tooth f_z

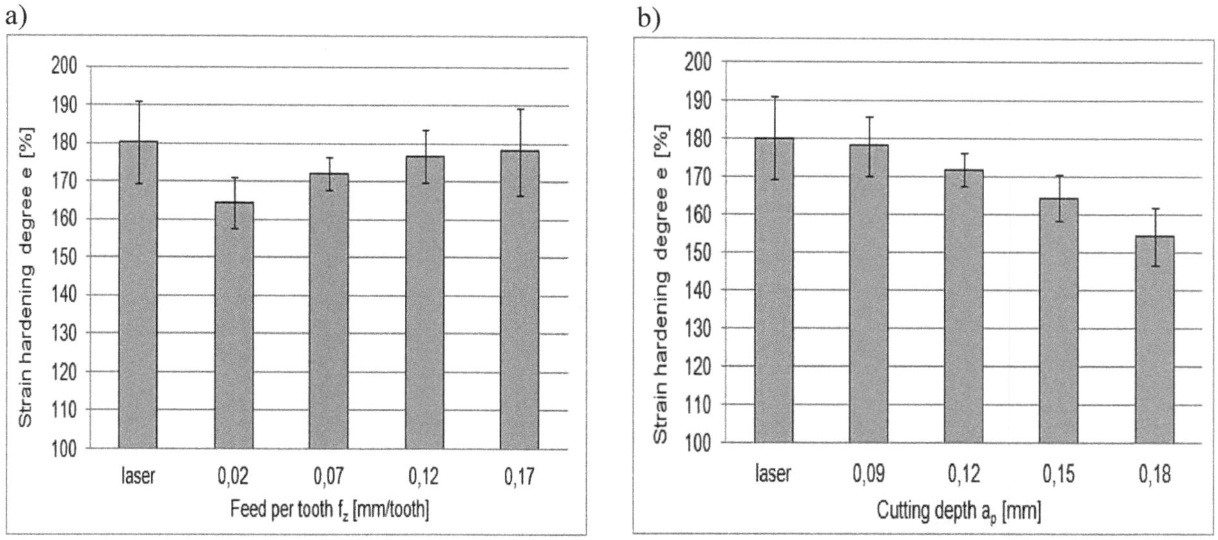

Fig. 8. Strain hardening degree as function a) feed per tooth f_z b) cutting depth a_p

The results of studies on the effects of milling parameters upon microhardness of surface layer are shown in Figure 8.

The increased feed per tooth causes slight increase of surface layer microhardness. That is probably caused by the increased chip thickness and higher passive forces (Fig. 8a). The strain hardening degree for samples after milling is lower than for a sample after laser cutting, because a part of the hardened layer was removed. Changes in the strain hardening degree are slight indeed, but that proves the mechanical effect of the tool teeth upon the machined surface after laser cutting.

Analyzing the influence of cutting depth upon the strain hardening degree (Fig. 8b), one can notice the decrease of strain hardening degree together with the increase of cutting depth. Milling with the increasing cutting depth causes removal of the thicker hardened layer as a result of laser cutting (the biggest hardening occurs down to the depth of about 120 μm), which leads to the decrease of surface layer hardness.

Microhardness of surface layer after laser cutting and milling probably results from the influence of two opposite phenomena: drawing the hardened layer, which causes hardness reduction and/or secondary reinforcing through the mechanical effect of teeth [9].

SUMMARY

In this study the effect of milling technological parameters upon surface roughness and microhardness of C45 steel after laser cutting were examined. Also the structure formed in the effect of laser cutting was examined. The following conclusions sum up the results of performed studies.

In the effect of milling the roughness was highly improved. Differences in surface roughness occurring after laser cutting (radius input and output zone) were eliminated. No significant effect of machining parameters upon surface roughness was found.

After laser cutting process edge zone was quenched. A non-easily treatable martensitic-bainitic zone was formed. Milling caused removal of a part of the layer, which was hardened in the laser cutting process.

Feed and depth of cut do not significantly affect the microhardness of surface layer. A slight accretion of microhardness is observed with the increased feed.

C45 steel milling after laser cutting makes it possible to remove the negative effects of cutting process (striae on the cutting surface, contour shape error, as well as rounding of the cutting edge), which allows for the improvement of the surface geometrical structure, simultaneously leaving a small layer of hardened material.

REFERENCES

1. Arif A.F.M., Yilbas B.S., Abdul Aleem B.J.: Laser cutting of thick sheet metals: Residual stress analysis. Optics & Laser Technology, 41, 2009, 224–232.
2. Bień A., Żarnowski R.: Optymalizacja parametrów cięcia laserowego. Przegląd Mechaniczny, 12, 2011, 37–40.

3. Bouzid Saï W., Ben Salah N., Lebrun J.L.: Influence of machining by finishing milling on surface characteristics. International Journal of Machine Tool & Manufacture, 41, 2001, 443–450.
4. Feldshtein E., Koman I.: Wycinanie laserowe elementów o dużej grubości w blachach ze stali nierdzewnej. Przegląd Mechaniczny, 4, 2010, 13–18.
5. Górka J., Skiba R.: Wpływ procesów cięcia termicznego i strumieniem wody na właściwości i jakość powierzchni ciętych stali niskostopowych o wysokiej granicy plastyczności. Przegląd Spawalnictwa, 2, 2013, 11–18.
6. Kalita W., Hoffman J., Radziejewska J.: Jakościowe efekty cięcia blachy wiązką lasera CO_2 z wysoko-ciśnieniowym nadmuchem gazów neutralnych. Przegląd Mechaniczny, 13-14, 1996, 22–27.
7. Kannatey-Asibu E., Jr.: Principles of laser materials processing. John Wiley & Sons, New Jersey, 2009
8. Kawalec M., Rybicki M.: Struktura geometryczna powierzchni po frezowaniu czołowym zahartowanej stali. Mechanik, 10, 2009, 780–785.
9. Kawalec M.: Mikrostruktura i twardość technologicznej warstwy wierzchniej po toczeniu zahartowanej stali narzędziowej. Postępy Technologii Maszyn i Urządzeń, 3, 1996, 23–35.
10. Klimpel A.: Technologie laserowe. Wydawnictwo Politechniki Śląskiej, Gliwice, 2012
11. Mäntyjävi K., Väisänen A., Karjalaninen J.A.: Cutting method influence on the fatigue resistance of ultra- high- strength steel. Journal of Material Forming, 2, 2009, 547–550.
12. Skoczylas A. Zaleski K.: Badania frezowania obwodowego warstwy utwardzonej wskutek cięcia laserem. Mechanik, 8-9, 2012, 249–256.
13. Trela S.: Badania wpływu cięcia laserowego na twardość stali nisko- i średnio węglowej. Mechanik, 12, 2004, 877–878.
14. Twardowski P.: Wpływ dynamiki procesu frezowania zahartowanej stali na chropowatość powierzchni obrobionej w warunkach HSM. Archiwum Technologii Maszyn i Automatyzacji, 2, 2008, 183–192.
15. Yilbas B.S., Arif A.F.M., Abdul Aleem B.J: Laser cutting of holes in thick sheet metals: Development of stress field. Optics and Lasers in Engineering, 49, 2009, 909–916.
16. Yilbas B.S., Arif A.F.M., Abdul Aleem B.J: Laser cutting of rectangular blanks in thick sheet steel: Effect of cutting speed on thermal stresses. Journal of Materials Engineering and Performance, 19, 2010, 177–184.
17. Zaborski S., Stechnij T.: Laserowe i plazmowe cięcie blach ze stali niestopowych i kwasoodpornych. Inżynieria Maszyn, 4, 2011, 109–116.
18. PN- EN ISO 9013:2008. Cięcie termiczne. Klasyfikacja ciecia termicznego . Specyfikacja geometrii wyrobu i tolerancje jakości
19. PN-EN ISO 12584: 2004. Niezgodności w procesach cięcia płomieniowego tlenowo- gazowego, cięcia wiązką laserową i cięcia plazmowego. Terminologia.

STACKING SEQUENCE OPTIMIZATION OF COMPOSITE BEAMS WITH DIFFERENT LAYER THICKNESSES

Fatih Karaçam[1], Taner Timarci[1]

[1] Department of Mechanical Engineering, Faculty of Engineering, Trakya University, Edirne 22030, Turkey, e-mail: fatihkar@trakya.edu.tr; tanert@trakya.edu.tr

ABSTRACT

In this study, stacking sequence optimization of composite beams with different layer thicknesses is investigated for various boundary conditions. A unified shear deformation theory is used for analytical solution. The optimization process is carried out in order to obtain the minimum deflection parameters for Clamped-Free (C-F), Clamped-Clamped (C-C) and simply supported (S-S) boundary conditions under a uniform distributed load by use of genetic algorithm for a specific number of population and generation. Finally, among all possible combinations of layer thicknesses, the one giving the minimum deflection parameter and corresponding stacking sequence is chosen. The minimum values and corresponding stacking sequences are presented for different boundary conditions.

Keywords: laminated composite beams, static analysis, genetic algorithm, layer thickness, stacking sequence optimization.

INTRODUCTION

Composite materials have gained popularity in modern structures where light-weight and high strength are sufficient to carry various loading types for different boundary conditions. Because of high elasticity modulus and strength, composite materials are more popular than any conventional materials, besides, the advantage of being designed with many parameters such as stacking sequence, layer thicknesses, material and geometrical properties, structures can be designed effectively in order to satisfy the desired mechanical properties. Walker and Smith [1] studied the ending analysis of laminated composite plates by minimizing a certain fitness function with genetic algorithm (GA). Due to the manufacturing constraints, they chose the sets of values from the ply angles and layer thicknesses. Conti et al. [2] described a method to optimize the thickness balance within a composite laminate with layers oriented according to a limited set of angles and used a finite element method in optimization process. Lee et al. [3], used a parallel evolutionary algorithm in multi-objective optimization of the stacking sequences in composite plates by choosing fiber type, thickness and orientation angles as design parameters for each layer. Brighenti [4], investigated the optimum fiber orientation of fiber reinforced composite materials in order to optimize a certain fitness function by use of GA. Khosravi and Sedaghati [5], considering the thickness of layers and fiber orientation angles as the design variables, presented two optimality criteria for optimum design of composite laminates using a finite element method. In the study, stacking sequence optimization of composite beams with different layer thicknesses is investigated for various boundary conditions. In the static analysis, a unified shear deformation beam theory is used which was first applied to composite shells by Soldatos and Tımarcı [6]. Initially, the layer thicknesses are generated so as to satisfy the total thickness of a four-layered composite beam. After the layer thicknesses are set, a random initial population of stacking sequences are used and evolved in each generation in order to obtain the minimum deflection parameter. Finally, among all the possible combinations of layer thicknesses, the one that has the minimum deflection parameter and corresponding stacking sequence is presented.

ANALYTICAL METHOD

In the study, the beam is assumed to be constructed of four linearly elastic layers. Various boundary conditions are applied to the beam under the effect of a uniform distributed transverse load. The beam has a unith width, length of L and thickness of h in a typical coordinate system where $0 \leq x \leq L$ and $-h/2 \leq z \leq h/2$ as shown in Figure 1.

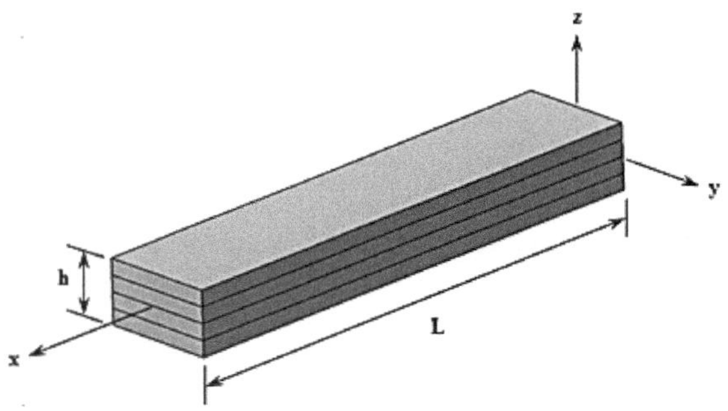

Fig. 1. Beam geometry and coordinate system

Uniform shear deformation beam theory is used in the static analysis where the other beam theories can be obtained as a special case [6]. The displacement fields are given by:

$$U(x,z) = u(x) - z\, w(x)_{,x} + \emptyset(z)\, u_1(x)$$
$$W(x,z) = w(x) \tag{1}$$

where: U, W are the displacement fields of the beam;
u, u_1, w are the displacement fields of the mid-plane;
$\emptyset(z)$ represents the shape function that varies depending on the beam theory chosen [7].

By the use of the stress-strain relations into force (N_x) and moment definitions (M_x, M_x^a), the constitutive equations are given as below:

$$\begin{bmatrix} N_x \\ M_x \\ M_x^a \end{bmatrix} = \begin{bmatrix} A_{11} & B_{11} & B_{111} \\ B_{11} & D_{11} & D_{111} \\ B_{111} & D_{111} & D_{1111} \end{bmatrix} \begin{bmatrix} u_{,x} \\ -w_{,xx} \\ u_{1,x} \end{bmatrix}, \tag{2}$$

$$Q_x^a = A_{55}\, u_1$$

where: A_{ij}, B_{ij} and D_{ij} denote the extensional, coupling and bending rigidities and Q_x^a denotes the shear force respectively.

The rigidities with more than two indices and the moment with superscript correspond to the shear deformation beam theory and subscript comma indicates the derivation with respect to the relevant axis. The governing equations of composite beam for static analysis under the effect of a uniform distributed load are given as follows:

$$N_{x,x} = 0$$
$$M_{x,xx} = q(x) \tag{3}$$
$$M_{x,x}^a - Q_x^a = 0$$

The boundary conditions are prescribed at the edges of the beam when $x = 0$ and L are given for simply supported, cantilever and free cases respectively as follows:

$$N_{x,x} = w = M_x = M_x^a = 0$$
$$u = w = w_{,x} = u_1 = 0 \tag{4}$$
$$N_x = M_{x,x} = M_x = M_x^a = 0$$

The use of constitutive equations into governing equations, three unknown displacement fields of the mid-plane with eight unknown integration coefficients (C_k) are obtained by integrating and solving the equations simultaneously as follows [8]:

$$u_1(x) = C_1 e^{-px} + C_2 e^{px} - (qx + C_3)\frac{D_{111}}{A_{55}D_{11}}$$

$$u(x) = -\frac{B_{111}}{A_{11}}\, u_1 + C_7 x + C_8$$

$$w(x) = \frac{D_{111}}{pD_{11}}\left[-C_1 e^{-px} + C_2 e^{px} - \frac{p}{A_{55}}\left(\frac{qx^2}{2} + C_3 x\right)\right] + \frac{1}{D_{11}}\left(\frac{qx^4}{24} + C_3\frac{x^3}{6}\right) + C_4\frac{x^2}{2} + C_5 x + C_6 \tag{5}$$

where:

$$p = \sqrt{\frac{-A_{55}A_{11}D_{11}}{D_{111}^2 A_{11} - D_{1111}A_{11}D_{11}}} \tag{6}$$

By use of a computational programme, the integration coefficients can easily be obtained by applying the boundary conditions at both edges of the beam.

OPTIMIZATION METHOD

Considering too many design parameters in an optimization problem will bring about too many solutions attendantly. Thus, instead of using all the possible solutions and obtain the optimum, using an evolutionary technique in the optimization process would be inevitable. Genetic algorithm is widely used in the design of laminated composite structures that have many variables and different fitness functions. Solving complex optimization problems by the use of conventional optimization techniques can be difficult in many fields. Genetic algorithm (GA) is an evolutionary optimization technique using Darwin's principal of survival of the fittest to improve a population of solutions [9]. The algorithm is generally based on natural selection, crossover and mutation which are also called genetic operators. Stacking sequences are the design variables that consist of fibre orientation angles ($\theta^{(k)}$) where k represents the number of layer. Therefore, the optimization problem may be written in a standard form such as:

$$\begin{aligned}&\text{Find: } [\theta^{(1)}/\theta^{(2)}/\theta^{(3)}/\theta^{(4)}]\\&\text{Minimize: } w\\&\text{Subject to: } 0° \le \theta \le 90°\end{aligned} \tag{7}$$

The problem considered in the study is to optimize the stacking sequences of laminated composite beam in order to obtain the minimum deflection parameter among all possible combinations of layer thicknesses for various boundary conditions. In Figure 2, the cross-section of composite beam with the same and various layer thicknesses satisfying the total thickness of are presented respectively.

The flowchart of genetic algorithm used in the study is illustrated in Figure 3. After an initial population of stacking sequences is generated, they are ranked from minimum to maximum with corresponding deflection parameters. New individuals representing the stacking sequences are generated by crossover so as to swap one or two fiber orientation angle of two consecutive stacking sequences and the genetic diversity from one generation to another is maintained by mutation,

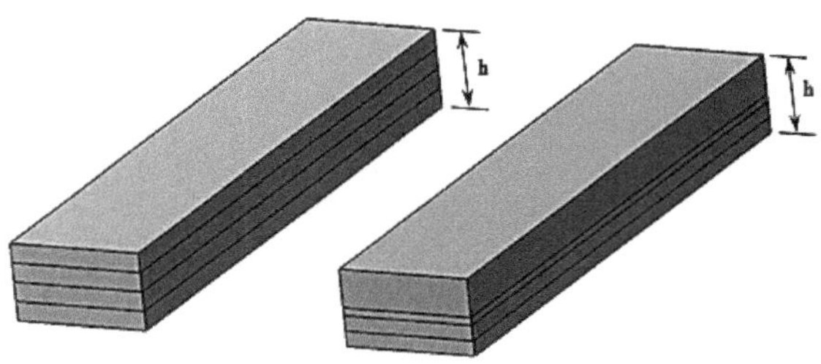

Fig. 2. Cross-section of the beam for the same and various layer thicknesses

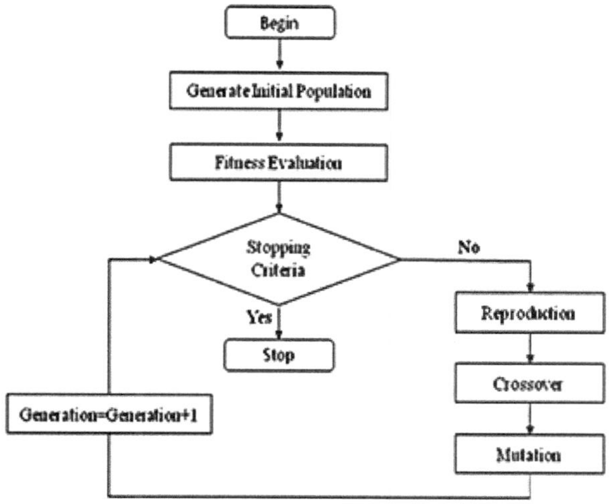

Fig. 3. The flowchart of genetic algorithm

altering one random fiber orientation angle of a random stacking sequence in each generation.

RESULTS AND DISCUSSION

The composite beam is assumed to be constructed of graphite/epoxy and under a uniform distributed load where $q(x) = 1000$ N/m, a length of $L = 1$ m and thickness of $h = 0{,}02$ m and the mechanical properties of graphite/epoxy are given as [10]:

$$\begin{gathered}E_{11} = 241.5 \text{ GPa}, E_{22} = E_{33} = 18.89 \text{ GPa}\\G_{23} = 3.45 \text{ GPa}, G_{12} = G_{13} = 5.18 \text{ GPa}\\\upsilon_{23} = 0.25, \upsilon_{12} = \upsilon_{13} = 0.24\end{gathered} \tag{8}$$

In Table 1, the minimum deflection parameters and corresponding stacking sequences are presented for various layer thicknesses and boundary conditions. The first term in layer thicknesses and stacking sequences indicates the first layer at the bottom of the beam where the last indicates the one at the top. There may be too many possible combinations of layer thicknesses for a composite beam depending on the variation and the number of layers. For instance, in a four layered beam, although there is only one combination for a variation of 0.25 h, there are 84 possible combinations for a variation of 0.1 h. Since there are more possible combinations of layer thicknesses with a variation of 0.1 h, the minimum deflection parameters are obtained for this case in all boundary conditions.

In Figure 4, Figure 5 and Figure 6, the variation of deflection parameters with respect to the number of generation for C-C, C-F and S-S boundary conditions are presented respectively. The algorithm is carried out for a population of 100 individuals and a number of generation of 50 in

Table 1. The minimum deflection parameters and corresponding stacking sequences for various layer thicknesses and boundary conditions

Boundary condition	Variation	Layer thicknesses (× *h*)	Deflection parameter [m]	Stacking sequence
C-C	0.25	0.25/0.25/0.25/0.25	$1.75226 \cdot 10^{-5}$	0°/0°/0°/0°
	0.2	0.2/0.2/0.2/0.4	$1.75226 \cdot 10^{-5}$	0°/0°/0°/0°
	0.1	0.1/0.1/0.2/0.6	$1.10796 \cdot 10^{-6}$	0°/0°/0°/40°
C-F	0.25	0.25/0.25/0.25/0.25	$3.87065 \cdot 10^{-4}$	0°/10°/50°/40°
	0.2	0.2/0.2/0.2/0.4	$3.19941 \cdot 10^{-4}$	20°/90°/70°/0°
	0.1	0.4/0.1/0.4/0.1	$2.89024 \cdot 10^{-4}$	0°/10°/80°/0°
S-S	0.25	0.25/0.25/0.25/0.25	$8.22771 \cdot 10^{-5}$	0°/0°/10°/0°
	0.2	0.2/0.2/0.2/0.4	$8.24556 \cdot 10^{-5}$	0°/0°/10°/0°
	0.1	0.2/0.2/0.1/0.5	$8.19918 \cdot 10^{-5}$	0°/0°/10°/0°

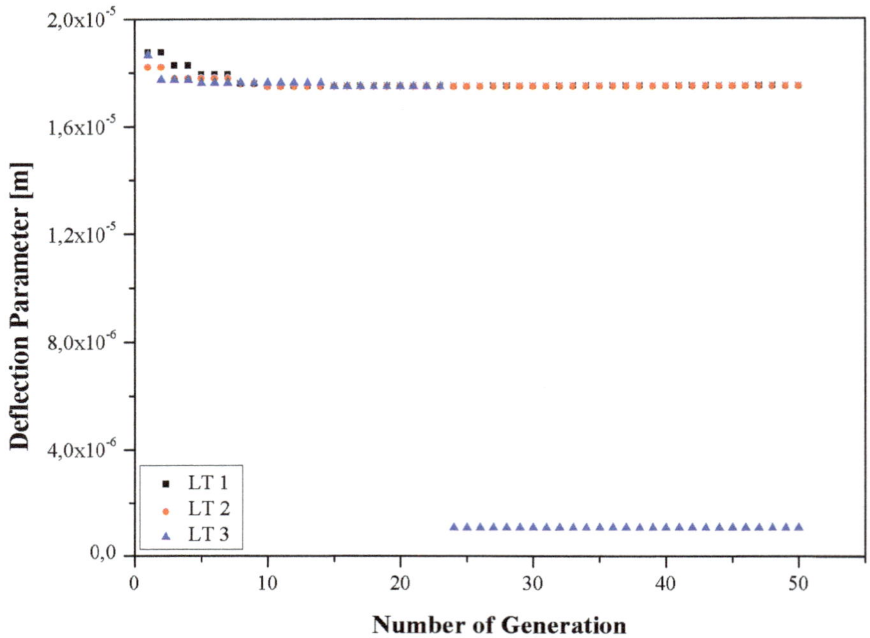

Fig. 4. The variation of deflection parameters with respect to generation for C-C

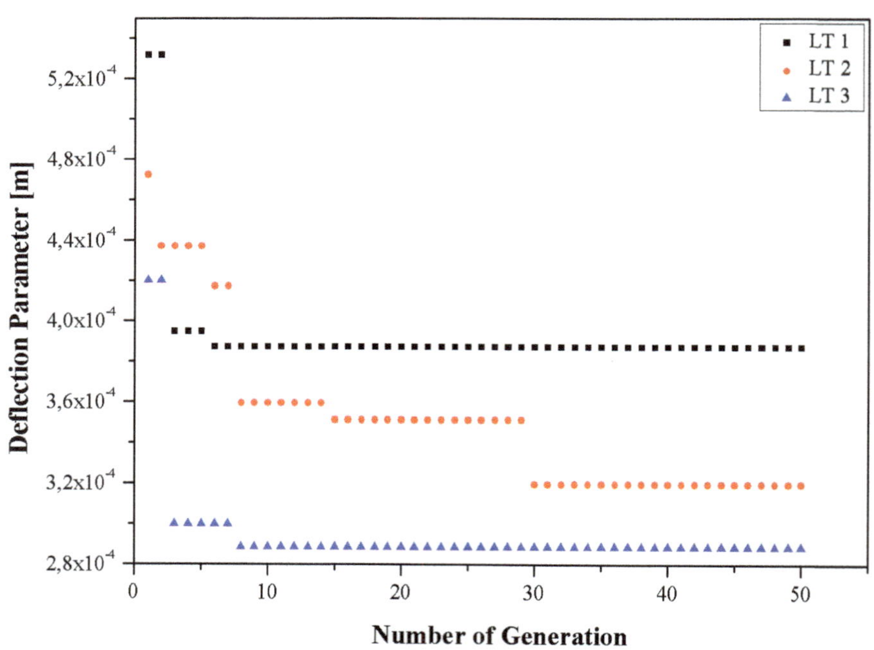

Fig. 5. The variation of deflection parameters with respect to generation for C-F

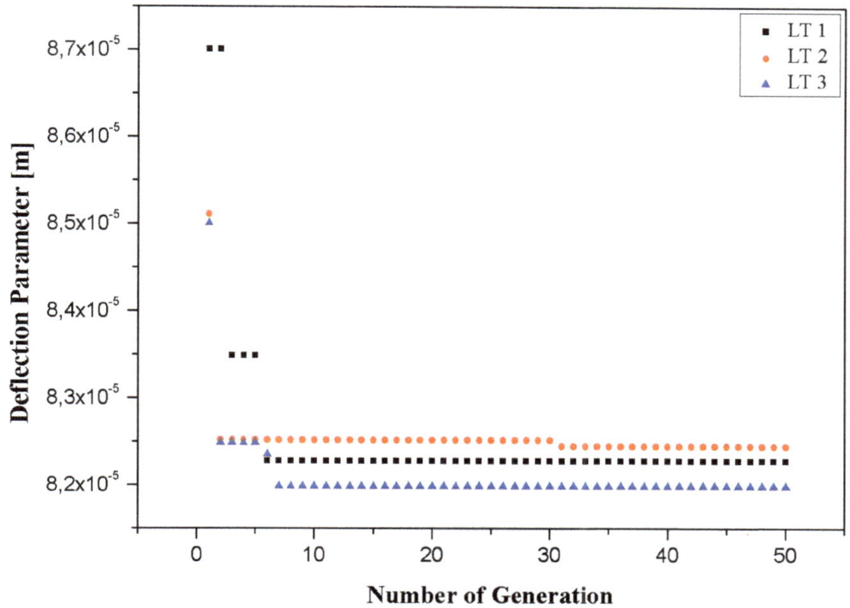

Fig. 6. The variation of deflection parameters with respect to generation for S-S

order not to stuck in a local minimum. The combination of layer thicknesses (*LT*), satisfying the total thickness of the beam together, are considered for three cases. In *LT*1, the layer thicknesses are obtained for a variation of 0.25 which correspond to four layers with a same thickness of 0.25 *h*. In *LT*2 and *LT*3 the variation of 0.2 and 0.1 are considered respectively. As it is obvious from the figures that the minimum values are obtained for a variation of 0.1 (*LT*3) in all cases. For C-C boundary condition, the deflection parameters are very close to each other but the minimum value is obtained in 24th generation. For C-F, the minimum value is obtained in 8th generation and it is obtained in 7th generation for S-S boundary condition.

CONCLUSIONS

In this study, the stacking sequence of a composite beam is optimized by the use of an evolutionary technique in order to obtain the minimum deflection parameters for various boundary conditions. Unlike the studies with the same layer thicknesses in the literature, the optimization process is carried out with different layer thicknesses. It is obvious from the results that, rather than the layers with the same thicknesses, the minimum values are obtained for different layer thicknesses due to the further number of possible combinations. In further works, the optimization problem can be expanded by taking new design parameters into account including strength, frequency and buckling.

REFERENCES

1. Walker M., Smith R.E.: A technique for the multiobjective optimisation of laminated composite structures using genetic algorithms and finite element analysis. Composite Structures, 62, 1, 2003, 123–128.
2. Conti P., Luparello S., Pasta A.: Layer thickness optimisation in a laminated composite. Composites Part B, 28B, 1997, 309–317.
3. Lee D.S., Morillo C., Bugeda G., Oller S., Onate E.: Multilayered composite structure design optimisation using distributed/parallel multi-objective evolutionary algorithms. Composite Structures, 94, 3, 2012, 1087–1096.
4. Brighenti R.: Fibre distribution optimisation in fibre-reinforced composites by a genetic algorithm. Composite Structures, 71, 2005, 1–15.
5. Khosravi P., Sedaghati R.: Design of laminated composite structures for optimum fiber direction and layer thickness, using optimality criteria. Structural and Multidisciplinary Optimization, 36, 2, 2008, 159–167.
6. Soldatos K.P., Tımarcı T.: A unified formulation of laminated composite, shear deformable five-degrees-of-freedom cylindrical shell theories. Composite Structures, 25, 1993, 165–171.
7. Timarci T., Soldatos K.P.: Comparative dynamic studies for symmetric cross-ply circular cylindrical shells on the basis of a unified shear deformable shell theory. Journal of Sound Vibration, 187, 4, 1995, 609–624.
8. Karaçam F.: MSc. Thesis (In Turkish), Trakya University, Institute of Science, Turkey, Edirne, 2005.
9. Goldberg D.E.: Genetic Algorithms in Search, Optimization and Machine Learning. Addison-Wesley Longman Publishing Co., Inc., USA, Boston, MA, 1989.
10. Karama K.S., Afaq S., Mistou S.: Mechanical behaviour of laminated composite beam by the new multi-layered laminated composite structures model with transverse shear stress continuity. International Journal of Solids and Structures, 40, 2003, 1525–1546.

15

THE CORROSION RESISTANCE OF THE BASE GEOPOLYMER FLY ASH

Paweł Mazur[1], Janusz Mikuła[1], Jerzy S. Kowalski[1]

[1] Institute of Materials Science and Engineering, Cracow University of Technology, Aleja Jana Pawła II 37, 31-155 Kraków, Poland, e-mail: pawelmazur89@interia.eu: jamikula@pk.edu.pl; jskowal@mech.pk.edu.pl

ABSTRACT
The article presents research on a new engineering material which becomes geopolymer. This is the material used to form a protective layer, to provide a high moisture absorption capacity, however, its composition is highly alkaline and, therefore, its absorption does not affect adversely the protected structures. The research included the effects of various corrosive environments for a change in weight of the test samples during the test, as well as the surface structure on a macroscopic scale.

Keywords: geopolymer, fly ash, corrosion resistance, new materials.

INTRODUCION

Geopolymer materials form a family of modern construction materials, exhibiting a number of properties that enable them to replace popular engineering plastics [2, 5]. Important researches have been conducted in both Poland and in the world to introduce geopolymer plastic for public use. The main proposed purpose is to replace them by not commonly used adhesives based on Portland cement-based materials. They look like the already mentioned Portland cement or natural stone [3]. Geopolymer mass can be used as a binder for various materials. In order to give them an aggregate-like structure commonly used cement concretes are used, however, the material using geopolymer has higher quality in every regard than popular concrete [2, 5].

The main advantages of plastic geopolymers is their high fire resistance, and superior mechanical strength than Portland cement. It is important to mention its strongly alkaline composition.

A basic geopolymer composition allows it to connect to construction steels widely used in the industry, while avoiding the formation of connections in the places of corrosion cells. Cracow University of Technology conducted extensive research on the properties of the material as well as on the development of new materials on the basis of and the search for new uses for them. One part of the study is devoted to geopolymer corrosion resistance to a number of chemical environments, as well as the possibility of combining the new plastic with steel surfaces to prevent their corrosion [1, 4, 6].

Geopolymer's high corrosion resistance to a number of highly corrosive environments present in the chemical industry as well as its resistance to abrasion and high temperatures allow new applications of the material. This material can be used in setting up protection for pipelines and industrial furnaces, as well as other large structures, not only industrial. Given the relatively low cost for construction of the geopolymer we can get effective protection against corrosion even at high temperatures without incurring high financial cost to provide thermal insulation.

Although most geopolymer plastics have a large absorption in environments containing electrolyte to the metal – geopolymer electrolyte can occur thanks to a strong geopolymer plastic composition of the basic charges at the interference between the surface of the cell is not corrosive. High moisture absorption can also be considered as an advantage, for example, for drying or room moisture control, where the walls are lined with it [1, 7, 8].

THE EXPERIMENT

The research material

Corrosion tests based on geopolymer ashes and coal-derived products were made (Table 1, Figure 1) to determine the resistance to various corrosive environments.

Table 1. Samples of the geopolymer

The composition of samples	Physical characteristics
• 100 g fly ash • 3 g NaOH • 25 ml of water glass • 10 ml H_2O • 2.5 ml plasticizer	• weight ≅15.2 g

Fig. 1. The example of the sample prior to testing

The study plan has taken into account the aspects of corrosion resistance in several environments of the geopolymer as well as the absorbency and concentration variations of NaOH as well as hydrochloric acid (Table 2).

Table 2. The solutions used in the study

Composition of solution	Concentration
H_2O (water from the water supply)	–
H_2O + NaCl	3%
H_2O + NaCl 75°C	3%
H_2O + NaOH	3%
	5%
	10%
Acetone	100%
H_2O + HCl	5%
	10%
	30%

The study was planned for a period of 49 days per test specimen in each of the solutions. During the course of the study sample weight was checked every two days between day 26 and day 49, after which the samples were removed from the solution and dried for two days in the open air.

Results and Discussion

To avoid errors in the conduct of research three control samples were placed in each solution to eliminate the possibility of accidental phenomena affecting the final results. When measuring the weight loss of all samples the same procedure was employed for the dried sample surface, with the exception of the samples placed in acetone. Material properties of geopolymer to donate stored moisture as well as highly volatile characteristics of acetone prevented repeatable mass measurement. Quick passing of the stored substance on the outer surface of the sample and highly volatile acetone properties enable rapid evaporation, which led to a rapid loss of weight and made it impossible to accurately measure the mass. Therefore, the standard procedure for drying the surface which evaporate after 40 minutes in the open air it is possible to measure the mass, but also it is not repeatable (Figure 2).

On the basis of the results of the samples placed in acetone, shows that geopolymer is resistant to the action of acetone

The solutions of NaOH for each concentration of the solution was observed to move from the solution into the material, which caused a significant increase in weight loss and surface hardness in the test samples (Figure 3).

After drying the samples to a piece of a 3% solution it retained their initial weight and structure. When the remaining the concentration of solution at 5% and 10% the saturation made geopolymer plastic outer surfaces increase the weight of the samples, leading to cracks in the material after its drying. The concentration of 5% resulted in a shallow surface cracks but at 10% breakage was so deep that the material was totally eliminated from use in this environment (Figure 4).

The concentration of the test samples with 10% NaOH solution at the area it was above 20% and decreased with distance from the outside area. Such a significant gain on the solution resulted in supersaturation which resulted in the destruction of the material (Figure 5, Table 3).

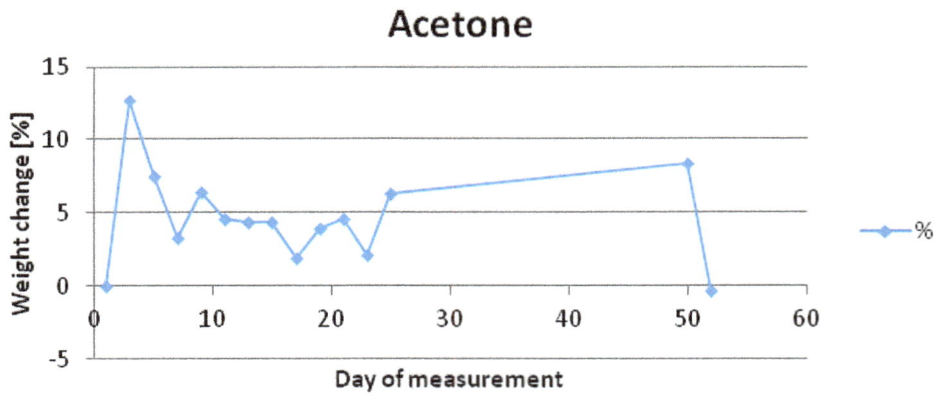

Fig. 2. The graph of the weight change of the measurement of samples immersed in acetone

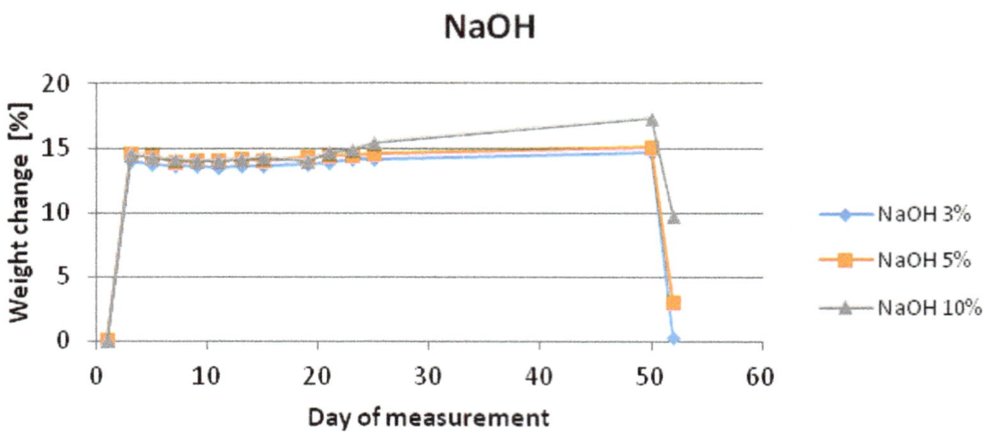

Fig. 3. The graph of the change in weight from the measurement for the sample immersed in NaOH

Fig. 4. Samples at 10% NaOH solution

As the only tested samples for which significant weight loss was recorded from HCl solutions. The increase in weight loss in the test samples was not associated with an increase in the concentration of hydrochloric acid (Figure 6).

The graphs present hydrochloric acid at a concentration of 5% and 30% had a similar loss in mass of about 2.5%, and the intermediate concentration of 10% hydrochloric acid weight loss is 8.2%.

Low weight loss of samples with 5% HCl solution is related to the geopolymer material resistance to hydrochloric acid in solutions containing low concentrations of the acid. In the case of samples in 30% HCl solution at the surface of the samples formed a tight adhering layer of solid well-inhibiting corrosion processes (Figure 7).

The solution on samples in 10% HCl also began creating a layer of pellet at a concentration of 30%, however, it does not have a tight layer of sludge on the surface of the sample. Under high concentrations of acid and the absence of a tight layer of the sediment the samples underwent rapid corrosion process. In other solutions, there was no case of corro-

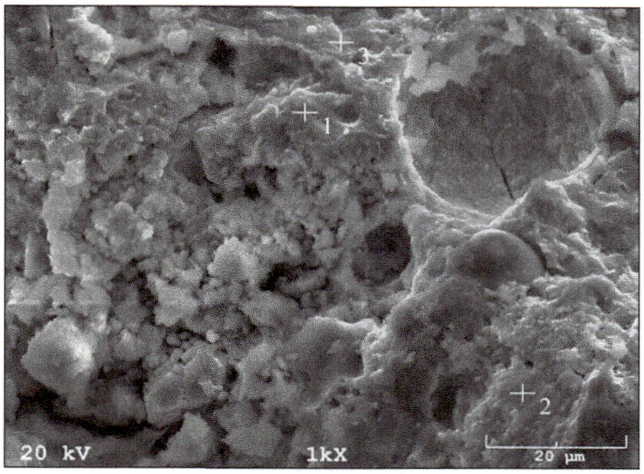

Fig. 5. Microstructure of the geopolymer in the 10%solution of NaOH

Table 3. The final composition of the material placed in 10% NaOH

Elt.	Line	Intensity (c/s)	Conc.	Units	Error 2-sig	MDL 3-sig
O	Ka	139.17	54.188	wt.%	1.743	0.728
Na	Ka	136.61	23.667	wt.%	0.777	0.364
Mg	Ka	2.33	0.401	wt.%	0.259	0.369
Al	Ka	43.80	5.835	wt.%	0.383	0.319
Si	Ka	116.70	13.568	wt.%	0.492	0.272
K	Ka	7.88	0.870	wt.%	0.166	0.185
Ca	Ka	9.58	1.097	wt.%	0.174	0.179
Fe	Ka	1.60	0.376	wt.%	0.224	0.299
Total			100.000	wt.%		

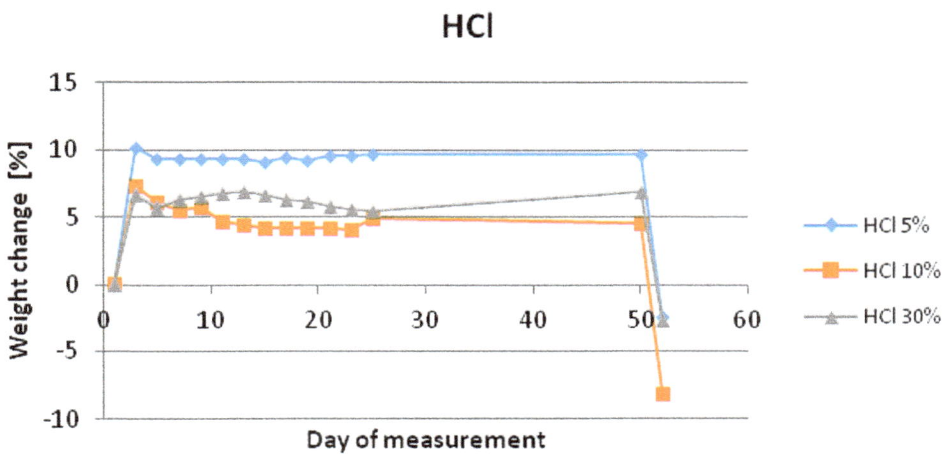

Fig. 6. The graph of weight change of the measuredsamples immersed in HCl

Fig. 7. Samples of the solution of HCl 30%

sion and only the formation of the precipitate formed from the mineral compounds contained in the solution (Figure 8).

From the results obtained for all the solutions it is showed that the material has excellent properties for enhancing the hygroscopic weight by up to 15% through the absorption of the solution. It should be noted, however, that depending on rates of solution absorbed differ by about 5%. The amount of solution absorbed determine the difference in mass between the first and second measurement of the mass of the test samples, in turn, the geopolymer material was able to absorb (moisture) to the ambient in less than 48 hours. The final weight is the weight of the initial sample plus a sample of the precipitate hiatus or mass acquired by the geopolymer ions from the solution.

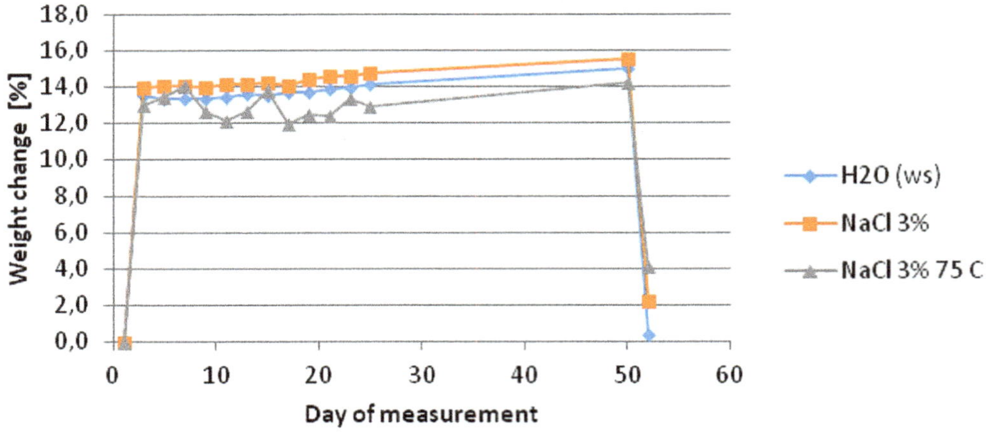

Fig. 8. The results of weight change

CONCLUSION

Research has shown a number of properties of geopolymer materials to enable its use as a material resistant to various aggressive environments. Thanks to its unique hygroscopic material that can be used to control humidity in chemical environments, by absorbing excess moisture from the products in chemical industry. Room or storage tanks whose walls were made of plastic allow geopolymer maintain the stored substances at constant level of humidity.

High resistance to the material at 30% hydrochloric acid. This points to the possibility to use such materials in industrial pipelines carrying concentrated acids, and its basic composition that would prevent the formation of corrosion on the steel walls, which would be covered with a geopolymer material. All the results are very promising, but require broader studies to confirm the correctness of the results.

REFERENCES

1. Bartoszewski M., Filipowicz S.: Wybrane zagadnienia ochrony katodowej rurociągów przed korozją. Prace Instytutu Elektrotechniki, z. 232, 2007.
2. Davidovits J.: Geopolymer – Chemistry & Applications. Institut Geopolymere, France 2008.
3. Davidovits J.: Geopolymeric Reactions in the Economic Future of Cements and Concretes. Geopolymer International Conference, France 1999.
4. Mikuła J.: Technika wytwarzania powłoki ochronnej. Wyd. Politechniki Rzeszowskiej, Rzeszów 1979.
5. Provis J., van Deventer J.: Geopolymers structure, processing, properties and industrial supplication. CRC Press.
6. Stansbury E.E., Buchanan R.A.: Fundamentals of Elektrochemical Corrosion. ASM International 2000.
7. Surowska B.: Wybrane zagadnienia z korozji i ochrony przed korozją. Wydawnictwo Politechniki Lubelskiej, Lublin 2002.
8. Young S., Feng D., Lukey G., van Deventer J.: Chemical characterisation of the steel-geopolymeric gel interface. Colloids and Surfaces A: Physicochem. Eng. Aspects, 302, 2007, 411-423.

ANALYSIS OF RUBBER FORMING PROCESS OF FIRE BARRIER FROM TITANIUM CP2 ALLOY FOR AW 139 HELICOPTER

Mariusz Krakowski[1, 2], **Jarosław Bartnicki**[1]

[1] Department of Computer Modelling and Metal Forming Technologies, Mechanical Engineering Faculty, Lublin University of Technology, 36 Nadbystrzycka Str., 20-618 Lublin, Poland, e-mail: j.bartnicki@pollub.pl

[2] The Communication Equipment Manufacturing „PZL-Świdnik" S.A., Lotników Polskich 1, 21-045 Świdnik, Poland, e-mail: mariusz.krakowski.gm@gmail.com

ABSTRACT

This paper presents conditions of forming products from titanium sheets by means of rubber stamping method. After introducing the issues connected with these elements manufacturing for aviation industry, considering standards and legal regulations, technological problems occurring during forming deep stiffening ribs. The results in the characteristic curvatures and distortion in the final products were generated in the process heterogeneous internal stress state. The Authors aimed at explaining and presenting of solutions limiting occurrence of chosen shape faults, which disqualify these products application in aviation industry. The proposed solutions significantly reduced the incidence of these unfavorable phenomena. The modified method of rubber stamping of firewall from titanium alloy CP-2 sheet was successfully implemented in manufacturing conditions of PZL Swidnik.

Keywords: titanium CP2, rubber forming.

INTRODUCTION

The forming process of sheet parts on hydraulic presses with rubber blanket has been applied in the industry for many years, being the most widely used in small series and unitary production such as aviation industry [1]. This is connected with relatively small costs of instrumentation. In classical approach to sheet parts stamping on hydraulic presses, production facilities use a device consisting of a punch and a die, which have to be ideally fitted and, because of that, ideally manufactured, hence, this instrumentation costs are very high and sometimes a small change of a part shape requires manufacturing of a completely new instrumentation. The stamping process on hydraulic presses with rubber blanket differs from classical stamping by means of a punch and die in that the role of the punch is played by a rubber blanket, which matches up to each die shape, hence, only a half of a classical instrument has to be made. Due to that fact this instrumentation costs are incomparably lower as the punch matches every shape [2, 3].

The development of plastics technology allowed for replacing used rubbers with polymer materials as well as classical rubber blanket with elastomeric bag, in which liquid is forced in under high pressure.

Presses of this type are widely used in aviation industry due to lower instrumentation costs and fast tool replacement in comparison with stamping by means of the punch and die. One of the companies in aviation industry applying such presses for sheet drawpieces production is PZL Świdnik from Augusta Westland group.

FIREWALL OF HELICOPTER AW139

For building engine decks firewalls were used in a form of flat surfaces with stiffening element guaranteeing stability and rigidity in a form of stiffening ribs. The shape of the considered stiffening was included in the factory standard obligatory for AW products, in which the stiffness depth was defined together with radii and permissible deviations from nominal values. The schema of the discussed firewall is shown in Figure 1.

The proposed part contains parts which are typical for this type of firewalls, that is on the plane stiffening elements were made, which have to guarantee rigidity and stability of the product without the necessity of connection with other stiffening elements.

ANALYSIS OF THE FORMING PROCESS BY MEANS OF A FLEXIBLE TOOL

Within the scope of forming by means of flexible tools in PZL Świdnik conditions, two hydraulic presses are presently used, one of them is equipped with a classical rubber blanket. A simplified schema of this forming method is shown in Figure 2, and a tunnel press for this operation is presented in Figure 3.

These machines are used for manufacturing of drawpieces, both from steel sheets and light al-

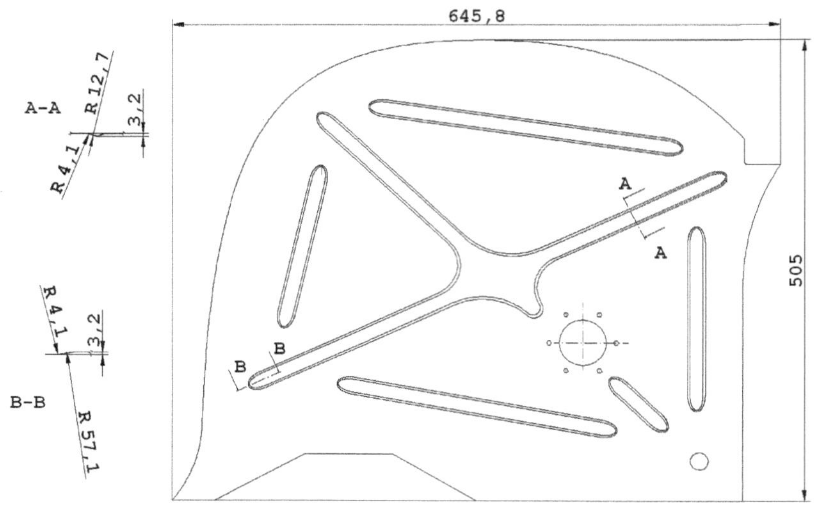

Fig. 1. Schema of firewall of helicopter AW139

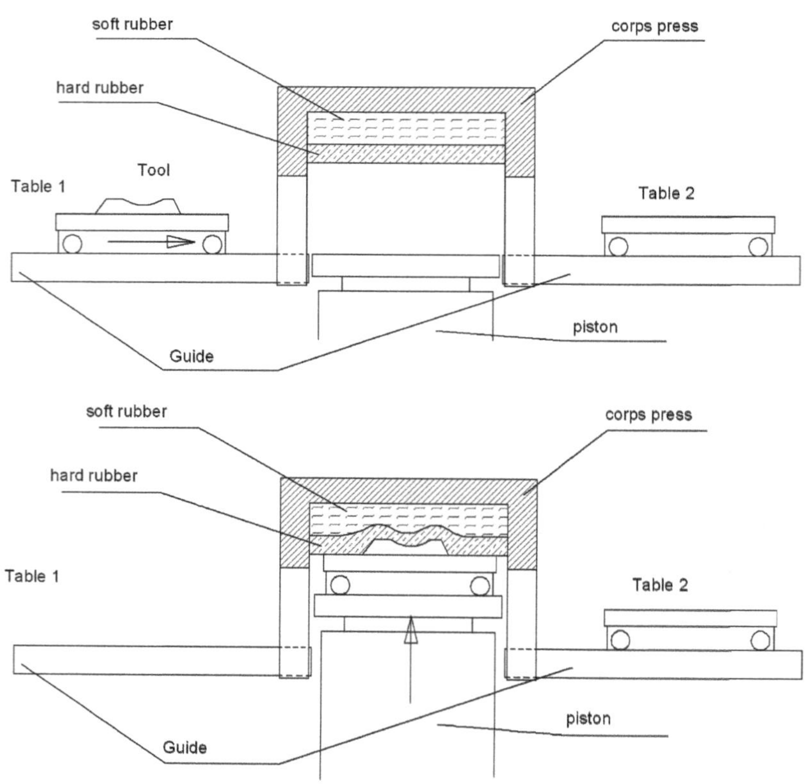

Fig. 2. Schema of rubber forming in the press with rubber blanket

Fig. 3. The photo of a tunnel press

loys, such as aluminum and titanium alloys. The scope of manufactured drawpieces is as wide as different is diversity of helicopter parts, starting from brackets and reinforcement by carrying girders and longerons to titanium firewalls of engine decks.

Increasing safety [4] and qualitative requirements resulted in abandoning the previously used alloy 3.7024.1 according to the standard LN9297 and its replacement by CP-2 material according to the standard MIL-T-9046, of higher mechanical properties and larger resistance to higher temperatures. The change of this material caused difficulties connected with metal forming of firewalls with stiffening ribs. Making stiffening ribs of the assumed shape in walls and tolerances led to the presence of internal stresses in the material, which resulted in curvatures forming and appearance of characteristic surface folding, named distortions (Fig. 4).

High qualitative requirements applied in Augusta Westland do not permit the part usage for further machining in such a state. This issue imposed the survey of the production process. The initial analysis led to conclusions that during stamping the formed material, flowing to stiffenings, caused stresses in the rest of the material, which cause drawpieces folding. One of the promising ideas is abandoning the previous approach, that is cutting off the part outline in development and deformation to the desired shape for stamping respectively larger material sheet on the device and, after this operation, outline cutting, which is presented in Figure 5.

Fig. 4. Firewall from titanium alloy CP-2 with characteristic folding of the surface

In order to balance the force causing material flow, additional element, making balancing stresses, was introduced in the stiffening. For that reason, a device from machine lingoform, modified by addition of forming threshold retreated from the part outline and enlarged by building over to rectangular shape was used. For designing of the threshold shape around the outline, a simplified parts shape (cuttings were omitted and radii were enlarged) was assumed, and next, trimming line was retreated of constant value from the cutting simplified edge, which is visible in Figure 6.

After initial cutting of the material excess together with stamped stiffening element the drawpiece fulfilled qualitative requirements concerning stiffening ribs depth, and a considerable improvement of the quality in the plane area took place, unfortunately, not on the whole surface. The first tests brought a certain improvement of the formed element geometry. The results of these works are presented in Figure 7, where the product stamped in a sheet is shown.

A considerable improvement of quality of flat surfaces was observed in the influence areas of forming threshold running on curves. Unfortunately, in the areas in which forming threshold ran in straight lines, constant compression of stresses causing material folding took place. The observed phenomenon suggested the idea of constructing an instrument into forming threshold, running on curves and without straight lines.

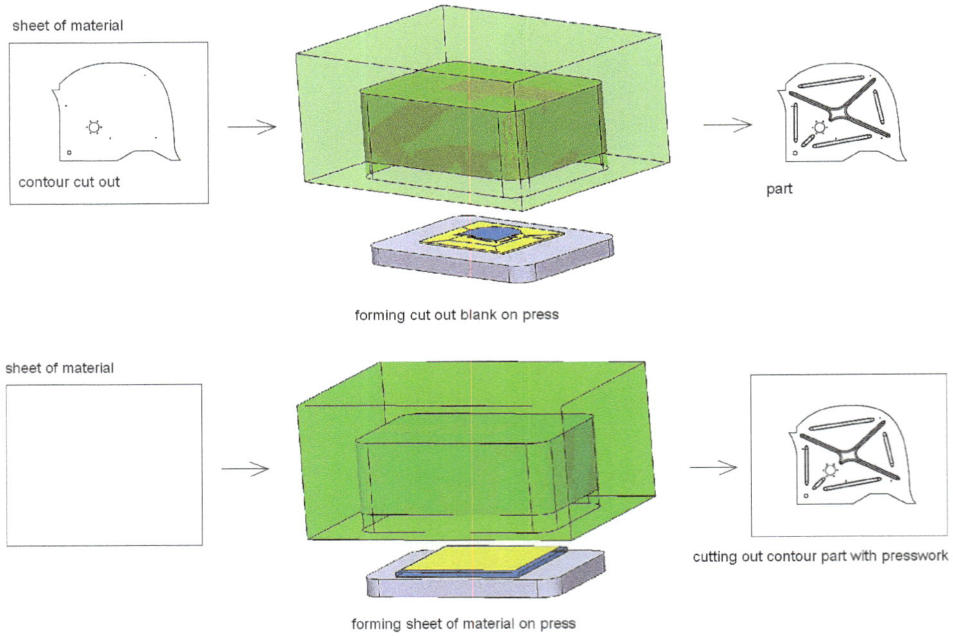

Fig. 5. Change of manufacturing schema of firewall drawpiece

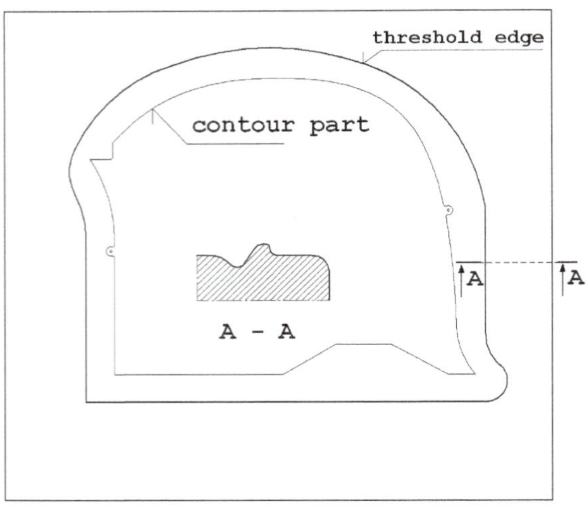

Fig. 6. Tool schema (on the left) and photo of the finished instrument (on the right)

Fig. 7. Results of stamping tests from sheet

This led to designing another tool modification, in which straight lines in the forming threshold geometry was replaced by curves. This solution is given in Figure 8, where the schema and photo of the finished tool are presented.

After stamping tests of the prepared material on the new instrument, a considerable improvement of surface quality was observed at obtaining required stiffening depth. In accordance with drawing requirements the depth was defined at 3.2 mm together with material depth equal 0.4 mm as it is presented in Figure 1 in sections A-A and B-B.

Considering requirements of the factory standard [5], the stamping depth should be between 2.55 mm to 3.05 mm, and considering the material depth, it should be from 2.95 to 3.45 mm. The manufactured drawpiece presented in Figure 9 shows a noticeable surface quality improvement in respect of deformations appearance in the plane area and elimination of surface folding phenomenon.

Application of simple short-time locksmith modifications allowed for obtaining acceptable adhesion of the part to the instrument shown in Figure 10.

The made drawpieces underwent measurement concerning stiffening depth. Measurements were made on 6 parts after 2 parts from following series. The areas of measurement are marked in Figure 11, yet their results are compared in the Table 1.

All obtained measurement results, for subsequent three series of finished products, remain within tolerance areas assumed in the factory standard [5]. This points to the rightness of the assumed solution of the technological problem and determine the further direction of works on sheet products with ribs forming.

CONCLUSION

Solution presented in the paper was successfully implemented in factory conditions of Augusta Westland in Świdnik. At present, the proposed method is used for manufacturing of the elements of firewalls type for helicopters

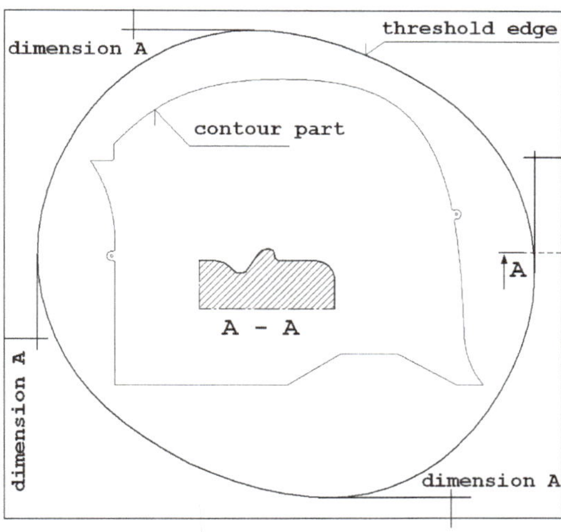

Fig. 8. Schema of the modified tool (on the left) and photo of the finished instrument (on the right)

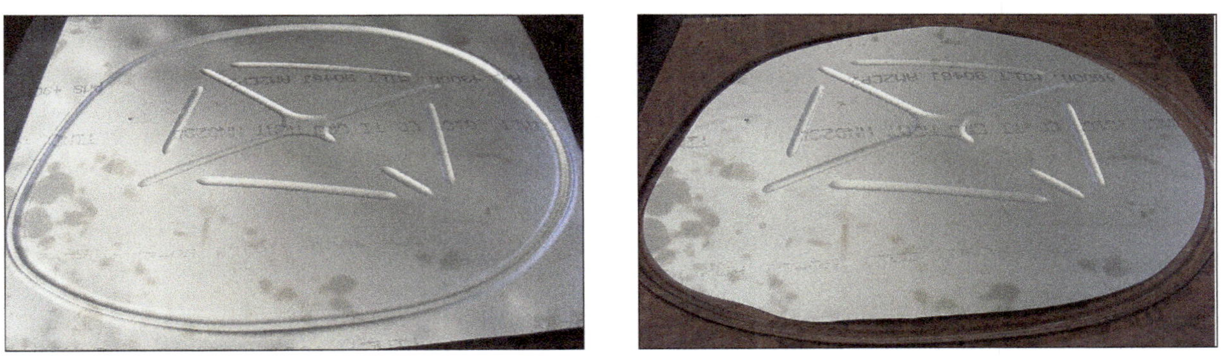

Fig. 9. Results of stamping tests from sheet in the modified tool

Fig. 10. Finished product in the form for geometry checking

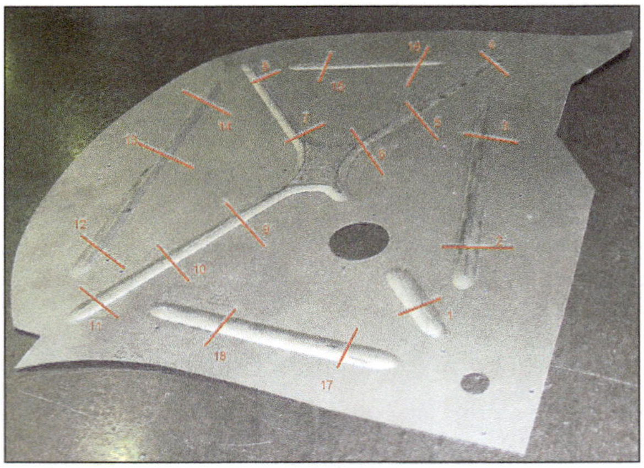

Fig. 11. Areas of measurement in the finished product

Table 1. Comparison of measurement of ribs depth according to Figure 11

The location of measurement	Part 1		Part 2		Part 3	
	piece no. 1	piece no. 2	piece no. 3	piece no. 4	piece no. 5	piece no. 6
1	2.57	2.58	2.58	2.56	2.56	2.57
2	2.56	2.60	2.59	2.58	2.59	2.60
3	2.57	2.59	2.58	2.59	2.59	2.59
4	2.57	2.57	2.58	2.58	2.57	2.57
5	2.58	2.60	2.59	2.61	2.58	2.59
6	2.60	2.58	2.59	2.58	2.57	2.58
7	2.59	2.58	2.57	2.59	2.58	2.57
8	2.57	2.57	2.58	2.58	2.57	2.58
9	2.60	2.61	2.60	2.61	2.58	2.59
10	2.60	2.58	2.59	2.58	2.57	2.58
11	2.57	2.58	2.58	2.56	2.56	2.57
12	2.56	2.58	2.59	2.58	2.59	2.58
13	2.59	2.60	2.59	2.61	2.58	2.59
14	2.58	2.57	2.58	2.58	2.57	2.58
15	2.57	2.60	2.58	2.59	2.58	2.60
16	2.56	2.59	2.59	2.59	2.59	2.59
17	2.56	2.58	2.59	2.58	2.59	2.58
18	2.57	2.59	2.58	2.59	2.58	2.69

AW139. Due to similar construction of these elements in all helicopters manufactured by this consortium (certificate requirements), it is planned to widen scope of application of the presented solution to further elements stamped from titanium and aluminum alloys. Obtaining the product in accordance with requirements do not eliminate the problem of undesired deformations of drawpiece. In order to improve the products quality and eliminate the necessity of hand-made straightening, the further tests and research will be carried out aiming at developing stamping technology of sheet products with ribs by means of flexible tools.

REFERENCES

1. Tekkaya A.E., Taylan A. Sheet metal forming. Processes and applications. Materials Park, ASM International, Ohio 2012.
2. Minghua Z., Qiang X.Y., Yu L., Wenliang C. Prediction algorithm of blank shape for rubber forming of aircraft sheet metal part. AIP Conference Proc. 22.8.2011, Vol. 1383, Iss. 1, 2011, 943–947.
3. Vilcans J., Torims T. Thin sheet metal stamping with elastic media. Annals of DAAAM & Proc. of the 21st DAAAM Symposium, Croatia, Zadar, (20-23.10.2010), 2010, 849–850.
4. Standards JAR/FAR 29.
5. Company standard Agusta Westland No 999-0010-08.

ASSESSMENT OF THE PRODUCTIBILITY OF HYBRID NODES USING THE MULTI-CRITERIA METHOD

Tomasz Urbański[1]

[1] Department of Ship Structure, Mechanics and Fabrication, Faculty of Maritime Engineering and Transport, West Pomeranian University of Technology in Szczecin, Poland, e-mail: tomasz.urbanski@zut.edu.pl

ABSTRACT

The article presents an assessment of the productibility of hybrid nodes. The hybrid node is a new structural element the implementation of which causes numerous (especially technological) problems that need to be solved. The most important element of the hybrid node is the so-called connector, the choice of which is a complex and difficult problem. It involves taking into account many aspects (constructional, strength-related, technological, economic) in order to make sure that the choice is as objective as possible. Therefore, an attempt to acquire a comprehensive view at the problem requires that a set of accurate criteria be used for assessment. In this paper, the author undertakes such an attempt. The expert method presented herein allows for choosing the right connector.

Keywords: innovative structural element, sandwich panel, hybrid node, expert method.

INTRODUCTION

In order to ensure unambiguous interpretation of the problems discussed in this paper, the following terms were introduced:

- hybrid node – a special fragment of a large-size steel structure within which two parts (distinguished from each other in structural and technological terms) of that structure are joined (in the case analyzed: an innovative structural element – a sandwich panel, and a conventional structural element – a stiffened plate). The fragments of the structure are joined with each other using an intermediate element – a connector [5],
- assembly suitability – the structure's ability to be joined with another structure, or a fragment thereof, preferably without the need for any further corrective procedures [5, 6].

The hybrid node constitutes an integral part of hybrid structures (Figure 1) and can be applied wherever sandwich panels are used [3, 4, 5].

Performing a hybrid node entails the need to:

- choose the shape of the intermediate joining element,
- choose the method for joining particular elements of the node,
- develop the technology for assembling the hybrid node at an acceptable quality level, i.e. its assembly suitability.

All of the aforementioned aspects are closely interrelated and create the broadly-defined area of the hybrid node's productibility. Since each of these aspects is so vast that it could be the topic of at least one separate article, the decision was made to narrow the scope of interest of this paper to the first of them, only.

The connector's shape has a substantial impact on the number and quality of the hybrid node's welding deformations (the effect that the welding deformation forms identified in the node have on its assembly suitability is shown in [7]).

HYBRID NODES ASSESSED

The objects (hybrid nodes) subjected to a multi-criteria assessment were developed on the basis of the author's knowledge of the phenomenon studied and his own material gathered, and

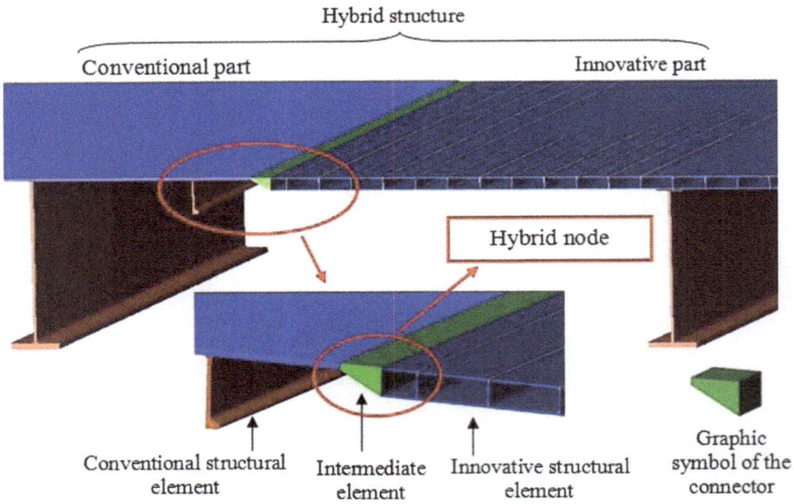

Fig. 1. Hybrid node – components [5, 6]

Table 1. Types of hybrid nodes under a multi-criteria assessment (on the basis of [5])

HN no.	Drawing of the hybrid node	Connector shape
1	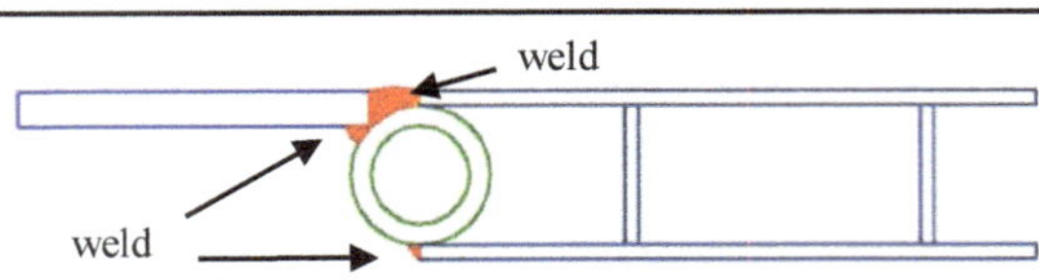	 simple*
2	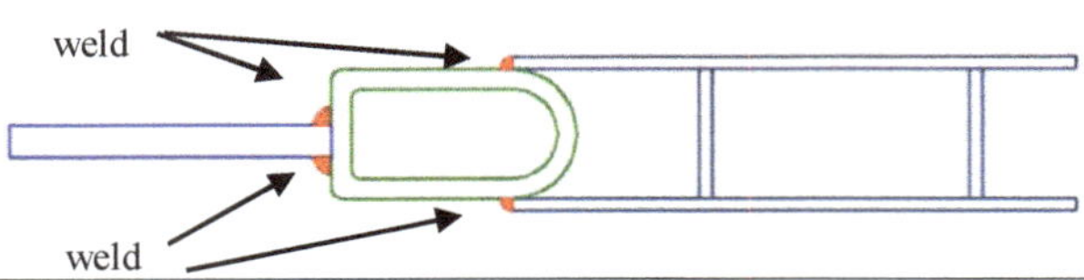	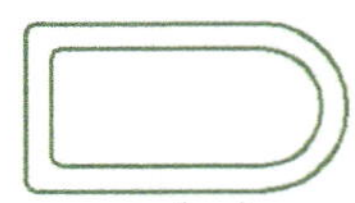 complex*
3	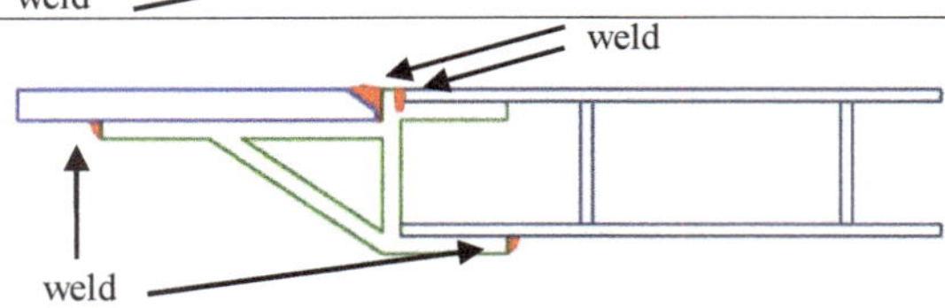	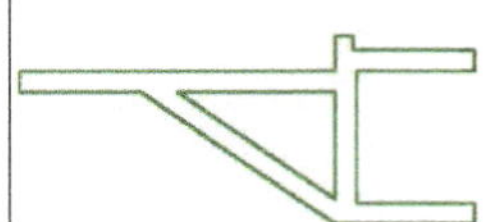 complex
4	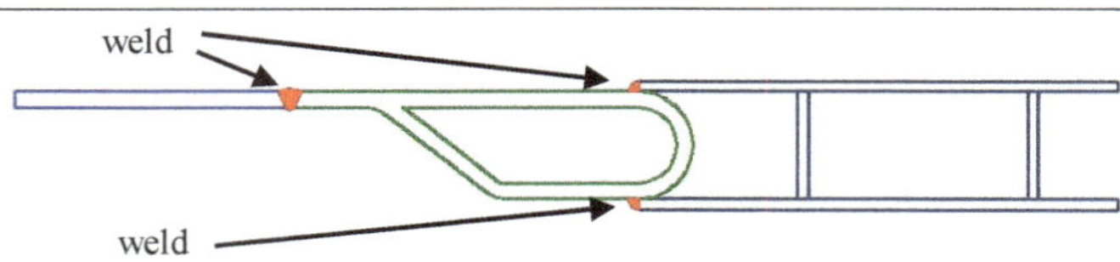	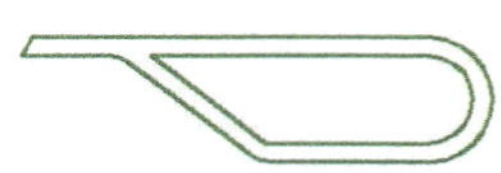 complex
5	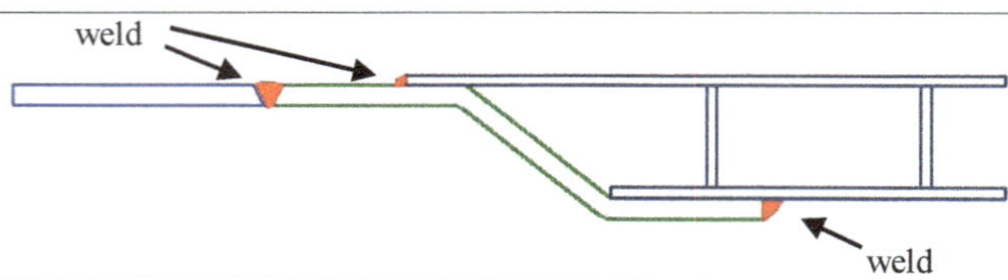	 simple
6	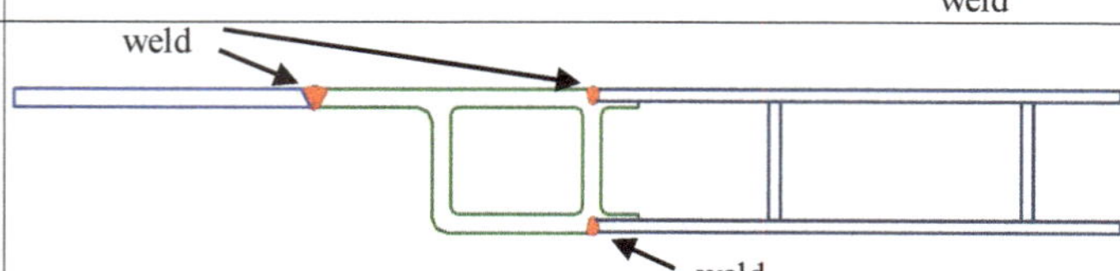	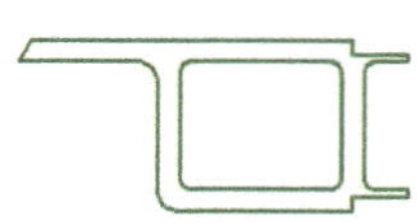 complex
7	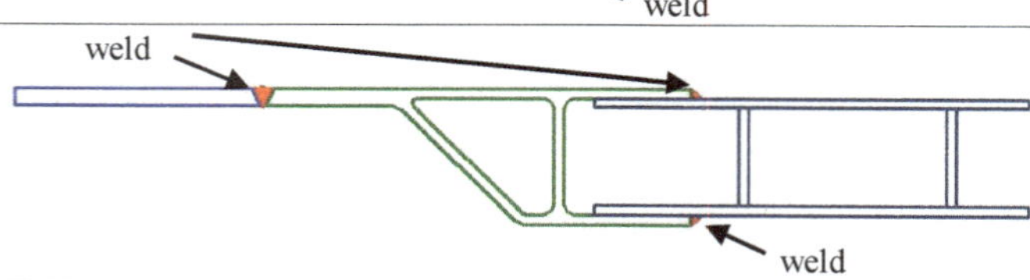	 complex

Denotations and notes: HN – hybrid node.
* – shape available in metallurgic products catalogues, e.g. [10].

illustrated in Table 1 (Table 1 shows the most representative node shapes, only).

PRODUCTIBILITY ASSESSMENT USING THE MULTI-CRITERIA METHOD

The purpose of the hybrid node productibility assessment is to show the direct impact that the node's geometry has on its assembly suitability, and thus to choose the connection variant that is most suitable in technological terms. The assessment was carried out using the so-called multi-criteria expert method. It follows a few basic steps:

- determination of the purpose of the analysis, and selection of the objects to be assessed.
- formulation of a set of criteria defining the group of features sought that describe the objects chosen.
- determination of assessment rules based on the criteria formulated.
- performance of the assessment for each of the objects analyzed, and selection of the best one/ ones of them.

The hybrid nodes shown in Table 1 were used as the objects of the multi-criteria assessment. Eleven criteria, shown in Table 2, were used for assessing particular joint variants (node geometries). In the author's opinion, these criteria make it possible to achieve the assessment's objective in a reliable way.

The hybrid nodes selected were assessed separately according to the criteria (Table 2) on a scale of 0 to 5. The higher the assessment value the better their properties, or their assembly suitability. All criteria were brought to a dimensionless form through dividing them by the maximum number of points attributable to the given criterion, i.e. by 5.

Table 2. Criteria for assessing hybrid node joints [5]

No.	Name	Description
1	(connector) Feasibility criterion	Refers to the degree of difficulty in manufacturing the connector by the manufacturing plant.
2	Adaptability (adjustment) criterion	Refers to the degree of assembly difficulty, i.e. of a possible need to apply force in order to adjust the connector to the panel as a result of, e.g., errors in connector manufacture.
3	(connector) Accessibility criterion	Refers to the necessity of performing additional procedures preparing the element to welding, e.g. scarfing * (of one edge or more), or thickness reduction.
4	(technological) Universality criterion	Refers to the possibility of using the connector at different stages of constructing a large-size steel structure, e.g. prefabrication and/or assembly.
5	(constructional) Universality criterion	Refers to the possibility of performing the hybrid node in various joint variants, i.e. according to the location of the stiffeners of the panel's core in relation to the main structural link system, without disturbing the inner structure of the core of the panel.
6	Instrumentality criterion (concerning tools)	Refers to the necessity of developing special tools useful for assembling the given node, i.e. the number of such tools and the degree of their technological advancement.
7	Size criterion	Covers the number of welds (thus the level of consumption of welding consumables, and labor intensity required the performance of the joint), the type of welds and their concentration.
8	(joint) Feasibility criterion	Refers to the degree of difficulty of performing the joint, i.e. mainly: access to the joint, welding position, etc., and thus the welder's qualifications required.
9	Deformability criterion	Refers to the expected (quantitative and qualitative) level of welding deformations formed within the node.
10	Correctibility criterion	Refers to the possibility (and degree of difficulty) of performing repairs, mainly thermal straightening.
11	Esthetics criterion	Refers to the general visual impression with regard to the entirety of the joint, from the point of view of the engineer.

Notes:
* – it should be remembered that if it is necessary to carry out the scarfing of sheet metal, the manner of preparing the edges will depend on the element's thickness and the welding method [1, 8, 9].

Table 3. Assessment of hybrid node joints depending on the radar chart area [5]

No.	Hybrid node productibility state	Value of the radar chart area [-]
1	Very good	$2.229 < a$
2	Good	$1.486 < a \leq 2.229$
3	Satisfactory	$0.743 < a \leq 1.486$
4	Poor	$a < 0.743$

For each of the nodes analyzed, an 11-parameter radar chart was drawn (Figure 2). The chart area is a criterion generalized to the assessment of the hybrid node's productibility state, i.e. its impact on assembly suitability. In the ideal case, each of the 11 criteria has a dimensionless value equal to 1. The radar chart area for an ideal object is 2.972. The assessment of the hybrid node's pro-

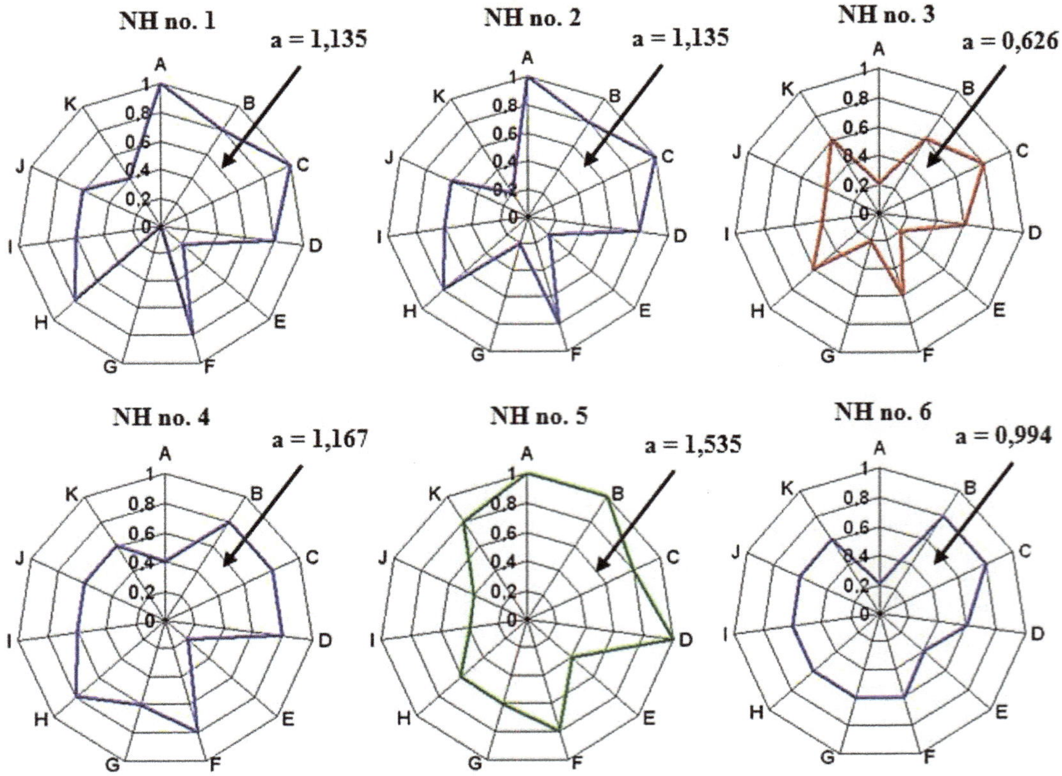

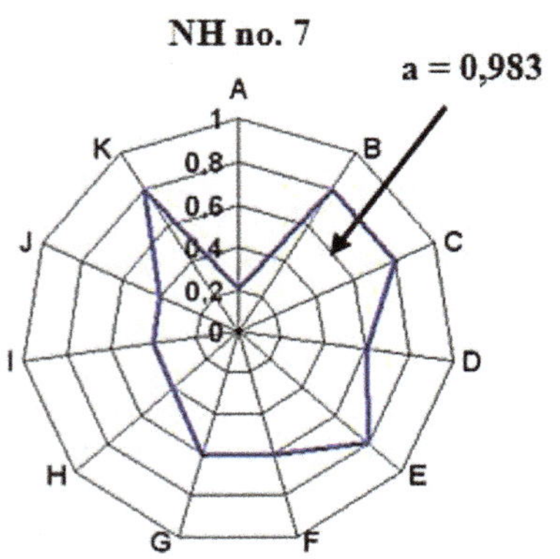

Letters from A to K in the charts denote the assessment criteria, i.e.:
A – (connector) feasibility,
B – adaptability,
C – accessibility,
D – (technological) universality,
E – (constructional) universality,
F – instrumentality,
G – size,
H – (node) feasibility,
I – deformability,
J – correctibility,
K - esthetics

a – radar chart area denotes the joint's productibility state

Fig. 2. A set of radar charts for the hybrid nodes assessed [5]

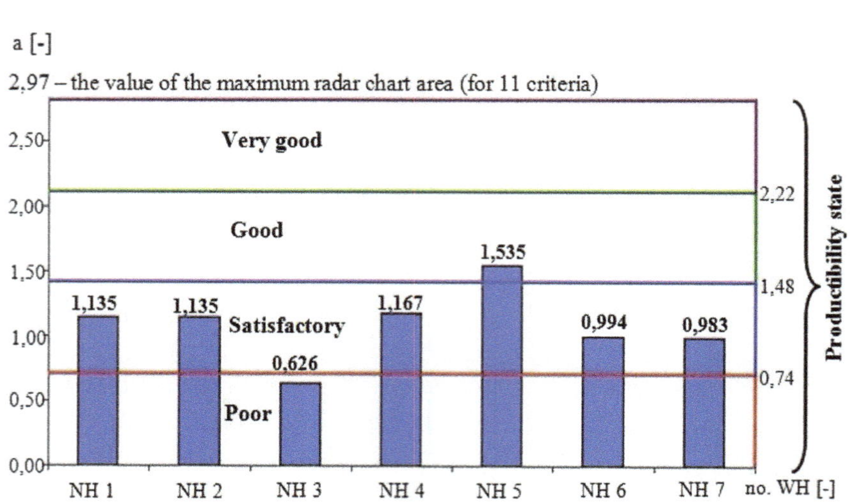

Fig. 3. A comparison of the radar chart area values describing the productibility of the hybrid nodes assessed (on the basis of [5])

ductibility depending on the radar chart area ("a") is shown in Table 3.

The resulting values of the multi-criteria assessment for this analysis are shown in Tab. 4 and illustrated in Figure 2 and Figure 3.

Fiugre 2 shows a set of radar charts for all of the objects assessed containing all the assessment states obtained (according to Table 4).

Figure 3 shows a set of radar chart areas describing the productibility state of the hybrid nodes assessed.

On the basis of the results obtained, it was found that:

- most of the nodes selected for the analysis (five out of the seven) were characterized by satisfactory productibility at comparable levels (the range of difference between the values in this area was 0.184),
- one of the nodes demonstrated poor productibility and one good productibility,
- none of the nodes reached the best, i.e. very good, productibility.

Table 4. Values of the multi-criteria assessment of the hybrid nodes analyzed (on the basis of [5])

No.	Assessment criterion	Assessment on a score scale (of 0 to 5) for particular nodes						
		HN1	HN2	HN3	HN4	HN5	HN6	HN7
1	(connector) Feasibility	5	5	1	2	5	1	1
2	Adaptability	4	4	3	4	5	4	4
3	Accessibility	5	5	4	4	4	4	4
4	(technological) Universality	4	4	3	4	5	3	3
5	(constructional) Universality	1	1	1	1	2	2	4
6	Instrumentality	4	4	3	4	4	3	3
7	Size	0	1	1	3	3	3	3
8	(node) Feasibility	4	4	3	4	3	3	2
9	Deformability	3	3	2	3	2	3	2
10	Correctibility	3	3	2	3	2	3	2
11	Esthetics	2	1	3	3	4	3	4
12	HN chart area [-]	1.135	1.135	0.626	1.167	1.535	0.994	0.983
13	Share of the HN chart area in the ideal chart area [%]	38	38	21	39	52	33	33

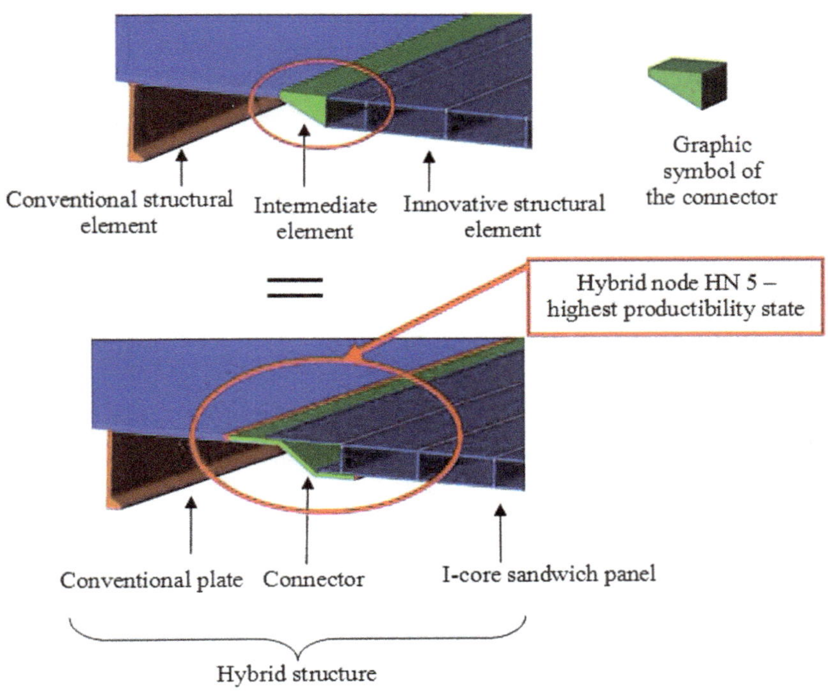

Fig. 4. The hybrid node characterized by the highest state of productibility (on the basis of [5])

The analysis performed in this paper allows for choosing the hybrid node and, at the same time, the connector shape that is characterized by the best (i.e. good) productibility. It is the node number 5 (Table 1). Therefore, the geometry of HN 5 was accepted as the most assembly suitable (Figure 4).

CONCLUSIONS

The multi-criteria expert method presented in this paper allows for determining the productibility of hybrid nodes, and thus for raising the objectivity of the choice. On the basis of this method, a hierarchy of the nodes' productibility states can be created.

Large-size steel structures are characterized by substantial structural and technological repeatability. Therefore, selecting the element that demonstrates potentially the highest assembly suitability can bring notable benefits at the stage of manufacturing the hybrid structure – namely smaller welding deformations, which in turn will help minimize the number of corrective procedures that entail considerable labor intensity and generate costs. It is estimated that the labor intensity involved in straightening large-size steel structures accounts for up to 30% of the total labor intensity required in order to perform the structure [1, 2].

That is why it is worth making the right choices at the earliest (conceptual) stages of developing the structure

REFERENCES

1. Kubera S.: Odkształcenia spawalnicze i zasady ich usuwania w okrętowych konstrukcjach cienkościennych, Doctoral Dissertation, Gdańsk University of Technology, Gdańsk 1964.
2. Metschkow B.: Współczesna technika utrzymania wysokiej jakości konstrukcji okrętowych, X Konferencja: Laboratoria technologiczne – aspekty utrzymania wysokiej jakości wyrobu, Wydawnictwo uczelniane Politechniki Szczecińskiej, Szczecin 1999.
3. Pyszko R.: Zastosowanie stalowych paneli typu sandwich w konstrukcjach okrętowych i oceanotechnicznych. Doctoral Dissertation, Department of Ship Technologies, Quality Systems and Materials Science, Faculty of Ocean Engineering and Shipbuilding of Gdańsk University of Technology, Gdańsk 2006.
4. Urbański T., Graczyk T.: Application of innovative materials in waterborne transport means - identification of technological problems, International Conference on Innovative Materials and Technologies for Surface Transport (INMAT 2005), Gdańsk, Poland, 7-8 November, 2005.
5. Urbański T.: Metoda prognozowania odkształceń spawalniczych węzła hybrydowego na podstawie badań eksperymentalnych, Doctoral Dissertation, Chair of Ship Structure, Mechanics and Fabrication, Faculty of Maritime Technology, West Pomeranian University of Technology in Szczecin, Szczecin 2009.
6. Urbański T.: Węzeł hybrydowy – technologiczność wielkogabarytowych konstrukcji spawanych – wprowadzenie, Przegląd Spawalnictwa, 11, 2010, 21-25.
7. Urbański T.: The influence of welding deformation types of hybrid node on assembly suitability, Advances in Science and Technology Research Journal 6(16), 2012, 42-47.
8. Company standard: T081-02, Spawanie łukowe w osłonie gazów ochronnych, część II – Instrukcje technologiczne spawania WPS, Stocznia Szczecińska S.A. 2001.
9. Company standard: T081-03, Spawanie automatyczne łukiem krytym, część II – Instrukcje technologiczne spawania WPS, Stocznia Szczecińska S.A. 2001.
10. The catalogue of Stalprodukt S.A. Kształtowniki stalowe gięte na zimno. (http://www.stalprodukt.com.pl).

18

CHARACTERISTICS OF INTERACTIONS BETWEEN SOME TEXTURE PROPERTIES AND COMPOSITION OF CARRAGEENAN GELS AS A RESULT OF ITS DEFINED DIVERSIFIED FREEZING AND THAWING TREATMENT

Katarzyna Kozłowicz[1]

[1] Department of Refrigeration and Food Industry Energetics, University of Life Sciences in Lublin, Doświadczalna 44, 20-280 Lublin, Poland, e-mail: katarzyna.kozlowicz@up.lublin.pl

ABSTRACT

Model samples of carrageenan gels based on water, milk and juice were air-blast frozen and frozen by immersion in glycol and in liquid nitrogen. The gel freezing rate was determined on the basis of the kinetics of freezing. Carrageenan gel samples were characterized by evaluation of its thawing drip loss and hardness determined with compression and penetration tests. Freezing in liquid nitrogen ensured the highest freezing rates. Thawing drip loss of gels significantly depended on the carrageenan content, pH of the solution, freezing method and freezing rate. The resulting relationships are linear functions with high determination coefficients. The results of compression and penetration tests prove the significant effect of the carrageenan content and pH on gel hardness. The higher carrageenan content in a sample, the higher compression force and penetration of the gel. Gel freezing resulted in lower hardness. Freezing conditions had a significant effect on the properties tested. The correlation between compression forces and penetration depending on the carrageenan content and the freezing method was described using regression equations with high determination coefficients. Gels based on milk and juice with 2.2% carrageenan content are recommended for immersion freezing at rates above 5.0 $cm \cdot h^{-1}$.

Keywords: freezing, carrageenan, gels, drip loss, hardness.

INTRODUCTION

Carrageenan is a sulfated polysaccharide extracted from red seaweed (*Rhodophyceae*), which is applied widely as a food additive. Carrageenan is classified into tree types as kappa (κ), iota (ι) – gel forming or lambda (λ) – which does not form a gel [25, 29, 30]. κ-Carrageenan consists of a repeating unit composed of the disaccharide, β-(1-3)-D-galactose-4-sulfate and α-(1-4)-3.6-anhydro-D-galactose. ι-Carrageenan possesses two sulfate groups in a disaccharide repeat unit; β-(1-3)-D-galactose-4-sulfate and α-(1-4)-3.6-anhydro-D-galactose-2-sulfate. λ-Carrageenan consists of β-(1-3)-D-galactose-2-sulfate and α-(1-4)-D-galactose-2.6-disulfate including three sulfate groups. All carrageenan types dissolve in hot water (above 70 °C). Carrageenans dissolve in hot milk yielding gels upon cooling whose strength and consistency depend on concentration and affinity to calcium ions [14, 30]. κ-Carrageenans are readily soluble in aqueous sucrose solutions, while ι-carrageenan is scarcely soluble [27]. The gelation mechanism of carrageenan is based on the formation of double helix structure [25, 29].

The stability of carrageenan gels depends on pH, temperature, time, presence of metal ions, protein content, presence of other colloids, common salt and sugars. Increased pH of the medium leads to higher gelation temperature of κ-carrageenan and higher hardness of gels [25, 27]. An example of this is the effect of pH and carrageenan concentration on whey protein rheology properties [12]. The viscosity of whey protein gels considerably increases after adding a carrageenan. The hardest gels were obtained at 0.3% polysaccharide concentration and pH 7. Gels formed from κ-carrageenan are hard, brittle, slightly opalescent and prone to syneresis. When gums, such as guar gum, are added to κ-carrageenan, its elasticity increases [11, 21, 29].

In the food industry, κ-carrageenan is used as a gelling, thickening, stabilizing and water-binding agent in various food products, such as instant products, dessert, sauces, milk, yogurt and meats. Carrageenans are used in meat processing as water-binding agents, fat-replacer, stabilisers for stuffing and meat emulsions, as functional additives for meat restructuring and in low-fat finely-ground sausages [1, 3, 25]. In dairy industry, they are used to stabilise condensed milk, creams and milk desserts, puddings, yoghurts [13, 31], processed cheese and emulsions [23, 25]. They are widely used in ice cream and other frozen food production in which they provide appropriate texture by reducing the growth of ice crystals and ensure higher product stability when storage temperature changes [25]. Also, they are used in baking and confectionary industry. The addition of carrageenan to dough in an amount of 0.1% improves the bread texture. The synergy of carrageenan and lecithin and milk proteins in dough leads to a higher strength of dough and improves loaf volume, shape and texture [19].

Positive effects of carrageenan as a cryoprotectant have been documented for frozen pork [8, 9] and poultry [20], ice cream [5], frozen dough [17], and survivability of lactic acid bacteria and yeasts [10].

The aim of this study was to analyse the effect of various freezing methods on selected properties of κ-carrageenan gels prepared in water, milk and juice. The scope of investigation included preparation of model gel samples, freezing them and determination of its hardness based on compression and penetration tests.

MATERIAL AND METHODS

In this study κ-carrageenan (E 407-*Satiagel™ of 10-Cargill, France)* was used. The model studies comprised carrageenan gels prepared on the basis of water (pH 7.0), milk (*Mlekpol,* protein 3.2 g, carbohydrates 4.7 g, fat 3.2 g /100 g milk, pH 6.6) and orange juice (*Sokpol* ekstrakt 10° brix, pH 3.7) with the carrageenan mass share of 1.0, 1.3, 1.6, 2.0, 2.2% to solvent weight ratio. The preparation of colloid solution (sol) consisted in measuring out adequate amount of cold solvent (juice, water or milk) and addition of carrageenan powder under vigorous agitation. The mixture was heated to temperature 80 °C. The prepared solutions were distributed into containers of 3.0 cm diameter and 14 cm^3 volume (for the compression tests) and the containers of 3.0 cm diameter and 10 cm^3 volume (for the penetration tests). The obtained solidified gel samples – not frozen (control – P_C) and those after the thermal treatment were investigated.

The gels were frozen using: air method (Freezer *Whirlpool*, Italy, natural convection conditions, temperature -33 °C) and immersion freezing (Immersion cryostat *Wiggen Hauser*, Germany, in glycol, temperature -35 °C) or liquid nitrogen immersion (Dewar *MVE Millennium 2000*, USA – Cryopreservation System, temperature -196 °C). The freezing process was continued until the temperature in the thermal centre of the prepared model was -18 °C. During the freezing, the sample thermal centre temperature was recorded by means of a multi channel digital thermometer equipped with NiCrNi thermoelements with measurement accuracy of ±0.05 K. All the measurements were preceded by the verification of the thermometer readings, taking the temperature of distilled water-ice bath. The recorded temperatures in time were converted to the Excel spreadsheet to obtain the freezing curves which served for determination of the mean linear freezing rate according to the Recommendations of the International Institute of Refrigeration [16, 18].

$$\overline{w} = \frac{\delta}{\tau} \tag{1}$$

where: $\overline{w}$ – mean linear freezing rate (cm·h^{-1}),
δ – thickness of frozen layer (cm),
τ – its freezing time (h).

After freezing, the samples were stored for 24 h in a cabinet freezer at temperature –33 °C. Then the samples were thawed at room temperature (+20 °C) until +10 °C was obtained in the centre of sample. Temperature of the sample center during thawing was measured and recorded with a multi channel digital thermometer equipped with NiCrNi thermoelements.

After thawing, the samples were dried with tissue paper and weighed each time to evaluate the amount of drip loss (L_D). The material weight changes caused by the drip loss depended on the freezing technique. Thawing drip loss was determined as a difference between the sample weight before and after the thawing process [2].

$$L_D = \frac{m_s - m_R}{m_s} \cdot 100\% \quad (2)$$

where: m_s – material mass before freezing (kg),
m_R – material mass after thawing (kg).

The unfrozen gel samples and those subjected to freezing process underwent the compression and penetration testing using the LFRA texture analyzer BROOKFIELD (*Brookfield Engineering Laboratories, Inc., Middleboro, Massachusetts*).

During the compression test the probe of 40 mm diameter cylinder was employed, mandrel displacement rate was 1.0 mm·s^{-1}, initiation force 0.5 N, the compression test depth up to 50% of the sample height. Whereas at the penetration test, a cylinder probe of 7 mm diameter with a cone base angle of 45° was used; mandrel displacement rate was 1.0 mm·s^{-1}, initiation force – 0.5 N, penetration depth was up to 50% of the sample height. The obtained research results were analyzed to establish the peak force of compression and penetration of gels [6].

The statistical analysis of the results was conducted by the variance analysis with Statistica 6 (*StatSoft*) software. Tukey tests was used to determine significant difference ($p \leq 0.05$). The results of the statistical analysis were presented as comments in the text. The data were fitted using regression equations. The degree of fit was judged by the R^2 coefficient.

RESULTS AND DISCUSSION

Significant differences were noted in the freezing time of samples with regards to a freezing technique employed and freezing conditions provided. It was also found that the freezing time of samples did not depend significantly on the type of the prepared gels. The analysis of freezing curves has proved that the linear freezing rate relies on the freezing conditions and its value is very similar for all the gels, irrespective of the type of solvent.

Freezing process of the gel samples appeared to be the slowest in the air freezing conditions at -33 °C ($\overline{w}$ = 0.6 cm·h^{-1}). The main cause of poor performance under these conditions was predominantly the low efficiency of heat transfer process between the surroundings and the material. The gel freezing process performed in the immersion cryostat with ethyl glycol as a refrigerating medium intensified the process and consequently, the freezing rate reached $\overline{w}$ = 5.0 cm·h^{-1}. Importantly, immersion freezing with liquid nitrogen ensured the highest freezing rate $\overline{w}$ = 35.7 cm·h^{-1}.

Drip loss is the synthetic indicator of the reversibility of gel freezing processes. Thawing drip loss determined by gel weight changes after thawing depending on the freezing method is shown in Table 1 (with significance of differences between means).

Changes in thawing drip loss, which indicate gel structure damage, were between 0 and 50.10% depending on the freezing method used. The largest significant ($p < 0.05$) weight losses due to thawing drip were noted for water gels (κ-carrageenan content 1.0%) subjected to air-blast freezing at a freezing rate of 0.6 cm·h^{-1} (50.10%) and to immersion freezing in liquid nitrogen at a freezing rate of 35.7 cm·h^{-1} (34.60%). The gels prepared in milk at pH 6.6 subjected to immersion freezing in glycol and in liquid nitrogen had the lowest free thawing drip loss. For a sample with a 2.2% κ-carrageenan content subjected to immersion freezing in liquid nitrogen, no free thawing drip loss was observed. The results are confirmed in a number of studies [4, 15, 24], in which thawing drip loss as a general indicator of gel quality after thawing depended on the freezing rate and its relationship with the nature and size of ice crystals being formed. Furthermore, in product with a high water content, such as gels, cracks and structural damage in surface product layers may form at very high rates (above 10 cm·h^{-1}), which may lead to increased thawing drip loss. According to the guidelines of the International Institute of Refrigeration [16], a freezing rate of 5.0 cm·h^{-1} is sufficient to maintain appropriate gel structure.

A significant effect was found ($p<0.05$) of the κ-carrageenan content, freezing method and media at various pH values on the occurrence of thawing drip loss. The higher the carrageenan content, the lower thawing drip loss in the gel. Milk-based gels with pH 6.6 were found to be the most favourable systems in terms of pH changes. The relations between changes of thawing drip loss versus carrageenan content and freezing method were expressed by linear relationships with high correlations (Table 2). Determination

Table 1. Dependence of drip loss (L_D) on freezing method and carrageenan addition

Addition of carrageenan (%)	Freezing in air	Freezing in glycol	Freezing in liquid nitrogen
Carrageenan – water			
1.0	50.10^{a} ± 0.79	3.70^{a} ± 0.45	34.60^{a} ± 1.53
1.3	45.07^{b} ± 0.32	1.40^{b} ± 0.00	21.37^{b} ± 0.92
1.6	41.03^{c} ± 0.66	0.97^{bc} ± 0.28	12.43^{c} ± 0.47
2.0	35.90^{d} ± 1.32	0.97^{bcd} ± 0.20	11.27^{cd} ± 0.85
2.2	33.37^{de} ± 1.59	0.40^{cde} ± 0.00	6.770^{e} ± 0.83
p	*p = 0.00*	*p = 0.0004*	*p = 0.00*
Carrageenan – juice			
1.0	7.10^{a} ± 0.26	2.83^{a} ± 0.40	2.47^{a} ± 0.15
1.3	4.67^{b} ± 0.64	2.40^{ab} ± 0.17	1.00^{b} ± 0.17
1.6	3.17^{c} ± 0.15	1.97^{bc} ± 0.15	0.50^{bc} ± 0.17
2.0	1.83^{d} ± 0.11	1.37^{d} ± 0.05	0.37^{cd} ± 0.05
2.2	0.77^{e} ± 0.05	0.50^{e} ± 0.17	0.27^{cde} ± 0.23
p	*p = 0.00*	*p = 0.73*	*p = 0.008*
Carrageenan – milk			
1.0	3.00^{a} ± 0.26	0.83^{a} ± 0.09	0.33^{a} ± 0.05
1.3	2.07^{b} ± 0.11	0.50^{ab} ± 0.10	0.27^{a} ± 0.23
1.6	1.07^{c} ± 0.05	0.30^{bc} ± 0.00	0.13^{a} ± 0.23
2.0	0.87^{cd} ± 0.20	0.33^{bcd} ± 0.03	0.10^{a} ± 0.17
2.2	0.40^{de} ± 0.00	0.23^{bcde} ± 0.12	0.00^{a} ± 0.00
p	*p = 0.59*	*p = 0.00*	*p = 0.00002*

^{abc}Mean values designated with different letters are statistically different in the columns ($p \leq 0{,}05$).

Table 2. Relations between the drip loss (L_D) (%) and the content of carrageenan (x) (%) depending on freezing method

Environmentgels / Freezing method	Carrageenan – water	R²	Carrageenan – juice	R²	Carrageenan – milk	R²
Freezing in air	L_D= 13.74x – 63.34	*-0.96*	L_D= 4.98x – 11.57	*-0.96*	L_D= 2.05x – 4.80	*-0.90*
Freezing in glycol	L_D= 2.23x – 5.12	*-0.71*	L_D= 1.84x – 4.79	*-0.90*	L_D= 0.42x – 1.13	*-0.61*
Freezing in liquid nitrogen	L_D= 21.01x – 51,33	*-0.86*	L_D= 1.63x – 3.55	*-0.74*	L_D= 0.26x – 0.59	*-0.64*

coefficients were between 0.61 and 0.96. Schmidt & Smith [26] defined the ability of polysaccharides, including carrageenan, to form more viscous solutions in milk than in water as milk reactivity. Similar interactions between carrageenan and milk and its effect on the rheological properties of carrageenan gels were also confirmed by other authors [7, 14, 21, 28, 31].

Hardness of all the formulated gels (the unfrozen samples and those thawed after earlier freezing at varying freezing rates) was examined on the grounds of the results of compression tests. The higher compression force applied, the higher gel hardness was displayed (Table 3).

The carrageenan content was found to determine the hardness of resulting gels in a statistically significant manner ($p<0.05$). Milk-based gels with a 2.2% carrageenan content had the maximum compression force (44.47 N). Water-based and juice-based gels had much lower compression force values with the same carrageenan content (18.31 N and 17.97 N, respectively). The available results of other authors [12, 21, 27, 29] confirm that the force needed to disrupt a carrageenan gel is a function of carrageenan concentration in a solution. Indeed, gel samples frozen at various rates had significantly ($p<0.05$) lower compression force values than those not subjected to freezing. Gels frozen in air at a rate of 0.6 $cm \cdot h^{-1}$ had the lowest compression force values in a range of 0.59 N to 4.90 N (medium: water, pH 7.0). Reduced compression force values, and thus the

Table 3. Compression force (N) in relation to gel freezing method

Addition of carrageenan (%)	Control unfrozen	Freezing in air	Freezing in glycol	Freezing in liquid nitrogen
Carrageenan – water				
1.0	$1.74^{a} \pm 0.30$	$0.59^{a} \pm 0.03$	$1.41^{a} \pm 0.09$	$1.11^{a} \pm 0.04$
1.3	$5.21^{b} \pm 0.21$	$0.74^{ab} \pm 0.06$	$6.19^{b} \pm 0.19$	$1.84^{ab} \pm 0.05$
1.6	$7.51^{c} \pm 0.61$	$1.34^{c} \pm 0.03$	$7.35^{c} \pm 0.27$	$2.12^{bc} \pm 0.16$
2.0	$12.72^{d} \pm 0.06$	$3.61^{d} \pm 0.42$	$12.75^{d} \pm 0.53$	$3.93^{d} \pm 0.16$
2.2	$18.31^{e} \pm 0.33$	$4.90^{e} \pm 0.17$	$15.15^{e} \pm 0.52$	$6.56^{e} \pm 0.56$
p	*p = 0.00005*	*p = 0.206*	*p = 0.00001*	*p = 0.01*
Carrageenan – juice				
1.0	$5.10^{a} \pm 0.53$	$3.43^{a} \pm 0.30$	$4.68^{a} \pm 0.47$	$2.33^{a} \pm 0.06$
1.3	$6.14^{ab} \pm 0.62$	$5.23^{b} \pm 0.28$	$6.61^{b} \pm 0.13$	$3.83^{b} \pm 0.05$
1.6	$8.48^{c} \pm 0.31$	$7.73^{c} \pm 0.50$	$8.59^{c} \pm 0.12$	$6.38^{c} \pm 0.15$
2.0	$14.69^{d} \pm 0.44$	$10.57^{d} \pm 0.45$	$11.72^{d} \pm 0.67$	$10.49^{d} \pm 0.49$
2.2	$17.97^{e} \pm 0.74$	$13.54^{e} \pm 0.38$	$13.95^{e} \pm 0.55$	$12.74^{e} \pm 0.43$
p	*p = 0.00*	*p = 0.00*	*p = 0.00*	*p = 0.00015*
Carrageenan – milk				
1.0	$12.60^{a} \pm 0.36$	$5.97^{a} \pm 0.15$	$10.70^{a} \pm 0.85$	$7.63^{a} \pm 0.15$
1.3	$16.67^{b} \pm 0.50$	$7.07^{b} \pm 0.25$	$11.57^{ab} \pm 0.20$	$10.17^{b} \pm 0.25$
1.6	$22.37^{c} \pm 0.86$	$9.27^{c} \pm 0.20$	$14.67^{c} \pm 0.60$	$13.13^{c} \pm 0.11$
2.0	$35.83^{d} \pm 1.83$	$16.07^{d} \pm 0.10$	$21.27^{d} \pm 0.45$	$20.13^{d} \pm 0.25$
2.2	$44.47^{e} \pm 0.58$	$20.30^{e} \pm 0.26$	$24.10^{e} \pm 0.10$	$22.83^{e} \pm 0.11$
p	*p = 0.00*	*p = 0.00*	*p = 0.00*	*p = 0.00*

[abc] Mean values designated with different letters are statistically different in the columns ($p \leq 0,05$).

hardness of frozen gels (compared to non-frozen ones) are related to changes due to the formation of crystalline ice structures. This may lead to changes in original properties of the gels [18, 22].

The relationships between the maximum compression force versus various carrageenan contents and freezing methods for carrageenan gels prepared in the media with various pH values are expressed by regression equations with high determination coefficients (Table 4).

A statistically significant ($p<0.05$) increase in penetration force with increasing carrageenan contents in samples irrespective of the medium in which they were prepared was found in the assessment of gel hardness, based on the maximum penetration force (Table 5). At the maximum carrageenan content (2.2%), the hardest gels were obtained in the carrageenan/juice (1.69 N) and carrageenan/milk (1.77 N) system. The freezing process had a significant effect ($p<0.05$) on penetration forces. Frozen gels had lower maximum penetration force values compared to the non-frozen ones.

The correlation between the carrageenan content and penetration force values for the gels tested depending on the freezing method is described using line equations (Table 6). The higher carrageenan content in a sample, the higher hardness of a gel (higher penetration force value). However, the dynamics of the changes varied for water, juice or milk based gels.

Table 4. Relations between the force compression of gels (F_C) (N) and the content of carrageenan (x) (%) depending on freezing method

Freezing method \ Environment gels	Carrageenan-water	R^2	Carrageenan - juice	R^2	Carrageenan - milk	R^2
Control unfrozen	$F_C = 12.97x - 11.91$	*0.96*	$F_C = 11.07x - 7.45$	*0.93*	$F_C = 26.67x - 16.81$	*0.96*
Freezing in air	$F_C = 3.71x - 3.77$	*0.89*	$F_C = 8.18x - 5.15$	*0.97*	$F_C = 12.12x - 7.90$	*0.93*
Freezing in glycol	$F_C = 10.97x - 9.21$	*0,97*	$F_C = 7.60x - 3.20$	*0.98*	$F_C = 11.81x - 2.67$	*0.94*
Freezing in liquid nitrogen	$F_C = 4.11x - 3.55$	*0.84*	$F_C = 8.86x - 7.21$	*0.98*	$F_C = 13.06x - 6.38$	*0.98*

Table 5. Penetration force (N) in relation to gel freezing method

Addition of carrageenan (%)	Control unfrozen	Freezing in air	Freezing in glycol	Freezing in liquid nitrogen
Carrageenan – water				
1.0	$0.05^{a} \pm 0.00$	$0.02^{a} \pm 0.002$	$0.04^{a} \pm 0.006$	$0.030^{a} \pm 0.003$
1.3	$0.14^{b} \pm 0.01$	$0.04^{b} \pm 0.001$	$0.07^{b} \pm 0.005$	$0.050^{ab} \pm 0.000$
1.6	$0.30^{c} \pm 0.01$	$0.06^{c} \pm 0.005$	$0.12^{c} \pm 0.010$	$0.083^{c} \pm 0.005$
2.0	$0.63^{d} \pm 0.01$	$0.09^{d} \pm 0.007$	$0.21^{d} \pm 0.002$	$0.087^{cd} \pm 0.010$
2.2	$1.03^{e} \pm 0.04$	$0.10^{de} \pm 0.004$	$0.35^{e} \pm 0.010$	$0.120^{e} \pm 0.010$
p	*p = 0.00001*	*p = 0.00001*	*p = 0.00001*	*p = 0.0102*
Carrageenan – juice				
1.0	$0.5^{a} \pm 0.01$	$0.18^{a} \pm 0.01$	$0.18^{a} \pm 0.005$	$0.24^{a} \pm 0.005$
1.3	$0.78^{b} \pm 0.01$	$0.31^{b} \pm 0.01$	$0.28^{b} \pm 0.005$	$0.65^{b} \pm 0.003$
1.6	$1.05^{c} \pm 0.08$	$0.46^{c} \pm 0.05$	$0.41^{c} \pm 0.008$	$0.96^{c} \pm 0.006$
2.0	$1.51^{d} \pm 0.04$	$0.63^{d} \pm 0.01$	$0.62^{d} \pm 0.006$	$1.41^{d} \pm 0.04$
2.2	$1.69^{e} \pm 0.01$	$0.78^{e} \pm 0.03$	$0.76^{e} \pm 0.006$	$1.63^{e} \pm 0.03$
p	*p = 0.00001*	*p = 0.00001*	*p = 0.00001*	*p = 0.00001*
Carrageenan – milk				
1.0	$0.7^{a} \pm 0.01$	$0.17^{a} \pm 0.05$	$0.20^{a} \pm 0.006$	$0.34^{a} \pm 0.02$
1.3	$1.0^{b} \pm 0.04$	$0.38^{b} \pm 0.01$	$0.39^{b} \pm 0.01$	$0.51^{b} \pm 0.006$
1.6	$1.27^{c} \pm 0.01$	$0.52^{c} \pm 0.01$	$0.61^{c} \pm 0.01$	$0.72^{c} \pm 0.02$
2.0	$1.53^{d} \pm 0.06$	$0.84^{d} \pm 0.01$	$0.98^{d} \pm 0.00$	$1.04^{d} \pm 0.02$
2.2	$1.77^{e} \pm 0.02$	$1.10^{e} \pm 0.07$	$1.24^{e} \pm 0.01$	$1.22^{e} \pm 0.02$
p	*p = 0.0012*	*p = 0.00001*	*p = 0.00001*	*p = 0.00001*

[abc] Mean values designated with different letters are statistically different in the columns ($p \leq 0,05$).

Table 6. Relations between the force penetration of gel (F_p) (N) and the content of gelatin (x) (%) depending on freezing method

Environment gels / Freezing method	Carrageenan -water	R^2	Carrageenan - juice	R^2	Carrageenan - milk	R^2
Control unfrozen	$F_p = 0.78x - 0.84$	*0.91*	$F_p = 1.01x - 0.53$	*0.99*	$F_p = 0.87x - 0.16$	*0.99*
Freezing in air	$F_p = 0.07x - 0.05$	*0.99*	$F_p = 0.49x - 0.32$	*0.99*	$F_p = 0.74x - 0.59$	*0.97*
Freezing in glycol	$F_p = 0.24x - 0.24$	*0.91*	$F_p = 0.48x - 0.32$	*0.98*	$F_p = 0.86x - 0.71$	*0.98*
Freezing in liquid nitrogen	$F_p = 0.07x - 0.03$	*0.89*	$F_p = 1.15x - 0.88$	*0.99*	$F_p = 0.73x - 0.42$	*0.99*

CONCLUSIONS

To conclude, it is noted that carrageenan gels are systems which form defined structures which undergo various changes during freezing. Various pH values and increased carrageenan contents had no significant effect on the gel freezing rate. Immersion freezing with liquid nitrogen ensured the highest freezing rate (35.7 cm·h^{-1}). Air-blast freezing (rate: 0.6 cm·h^{-1}) led to the greatest damage of gel structure, resulting in the largest thawing drip loss.

Gels prepared in milk at pH 6.6 and in juice at pH 3.7 with a 2.2% carrageenan content subjected to immersion freezing in glycol and in liquid nitrogen had the lowest thawing drip loss. Such factors as carrageenan content, medium type and pH (juice: 3.7; milk: 6.6; water: 7.0) and freezing conditions, had a significant ($p<0.05$) effect on the hardness of carrageenan gels in terms of compression forces and penetration. Frozen gels revealed a significant decrease in hardness, that is, lower compression forces and penetration, compared to non-frozen gels.

Gels prepared in milk at pH 6.6 and in juice at pH 3.7 with a 2.2% carrageenan content subjected to immersion freezing in glycol and in liquid nitrogen best retained their properties compared to non-frozen samples.

REFERENCES

1. Adamczak L., Słowiński M., Ruciński M., Wpływ dodatku κ karagenu, izolatu białka sojowego i błonnika pszennego na jakość technologiczną niskotłuszczowych kiełbas drobno rozdrobnionych. Acta Sci. Polonorum. Technologia Alim., 2003, 2, 85-93.
2. Official method of analysis (16th ed.). Association of the Official Analytical Chemists. Washington, DC: 1995.
3. Brewer M.S., Reducing the fat content in ground beef without sacrificing quality: A review. Meat Science, 2012, 91, 385-395.
4. Chevalier D., Bail A. Le., Ghoul M., Freezing and ice crystal formed in a cylindrical food model: Part I. Freezing at atmospheric pressure. J. Food Eng., 2000, 46(4), 277-285.
5. Damodaran S., Inhibitation of ice crystal growth in ice cream mix by gelatin hydrolysate. J. Agric. Food Chem., 2007, 55, 10918-10923.
6. Dobrzycki J.H., Baryłko-Pikielna N., Instrumentalne metody pomiaru tekstury żywności. IŻŻ., Warszawa, 1986, 22-33.
7. Drohan D.D., Tziboula A., McNulty D., Horne D.S., Milk protein-carrageenan interaction. Food Hydrocoll., 1997, 11(1), 101-107.
8. Dziomdziora M., Termostabilność białek mięsa wieprzowego mrożonego z dodatkiem krioprotektantów. Chłodnictwo, 2006, 1-2, 74-78.
9. Dziomdziora M., Krala L., The effect of hydrocolloids mixtures on frozen pork properties. Pol. J. Food Nutr. Sci., 2003, 12(53), 4, 55-58.
10. Dziugan P., Dziomdziora M., Wpływ wybranych krioprotektantów na przeżywalność bakterii fermentacji mlekowej i drożdży podczas zamrażania. Chłodnictwo, 2006, 9.
11. Gustaw W., Mleko S., Glibowski P., Synergiczne interakcje występujące pomiędzy polisacharydami w ich mieszaninach. Żywność. Nauka. Technologia. Jakość, 2001, 3(28), 5-15.
12. Gustaw W., Mleko S., The effect of pH and carrageenan concentration on the rheological properties of whey protein gels. Pol. J. Food Nutrition, 2003, 12(53), 4, 39-44.
13. Gustaw W., Nastaj M., Sołowiej B., Wpływ wybranych hydrokoloidów na właściwości reologiczne jogurtu stałego. Żywność. Nauka. Technologia. Jakość, 2007, 5(54).
14. Gustaw W., Nastaj M., Żelowanie mieszanin białek serwatkowych z kappa karagenem. Acta Agrophysica, 2009, 13(3), 651-660.
15. Hung Y.C., Kim N.K., Fundamental aspects of freeze-cracking. Food Technology, 1996, 50(12), 59-61.
16. Recommendations for the processing and handling of frozen foods. International Institute of Refrigeration, Denmark 2006.
17. Kozłowicz K., Kluza F., Experimental characteristics of dough freezing and products obtained from frozen doughs. EJPAU. Series: Agricultural Engineering., 2001, 4(1).
18. Kozłowicz K., Kluza F., Gel products properties influenced by freezing in different conditions. Int. J. Refrig. 2012, 35, 1715-1721.
19. Kowalski S., Łukasiewicz M., Juszczak L., Charakterystyka teksturalna i sensoryczna mas cukierniczych otrzymanych na bazie miodu naturalnego i wybranych hydrokoloidów. Żywność. Nauka. Technologia. Jakość, 2011, 3 (76), 40-52.
20. Kovačević D., Mastanjević K., Kordić J., Cryoprotective effect of polydextrose on chicken surimi. Czech J. Food Sci., 2011, Vol. 29, No. 3, 226-231.
21. Mleko S., Otrzymywanie mieszanych żeli białek serwatkowych z wybranymi polisacharydami i badanie ich właściwości mechanicznych. Żywność. Nauka. Tech. Jakość, 2004, 1(38), 34-40.
22. Olivera D.F., Salvadori V.O., 2009. Effect of freezing rate in textural and rheological characteristics of frozen cooked organic pasta. J. Food Eng. 90: 271-276.
23. Perrechil F.A., Cunha R.L., Stabilization of multilayered emulsions by sodium caseinate and κ-carrageenan. Food Hydrocoll., 2013, 30, 606-613.
24. Petrović L., Grujić R., Petrović M., Definition of the optimal freezing rate. 2. Investigation of the physico-chemical properties of beef M. Longissimus dorsi frozen at different freezing rates. Meat Sci. 1993, 33 (3): 319-331.
25. Phillips G.O., Williams P.A., Handbook of hydrocolloids. Woodhead Publishing Ltd., 2009, Cambridge.
26. Schmidt K.A., Smith D.E., Milk reactivity of gum and milk protein solutions. J. Dairy Sci., 1992, 75, 3290-3295.
27. Sołowiej B., Gustaw W., Mleko S., Wpływ wybranych czynników na żelowanie κ- i ι-karagenu. Żywność. Nauka. Technologia. Jakość, 2005, 2(43) Supl., 193-202.
28. Tziboula A., Horne D.S., Influence of milk proteins on κ-carrageenan gelation. International Dairy Journal, 1999, 9(3-6), 359-364.
29. Yuguchi Y., Thuy T.T., Urakawa H., Kajiwara K., Structural characteristics of carrageenan gels: temperature and concentration dependence. Food Hydrocoll., 2002, 16.
30. Yuguchi Y., Urakawa H., Kajiwara K., Structural characteristics of carrageenan gels: various types of counter ions. Food Hydrocoll., 2003, 17, 481-485.
31. Verbeken D., Bael K., Thas O., Dewettinck K., Interactions between κ-carrageenan, milk proteins and modified starch in sterilized dairy desserts. J. Dairy 2006, 16, 482-488.

FEM SIMULATION OF THE TUBE ROLLING PROCESS IN DIESCHER'S MILL

Zbigniew Pater[1], Arkadiusz Tofil[1]

[1] Mechanical Engineering Faculty, Lublin University of Technology, 36 Nadbystrzycka Str., 20-618 Lublin, Poland, e-mail: z.pater@pollub.pl

ABSTRACT

This paper deals with the issue of numerical modelling of the piercing process of a thick–walled bush in a two-rolled skew rolling mill, equipped with guiding devices of Diescher's type. After a short characteristic of the subject matter, the developed geometric model of the process was discussed, with taking kinematics of tools movement and thermal phenomena present in metal during forming into consideration. Next, the results of calculations were presented in a form of fields of strain, damage criterion and temperature. Distributions of force parameters acting on particular tools during the process of bush rolling were also given.

Keywords: tube rolling, FEM, state of stress and strain.

INTRODUCTION

Thick-walled bushes, from which later seamless pipes are manufactured, are made in skew piercing rolling mills. Depending on the number of rolls, the skew rolling mills are divided into [1]:

- Two-roll rolling mills with two working rolls and two guiding devices, which aim at holding of the rolled bush along the rolling axis. These guiding devices are manufactured as rolls of Mannesmann's type, flat of Stiefel's type and disk of Diescher's type.
- Three-roll mills with three working rolls, which work without additional guiding devices.

Although the piercing process in skew rolling mills has been used for more than a hundred years, this technology has not been so far analyzed theoretically in a satisfactory way, which is the result of a complex kinematics of metal flow. The improvement of this situation can be seen in possibilities provided by the numerical modelling application, based on FEM.

The first works concerning numerical modelling in skew piercing rolling mills made in the 90's of the XXth century based on two-dimensional FEM models. Due to the complex character of material flow, piercing processes should be modeled in three-dimensional state of strain. Pietsh and Thievien [2], Cereti et al. [3] and Komori [4] were the first researchers who tried modelling the piercing process. However, numerical models developed by these researchers, based on considerable simplifications, e.g.: omitting of thermal phenomena present in the formed material, considering only the initial stage or given stage of the piercing process during calculations or assuming that the formed material is perfectly plastic. More complex models of the piercing process, taking into thermal phenomena account, were presented by Pater and et al. [5–7] and Lu and et al. [8].

The results of calculations obtained by means of these models remained in good convergence with the results of experiments. A disadvantage of the above mentioned works was, however, omitting rotational movement of the piercing plug and friction forces present at the workpiece – plug interface.

This paper presents a new numerical model of the piercing process of a bush in the skew rolling mill with Deischer's guiding devices, in which all tools' kinematics of movement was considered.

NUMERICAL MODEL OF THE PIERCING PROCESS

The numerical analysis of the piercing process in the skew rolling mill was made by applying the commercial software Simufact.Forming 10.0, which uses FEM displacement representation. The schema of the analyzed case is shown in Figure 1, in which basic parameters describing the used tools and their reciprocal position are also marked.

It was assumed in simulation that cylindrical billet (of dimensions Ø60×150 mm) from 100Cr6 steel would be used. The material model was taken from the applied software library. It was also assumed that rolls moved in the same direction with a constant rotational velocity n = 60 rpm, disks rotated (in the opposite direction) with velocity 6.8 rpm, and the plug rotational movement was caused by the formed semi-finished part.

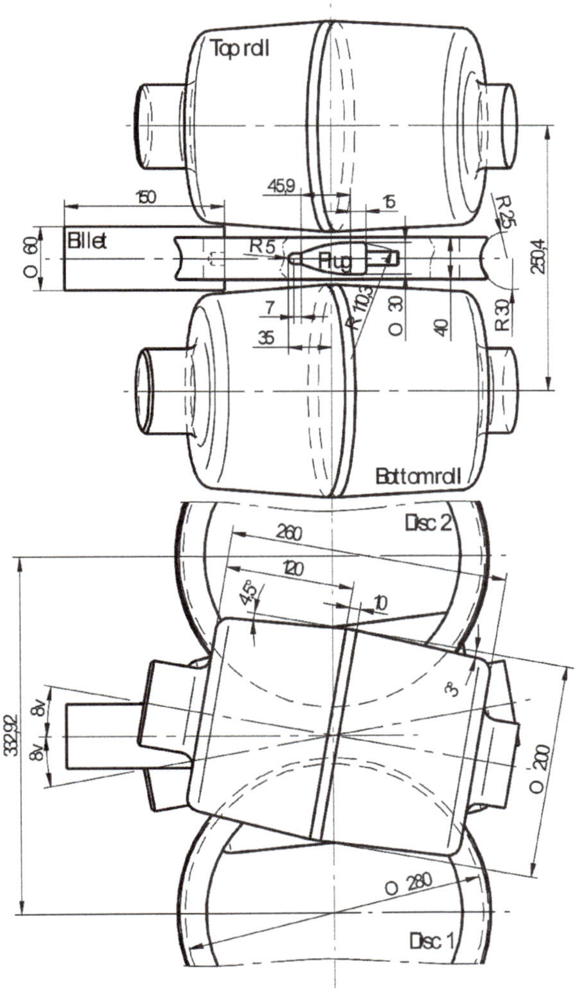

Fig. 1. Schematic view a of the analyzed numerically piercing process of a thick-walled bush

Thermo-mechanical schema of the calculations was considered. It was assumed, that the billet was heated in its whole volume to the temperature 1180 °C, tools temperature was constant during the whole forming process and equal 900 °C for the piercing plug and 50 °C for rolls and disk guiding devices. On a basis of the reference [9], it was presumed that the coefficient of heat exchange between tools and the workpiece was 4000 W/m^{2}K. Yet, the coefficient determining heat exchange between workpiece and the environment was equal 200 W/m^{2}K.

Due to the change of direction of friction forces acting on the tool-workpiece interface, a model of constant friction dependent on metal slide velocity in relation to tool was assumed in calculations, and described by the equation:

$$\tau = -m\,k\,\arctan\left(\frac{v_p}{a_p}\right)\frac{v_p}{|v_p|} \tag{1}$$

where: m – friction factor,
k – shear yield stress,
v_p – slide velocity vector,
a_p – coefficient a few grades lower than slide velocity (it was assumed in calculations that a_p = 0,1% of roll velocity).

Due to the lack of lubrication and purposeful rolls coarseness (for an easier billet implementation into the space between rolls), it was assumed that friction factor for rolls reached limiting value m = 1.0. In the case of other tools (discs and piercing plug), a lower value of the friction factor m = 0.7 was considered.

Figure 2 shows FEM model of the piercing process. This model consists of: two barrel rolls,

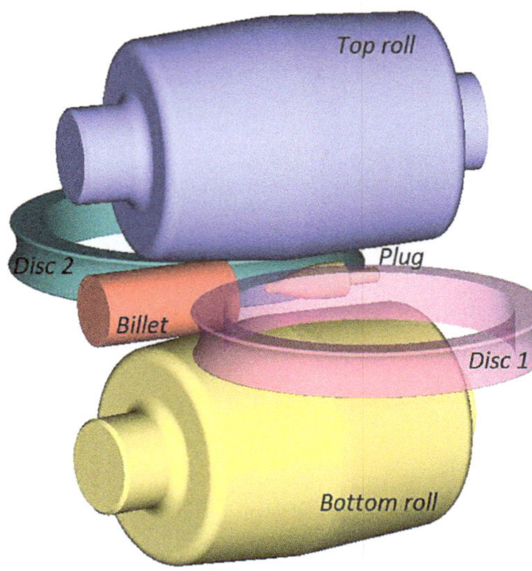

Fig. 2. FEM model of the piercing process of a thick-walled bush

two disc guiding devices, a piercing plug and a workpiece. All tools were modelled as rigid bodies. Hexahedral elements were used for billet modelling.

RESULTS OF NUMERICAL ANALYSIS

The application of FEM for the analysis of the piercing process in the skew rolling mill allows for precise observations of the workpiece shape changes during rolling. These changes are shown in Figure 3, in which, in order to improve visibility, one of the disc guiding devices was made transparent. At the beginning of the process, the billet is grabbed by rolls which rotate and displace the workpiece in the axial direction. At this stage of forming, rotary compression process is realized, which lasts until the plug nose contacts the workpiece surface. Then the hollow forming begins. The hole diameter depends mainly on the applied piercing plug. During the hollow forming, the disc guiding devices also contact the metal and their rotation increases the axial forces drawing the workpiece on the piercing plug. After

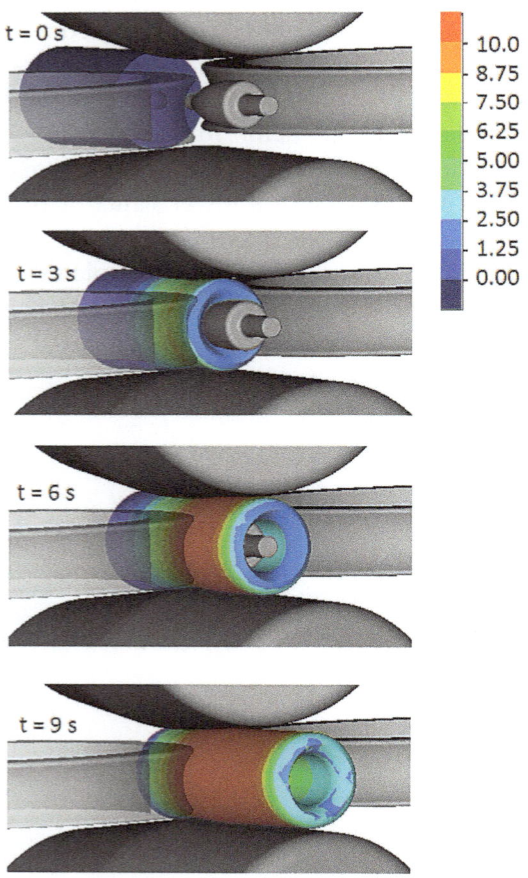

Fig. 3. Progression of the workpiece shape in the piercing process with marked distribution of effective strain

the whole head goes into the metal the forming process becomes stable and is characterized by relatively stable forming forces. At the end of the process, the plug pierces (in the axial direction) through the material and a thick-walled bush is formed.

During rolling the formed element is put into rotation, which dominates. Tool forces acting on the formed element cause also its displacement in the axial direction. In consequence, the material in the piercing process displaces in screw motion. The most intensive axial flow of the material is localized at the surface area adhering the rolls, the smallest flow is before the piercing plug nose.

The analysis of strain distributions presented in Figure 3 shows that they reach relatively large values. This is the result of the presence (during forming) of internal shearing of the metal structure, which does not influence the desired change of shape and causes unnecessary metal deformation. It is the most intensive at the bush external surface, which is influenced by friction forces causing material flow mainly in the circumferential (tangent) direction. The appearance of large deformations is a characteristic feature of skew and cross rolling processes [10–13].

Figure 4 presents the distribution of damage function according to Cockroft-Latham [14] criterion in the semi- finished product axial section at the determined stage of the piercing process. This function assumes the largest values in the area of the piercing plug nose, where the hole forming begins. However, these values are smaller than critical (being about 0.8), at which the material cracks. This means that the piercing process has been properly designed as the bush internal surface is formed in the result of plastic flow of material and is free from surface faults caused by cracking.

Figure 5 presents distributions of temperature at the surface and in cross section of the formed element in the piercing process. Although the

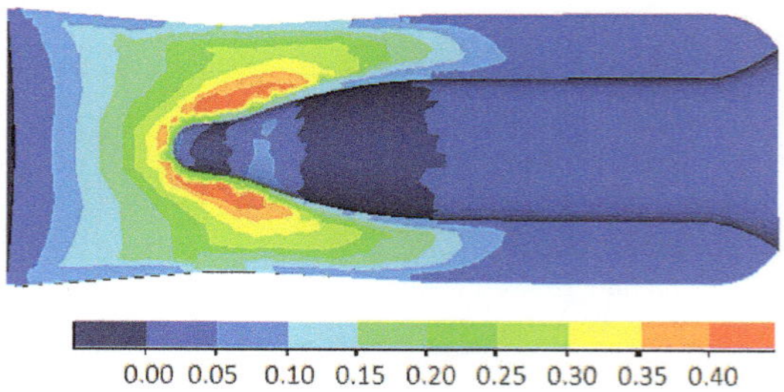

Fig. 4. Distribution of damage function according to Cockroft-Latham criterion for time t = 7 s

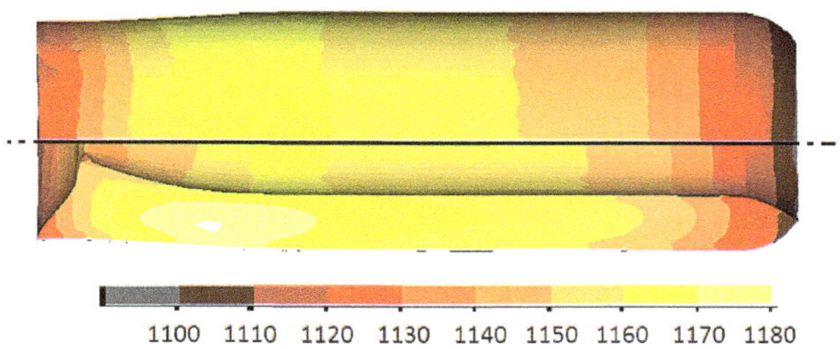

Fig. 5. Distribution of temperature in the bush at the end of piercing process (for t=10 s)

forming time is relatively long, the metal temperature in the wall of the rolled bush remains at the level close to the billet temperature (1180 °C). This is the result of compensation of loss of heat abstraction into tools by heat produced by plastic strain work and friction work. The smallest temperatures are observed at the formed product edges, where the best conditions of heat abstraction into the environment take place.

During the piercing process the forming tools rotate. The rolls and guiding devices are driven and they move with constant rotary velocity. Hence, the piercing plug rotates due to the influence of the workpiece which displaces in screw motion. Yet, at the determined forming stage, rotary velocity of this tool is the largest and it remains at relatively constant level – Figure 6.

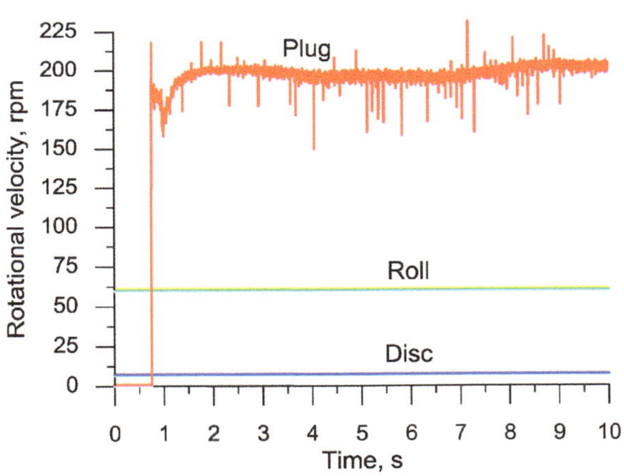

Fig. 6. Rotational velocities of tools in the analyzed piercing process

The application of FEM allows for precise determining of loads of particular tools during piercing. Distributions of forces obtained for the analyzed piercing process are shown in Figure 7. It can be seen that at the beginning all forces increase gradually until the bush contacts the piercing plug. After that, the process reaches the determined stage, at which forces are kept at constant level. It should be considered that loads of both rolls are almost identical. Certain differences can be observed in loads distributions of particular disc guiding devices, which is the result of irregular pressure of the semi-finished product to these tools. The forces acting on rolls are the largest; yet, the disc guiding devices have the smallest loads during piercing. At the end of the process all forces undergo a gradual decrease.

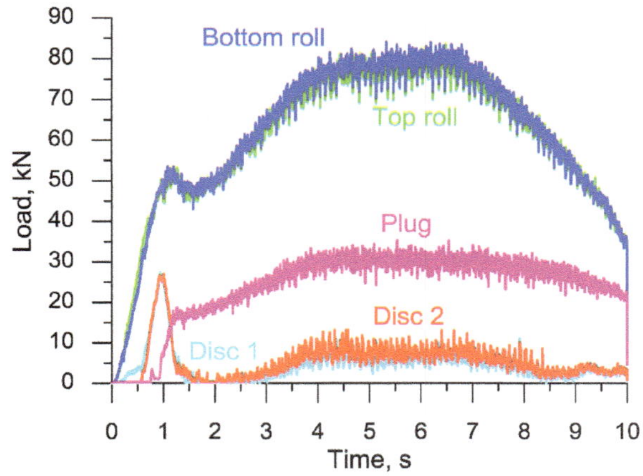

Fig. 7. Forces acting on tools In the piercing process, marked according to Figure 1

Distribution of rolls and discs rotary moments, shown in Figure 8, can be used for determining the rolling mill energy requirements. It is visible that a decisive role within this scope plays the rolls drive. The discs rotary moment is much lower, which is mainly the result of the resultant force placement acting on the disc placed angularly to the disc rotation axis. Considering the character of moments distribution it can be stated that it is similar to the course of forces shown in Figure 7.

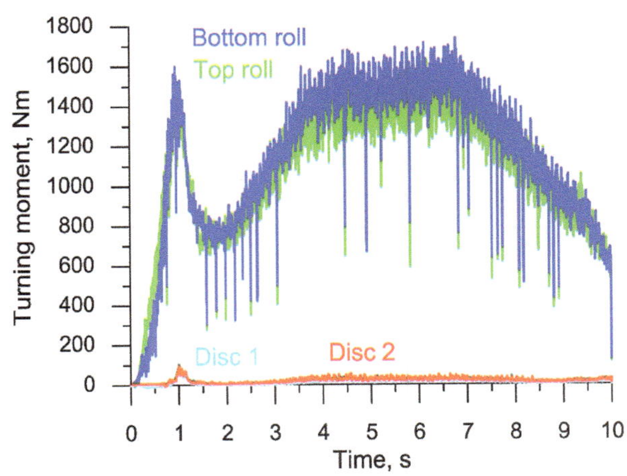

Fig. 8. Turning moments acting on tools in the piercing process, marked according to Figure 1

SUMMARY

This paper presents the thermo-mechanical analysis of the piercing process in the skew rolling mill made by means of finite element method and based on commercial software Simufact. Forming 10.0. The development model of the piercing process deals with real kinematics of all tools movement (piercing plug for the first time) and allows for the analysis of the stress, strain and temperature distributions in the workpiece as well as force-energetic parameters. Basing on this model, it is planned to make numerical simulations of a number of piercing processes in order to determine the process basic parameters influence on the diametric dimensions of the manufactured bush and particular tools loads. It is also possible to develop the worked out model FEM of the piercing process in the direction of stresses calculations in forming tools, assumed as elastic bodies.

REFERENCES

1. Kazanecki J. Production of seamless pipes. Ed. AGH, Krakow 2003 (in Polish).
2. Pietsch J., Thievien P. MPT International, Vol. 2, 2003, 52-60.
3. Ceretti E., Giardini C., Attanasio A., Brisotto F., Capoferri G.: Tube & Pipe Technology, 2004, March/April, 155-159.
4. Komori K.: International Journal of Mechanical Sciences, Vol. 47, 2005, 1838-1853.
5. Pater Z., Kazanecki J., Bartnicki J.: Journal of Materials Processing Technology, Vol. 177, 2006, 167-170.
6. Pater Z., Kazanecki J.: Metallurgy and Foundry Engineering, 32(1), 2006, 31-40.
7. Kazanecki J., Pater Z., Bartnicki J.: Computer Methods in Materials Science, 7(2), 2007, 299-304.
8. Lu L., Wang Z., Wang F., Zhu G., Zhang X.: J. Shanghai Jiaotong Univ. (Sci), 16(3), 2011, 281-185.
9. Prince J.C., Marono R., Leon F.: Proceedings of the Institution of Mechanical Engineers, Part E: Journal of Process Mechanical Engineering, Vol. 217, 2003, 337-340.
10. Pater Z. Cross-wedge rolling. Ed. Lublin University of Technology, Lublin 2009 (in Polish).
11. Fu X.P., Dean T.A.: International Journal of Machine Tools and Manufacture, 33(3), 1993, 367-400.
12. Pater Z.: Archives of Metallurgy, 48(1), 2003, 21-35.
13. Pater Z., Tofil A.: Archives of Metallurgy and Materials, 52(2), 2007, 289-297.
14. Gouveia B.P.P.A, Rodrigues J.M.C, Martins P.A.F: Journal of Materials Processing Technology, Vol. 101, 2000, 52-63.

CAVITATION EROSION RESISTANCE OF Ni-Co BASED COATINGS

Mirosław Szala[1], Tadeusz Hejwowski[2], Iwona Lenart[3]

[1] Department of Materials Engineering, Faculty of Mechanical Engineering, Lublin University of Technology, Nadbystrzycka 36, 20-618 Lublin, Poland, e-mail: m.szala@pollub.pl

[2] Department of Materials Engineering, Faculty of Mechanical Engineering, Lublin University of Technology, Nadbystrzycka 36, 20-618 Lublin, Poland, e-mail: t.hejwowski@pollub.pl

[3] Student of Mechanical Engineering Faculty, Lublin University of Technology, Nadbystrzycka 36, 20-618 Lublin, Poland, e-mail: iwonalenart29@wp.pl

ABSTRACT

The Ni-Co based coatings were performed by means of PTA method from Ni-based and Co-based powders. The cavitation erosion resistance of Ni-Co specimens and reference sample in deionized water was determined by the ultrasonic vibration system. The microstructure, as well as mechanism of cavitation erosion of coating was examined by using SEM and LOM. The main mechanism of erosion is cracking and spalling at the boundary of dendrites and eutectics. Coatings exhibit higher cavitation erosion resistance than the reference sample. Therefore, they could be applied to repair and protect pump elements from cavitation wear.

Keywords: cavitation erosion, Ni-Co based coating, PTA deposition.

INTRODUCTION

Cavitation can be defined as the formation and subsequent collapse, within a liquid, of cavities or bubbles that contain vapor or gas, or both. It is a common phenomenon in the hydraulic components such as valves, impellers, hydraulic systems, propellers, engine cylinder sleeves and high-frequency ultrasonic devices. Cavitation erosion is a progressive loss of original material from a solid surface due to continued exposure to cavitation. Bubble collapse is a violent process that generates highly localized, large-amplitude shock waves and microjets in the fluid at the point of collapse. Cavitation bubble collapse loads act repeatedly on a material surface to produce plastic deformation, crack initiation, crack growth and material removal [1–6]. Erosion reduces the machine performance and lifetime. Figure 1 presents pump impeller with the cavitation erosion damage located at the edge of impeller blades.

According to the data included in [6–9], modern materials, such as PTA, deposited coatings based on Ni and Co matrix are applied for regeneration and protection of machines and their components against cavitation erosion wear. In addition, application of coatings is usually easy and fast, a wide variety of materials can be deposited and minimal surface preparation is needed [10]. In the literature, cavitation erosion resistance of Ni-Co based coatings is not so widely presented as for Co [11] or Ni [12] based ones. Hence the cavitation damage mechanism of Ni-Co coatings is not recognized and scarce information is available in contradiction to those relating to Co or Ni. This paper describes the results of research into the cavitation wear behavior of Ni-Co coatings. The presented results of investigations will

Fig. 1. Radial impeller worn by cavitation

be useful in elaboration of protection technology and repair technology of pump elements damaged by cavitation.

The aim of the present study was to determine the cavitation erosion resistance of Ni-Co based coatings PTA deposited on a steel substrate combined with the research into wear mechanism.

MATERIALS AND METHODS

In the present work, the Ni-Co based coatings were deposited by means of NP1250 PTA deposition unit from the mixtures of Ni and Co based powders (Table 1). The substrate was made of a steel plate grade S235JR (C = 0.22%; Si = 0.35%; Mn =1.1%; Cr = 0.3%; Fe = bal.). Coating 1 (Table 2) was deposited by using only PMNi50p powder. Coatings 2–10 were synthesized from the PMNi50p and PMCo45 powders mixed in various proportions. The chemical compositions of the 2 layered coatings (1, 5, 6 and 8) were determined by weight percentage of powders used for coatings deposition whereas chemical compositions of the 1 layer coatings (24, 7, 9, 10) were identified by using EDS (energy-dispersive X-ray spectroscopy) method on scanning microscope Zeiss Ultra Plus. Results of chemical analysis are given in Table 2. Microhardness HV0,05 was measured by means of the Hanemann microhardness head. The head was mounted on the Neophot 2 optical microscope. Sixty indentations were performed on the coating cross-section at accidental locations.

From the padded steel plates, the samples were cut off with dimensions of 8×30×30 mm. The reference sample was made of normalized carbon steel, grade C45 (C = 0.42–0.5%; Mn = 0.5–0.8%; Si = 0.10,4%; P = max. 0.4%; S = max. 0.4%; Cr = max. 0.3%; Ni = max. 0.3%; Cu = max. 0.3%; Fe = bal.). Before the cavitation tests, surfaces of samples were grinded and polished with 1 μm oxide suspension.

Cavitation erosion tests were carried out for different coatings by using the stationary specimen method with a vibratory apparatus, shown in Figure 2. Tests were conducted in accordance with the procedure described in ASTM G32 standard [2]. An acoustic power output equaled 500 W. The distance between the horn tip and the sample was 1.5 mm. The resonance frequency of the oscillator was 20 kHz and the amplitude of the tip was 50 μm. Deionized water was used as test liquid, which was kept at 25±4°C with a temperature

Table 1. Nominal chemical composition of powders

Powder	Nominal chemical composition, weight %
PMNi50p	Ni = 87%, Cr = 6%, Fe = 2.1%, Si = 3.5%, B = 1.8%, C = 0.1%
PMCo45	Co = 63.2%, Cr = 30%, W = 4%, Si = 1.3%, B = 0.8%, C = 0.7%

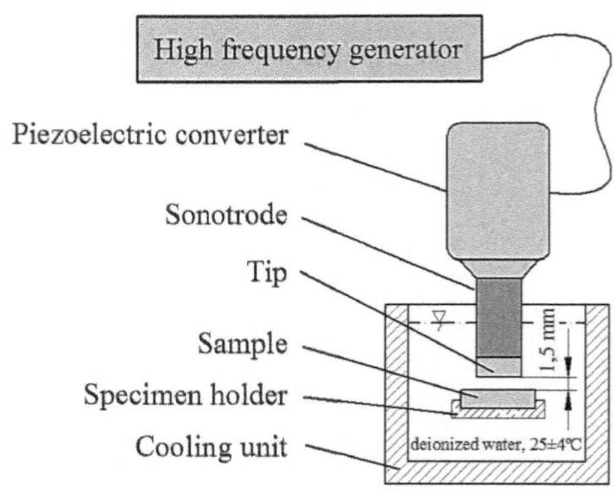

Fig. 2. Schematic diagram of the ultrasonic equipment

Table 2. Chemical composition of coatings

Coating number	Chemical composition, weight %							
	Co	Ni	Cr	W	Fe	Si	C	B
1	–	87.00	6.00	–	2.10	3.50	0.10	1.80
2	3.84	48.45	14.12	0.22	29.88	2.27	0.15	0.97
3	10.97	38.12	14.05	0.73	33.22	1.74	0.22	0.86
4	24.91	5.64	11.52	3.10	51.16	2.84	0.38	0.41
5	27.09	38.18	10.28	1.89	18.21	2.55	0.40	1.35
6	32.12	42.58	18.98	2.15	–	2.46	0.39	1.27
7	33.26	41.06	19.40	2.22	–	2.36	0.40	1.25
8	34.76	39.30	20.01	2.27	–	1.96	0.42	1.23
9	44.21	7.39	24.47	2.93	17.58	2.17	0.50	0.66
10	48.67	20.66	25.80	3.25	–	–	0.54	1.00

control device. Total test time for each sample was equal to 120 minutes (tests were termineted after 2 hours because of the detachment of some coatings). After each test step, which last 30 minutes, the specimen was weighed by means of the analytical balance with the accuracy of 0.2 mg to determine mass loss.

A quantitative assessment of cavitation erosion resistance of samples based on the evaluation of mean depth of erosion (MDE) parameter, in accordance with [2]. MDE is calculated by dividing the measured mass loss by the product of material density and the tested area. To assess qualitatively cavitation erosion and to study the mechanism of cavitation erosion, SEM microscope VEGA 3XMH TESCAN as well as light optical microscopes Nikon MA200 and Nikon SMZ1500 were used.

DESCRIPTION OF RESULTS

Typical microstructures of Ni-Co based coatings are illustrated in Figure 3. According to literature [13–15], at first stage of solidification dendrites are formed from the liquid phase at the boundary substrate/coating (Figure 3A), and then eutectics solidifies. The basic microstructure of coatings contains dendrites (γ phase) and interdendritic eutectic regions (dark area). The eutectic regions consisted of γ phase – boride eutectics as well as other eutectics. The γ phase is composed of alloying austenite based on cobalt which incorporated elements such as chromium, nickel, iron, tungsten and molybdenum.

Diagram with the MDE parameters of coatings and the reference sample is presented in Figure 4. The highest values of MDE parameter were for coatings 6 and 1, whereas the lowest ones were for coatings 4 and 9. Ni-Co based coatings compared to reference mild steel sample reveal higher cavitation erosion resistance (Figure 4) which is in agreement with literature, e.g. [6, 16].

In the works [16–18] it was stated that the presence of Co in metal matrix lowered the stacking-fault energy, resulting in planar slip, high strain-hardening and enhanced strain-induced martensitic transformation, all of which favored an increase of resistance to cavitation erosion. In the presented work no evident relationship between Co content in matrix of coatings and cavitation erosion resistance was found (Figure 5). Moreover, some influence of nickel on cavitation erosion wear was observed. Ni-Co based coatings (4 and 9) that contain the smallest amount of nickel (according to Table 2) exhibit lower values of MDE parameter and relatively higher cavitation erosion resistance than nickel rich coatings (Figure 5).

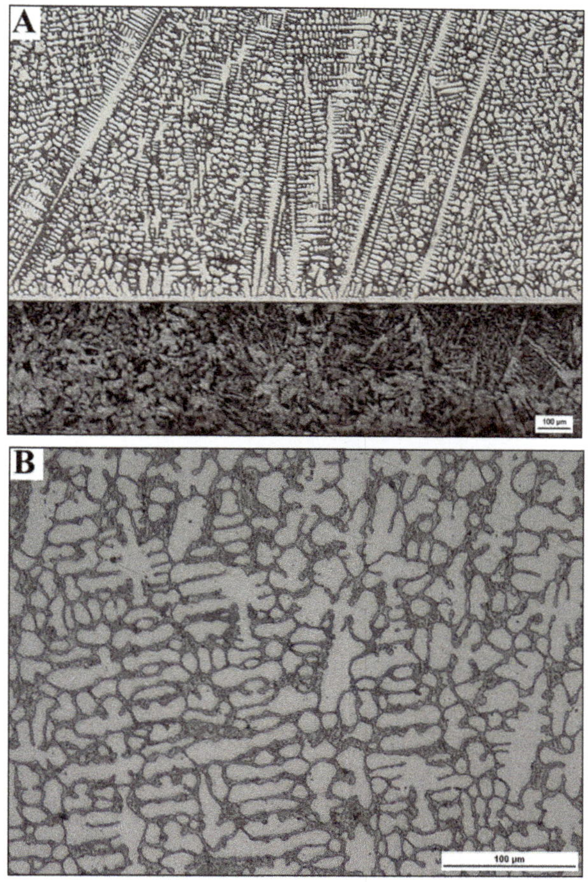

Fig. 3. LOM micrographs. The cross-section of sample 3 (A) and microstructure of coating 9 (B)

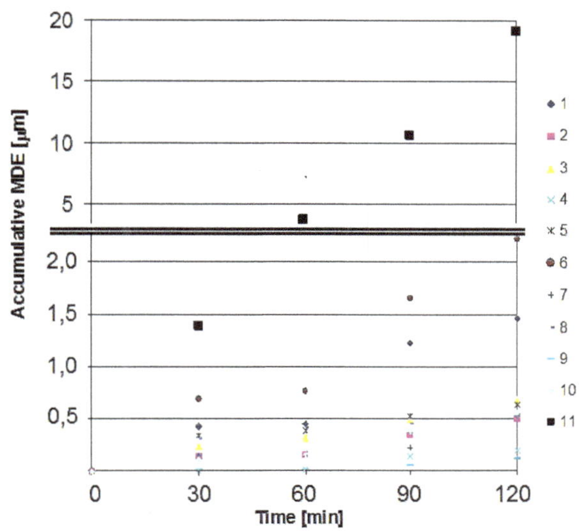

Fig. 4. Cavitation erosion resistance of coatings (1–10) and reference sample (11)

After 120 minutes of test, coatings 3, 1 and 6 show relatively high values of MDE parameter equal 0.69; 1.49 and 2.22, respectively. Their cavitation wear mechanism was mainly the detachment of the whole coating grains. Coating 6 surface with pits after 30 minutes of cavitation is presented in Figure 6A. After 120 minutes of cavitation these worn regions increased and produced higher mass loss, Figure 6B. In contrast, coating 5 after 30 minutes exhibits relatively low value of MDE parameter, 0.34. However after 120 minutes of cavitation the matt-surface area increased and small pits occurrence was observed (Figure 7). Matt-surface appearance is a typical

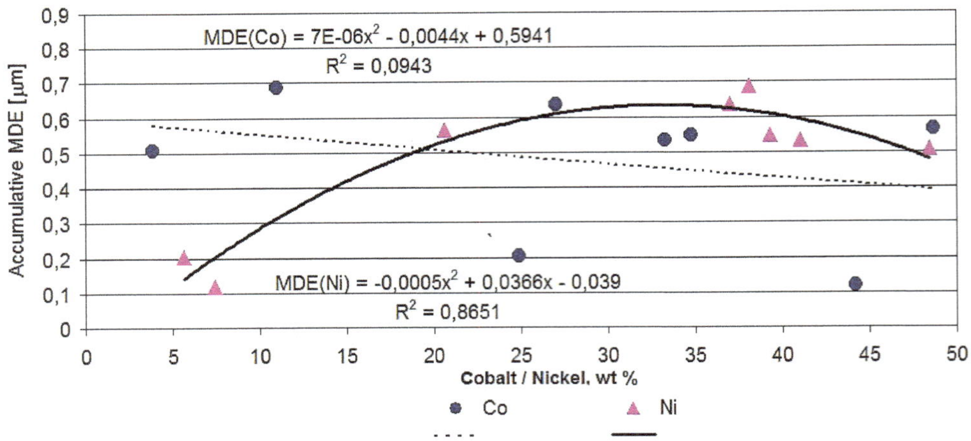

Fig. 5. Accumulative MDE parameters of coatings 2–5, 7–10, after 120 minutes of cavitation, with relation to their main matrix elements

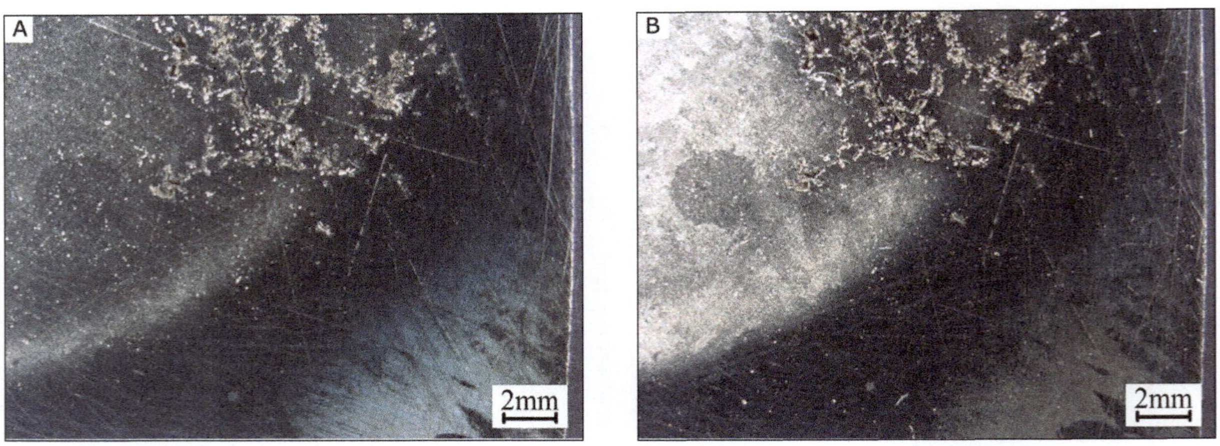

Fig. 6. Coating 6: A – after 30 minutes and B – after 120 minutes of cavitation test

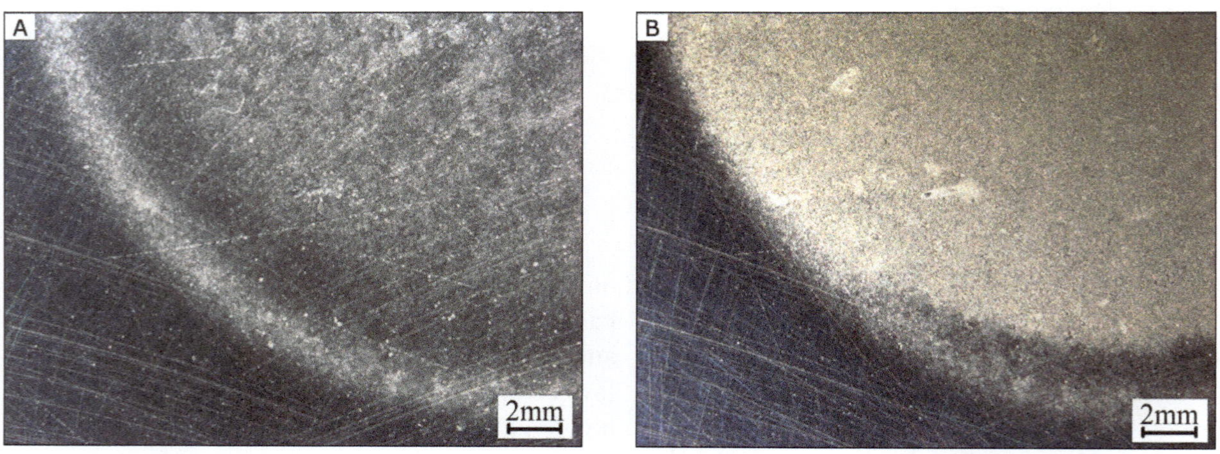

Fig. 7. Coating 5: A – after 30 minutes and B – after 120 minutes of cavitation test

phenomenon observed for polished surface of metals and alloys in incubation period of cavitation wear. Some coatings reveal matt-areas and very little mass loss even after 120 minutes of cavitation (coatings 9 and 4; MDE = 0.12 and 0.20, respectively). The genesis of matt-effect is a plastic deformation of surface and it is clarified in literature [19, 20]. In Figure 8A, the areas of cavitation damage of coating 3 are clearly visible. Figure 8B presents matt-surface area of coating 2. Surface particle detachments may be accelerated by coating non-uniform structure (occurrence of cracks and pores). The presence of pores in coatings was revealed in microscopic examinations. Surface areas with inhomogenities are less resistant to brittle fracture and to plastic shearing caused by a repeated collapse of cavitational bubbles [16].

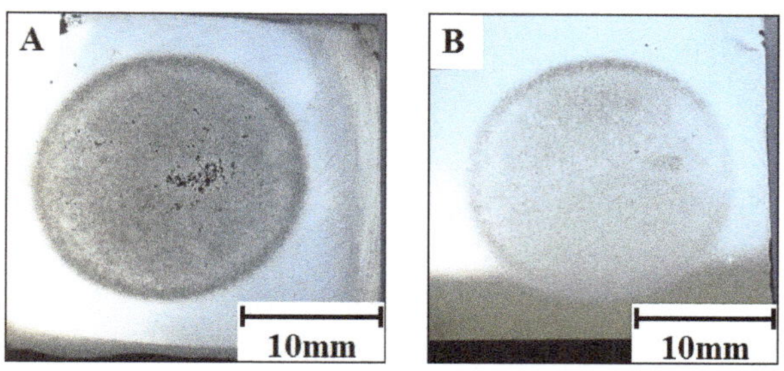

Fig. 8. Specimen surface after 120 minutes of cavitation: A – coating 3; B – coating 2

The results of the microhardness measurements are given in Table 3. Some of researchers [11, 21, 22] assert that mechanical properties such as hardness influence the cavitation resistance of coatings. The range of microhardness (Table 3) confirmed the presence of hard phases (carbides, borides and silicides) in austenitic matrix. However, the influence of hard phases presence in coating microstructure on cavitation erosion resistance after 120 minutes was not recognized. In our investigation, relationship between microhardness and MDE parameter was not found, which is presented in Figure 9. This is rough agreement with previous work [19] in which, after 90 minutes of cavitation no effect of C or B content on cavitation resistance of Ni-Co coatings was found.

Table 3. Results of microhardness measurements

Coating	Microhardness, HV 0.05	
	Average	Range
1	791	380 – 1356
2	946	535 – 1357
3	660	471 – 954
4	1015	855 – 1261
5	606	447 – 885
6	500	357 – 771
7	839	575 – 1274
8	935	588 – 1357
9	676	498 – 954
10	539	413 – 885

Micrographs of the less (Figure 10) and the most worn (Figure 11) surfaces of coatings at different magnifications are presented. Generally cavitation pits arose close to the regions of surface defects such as micro-scratches (Figure 10A and B), also at the interfaces between hard and soft phases. Thus in our case the mechanism of erosion of Ni-Co based coatings depends on cracking that starts at the borders of dendrites with interdendritic eutectics which incorporated hard phases. Some coatings demonstrate advanced cavitation (Figure 11) so in that case arrays of dendrites are clearly visible. The observed mechanism of cavitation erosion of the Ni-Co based coatings was similar to the already published in the literature for Co-, Ni- and Fe- based coatings [11, 22, 23]. Hard phases are the areas which act as initiating site for erosion damage, whereas the matrix was

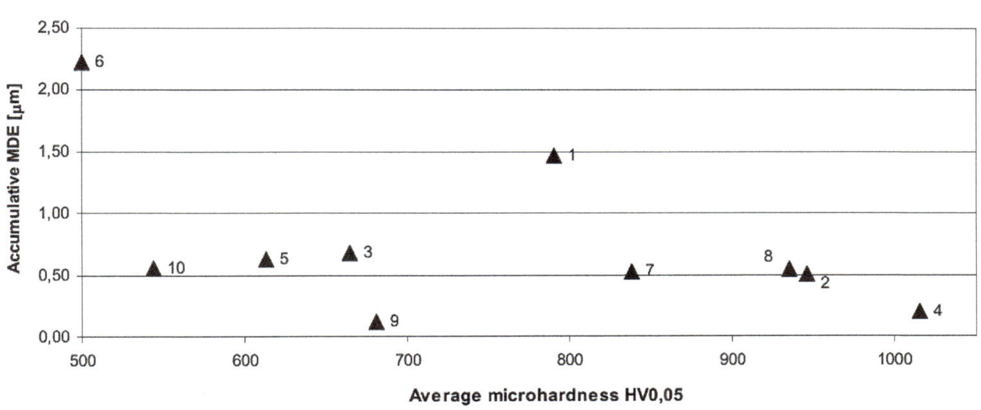

Fig. 9. Accumulative MDE after 120 minutes of cavitation versus the range of microhardness of coatings (1-10)

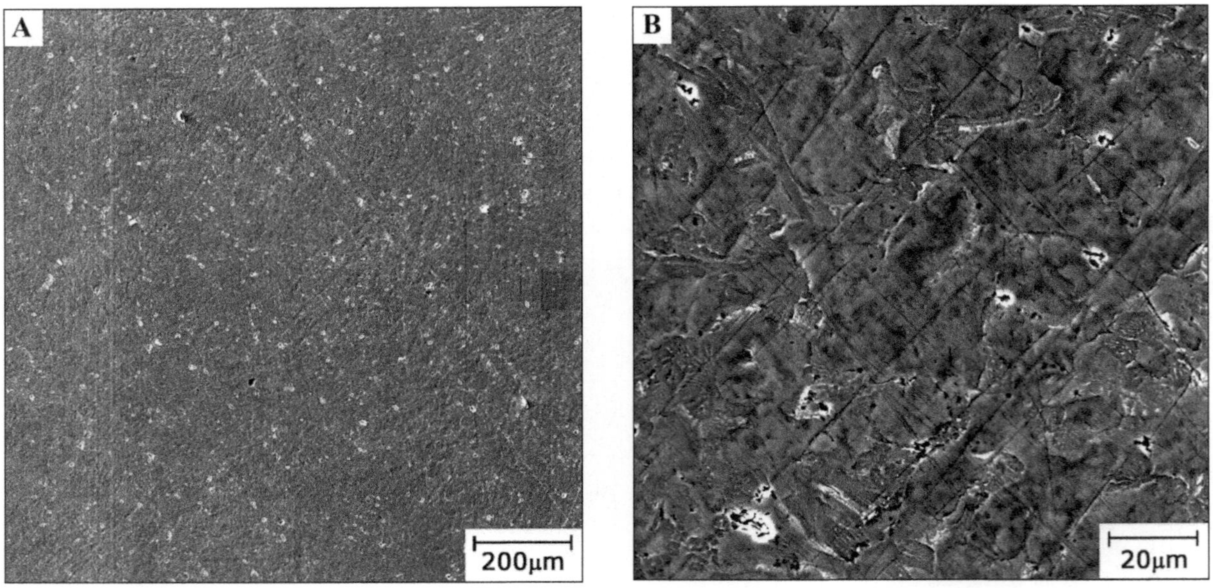

Fig. 10. SEM micrographs of coating 9, after 120 minutes of cavitation test

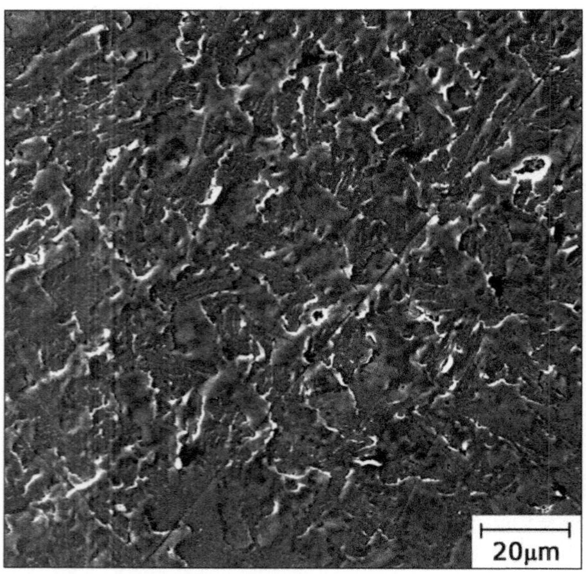

Fig. 11. SEM micrograph of coating 6 after 120 minutes of cavitation test

plastically strained which could be accompanied by strain-induced transformation.

After 2 hours of test, cavitation erosion wear of some coatings (9, 4) may be classified as incubation period of the cavitation erosion [1, 4, 16]. This period of erosion is characterized by a very little mass loss. Coatings 1, 3, 6 exhibit high mass losses. They could experienced accelerated period of erosion [1, 4, 16]. After 2 hours of cavitation, the reference sample reveals considerably higher value of MDE parameter compared with Ni-Co coating (Figure 4). Cavitation wear of reference carbon steel may by allocated to a stationary period of cavitation erosion wear [1, 4, 16].

CONCLUSIONS

Ni-Co based coatings exhibit nearly 10 times higher resistance to cavitation erosion than reference sample made of mild steel, therefore, they could be suitable to protect and repair machine parts applied under conditions of cavitation erosion.

The main mass removal mechanism observed in Ni-Co overlay coatings in cavitation tests was cracking and spalling of phases located at the boundary of dendrites and eutectics.

The influence of microhardness on cavitation erosion resistance of coatings was not stated. Ni-Co based coatings that contain relatively high percentage of nickel in their matrix exhibit relatively lower cavitation erosion resistance.

The presence of inhomogenities at the coating surface reduced the resistance of coatings to cavitation erosion.

REFERENCES

1. Franc J.-P. , Michel J.-M., Fundamentals of Cavitation. Softcover reprint of hardcover 2004, Springer, 2011.
2. Standard Test Method for Cavitation Erosion using Vibratory Apparatus. ASTM Standard G32-98, ASTM International, 1998.
3. Bagieński J., Kawitacja w urządzeniach wodociągowych i ciepłowniczych, Wydaw. Politechniki Poznańskiej, 1998.
4. Steller K., O mechanizmie niszczenia materiałów podczas kawitacji. IMP PAN, 1983.

5. Szala M., Rafalski K., Erozja kawitacyjna wybranych materiałów konstrukcyjnych, in: J. Caban, M. Szala (Eds.), Drugi Lubelski Kongres Studenckich Kół Naukowych TYGIEL 2010: czyli jak rozwijać naukę?, Biblioteka Politechniki Lubelskiej, 2011, 230–240.

6. Karassik I.J., Pump handbook. McGraw-Hill, New York 2001.

7. Klimpel A., Napawanie i natryskiwanie cieplne: technologie, WNT, Warszawa 2009.

8. Kumar A., Boy J., Zatorski R., Stephenson L.D., Thermal spray and weld repair alloys for the repair of cavitation damage in turbines and pumps: A technical note, 2005, 177–182.

9. Hejwowski T., Nowoczesne powłoki nakładane cieplnie odporne na zużycie ścierne i erozyjne. Biblioteka Politechniki Lubelskiej, 2013.

10. Espitia L.A., Toro A., Cavitation resistance, microstructure and surface topography of materials used for hydraulic components. Tribology International, 43, 2010, 2037–2045.

11. Hattori S., Mikami N., Cavitation erosion resistance of stellite alloy weld overlays. Wear, 267, 2009, 1954–1960.

12. Santa J.F., Espitia L.A., Blanco J.A., Romo S.A., Toro A., Slurry and cavitation erosion resistance of thermal spray coatings. Wear, 267, 2009, 160–167.

13. Tasak E., Metalurgia spawania. Wydawnictwo JAK Andrzej Choczewski, Kraków 2008.

14. Blicharski M., Zmiany mikrostruktury w połączeniach spawanych różnoimiennych materiałów stosowanych w energetyce. Przegląd Spawalnictwa, 85, 2013, 2–13.

15. Madadi F., Shamanian M., Ashrafizadeh F., Effect of pulse current on microstructure and wear resistance of Stellite6/tungsten carbide claddings produced by tungsten inert gas process. Surface and Coatings Technology, 205, 2011, 4320–4328.

16. Szkodo M., Erozja kawitacyjna materiałów konstrukcyjnych metalowych. Wydawn. Politechniki Gdańskiej, 2008.

17. Feller H.G., Kharrazi Y., Cavitation erosion of metals and alloys. Wear, 93, 1984, 249–260.

18. Heathcock C.J., Ball A., Protheroe B.E., Cavitation erosion of cobalt-based Stellite® alloys, cemented carbides and surface-treated low alloy steels. Wear, 74, 1981, 11–26.

19. Hejwowski T., Szala M., Cavitation resistance of plasma Ni-Co based overlay coatings. In: A. Świc, J. Lipski (Eds.), Systemy technologiczne w inżynierii produkcji, Biblioteka Politechniki Lubelskiej, 2013, 126–138.

20. Szkodo M., Image analysis to define cavitation properties of eroded materials. Problemy Eksploatacji. 1, 2006, 225–231.

21. Sang K., Li Y., Cavitation erosion of flame spray weld coating of nickel-base alloy powder. Wear, 189, 1995, 20–24.

22. Kwok C., Cheng F., Man H., Laser surface modification of UNS S31603 stainless steel. Part II: Cavitation erosion characteristics. Materials Science and Engineering: A, 290, 2000, 74–88.

23. Kim J.H., Na K.S., Kim G.G., Oh J.Y., Yoon C.S., Kim S.J., The effects of Mn and B on the cavitation erosion resistance of austenitic Fe-base hardfacing alloys. Mater. Sci. and Engin.: A, 477, 2008, 204–207.

OPTIMIZATION OF THE HEAT TREATMENT PROCESS OF A STEEL POROUS CHARGE USING AN INTEGRATED MODELLING

Rafał Wyczółkowski[1], Agnieszka Benduch[1]

[1] Department of Industrial Furnaces and Environmental Protection, Częstochowa University of Technology, Al. Armii Krajowej 19, 42-200 Częstochowa, Poland, e-mail: abenduch@wip.pcz.pl

ABSTRACT

The paper discusses the structure and principle of operation of programs for integrated modelling of the processes of heat treatment of porous-structure steel charges, such as long product bundles or strip or wire coils. Consideration is given to the specificity of these models in respect to porous charges. This is associated with their untypical thermal properties, which are expressed using the concept of effective thermal conductivity.

Keywords: porous charge, heat treatment, integrated modelling, effective thermal conductivity.

INTRODUCTION

Technological operations involving the heat treatment of steel products may significantly influence the efficiency and energy intensity of production, the emission of pollutants and the quality of finished products [1, 2]. Therefore, efforts are being made to achieve the capability to control the technological parameters of the process of a particular heat treatment operation in a precise and, at the same time, continuous manner. The main difficulty in such control is the lack of opportunity to directly monitor changes in the microstructure of material being treated. Those changes are determined indirectly, based on the data from temperature measurements taken at selected points in the furnace. This approach is quite inconvenient, because thus measured temperature may considerably deviate from the actual temperature of the charge being either heated or cooled.

An extremely attractive alternative in this respect is mathematical modelling. Computer simulation of heat treatment processes using numerical methods is a relatively cheap and effective means of their optimization. The development of information technology and, above all, the increase in computing power, offers a great potential to these methods [3]. In the simplest case, a mathematical model of a process can apply to thermal problems, being limited to computing only the temperature field in an element being treated. Models of this type have been very successfully used in the metallurgical processes for over three decades [4, 5]. The most developed in this respect are integrated models, which, in addition to the temperature, determine also the microstructure and mechanical properties of an element being heated [6, 7, 8]. Computation results yielded by these models provide primarily the knowledge of the current mechanical properties of the treated elements. Owing to this, it is possible to effectively control the technological process by determining the optimal values of its parameters by computational means [9].

THE SPECIFICITY OF HEATING A POROUS CHARGE

In numerous cases of heat treatment of steel elements, the charge introduced to the furnace is characterized by a porous structure. It can be divided into two basic categories: coils and bundles

[10]. In the form of coils, sheet, strip, wire rod and wire are heated. Whereas, different types of long elements, such as bars or tubes, are heated in bundles. Porous charges are distinguished by the existence of a steel solid phase skeleton, within which voids (pores) filled with gas occur. In coils, this skeleton is made up of wire or strip turns, while in bundles, individual long products. Another important feature of these elements is lack of the continuity of the solid phase in specific directions. For this reason, the charges under discussion can be regarded as porous media with a grainy structure.

Due to the structure of porous charges described above, a complex heat transfer occurs within them during heating. For example, the following can be distinguished in a bundle of bars: conduction in the bar cross-section, conduction in the gaseous phase region, contact conduction in the bar contact location, and radiation between bar surfaces. For this reason, the ability of porous charges to transfer heat is an effective quantity and it is therefore termed the effective thermal conductivity [11].

It should be emphasized that, in contrast to the thermal conductivity l of solid materials, the effective conductivity is not a material's property. Due to the lack of the continuity of porous charge's solid phase and the presence of pores filled with gas, the values of the coefficient l_{ef} of those elements are much lower than the thermal conductivity of the steel alone. This is highlighted by the diagrams in Figure 1, representing the effective thermal conductivity measurement results obtained for flat beds made up of round bars of low-carbon steel with a carbon content of 0.2% C. From the point of view of charge structure, these beds correspond to bundles of bars. The data in these diagrams are expressed in the form of the reduced K_{ef}, as the quotient of the effective thermal conductivity l_{ef} of samples by the thermal conductivity l of their metallic phase. As can be seen, the effective conductivity of these beds is smaller by approximately an order of magnitudes than the thermal conductivity of steel. What is characteristic is that it increases with the increase in the lateral dimension of the bars. The observed relationships indicate also that, in contrast to the conductivity of steel, the l_{ef} coefficient increases as a function of temperature. This can be explained by the intensity of thermal radiation in charge pores increasing with temperature and by the increase in the thermal conductivity of the gaseous phase.

Therefore, relatively small values of thermal conductivity cause considerable temperature differences to occur in the cross-section of those bundles during heating. This is an unfavourable phenomenon, as it contributes to a considerable spatial diversification of the microstructure, and thus also the mechanical properties. To eliminate this undesirable effect, soaking operations are used. The correct definition of the duration of those operations for a given type of porous charge is a complex and difficult problem. Very helpful to this end is integrated modelling, whereby the heat treatment parameters, including also the soaking duration, are determined based on the required final mechanical properties of the material to be treated.

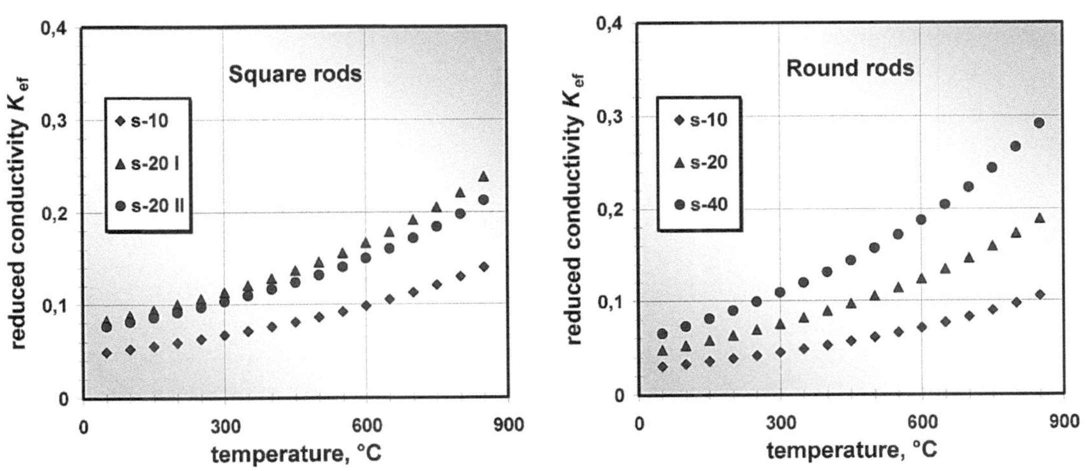

Fig. 1. The results of measurements of the effective thermal conductivity of beds of round and square bars in a reduced form – the numerical value in the sample symbol denotes the lateral dimension of bars in mm

THE GENERAL STRUCTURE OF INTEGRATED MODELS

The numerical algorithms of integrated models rely on the equations of heat transfer, the kinetics of recrystallization and grain growth, and on the correlations between the microstructure of a given material and its mechanical properties [6-8, 12]. In these models, based on input data such as material grade, charge dimensions, furnace geometry and operation parameters, and planned heat treatment cycles, the time and spatial distribution of temperature within the charge is determined in the first place. Then, based on the information on the formation of the temperature field, the grain size distribution and mechanical properties of the material are estimated.

In view of the described procedure, three separate modules can be distinguished within integrated models: the thermal, the microstructural transformation, and the mechanical properties modules. In the thermal module, calculations of the time and spatial temperature changes within the charge are made. The results of these calculations are then used as input data for the microstructural transformation module, in which the sizes of grains formed during heating of metal are determined. The knowledge of the microstructure of the material enables its mechanical properties to be determined in the last module.

THE CONFIGURATION AND PRINCIPLE OF OPERATION OF THE THERMAL MODULE

The design of the thermal module must allow for thermal interactions occurring between the charge and various elements of the furnace, which are associated with the complex mechanisms of conduction, convection and radiation. Information on changes in the temperature field within the charge is obtained by solving the Fourier-Kirchhoff (F-K) heat equation [13]. The form of this equation depends on the geometry of charge being heated. As the porous charge has most commonly a cylindrical shape (coils being regarded as rings, while bundles as solid cylinders), thermal modules use the F-K equation as written for the cylindrical coordinate system.

Obtaining the analytical solution of the F-K equation is only possible for identical uniqueness conditions [13]. For processes showing variability of these conditions, the solution can only be achieved using numerical methods. One of the methods used for the calculation of the temperature field of heated charge in computerized implementations is the finite difference method [14]. In this method, the starting differential heat equation is expressed in the form of finite differences, which involves the division of the region under analysis into elementary parts using a difference division grid [3, 13, 15]. Substituting the forward difference quotient for the time derivative yields a system of equations in which only one unknown occurs, which is temperature in a given difference grid node i for the successive time step $k + 1$, denoted by the symbol $T_{i,\,k+1}$. The obtained system of equations needs to be complemented with boundary and initial conditions. Their form depends on the technological conditions being associated, e.g., with the type and design of the furnace used.

In modelling of porous charge heating, it is crucial to determine the value of the l_{ef} coefficient, as one of the uniqueness conditions. These issues have not been sufficiently extensively covered so far in the literature on the subject [16]. The most appropriate solution of this problem is to incorporate a suitable numerical procedure for determining the current value of the l_{ef} coefficient in the thermal model. These computations are performed by analyzing the thermal resistance R_t of an elementary cell isolated from the charge concerned. This approach takes advantage of the analogy between the electric current flow and the heat transfer phenomena. The effective conductivity searched for is calculated as the quotient of the resistance R_t to the characteristic dimension of the cell in the heat transfer direction. This problem is discussed in a greater detail in study [17] on the example of several types of porous charge.

THE MICROSTRUCTURAL TRANSFORMATION AND MECHANICAL PROPERTY MODULES

This module uses primarily the Avrami equation that describes the variations in the fraction of recrystallized grains X as a function of time [12]:

$$X = 1 - \exp(-K_{re}\tau^{n_{re}}) \tag{1}$$

where: K_{re} – a temperature-dependent constant,
n_{re} – the Avrami exponent.

The values of these parameters for a specific steel grade are determined based on experimental tests that involve the determination of the value of the X parameter in isothermally soaked samples for different time intervals, ranging from several to several hundred minutes.

After transforming Equation (1) into the form of $\ln[\ln(1/(1-X))] = \ln K_{re} + n_{re} \ln t$, the values of the exponent n_{re} and the constant K_{re} are determined based on the analysis of the graph, in which the experimental data are approximated with a linear relationship (Figure 2a) [12]. The values of the parameter $\ln K_{re}$ are the absolute terms, whereas the exponent n_{re} corresponds to the slopes of the lines.

Thus determined K_{re} constants are a function of the crystal nucleation and growth rate, as described by the relationship [12, 19]:

$$K_{re}^{\frac{1}{n_{re}}} = k_0 \exp\left(-\frac{Q_{re}}{RT}\right) \quad (2)$$

where: Q_{re} – the activation energy characteristic of the process of transformations of the crystalline structure in question.

The values of this parameter are determined based on the analysis of the experimental data summarized in the form of the diagram shown in Figure 2b [12].

In the same way, the parameters describing the grain growth kinetics are determined. To this end, grain size measurements should be taken on samples isothermally soaked at different temperatures. The mean grain growth rate is described by the relationship [12]:

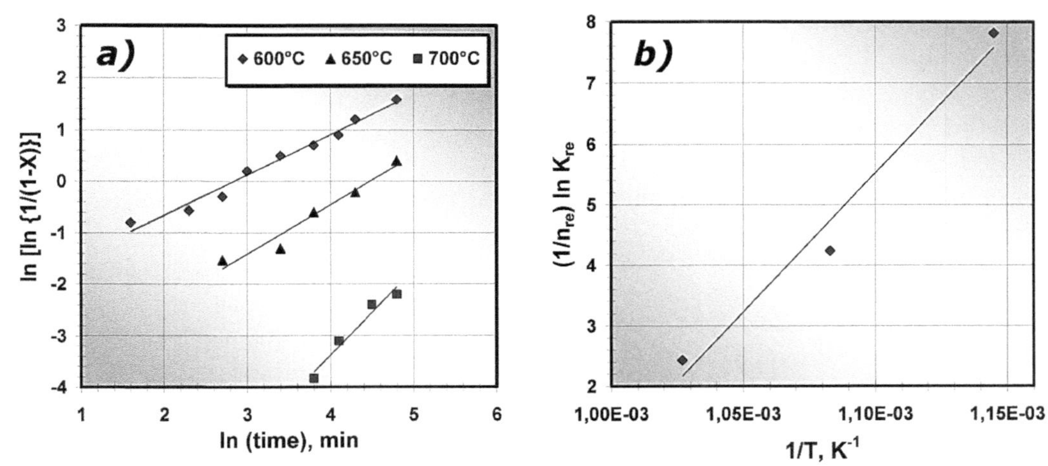

Fig. 2. a) the graph of the parameter X as a function of the logarithm of time, b) the graph for the determination of the activation energy for the recrystallization process [12]

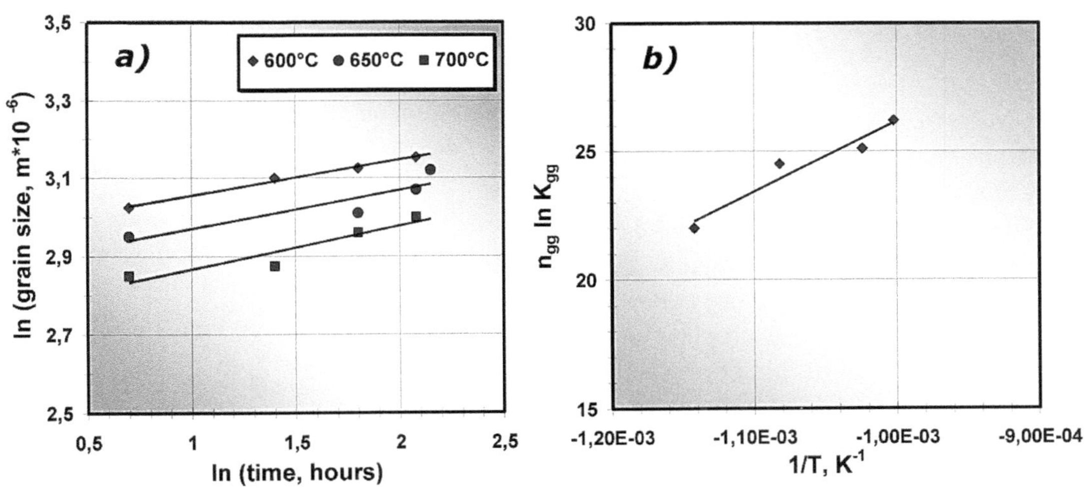

Fig. 3. a) the graph of the grain size d_{gg} as a function of the logarithm of time, b) the graph for the determination of the activation energy for the grain growth process [12]

$$d_{gg} = K_{gg}\tau^{\frac{1}{n_{gg}}} \tag{3}$$

where: d_{gg} – the mean metal grain size after the time t.

The values of the coefficient K_{gg} and the exponent n_{gg} are determined from the logarithmic graphs representing the relationship of grain size as a function of annealing duration (Figure 3a). The value of the constant K_{gg} is described by the relationship [12]:

$$K_{gg}^{n_{gg}} = k_0 \exp\left(-Q_{gg}/(RT)\right) \tag{4}$$

The value of grain growth activation energy Q_{gg}, occurring in Equation (4), as illustrated in Figure 3b, is obtained from the analysis of the graph representing the relationship of n_{gg} ln K_{gg} as a function of the inverse of absolute temperature [12].

After determining the grain size value, further computations are performed in the mechanical property module. First of all, the yield strength of the material is determined. For the determination of this parameter, the Hall-Petch relationship is used, which describes the effect of the mean grain size d_{gg} on the value of the lower yield stress σ_d [18]:

$$\sigma_d = \sigma_0 + k \cdot d_{gg}^{-0,5} \tag{5}$$

where: σ_0, k are constants, whose values for a specific steel grade are determined based on experimental tests.

Similar procedures are used for determining the tensile strength.

CONCLUSIONS

Integrated heat treatment process models, operating according to the description given above, are successfully used on industrial lines for various types of porous charge. They are most often employed for controlling the annealing of sheet coils in bell-type furnaces [6, 19] and bundles of bars in continuous furnaces [7, 8]. In view of their thermal properties, these elements require soaking. Integrated models make it possible to establish the optimal duration of this operation. Thanks to this, these tools contribute to the reduction of production costs by reducing energy consumption, increasing productivity, minimizing the costly industrial tests and improving the quality of manufactured products [19].

It should be emphasized that prior to implementing it for the control and monitoring of a specific industrial process, an integrated program requires careful validation of each of its component modules. The validation procedures are carried out based on the data obtained both from industrial trials and from laboratory tests [12, 19]. This issue, though very important from the point of view of the correctness of operation, has not been discussed in this paper for the reasons of its space.

REFERENCES

1. Mehta R., Sahay S.S.: Heat transfer mechanism and furnace productivity during coil annealing: aluminium vs. steel. Journal of Materials Engineering and Performance, Vol. 18(1), 2009, 8-15.
2. Rudnik S.: Physical metallurgy. PWN, Warszawa 1996 (in Polish).
3. Adrian H.: Numerical modeling heat treatment process. Wydawnictwa AGH, Kraków 2011 (in Polish).
4. Jaluria Y.: Numerical simulation of the transport process in a heat treatment furnace. International Journal for Numerical Methods in Engineering, 25, 1988, 387-399.
5. Rao T.R., Barth G.J., Miller J.R.: Computer model prediction of heating, soaking and cooking times in batch coil annealing. Iron and Steel Engineering, 60 (9), 1983, 22–33.
6. Sahay S.S., Kumar A.M.: Application of integrated batch annealing furnace simulator. Materials and Manufacturing Processes, Vol. 17 (4), 2002, 439–453.
7. Sahay S.S., Krishnan K.: Model based optimization of continuous annealing operation for bundle of packed rods. Ironmaking and Steelmaking, Vol. 34 (1), 2007, 89–94.
8. Sahay S., Mehta R., Krishnan K.: Genetic-algorithm-based optimization of an industrial age-hardening operation for packed bundles of aluminum rods. Materials and Manufacturing Processes, 22, 2007, 615–622.
9. Wyczółkowski R.: Integrated modeling heat treatment process of porous charge. In: A. Kawałek (ed.). Metallurgy, New technologies and achievements. Series: Monographs No 25, Częstochowa 2012, 40–50 (in Polish).
10. Wyczółkowski R.: Classification and characterization of porous charge found in the industrial heat treatment practice. Hutnik-Wiadomości Hutnicze, Vol. 12, 2012, 877–879 (in Polish).
11. Wyczółkowski R., Benduch A.: Effective thermal conductivity as primary thermal parameter

of porous charge. In: XII Międzynarodowa Konferencja Naukowa „Nowe technologie w metalurgii i inżynierii materiałowej. Monografie nr 15, Częstochowa 2011 (in Polish).

12. Sahay S., Kumar M., Chatterjee A.: Development of integrated model for batch annealing of cold rolled steels. Ironmaking and Steelmaking, 2004, Vol. 31 (2), 144–152.
13. Wiśniewski S., Wiśniewski S.T.: Heat transfer. WNT, Warszawa 2012 (in Polish).
14. Piechowicz Ł., Wyczółkowski R., Łukaszek D.: Mathematical model of thermal process on the basis of installation to annealing coils of aluminum sheet. Report from investigatory project order by SECO/WARWICK ThermAL S.A., Świebodzin 2012.
15. Gdula S.J. (ed.): Heat conduction. PWN, Warszawa 1984 (in Polish).
16. Wyczółkowski R.: Selected problems connected with hating up of steel charge of porous structure. In: Metalurgia. Nowe technologie i osiągnięcia, Częstochowa 2009, 337–359 (in Polish).
17. Wyczółkowski R.: Modeling of the thermal properties of porous charge by application of the elementary cell concept. In: Wybrane zagadnienia produkcji i zarządzania w przedsiębiorstwie. Seria: Monografie nr 29, Częstochowa 2012, 54–65.
18. Przybyłowicz K.: Physical metallurgy. PWN, Warszawa 2007 (in Polish).
19. Sahay S.S., Kumar A.M.: Applications of integrated batch annealing furnace simulator. Materials and Manufacturing Processes, Vol. 17(4), 2002, 439–453.

NEW CONCRETE ADDITIVE BY CHEMICAL RECYCLING OF PET

José Miguel Mendivil-Escalante[1], Jorge Luis Almaral-Sánchez[1], José Manuel Gómez-Soberón[2], Susana Paola Arredondo-Rea[1], Ramón Corral-Higuera[1], Andrés Castro-Beltrán[1], Francisca Guadalupe Cabrera-Covarrubias[3]

[1] Universidad Autónoma de Sinaloa, Facultad de Ingeniería Mochis, Los Mochis, Sinaloa, Mexico, e-mail: josemiguelmendivil@hotmail.com; jalmaral@gmail.com; susypao79@gmail.com; rmn1779@gmail.com; andrescb04@gmail.com

[2] Universidad Politécnica de Cataluña, Departamento de Construcciones II-EPSEB, Barcelona, Spain, e-mail: josemanuel.gomez@upc.edu

[3] Universidad Politécnica de Cataluña, Escuela de Ingenieros Caminos Canales y Puertos, Barcelona, Spain, e-mail: lupita_4418@hotmail.com

ABSTRACT

Due to the problem that represents the accelerated production of waste from the consumption of polyethylene terephthalate (PET), it becomes necessary to look for alternative solutions; chemical recycling is a suitable method for conversion into a material with potential application in concrete additive, such as unsaturated polyester resins; with the above improvements and conservation of non-renewable raw materials can reduce the environmental impact. This investigation was focused on obtaining the synthesis of a polymeric resin by the method of glycolysis, and their characterization by Infrared Spectroscopy (FT-IR); the results show that the recycled materials (PET and its processing) are similar to reagents made from primary materials so that it is considered feasible to use in the intended application.

Keywords: concrete admixtures, glycolysis, polymeric resins, polyethylene terephthalate.

INTRODUCTION

As a globalization in today's lifestyle, and looking for new adaptations in the environment, perception and the archetypes of society are adapting; among these, new consumer behaviors are generating waste with significant figures, especially in the domestic scope. Inadequate treatment of these, partly caused by the lack of awareness on correct production, consumption, waste and responsible use; and secondly, by little or no social system oriented environmental conservation, require actions to foster this waste minimization, incorporating in new products, where possible [1].

One of the residues with significant representation in our society, with a generation of 13 MT/year [7], is PET, which belongs to the group of polyesters called thermoplastics.

PET is a polymer derived from a combination of two monomers: dimethyl terephthalate (DMT) and ethylene glycol (EG) whose molecules are configured in long chains of organic elements, such as carbon (C), oxygen (O) and hydrogen (H); and which was patented for the purpose of manufacture of fibers, by Whinfield and Dickson (1941) [2]; this material has been in use with wide dissemination throughout the world since the 80's, mainly for bottling beverages [4]. The Society of the Plastics Industry (SPI) identifies the material with a logo of the three arrows in the center and the number one (100% recyclable) or the acronym PET (PETE English) in the bottom of containers made of this material [3].

The benefits of its various uses are countless, however, the material with a chemical contamination, identical to the original, at the end of PET use,; is difficult in degradation (between 100 and

1000 years), therefore, it is important to reintegrate PET into a new cycle of use [5]. Because of this, we are currently working on finding techniques for recycling these plastics as alternatives to decrease the harmful effects to the environment, and likewise also reducing consumption and extraction of non-renewable raw materials for your manufacturing [6, 7].

In 1977, PET was recycled and used as the basis for a new bottle; today much of European and American cities have a protocol for collection and recovery of recyclables; generally plastics being one of the most popular representative, which after sorting, crushing, smelting and manufacturing, can be re-incorporated as a new material [8]. After this process, waste PET may become modified to produce different materials; including unsaturated polyester resins [9] relevant to this work for the feasibility of implementation as an additive to the polymer-modified concrete, in order to improve its properties. As it is necessary to establish the process that allows such an application, this requires the study of thermal sequencing chemistry that enables generating the said additive, thus enhancing the industrial application in the construction and especially in the use of concrete.

MATERIALS

The materials used for the development of the experimental phase were:

1. Study base materials: PET bottles used: transparent color, origin of waste and beverage containers of different capacities. Their characteristics and properties are typical [10].
2. Chemical reagents: Sodium Hydroxide (NaOH), propylene glycol (PG), Zinc Acetate (AZ), Maleic Anhydride (MA), Adipic Acid (AA), Bis-(2-HidroxiPropilTereftalato) (BHPT), dimethyl terephthalate (DMT) and styrene (all with 99% purity of the Aldrich brand).

METHODOLOGY

Obtaining recycled PET flakes

As a preliminary step to the chemical recycling of PET requires the collection and segregation at colors and other solids present; then the bottle labels and rings were removed to avoid contamination to include other polymeric materials, other than the studied; after that a mechanical treatment for the reduction and homogenization of size was carried out, using a not standardized mill, with five HP power and equipped with a scissor cutting knife (the process was performed twice, to obtain small flakes, smaller than than ½″); in order to remove impurities, the flakes were washed with a solution of sodium hydroxide and 50% water; and finally, the resulting product was dried in an oven at 60 °C for 8 hours to remove any moisture present therein (prevent unwanted chemical reactions).

Process of Glycolysis

In accordance with the principles of sustainable development for the choice of the most appropriate recycling technique, chemical recycling is optimal [11]; it consists of complete or partial depolymerization of PET to degradation into its monomers or oligomers; its main virtue is that basic raw materials that could be re-used for the production of PET or other materials are obtained (such as polymeric resins). The process may represent a saving of up to 50–60% energy, when compared with manufacturing the same product, but on the basis of laboratory reagents [7, 12].

The chemical processes used are: methanolysis, hydrolysis and glycolysis, the first two are carried out under conditions of pressure and temperature higher than the third; and the hydrolysis can generate environmental problems due to acidic or basic conditions requiring process [3, 12], therefore, glycolysis was selected as the most suitable method for this research. It is based on the depolymerization of the components by chain scission of the polymer [13], which in the presence of a catalyst facilitates the fission of the glycol possessing free radicals (oxygen and glycol ester), thus causing a higher concentration of electron density and an attraction between the carbonyl oxygen and, thus arriving to create a single bond with partial loads. This occurs after an electronic rearrangement in which again forms a double C = O of ester, caused by the electrostatic attraction between the O and H gives glycol, and thus producing the monomers plus a small remnant portion of glycol [13].

The glycolysis process is carried out in a vessel type reactor (Fig. 1); which consists of the following elements: a glass cup double layer resistant to temperatures between -40 ° C to 200 ° C, a motor with a digital display for the control of agitation and a system for recirculating the fluid (silicone) with passageways allowing deposit and maintain temperature.

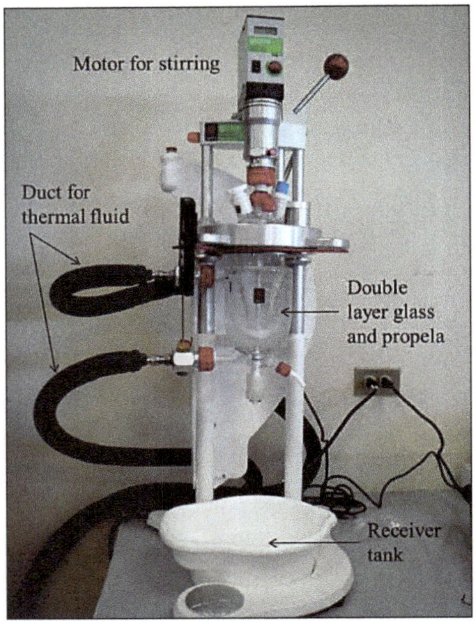

Fig. 1. Type reactor Vessel

The process involves adding PET, PG solvent (50% by weight of PET) and AZ catalyst (0.5% by weight of PET); the latter being essential in the process of degradation, which allows the breaking of the polymer chain.

Then a sequence of increases in temperature (25 °C every 20 minutes up to 200 °C) was performed, keeping in agitation (165 rpm) for eight hours. The material resulting from this process showed a viscous consistency and transparent gray color, which upon reaching room temperature passed into semi-solid and opaque gray (BHET-R).

Synthesis of the Resin

In the synthesis of the resin (R-R) BHET-R, AM and AA were used at a molar ratio of 1.1:0.5:0.5, these compounds were dosed by weight and were introduced them into the reactor; which followed the same programming sequence of temperature and duration as the previous process. The resulting product showed a viscous consistency and a gray color, which at room temperature became semi-solid and amber. Finally, a curing process, which consisted of dissolving the RR (40%) styrene (60%) until obtaining the material in liquid form, which is a desired material as a sustainable alternative application of PET in a densifying additive of the cementitious matrix of concrete.

To make a comparison between the obtained materials (BHET-R and R-R), comparable products were synthesized from primary reagents (BHPT and DMT). The synthesis procedures and required equipment for this new process were the same as the ones used for the preparation of the R-R; these new materials resulting from the primary reagents were appointed as BHET-P and R-P.

Characterization

For the comparative analysis of the polymeric materials created, Infrared Spectroscopy Fourier Transform (FT-IR) using spectrometer Nicolet 6700 was used, comprising an information processor and screen for viewing and manipulating the spectra. Solid samples were prepared by the compressed method with potassium bromide (KBr) [15], which comprises milling the polymeric material over the portion of KBr. The obtained material was introduced into metal capsules and was placed in the press, in order to make the tablet, and form a thin wafer; which is introduced into the spectrometer for evaluation, after calibration using reference test (blank test) only with a KBr pellet. The samples were analyzed after of 32 scans wavenumber in the range of 600–4000 cm^{-1} and a resolution of 4 cm^{-1}.

RESULTS

FT-IR Spectroscopy

The spectra of FT-IR were obtained in the studied materials: PET, BHET-R and RR.

In Figure 2-a (PET spectrum), the following bands are reported: OH glycol stretching, C = O of carbonyl, CH of aromatic ring, type bonds $-CH_2-$, at frequencies of 3539 cm^{-1}, 3435 cm^{-1}, 3098 cm^{-1}, 2958 cm^{-1} and 2885 cm^{-1}, respectively; bands of aromatic CH bending out of plane with a 1948 cm^{-1}; a peak at 1710 cm^{-1} corresponding to C = O stretching carbonyl 1458 cm^{-1} and link $-CH_2-$ type in-plane bending; bonds to 1100 and 1017 cm^{-1} asymmetric CO ester, 870 cm^{-1} CH out of plane and 743 cm^{-1} aromatic CH symmetric stretch. In Figures 2-b and 2-c, corresponding to BHPT-R and RR, can be observed very similar spectra to that of PET, since the same bands appear with no significant displacements; passes as

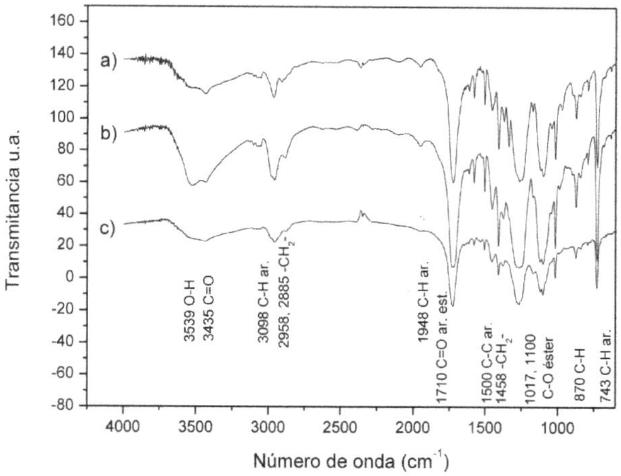

Fig. 2. Spectrum FT-IR: a) PET, b) BHET-R, c) R-R

well as (with reference to the frequency of 3500 cm^{-1}), which for the BHET-R peak is located at 3539 cm^{-1} being only sharper and more distant than that of PET.

Likewise, analyzing the polymeric materials obtained with the primary reagents (BHET-P and R-P), the Figure 3, indicates that the band at 3529 cm^{-1} corresponds to bonds of glycol O-H, the 3420 cm^{-1} is C=O of ester carbonyl, the 3055 cm^{-1} is C-H aromatic ring and the 2960 cm^{-1} and 2875 cm^{-1} is-CH_2- bonds, all the above of stretching vibration, likewise, the band 1955 cm^{-1} is aromatic C-H bending out of plane with the 1718 cm^{-1} C=O aromatic aliphatic ester carbonyl stretching a 1500 cm^{-1}, stretching of C-C on the aromatic ring, the 1450 cm $^{-1}$ -CH_2- bending in the plane, 1100 and 1017 cm^{-1} ester C-O bonds, the 873 cm^{-1} C-H out of plane, and 735 cm^{-1} symmetric stretch aromatic C-H . With respect to Figure 3-b (BHPT-P), has similar spectrum bands, with only small differences, as is the case of the 3098 cm^{-1} band of C-H and 2958 cm^{-1} -CH_2-, moving a negligible extent, and varying intensity.

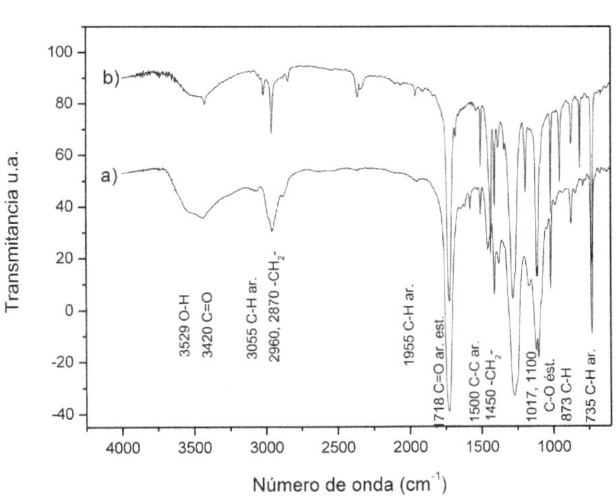

Fig. 3. .Spectrum FT-IR: a) R-P and b) BHET-P

When compared the spectra of BHET-P (Fig. 4a) and BHET-R (Fig. 4-b) the following small variations of bands: 3539 cm^{-1}, 3435 cm^{-1} and 3098 cm^{-1} (BHET-R) and 3529 cm^{-1}, 3420 cm^{-1} and 3055 cm^{-1} (BHET-P) were detected, on the other hand, also shows that the bands 2960 cm^{-1} and 2870 cm^{-1} are slightly wider to BHET-R. This must be interpreted as insignificant, given that their displacement and intensity are reduced and attributed to the change of the chemical environment of the reaction and small differences in their molecular weight.

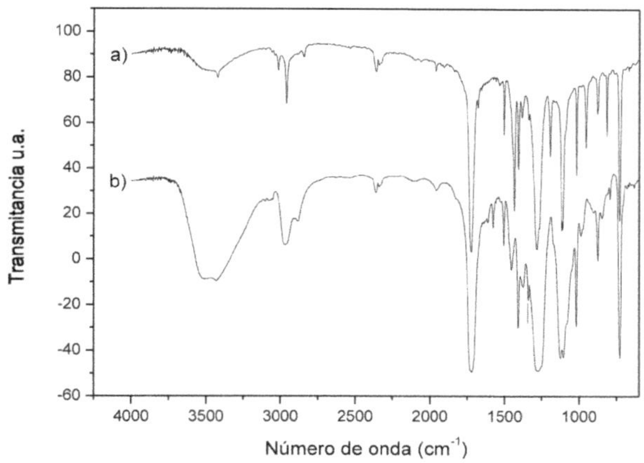

Fig. 4. Spectrum FT-IR: a) BHET-P and b) BHET-R

Comparing R-P (Fig. 5-a) and R-R (Fig. 5-b), one can see the following differences: for R-P the band at 1109 cm^{-1} is shifted to 1115 cm^{-1} of the R-R; while in the latter, a band 1955 cm^{-1} disappears. Furthermore, bands are also shown in 1115 cm^{-1} and 1090 cm^{-1} respectively, which are displaced between them slightly. The change is related to the fact that wider bands are shown in the higher molecular weight materials.

The FT-IR results shown above are consistent with those reported in similar studies by other authors [13–16].

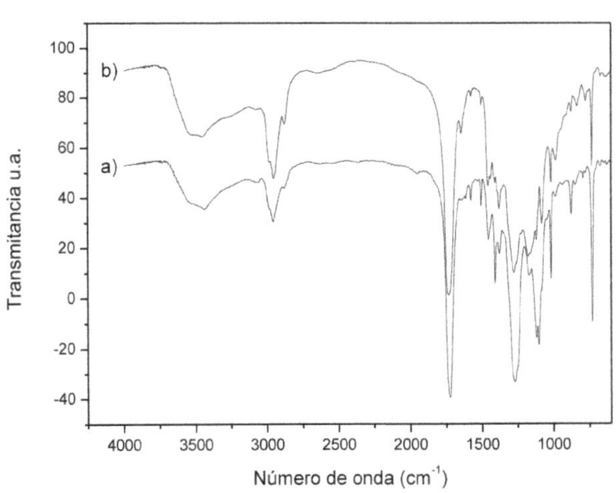

Fig. 5. Spectrum FT-IR: a) R-P and b) R-R

CONCLUSIONS

The features found in the FT-IR technique in the studied materials, both PET product origin and the primary reagents, revealed the same typical bands; and therefore, the trends are similar and well defined there.

By FT-IR technique is feasible define the molecular structure of each polymeric material from the chemical recycling, showed no significant difference in the reference material (primary reagents); therefore considered feasible the processing of the R-R for use as an additive for making CMP with sustainable uses. However, it will require further research in the mechanical and durability performance in these CMP.

Acknowledgment

The authors express their gratitude to the scholarship program for doctoral studies by CONACYT, the FIM-UAS and EPSEB-UPC.

REFERENCES

1. Lecitra M., Reducir, reutilizar y reciclar: el problema de los residuos sólidos urbanos. Grupo de Estudios Internacionales, 2010.
2. Berho M., Pisoni M., Recomendaciones para reciclar: una introducción al reciclado a gran escala de PET. Metodología Diseño Industrial Fadu-Uba, 2007.
3. Hurd D.J., Best Practices and Industry Standards in PET plastic recycling. Washington State Department of Community, Trade and Economic Development's Clean Washington Center. [En línea, 07-07-2014]. http://www.napcor.com/pdf/Master.pdf.
4. Abdel-Azim A., Unsaturated polyester resin from PET waste for polymer concrete. Polymer Engineering & Science, vol. 36, 1996, p. 2973.
5. Abdel-Azim A., Abdel-Azim I., Attia A., and El-Saids S.N., Characterization of some aliphatic unsaturated polyesters synthesized for gigantic castings. Polymer-Plastics Technology and Engineering, vol. 34, 1995, p. 79.
6. Aslan S., Immirzi B. and Laurienzo P., Unsaturated polyester resin from glycolyzed polyethylene terephthalate: Synthesis and comparison of properties and performance with virgin resin. Journal of Materials Science, vol., 32, 1997, p. 2329.
7. Paszun D. and Spychaj T. Chemical recycling of polyethylene terephthalate. Industrial & Engineering Chemistry Research, vol. 36, 1997, p. 1373,.
8. Reyes C., Estudio de factibilidad para la instalación de una planta recicladora de envases de PET. México, 2009, Tesis de grado (Ingeniería industrial), Instituto Politécnico Nacional.
9. Viksne A., Kalnins M., Rence L., and Berzina R., Unsaturated polyester resins based on PET waste products from glycolysis by ethylene, propylene, and diethylene glycols and their mixtures. Arabian Journal for Science and Engineering, vol. 27, 1C, 2002.
10. Awaja F. and Pavel D., Recycling of PET. European Polymer Journal, vol. 41, 2005, 1453–1477.
11. Abu-Bakar D.R., Ahmad I., Ramli A., Chemical recycling of PET from soft drink bottles to produce a thermosetting polyester resin. Malaysian Journal of Chemistry, vol. 8, 1, 2006, 22–26.
12. Achilias D.S. and Karayannidis G.P. The chemical recycling of PET in the framework of sustainable development. Water, Air and Soil Pollution: Focus, vol. 4, 2004, 385–396.
13. Carta D., Cao G., D'Angeli C., Chemical recycling of poly (ethylene terephthalate) (PET) by hydrolysis and glycolysis. Research Article-INCA, vol. 10, 6, 2003, 390–394.
14. Mehrabzadehl M., Shodjaei S.T. and Khosravi M., Chemical recycling of polyethylene terephthalate. Iranian Polymer Journal, vol. 9, 1, 2000.
15. Pickering W.F., Modern analytical chemistry. Ed. Reverté, 1980.
16. Karayannidis G.P., Achilias D.S., Sideridou I. and Bikiaris D.N., Chemical recycling of PET by glycolysis. 8th. International Conference on Environmental Science and Technology, Lemnos Island, 2003.

23

FLOW STRESS MODEL FOR COLD-FORMED 40HM CONSTRUCTIONAL STEEL

Anna Dziubińska[1], Andrzej Gontarz[1], Grzegorz Winiarski[1]

[1] Faculty of Mechanical Engineering, Lublin University of Technology, 36 Nadbystrzycka Str., 20-618 Lublin, Poland, e-mail: a.dziubinska@pollub.pl; a.gontarz@pollub.pl; g.winiarski@pollub.pl

ABSTRACT

The paper presents the results of research undertaken to investigate cold forming process for 40HM constructional steel suitable for heat treatment. In the first part of the paper, mechanical properties of this steel and its industrial applications are described. The second part of the paper presents the results of the analysis of flow curves for two kinds of steel specimens: those that were subjected to annealing and those that did not undergo any heat treatment. It was found that the application of heat treatment had a significant effect on improving the forming conditions for this steel at room temperature. The experimental flow curves obtained in a compression test were described by constitutive equations illustrating the dependence between flow stresses and strain value. In order to determine the equation coefficients, the Generalized Reduced Gradient method implemented in Microsoft Excel was used. Based on the obtained equations, a material model will be developed to perform numerical simulations of cold forming for 40HM steel, using FEM-based software that aids the design of metal forming technologies.

Keyword: flow stress, 42CrMo4 constructional steel, cold forming.

INTRODUCTION

The research was undertaken to investigate constructional steel type 40HM (42CrMo4), classified under the PN-EN 10083-1+A1:1999 norm. One of the machinery and alloy steels suitable for heat treatment, this steel type has a carbon content of approximately 0.3–0.5% C, which determines its strength properties, and contains alloy additions, such as Mn, Cr, Ni, Mo, Si that provide the steel with a particular hardening capacity [1]. 40HM steel is used to produce most machinery, vehicle and construction parts. The steel can be heat treated due to a high Re/Rm ratio that ensures high ductility, while the alloy additions ensure a better hardening capacity [2].

The minimum mechanical properties of the investigated steel after heat treatment are as follows: tensile strength Rm of 1000 MPa, yield strength R0.2 of 750 MPa, elongation A5 of 11%. The steel is used for producing machinery elements with very high strength and ductility as well as parts exposed to changing loads, such as axles, crankshafts, gears, disks, rotors, levers, pushers, connecting rods, fasteners, casing elements and gripping elements of tools. 40HM steel is suitable for hot forming. Among others, it is used in processes, such as forging, extrusion and rolling in the temperature range of 850–1050 °C [3].

The analysis of the specialist literature demonstrates that there are no studies providing information on flow curves for 40HM steel subjected to cold forming. Therefore, it is justified that such research be performed to investigate the suitability of this steel for cold forming. The present study offers the results of the investigation undertaken to determine the flow stress model for cold-formed 40HM steel, examining the effect of applied heat treatment on the properties of steel formed at room temperature. The results can find application in the design and analysis of cold forming processes for this steel type.

EXPERIMENTAL DETAIL AND RESULTS

The research objective was to determine flow curves for 40HM steel. To this aim, a compression test method was employed [4, 5]. The forming process for 40HM steel specimens with a diameter of 10 mm and length of 20 mm (Figure 1) was run with lubrication at room temperature using a hydraulic press, at a mean strain rate of. The chemical composition of the investigated steel is listed in Table 1.

Prior to the forming process, some specimens (5 items) were subjected to soft annealing at a temperature of 680 °C for 0.5 h and then furnace cooled until total cool-down [8–10]. The design of the forming process and the view of the specimens after compression are shown in Figure 2 and Figure 3.

During the experimental tests, the specimen compression time, pressure and tool motion were determined. Based on the results of the measurements made, the values of flow stresses and equivalent strains were calculated. The flow stresses as a function of strain for the specimens made from annealed steel and for those formed without heat treatment are shown in Figure 4.

The flow curves determined in the experiments were described by functions (constitutive equations), making the flow stress values dependent on strain parameters. In order to describe the data, the following function was selected:

$$\sigma_p = A \cdot \varepsilon^B \tag{1}$$

where: σ_p – denotes the flow stresses (MPa),
ε – denotes the equivalent strains,
A, B, C – denote the coefficients [6].

In order to determine the constant function coefficients, the Generalized Reduced Gradient (GRG2) method implemented in Microsoft Excel Solver was used. To this aim, the objective function described by the below equation was minimized:

$$\Phi_\sigma = \frac{1}{k}\sum_{i=1}^{k}\frac{(\sigma_{pt}-\sigma_{pe})^2}{\sigma_{pe}^2}\cdot 100\% \tag{2}$$

where:
σ_{pt} – denotes the value of flow stresses calculated on the basis of function (1),
σ_{pe} – denotes the experimental value of flow stresses,
k – denotes the number of measuring points [7].

The results of the conducted optimization calculations are listed in Table 2.

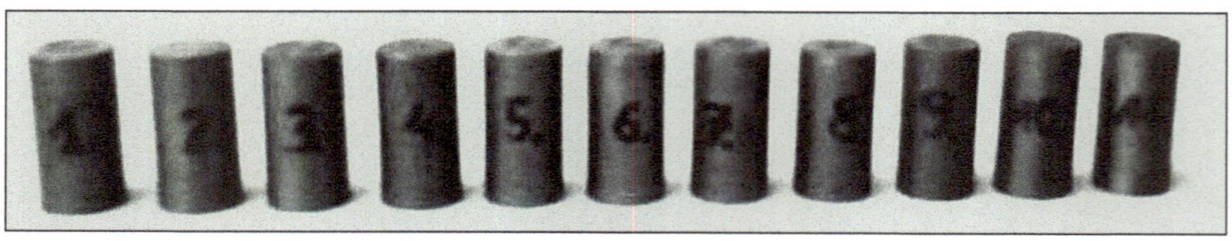

Fig. 1. Cylindrical specimens prepared for rolling

Table 1. Chemical composition of 40 HM steel [%]

C	Mn	Si	P	S	Cr	Ni	Mo	W	V	Cu	C
0.38–0.45	0.4–0.7	0.17–0.37	max 0.035	max 0.035	0.8–1.2	max 0.3	0.15–0.25	max 0.2	max 0.05	max 0.25	0.38–0.45

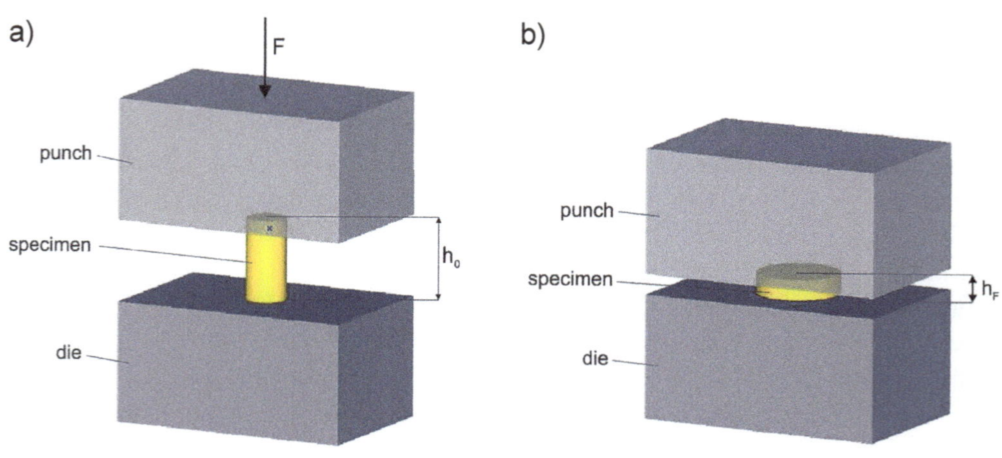

Fig. 2. Design of the compression test

Fig. 3. Surfaces of steel specimens formed: a) without prior heat treatment, b) by annealing

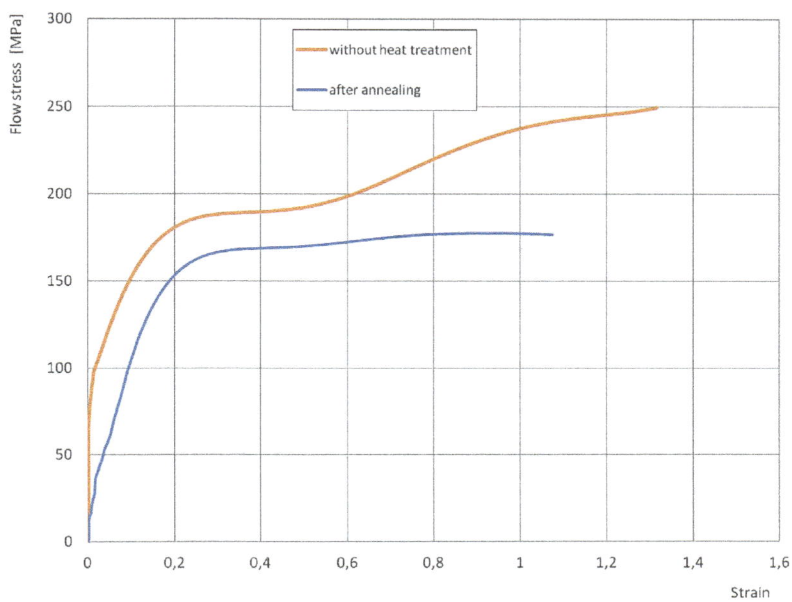

Fig. 4. Experimentally determined flow curves for cold formed 40HM steel: a) without heat treatment, b) after annealing

Table 2. Calculated coefficients of function (1)

Function (1)	Coefficients		Objective function Φ_σ, %
	A	B	
For 40 HM steel without heat treatment	230.7450	0.1903	0.0017
For 40 HM steel after annealing	224.8095	0.4941	0.0583

The following constitutive equations were obtained:

a) for 40HM steel specimens formed without annealing

$$\sigma_p = 230{,}7450 * \varepsilon^{0{,}1903} \quad (3)$$

b) for 40HM steel specimens formed by annealing

$$\sigma_p = 224{,}8095 * \varepsilon^{0{,}4941} \quad (4)$$

Figure 5 presents the comparison of the flow curves described by the determined equations (3) and (4). The curves illustrating the changes in the flow stress of steel after heat treatment (Figure 5) manifest different behaviors. The experimental tests demonstrated that heat-treated 40HM steel specimens were characterized by a much lower flow stress compared to the specimens that were not subjected to annealing, which proves that this steel exhibits better formability after heat treatment.

The tests undertaken to investigate whether 40HM steel can be formed at room temperature confirmed that the non-heat treated material was less suitable for cold forming, which was

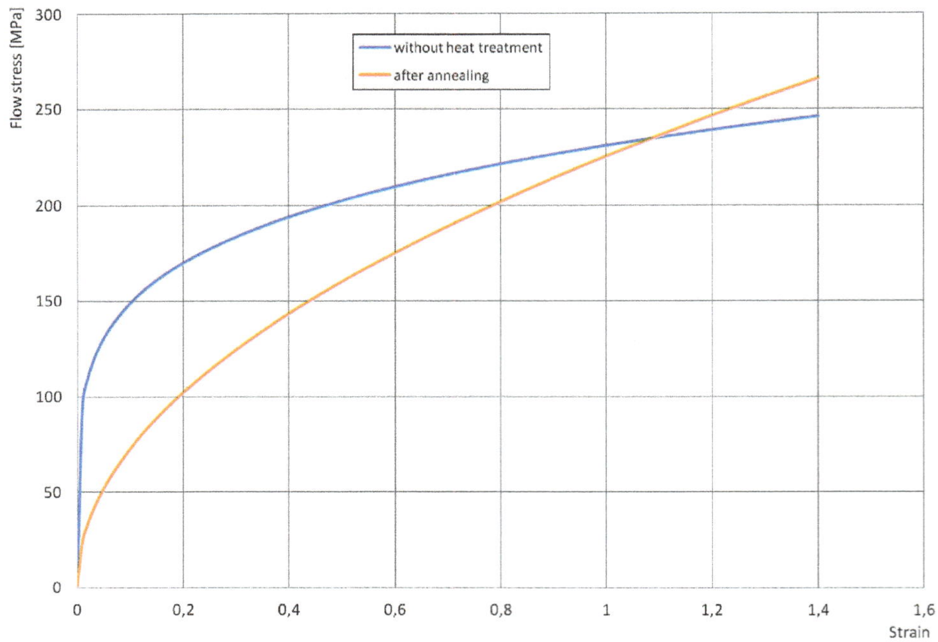

Fig. 5. Comparison of the flow curves described by the determined equations (3) and (4)

proved by the surface of the compressed specimens (Figure 3). The flanks and flat surfaces of the compressed non-heat treated 40 HM steel specimens had cracks, while the specimens subjected to annealing were crack-free. The heat treated steel specimens had a better surface quality, too.

CONCLUSIONS

Based on the conducted theoretical and experimental tests, the following conclusions can be drawn:

1. The performed plastometric compression tests allowed for the determination of flow curves for 40 HM constructional steel subjected to cold forming. The obtained flow curves were described by the function (1), making the flow stress values dependent on the equivalent strains.
2. The results of the conducted tests were used to design a flow stress model for this steel prior to and after heat treatment. The constitutive equations (3) and (4) were defined, and their parameters were determined using the GRG2 optimization method. The experimental flow curves show a good agreement with their values calculated based on the determined constitutive equations.
3. The results demonstrate that heat treatment has a significant effect on both the flow curves for cold-formed 40HM steel and flow stress values. The specimens that did not undergo heat treatment have a higher flow stress value. The investigated steel type exhibits good formability at compression at room temperature after the prior application of annealing. For the parameters applied, only the specimens not subjected to heat treatment had their material cohesion broken. Owing to that, it is recommended that 40Hm steel be heat treated in order to reduce deformation resistance in the cold forming process.
4. The knowledge of the flow characteristics of 40 HM steel under the investigated forming conditions is indispensable for developing a material model that will constitute the basis when designing technological processes for cold forming of this material. The results will be used to perform numerical simulations of forming 40HM steel by the finite element method.

Acknowledgments

MSc Anna Dziubińska and MSc Grzegorz Winiarski are participants of the project: "Qualifications for the labour market – employer friendly university", cofinanced by European Union from European Social Fund.

REFERENCES

1. Dobrzański L.A.: Materiały inżynierskie i projektowanie materiałowe: podstawy nauki o materiałach i metaloznawstwo. Wydawnictwo Naukowo-Techniczne, Warszawa 2006.
2. Kazanecki J., Kajtoch J., Kuźmiński Z.: Badania plastometryczne półwyrobów ze stali 100Cr6. Hutnik – Wiadomości Hutnicze, Wydawnictwo SIGMA-NOT, 10, 2007, 525-529.
3. Gojić M.: The Effect of Tempering on Mechanical Properties of Cold-Rolled 42CrMo4 Steel Pipes. Kovoné Materiály, 39, 2001, 85-92.
4. Głowacki M., Jędrzejczyk D., Hojny M.: Własności plastyczne próbek stalowych w aspekcie komputerowego modelowania próby ściskania w stanie półciekłym. Hutnik – Wiadomości Hutnicze, Wydawnictwo SIGMA-NOT, 3, 2004, 134-142.
5. Samołyk G.: Model płynięcia stopu Al. Z serii 6000 odkształcanego na zimno. Rudy Metale, 6, 2010, 326-329.
6. Połuchin P.I., Gun G., Gałkin A.K.: Soprotivlenije płasticzeskoj dieformacji mietałłov i spłavov. Spravocznik. Mietałłurgija, Moskva 1983.
7. Gontarz A., Dziubińska A.: Properties of MA2 magnesium alloy in hot forming conditions. Rudy i metale nieżelazne, Wydawnictwo SIGMA-NOT, 6, 2010, 340-344.
8. Bayrak M., Ozturk F., Demirezen M., Evis Z.: Analysis of Tempering Treatment on Material Properties of DIN 41Cr4 and DIN 42CrMo4 Steels. Journal of Materials Engineering and Performance, 16(5), 2007, 597-600.
9. Chaoucha D., Guessasmab S., Sadoka A.: Finite Element simulation coupled to optimisation stochastic process to assess the effect of heat treatment on the mechanical properties of 42CrMo4 steel. Materials & Design, 34, 2012, 679–684.
10. Živković D., Gabrić I., Šitić S.: The influence of low and high annealing on the hardness of steel EN 42CRMO4. Tehnički Glasnik, 6(2), 2012, 171-177.

FE-MODELLING OF A CONTACT LAYER BETWEEN ELEMENTS JOINED IN PRELOADED BOLTED CONNECTIONS FOR THE OPERATIONAL CONDITION

Rafał Grzejda[1]

[1] Faculty of Mechanical Engineering and Mechatronics, West Pomeranian University of Technology, Szczecin, 19 Piastów Ave., 70-310 Szczecin, Poland, e-mail: rafal.grzejda@zut.edu.pl

ABSTRACT
Modelling and calculations of a contact layer between elements joined in a preloaded bolted connection for the operational condition are presented. The physical model of the bolted connection is based on a flexible flange element that is joined with a rigid support by means of the no-bolt model (at the assembly stage) or the rigid body bolt model (at the operational stage). The contact layer between joined elements is described as the nonlinear Winkler model. The contact joint model considering an experimental normal elastic characteristic is presented. Examples of normal contact pressure and normal contact deformations distributions are contained.

Keywords: bolted connection, contact layer, assembly condition, operational condition.

INTRODUCTION

Preloaded bolted connections are an excellent alternative to other structural joints used in mechanical engineering. By the suitably carried out pretensioning process it is possible to prevent both the phenomenon of loosening the connection [1] or its leakage phenomenon [2]. Therefore, works on modelling of the contact layer between elements joined in such connections are very important.

Currently, in modelling of the contact joint in bolted connections the finite element systems are often used [3–5]. However, using standard contact elements available in these systems, only constant stiffness coefficients for each contact element at the contact surface can be taken into account. Meanwhile, in the analysis of the contact of elements joined in the bolted connection with an experimentally defined characteristic [6], it is necessary to take into consideration variations of stiffness coefficients for each element of the contact layer. Then creation of specific calculation procedures and using them in conjunction with calculations carried out in the finite element computer system is required.

In the present paper the method for individual modelling of elastic properties of the contact layer's elements is proposed. The study is conducted on the example of the contact joint of elements in the preloaded bolted connection loaded externally by a normal force.

PHYSICAL MODEL OF THE BOLTED CONNECTION

A general structure of the bolted connection model is illustrated in Figure 1. The model is based on a fragment of a flexible flange element that is fastened to a rigid support by means of a bolt. Calculations are carried out in two stages. During tightening the connection, the bolt is modeled only as an input force (Figure 1b). However, in the operational condition, the bolt is treated as the rigid body bolt model [7] (Figure 1c).

A nonlinear contact layer between the flange element and the support is modeled as the Winkler model, which is defined by means of a set of j ($j = 1, 2, \ldots, l$) one-sided spring elements, described by the following relationship:

$$R_j = A_j \cdot p_{nj} = A_j \cdot f(u_{nj}) \qquad (1)$$

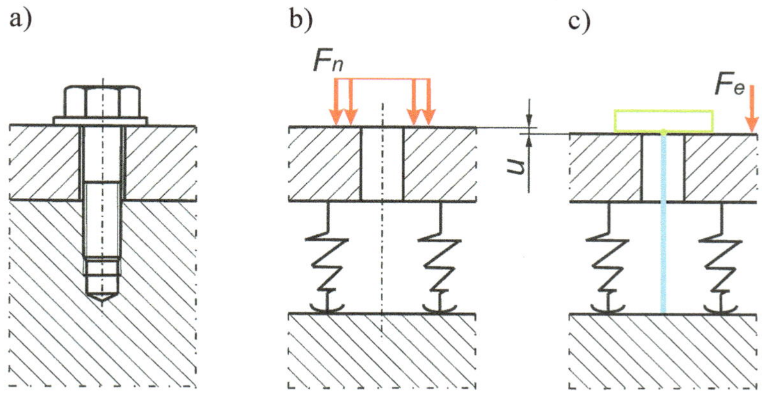

Fig. 1. Fragment of a bolted flange connection: a) diagram, b) physical model for the assembly condition, c) physical model for the operational condition (F_n – preload, F_e – external normal load, u – initial deformation of the bolted connection)

where: R_j – the force in the centre of the j-th elementary contact area,
p_{nj} – normal contact pressure at the j-th elementary contact area,
u_{nj} – normal deformation of the j-th nonlinear spring element,
A_j – the j-th elementary contact area.

Structuring the contact layer model is run in the following steps:

- division of the contact surface between the flange and the support (Figure 2a) into elementary contact areas (Figure 2b),
- assignation mesh nodes in the centers of gravity of elementary contact areas (Figure 2b),
- insertion of nonlinear springs at the nodes identified in the previous step,
- creation of the 2D finite element mesh on the contact surface (Figure 2c).

Based on the 2D finite element mesh on the contact surface between the flange and the support, a uniform 3D finite element mesh for all the volume of the flange is generated.

The equation of system equilibrium (Figure 1) can be written in the form:

$$\boldsymbol{K} \cdot \boldsymbol{q} = \boldsymbol{p} \tag{2}$$

where: $\boldsymbol{K}$ – the stiffness matrix,
$\boldsymbol{q}$ – the displacements vector,
$\boldsymbol{p}$ – the loads vector.

Assuming the division of the model into three subsystems (***B*** – the bolt, ***F*** – the flange element, ***C*** – the contact layer), the equation (2) can be rewritten as follows:

$$\begin{bmatrix} \boldsymbol{K}_{BB} & \boldsymbol{K}_{BF} & \boldsymbol{0} \\ \boldsymbol{K}_{FB} & \boldsymbol{K}_{FF} & \boldsymbol{K}_{FC} \\ \boldsymbol{0} & \boldsymbol{K}_{CF} & \boldsymbol{K}_{CC} \end{bmatrix} \cdot \begin{bmatrix} \boldsymbol{q}_B \\ \boldsymbol{q}_F \\ \boldsymbol{q}_C \end{bmatrix} = \boldsymbol{p} \tag{3}$$

where: $\boldsymbol{K}_{BB}$, $\boldsymbol{K}_{FF}$, $\boldsymbol{K}_{CC}$ – the stiffness matrices of subsystems ***B***, ***F*** and ***C***,
$\boldsymbol{K}_{BF}$, $\boldsymbol{K}_{FB}$, $\boldsymbol{K}_{FC}$, $\boldsymbol{K}_{CF}$ – the matrices of elastic couplings among subsystems ***B***, ***F*** and ***C***,
$\boldsymbol{q}_B$, $\boldsymbol{q}_F$, $\boldsymbol{q}_C$ – the displacements vectors of subsystems ***B***, ***F*** and ***C***.

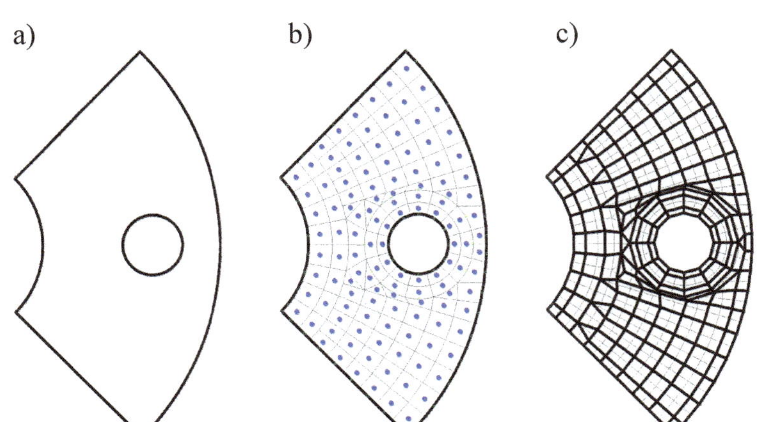

Fig. 2. Creating of the finite element mesh at the contact surface: a) real contact area, b) elementary contact areas, c) scheme of the mesh

At the assembly stage, the system is composed of only two subsystems (***F*** and ***C***), preloaded by the force F_n (Figure 1b). Then the equation (3) is reduced to the form:

$$\begin{bmatrix} \boldsymbol{K}_{FF} & \boldsymbol{K}_{FC} \\ \boldsymbol{K}_{CF} & \boldsymbol{K}_{CC} \end{bmatrix} \cdot \begin{bmatrix} \boldsymbol{q}_F \\ \boldsymbol{q}_C \end{bmatrix} = \begin{bmatrix} \boldsymbol{F}_n \\ \boldsymbol{0} \end{bmatrix} \tag{4}$$

At the operational stage, the preloaded system is loaded by a normal external force F_e. Then the loads vector ***p*** can be written as:

$$\boldsymbol{p} = \begin{bmatrix} \boldsymbol{0} \\ \boldsymbol{F}_e \\ \boldsymbol{0} \end{bmatrix} \tag{5}$$

Thus, the equation (3) can be converted to the formula:

$$\begin{bmatrix} \boldsymbol{K}_{BB} & \boldsymbol{K}_{BF} & \boldsymbol{0} \\ \boldsymbol{K}_{FB} & \boldsymbol{K}_{FF} & \boldsymbol{K}_{FC} \\ \boldsymbol{0} & \boldsymbol{K}_{CF} & \boldsymbol{K}_{CC} \end{bmatrix} \cdot \begin{bmatrix} \boldsymbol{q}_B \\ \boldsymbol{q}_F \\ \boldsymbol{q}_C \end{bmatrix} = \begin{bmatrix} \boldsymbol{0} \\ \boldsymbol{F}_e \\ \boldsymbol{0} \end{bmatrix} \tag{6}$$

Both in the assembly and the operational condition, by solving the system of equations one obtains the displacements vector $\boldsymbol{q}_C$:

$$\boldsymbol{q}_C = col(u_{n1}, u_{n2}, ..., u_{nj}, ..., u_{nl}) \tag{7}$$

In the next step, reactions R_j can be computed based on the relationship (1).

Solution of equations (4) and (6) is conducted in an iterative process using the secant method. In the case of the bolt tightening, the linearization runs according to the way shown in Figure 3a, starting from the origin of the coordinate system. And in the case of the operational stage, it starts with the working points WP_j corresponding to the preload of the j-th nonlinear spring element.

The linearization process is kept running until the following condition has been met:

$$\left| \frac{R'_\alpha - R_\alpha}{R_\alpha} \right| \le \varepsilon \tag{8}$$

where: R'_α – the reaction in the j-th nonlinear spring obtained from the linearization,
α – the index dependent on the case of the calculation process ($\alpha \in \{j, mj\}$),
ε – the admissible error of the linearization.

The diagram of iterative calculations of the bolted connection is shown in Figure 4.

EXAMPLE CALCULATIONS

According to the presented method, computations of a bolted connection of a flange shown in Figure 5 with a rigid support were realized. The subject of the research is a connection of a quarter of the flange EN 1092-1/01/DN25/PN6/S235JR [8] with the rigid support performed by means of the bolt M10 of mechanical property class 5.6. The preload of the bolt F_n is equal to 14,9 kN and it is set down on the base of Polish Standard [8]. The surface area of preload influence A_n is equal to 69,75π mm² and it is set down on the base of Polish Standard [9]. After the preloading process, the connection is loaded by a normal force F_e equal to 10 kN as shown in Figure 5.

Calculations were carried out by means of the Midas NFX 2014 finite element program. The bolt is modeled with two types of elements:

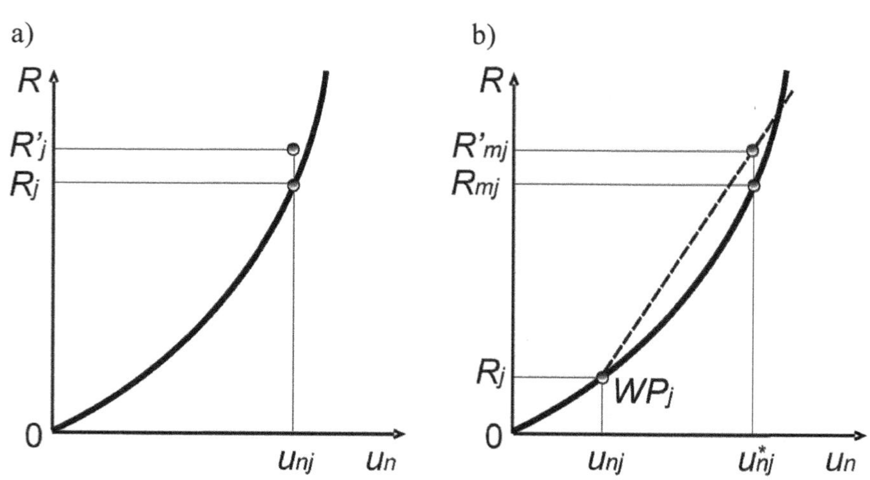

Fig. 3. Linearization of a curve by the secant method: a) in the case of the assembly condition, b) in the case of the operational condition

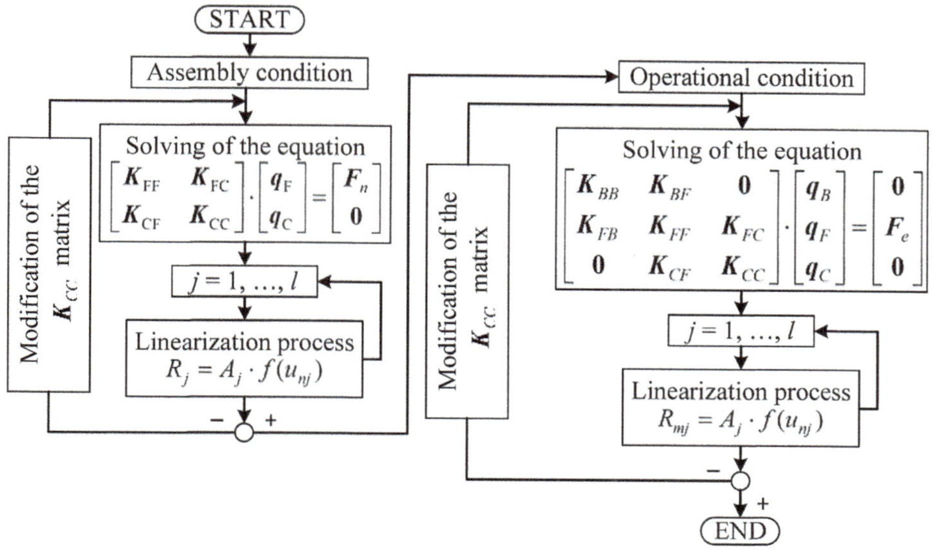

Fig. 4. Block diagram of the calculation process

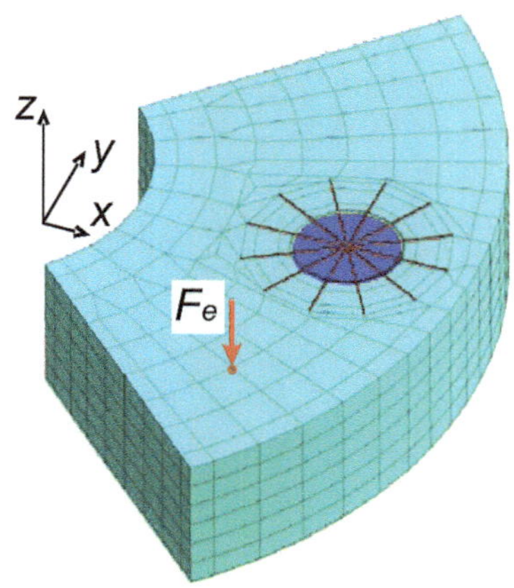

Fig. 5. FE-model of the tested bolted connection

- the plain part of the bolt is treated as a bar element,
- the bolt head is treated as a rigid element.

Elastic properties of the nonlinear contact layer are described as the following power function (obtained for a sample contact of elements joined in a multi-bolted connection [6]):

$$R_j = A_j \cdot \left(3{,}428 \cdot u_{nj}^{1{,}657}\right) \tag{9}$$

As results of calculations, normal contact pressure and normal contact deformations distributions for two tested conditions were received. The graphs were presented in Table 1.

Table 1. Results of calculations

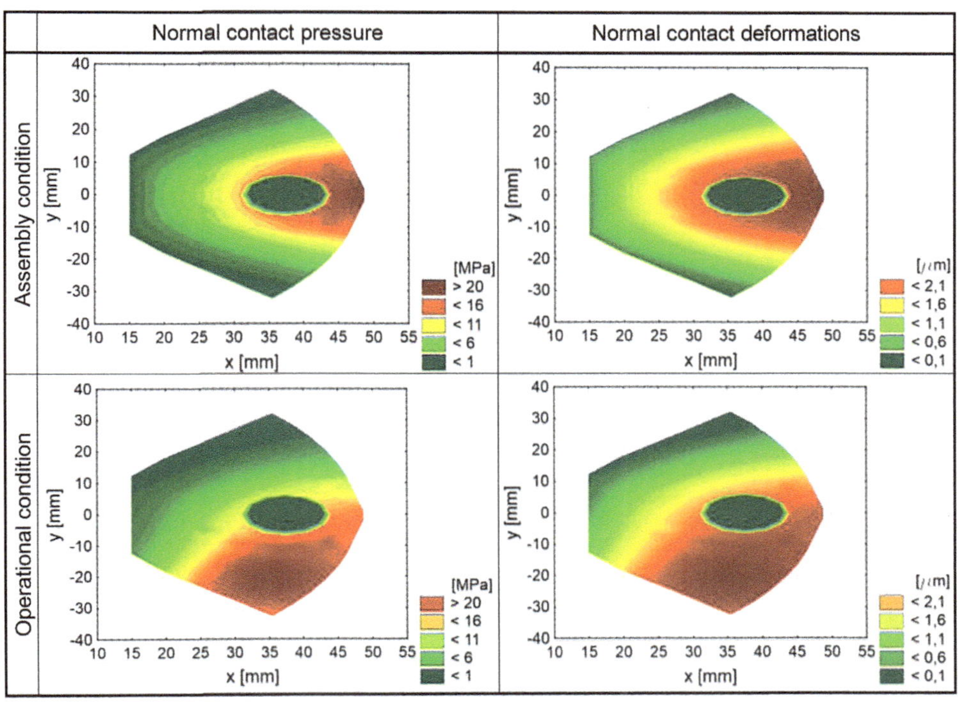

CONCLUSIONS

In the paper a procedure of computations of a contact joint between elements in a bolted connection with the use of its experimental normal characteristic is presented. Calculations were related to the assembly stage and the operational stage of the connection. In the model only contact stiffness in a direction normal to the contact surface were taken into account.

However, it is also possible to accommodate tangential contact stiffness and friction into the model. Results of works on development of the model will be published in future articles.

REFERENCES

1. Gardyński L., Nieoczym A.: Workplace research on the quality of screwing-in head (in Polish). Postępy Nauki i Techniki, 7, 2011, 120–126.
2. Izvolenskii E.V., Petrov R.A.: Determination of tightening parameters of threaded leak-tight joints. Chemical and Petroleum Engineering, 41 (1-2), 2005, 85–86.
3. Caliskan M.: Evaluation of bonded and bolted repair techniques with finite element method. Materials & Design, 27 (10), 2006, 811–820.
4. Chakherlou T.N., Razavi M.J., Aghdam A.B.: On the variation of clamping force in bolted double lap joints subjected to longitudinal loading: A numerical and experimental investigation. Strain, 48 (1), 2012, 21–29.
5. Shi G., Shi Y., Wang Y., Bradford M.A.: Numerical simulation of steel pretensioned bolted end-plate connections of different types and details. Engineering Structures, 30 (10), 2008, 2677–2686.
6. Grzejda R., Cieloszyk J.: Determination of normal flexibility characteristics of surfaces' contact with the specified surface texture (in Polish). Bulletin of the Military University of Technology, 63 (3), 2014, 175–187.
7. Grzejda R.: FE-modelling of bolts in bolted joints for the assembly condition (in Polish). Mechanik, 87 (8-9), 2014, 672–675.
8. PN-EN-1092-1:2010: Flanges and their joints – Circular flanges for pipes, valves, fittings and accessories, PN designated – Part 1: Steel flanges.
9. PN-EN ISO 7091:2003: Plain washers – Normal series – Product grade C.

PRESSURE DURING CURING AND THE STRENGTH OF 2024, 2017A AND 1050 ALUMINIUM ALLOY SHEET ADHESIVE JOINTS

Anna Rudawska[1]

[1] Department of Production Engineering, Faculty of Mechanical Engineering, Lublin University of Technology Nadbystrzycka 36, 20-618 Lublin, Poland, e-mail: a.rudawska@pollub.pl

ABSTRACT

This paper is primarily devoted to one of adhesive curing process parameters – pressure applied during curing, as well as the type of adhesive. The aim of this work is to determine the influence of pressure applied during curing process of aluminium alloy sheet adhesive joints on the strength of adhesive joints formed with two epoxy adhesives: Loctite 9484 and Loctite 9466. Three types of aluminium alloy sheets were tested, namely AW 2024, AW 2017A and AW 1050, applied in many branches of engineering. During formation of adhesive joints, other conditions remained unchanged, and surface preparation operations included mechanical processing and degreasing.

Keywords: adhesive joint strength, pressure, curing, aluminium alloys.

INTRODUCTION

Adhesive joining significantly supplements other methods of joining structural materials. Nowadays, adhesive technology is used in multiple branches of industry: building engineering, automotive and aircraft industries, machine construction, packaging production and many more. Joining metals by adhesive joining is becoming increasingly commonplace due to several factors, such as: considerable strength of joints, vibration damping, joining different structural materials, frequently of substantially disproportionate in geometric dimensions [1–4].

The adhesive bonding technology comprises several consecutive technological operations. Individual operations may consist of a number of stages, be characterised by specific technological parameters, involve diverse equipment and instrumentation. Detailed conditions of adhesive bonding operation are chosen based on, *inter alia*, the type of adherends, the type of adhesive, geometry of elements and structures, joint forming conditions etc. [2, 3, 5–7].

The article presents some aspects of curing process: the influence of pressure on the strength of adhesive joint strength. A lot of authors are presenting various kinds of problems of curing process, adhesives, ageing in various environments, etc. [5, 8]. S.G. Prolongo and others [5] presented the study of adhesive properties of different epoxy resin and among others things analyse of the curing reaction. Borsellino and others [9] analysed for example the effect of curing time of resins on the mechanical properties of the composite joints. Curing constitutes one of technological factors considerably influencing adhesive joint strength. Adams and others [10] presented the effect of temperature on the strength of adhesive. Stewart, Chambers and Gordon [11] described the cohesive performance of an epoxy adhesive as a function of cure level.

PRESSURE DURING CURING PROCESS

Pressure is another technological parameter of adhesive layer curing. Pressure in adhesive bonding process enhances wetting process, particularly when filling micropores and microirreguralities of the surface is involved. It facilitates spreading the adhesive and fixing the position of adhered elements. The adjustment of curing pressure depends on the adhesive (its thickness, viscosity, thermal shrinkage, chemical shrinkage), the type of joined materials and surface preparation methods. Additionally, size, shape, thickness and position of the adhesive joint itself plays a crucial role in adhesive joint construction [2, 3, 12].

Pressure applied during adhesive bonding should ensure a close contact between the joined surfaces as well as optimal thickness of adhesive layer (0.05 – 0.20 mm) [1, 2]. Pressure conditions thickness of adhesive layer, which in high-reliability constructions is a critical requirement.

In many instances, excessive pressure force or inappropriate arrangement of pressure pads on the adhesive bonded joint may lead to deformation of joined structures. The value of the applied pressure depends on whether the adhesive solidifies as a consequence of condensation polymerisation (pressure of 1 MPa), or in another way (pressure of 0.05 MPa) [2].

Required pressure may be applied with the use of different equipment. In aircraft structures, different types of joints may be cured in autoclaves of different structure and overall dimensions depending on the dimensions of the joined structure. The curing method depends on the type of adhesive, the type of structure, applied pressure and operating conditions [2, 3].

Adhesives consisting of epoxy resins cured with different types of curing agents are among the most common adhesive compositions [2, 11]. Epoxy resins cross-link (set) as epoxy or hydroxyl groups in the resins react with amines or acid anhydrides. Epoxy resin curing agents are chemical compounds, which, as a result of chemical reactions, cause spatial cross-linking of resins imparting properties of adhesive, saturant or chemosetting plastics to them [13].

EXPERIMENTAL TESTS

Experimental tests consisted in forming 6 variants adhesive single-lap joints of AW2024, AW2017A and AW1050 aluminium alloy sheets. Two types of epoxy adhesives of different properties were used. Furthermore, for each adhesive, five variants of pressure value applied were tested. Following specified seasoning time adhesive joints were subjected to strength tests.

Materials and adhesive joints characteristic

The research consisted in testing single-lap adhesive joints of AW2024, AW2017A and AW1050 aluminium alloys, the geometry of which is presented in Figure 1.

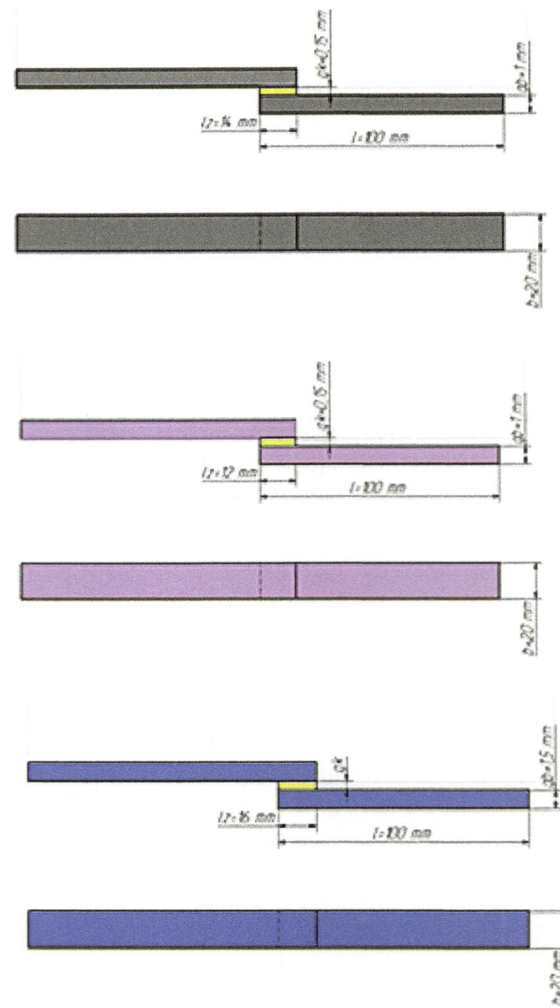

Fig. 1. Size and dimensions of tested adhesive specimens: a) AW 2024, b) AW 2017A, c) AW1050

Thickness and length of adherends were identical for all tests and equal to $b = 20\pm0{,}11$ mm and $l = 100\pm0{,}8$ mm respectively. The thickness of adhesive layer was determined in experimental tests.

Adhesives characteristic

Adhesive joints of aluminium alloys were formed with two types of two-component epoxy adhesives: Loctite 9466 and Loctite 9484, the selected properties of which are presented in Table 1 (for cured material) [14, 15].

Adhesives Loctite 9466 and 9484 require 24 hours in room temperature in order to produce high strength. The process, however, can be accelerated by applying higher temperature.

Adhesive compositions tested in experimental study contain an identical chemical resin compound – epoxy, yet differ in terms of the curing

Table 1. Selected properties of applied adhesive (cured material)

No.	Properties	Adhesive	
		Loctite 9466	Loctite 9484
1	Tensile Strength, ISO 527-3, N/mm²	32	15
2	Tensile Modulus, ISO 527-3, N/mm²	1718	161
3	Shore Hardness, ISO 868, Durometer D	60	55
4	Elongation, ISO 527-3, %	3	50

agent chemical type, which results in their different properties. Different curing agent requires different adhesive components mixing ratio by volume and by weight. Another factor taken into particular consideration was the working time of the adhesive, which necessitated performing joints within a couple of minutes after mixing the components of adhesives. Moreover, attention was given to the amount of adhesive mixed, due to the fact that excessive amounts of adhesive and ambient temperature above 22 °C contribute to shortening the working time, owing to, *inter alia*, emitting excessive amounts of heat.

Adhesives used in the tests are delivered in dual packages (cartridges), facilitating dosing and preparation. Mixing and dosing was performed with special dispensing device – a double syringe and a static mixer. After mixing the adhesive was applied on one of the adhered surfaces.

Joint forming conditions

The surface of adhered alloy sheets was prepared as follows:

- the first stage consisted in degreasing with acetone in order to remove any surface contaminants,
- the second stage was mechanical working with P400 grit abrasive discs,
- finally, degreasing of the surfaces with acetone was repeated.

Degreasing agent evaporation time equalled 3 minutes and was followed by applying the prepared adhesives. The cold-setting process was conducted at 21 °C. Five variants of pressure were applied during curing, respectively: 0.016 MPa; 0.032 MPa; 0.048 MPa; 0.064 MPa and 0.080 MPa. Adhesive joints were formed at relative air humidity between 22–23%. Prepared adhesive joints of the analysed sheets are presented in Figure 2.

Adhesive joints conditioning time was 6 days (at temperature and humidity conditions identical to joint forming), followed by strength tests, allowing to determine joint failure force values, thus adhesive joint strength.

Tests were conducted on a test stand – materials testing machine Zwick/ Roell Z150 according to standard DIN EN 1465. Figure 3 shows a model view of a tested adhesive joint test sample.

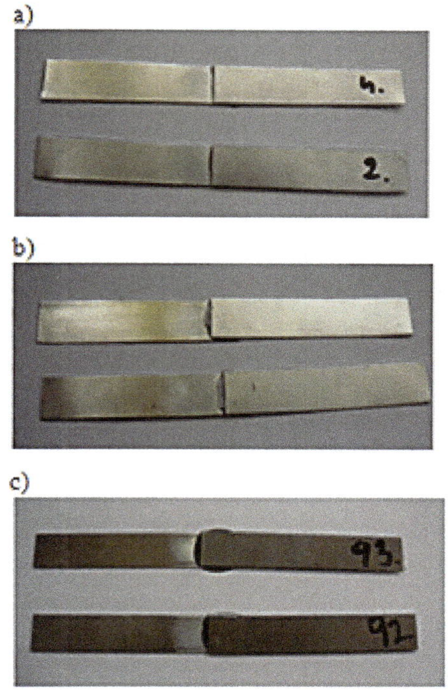

Fig. 2. Aluminium alloy sheet adhesive joints: a) AW 2024, c) AW 2017A, b) AW 1050

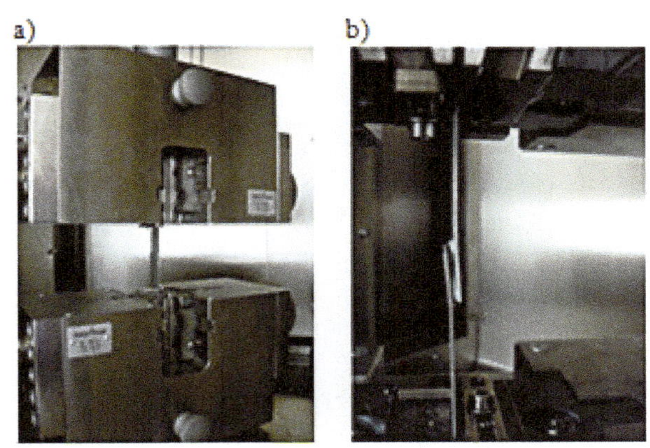

Fig. 3. The view of: a) wedge-screw grips with a test sample prior to destructive tests, b) grips with a test sample after failure

TEST RESULTS – ADHESIVE JOINTS STRENGTH

AW 2024 aluminium alloy

Strength test results of AW 2024 aluminium alloy sheet adhesive joints, formed with the analysed epoxy adhesives, at different pressure applied during curing, are shown in Figures 4 and 5.

Based on test results presented in Figure 8, it was observed that the highest joint strength (23 MPa) was achieved in the case of 0.048 MPa pressure applied during curing. For the pressure of 0.080 MPa, the second highest strength value was achieved – 21.50 MPa. In the remaining cases strength values obtained varied insignificantly, therefore the value of pressure applied during curing is of little influence on strength.

The study of test results (Fig. 5) reveals that in the case of joints formed with Loctite 9484 no significant impact of pressure applied during curing on joint strength can be observed for the first four pressure variants. The last variant (0.080 MPa) produced the highest joint strength – 9.66 MPa.

Based on the conducted tests it may be concluded that Loctite 9466 adhesive provides higher joint strength than 9489 (Fig. 4 and Fig. 5). In each variant of pressure applied during curing the strength manifested by joints formed with Loctite 9466 proved twice, and for two variants even three times as high as an analogical 9484 adhesive joint. The differences for tested pressure values are as follows: 0.016 MPa – 127 %; 0.032 MPa – 128%; 0.048 MPa – 215%; 0.064 MPa – 108%; 0.080 MPa – 123%.

AW 2017A aluminium alloy

Strength test results of AW 2017A aluminium alloy sheet adhesive joints, formed with the analysed epoxy adhesives, at different pressures applied during curing, are shown in Figures 6 and 7.

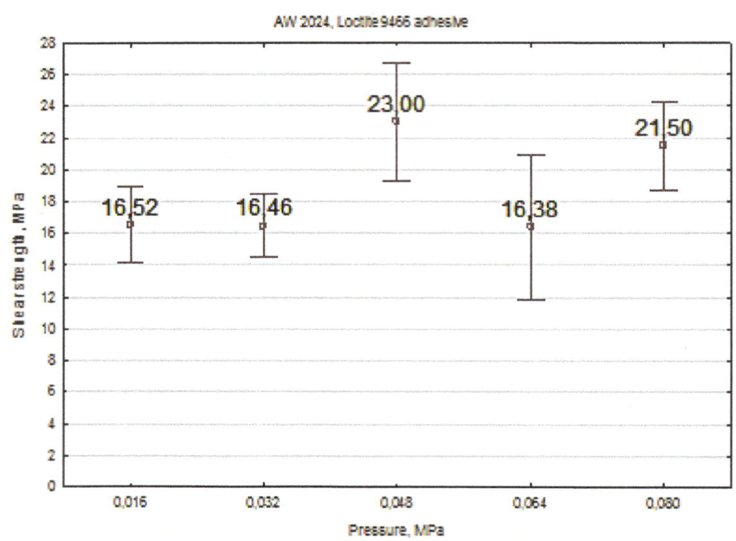

Fig. 4. Shear strength of 2024 aluminium alloy sheet adhesive joints formed with 9466 adhesive, depending on pressure value

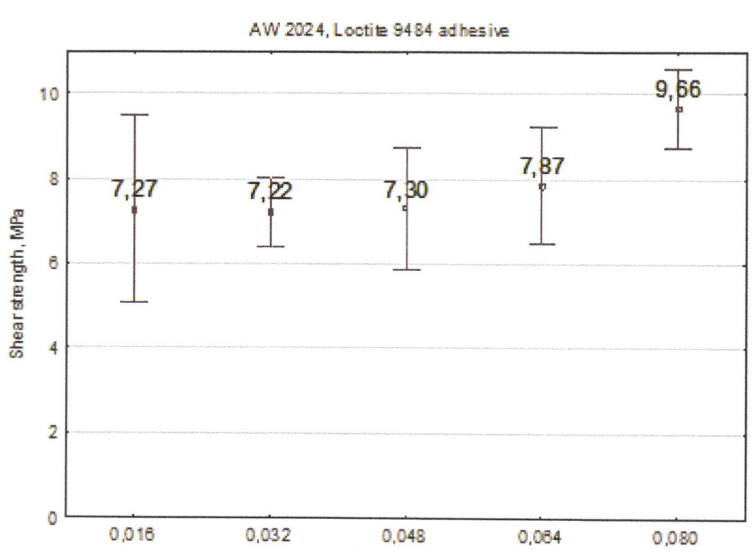

Fig. 5. Shear strength of adhesive joints formed with 9484 adhesive, depending on pressure value

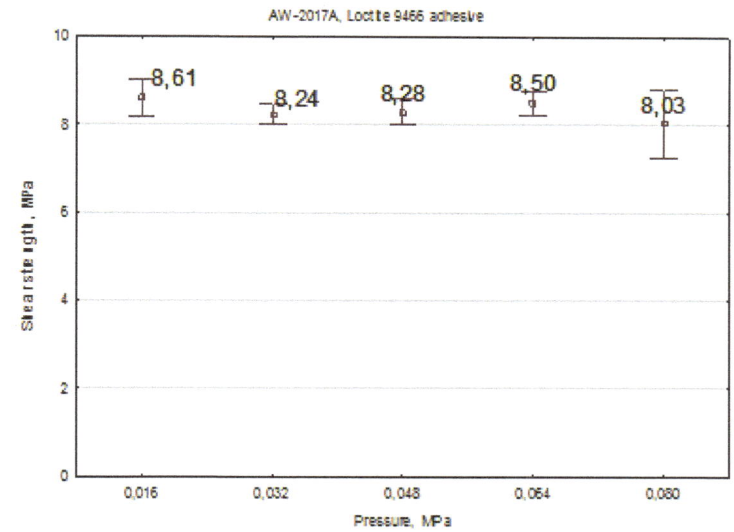

Fig. 6. Shear strength of AW 2017A aluminium alloy sheet adhesive joints formed with 9466 adhesive, depending on pressure value

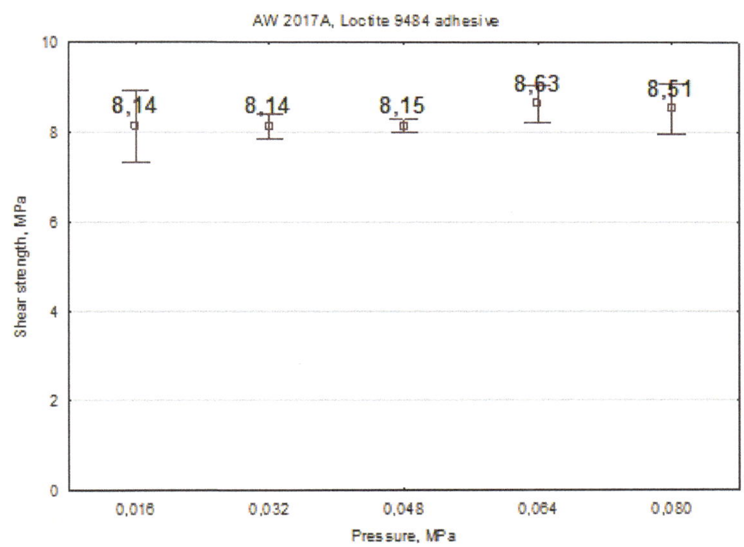

Fig. 7. Shear strength of AW 2017A aluminium alloy sheet adhesive joints formed with 9484 adhesive, depending on pressure value

Based on test results (Fig. 6 and Fig. 7), it was noticed that applying both epoxy adhesives produces similar joint strength values. Moreover, no influence of pressure applied during curing on the strength of 2017A aluminium alloy sheet adhesive joint was observed.

AW 1050 aluminium alloy

Strength test results of AW 1050 aluminium alloy sheet adhesive joints, formed with the analysed epoxy adhesives, at different pressures applied during curing, are shown in Figures 8 and 9.

Analysing the test results (Fig. 9) it was observed that the highest joint strength (20.25 MPa) was obtained for the pressure of 0.032 MPa, and the lowest strength in the case of the highest pressure applied – 0.080 MPa. Moreover, it was noticed that joint strength increases up to the pressure value of 0.032 MPa after which point it decreases with the increase of pressure applied during curing. It may be concluded therefore that there exists an optimum pressure value where the strength of the joint is the highest.

Examples of joint failure

Examples of test samples following failure tests with two epoxy adhesives are presented in Figures 10–12. Analysis proves that it can be observed adhesive-cohesive failure of adhesive layer.

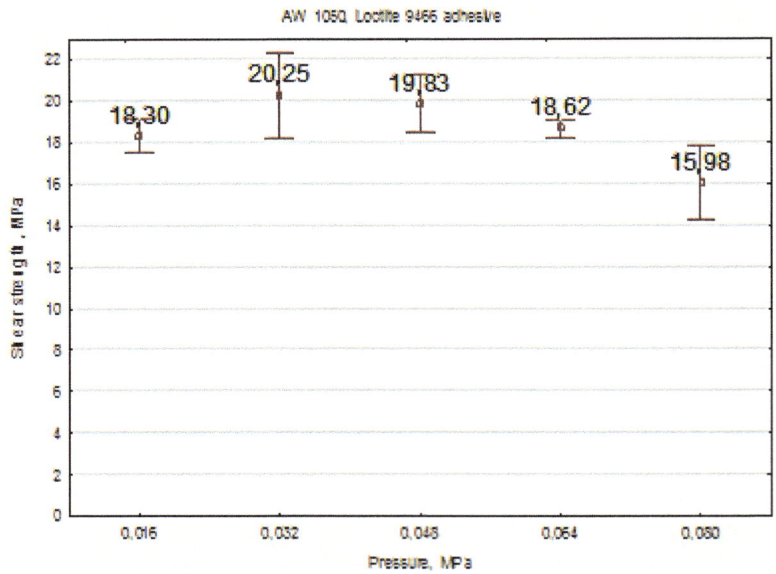

Fig. 8. Shear strength of AW 1050 aluminium alloy sheet adhesive joints formed with 9466 adhesive, depending on pressure value

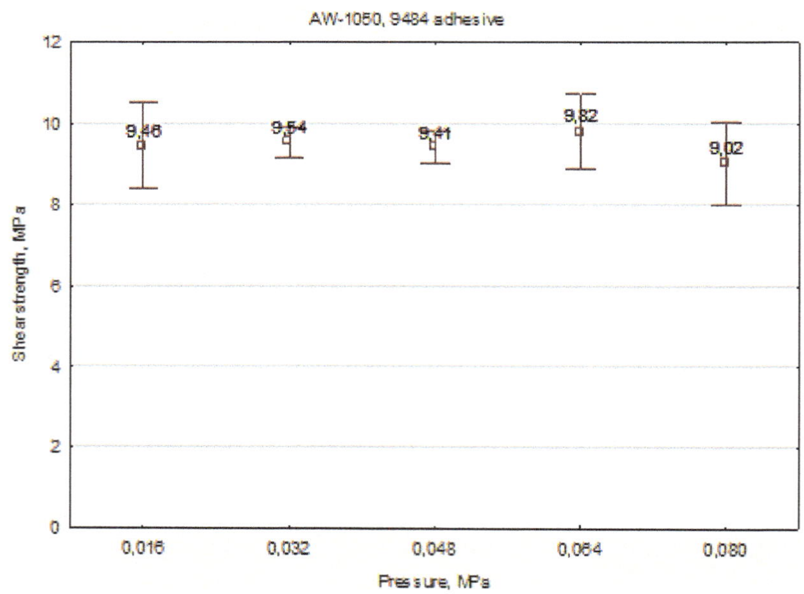

Fig. 9. Shear strength of 1050 aluminium alloy sheet adhesive joints formed with 9484 adhesive, depending on pressure value

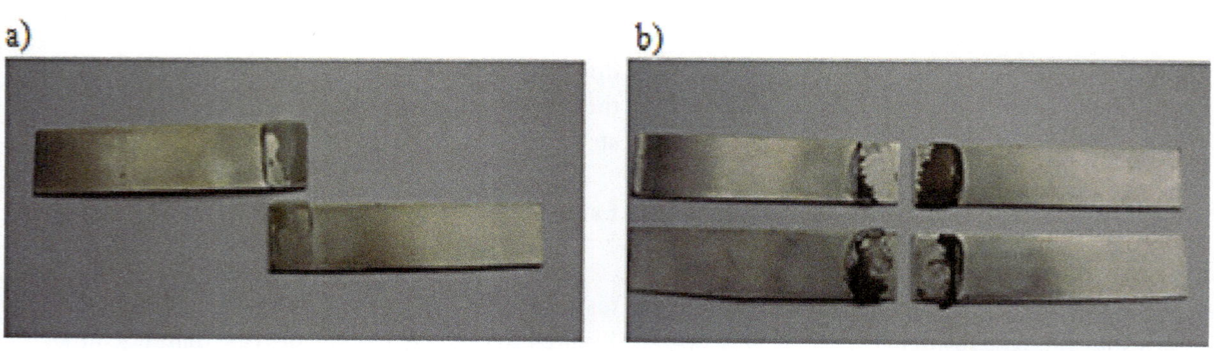

Fig. 10. Failure adhesive joints of AW 2024 aluminium alloy sheets formed with adhesives: a) Loctite 9466, b) Loctite 9484

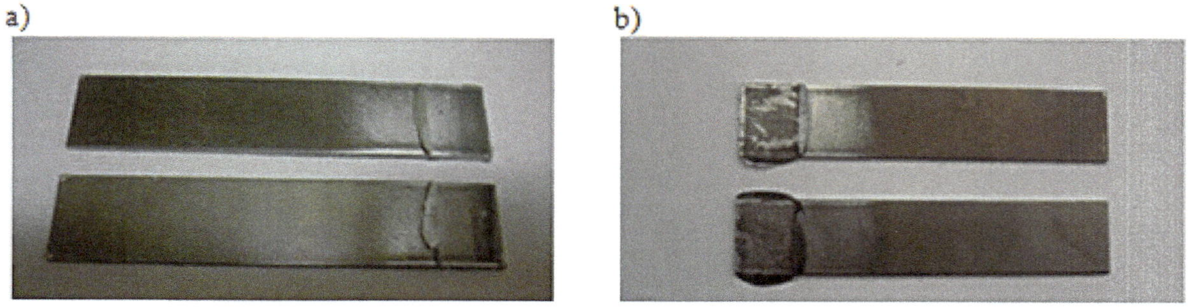

Fig. 11. Failure adhesive joints of AW 1050 aluminium alloy sheets formed with adhesives: a) Loctite 9466, b) Loctite 9484

Fig. 12. Failure adhesive joints of AW 2017A aluminium alloy sheets

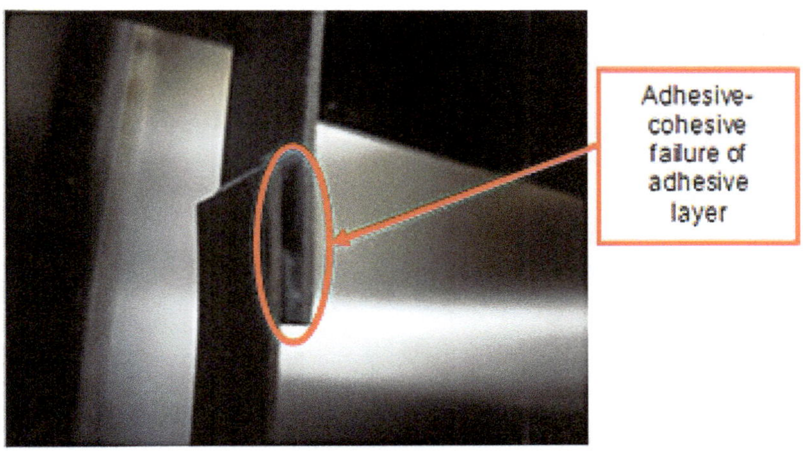

Fig. 13. AW1050 aluminium alloy sheet adhesive joints following strength test

Figure 13 shows adhesively joined AW 1050 aluminium alloy sheets in the materials testing machine clamping jaws, after a strength test. Characteristic shape of failure samples as well as adhesive and cohesive adhesive layer failure can be seen.

CONCLUSIONS

The study presented in the paper was oriented towards determining the impact of pressure value on the strength of adhesive joints of AW 2024, AW 2017A and AW1050 aluminium alloy sheets, formed with epoxy adhesives.

Analysing the obtained aluminium alloy sheet adhesive joints strength, nearly double difference in strength values of the analysed sheet joints formed with tested adhesives can be observed, *nota bene*, these were the joints formed with Loctite 9466 which were characterised by higher strength throughout the research. This applies to AW 2024 and AW 1050 aluminium alloy sheets joints. The highest strength values for the materials in question were noted for the pressure of 0.048 MPa and 0.032 MPa respectively. It may be therefore concluded that in the case of joining analysed aluminium alloy sheets the preferred choice of adhesive should be epoxy adhesive Loctite 9466, due to higher strength bond it creates. For these joints bigger differences in adhesive strength values were observed than for Loctite 9484, for which neither pressure value nor adhesive joint thickness has a significant impact on obtained joint strength. Finally, in the case of AW 2017A aluminium alloy sheets samples, the selection of adhesive was not a factor, as strength values for both tested adhesives were comparable. The presented conclusion is based on the analysed scope of pressure applied in tests.

Application of Loctite 9484 is advantageous in as much as this adhesive offers similar joint strength regardless of pressure applied during curing. This concerns all tested aluminium alloy sheets: AW 2024, AW 2017A and AW 1050. Moreover, it has been observed that the strength of adhesive joints formed with Loctite 9484 of three analysed aluminium alloy sheets was comparable regardless of tested adherend. It appears reasonable to conclude that the application of this adhesive to form aluminium alloy sheet joints will ensure little impact of the aforementioned factors (pressure) on the strength of these joints.

In conclusion, it is possible to select optimum pressure value which will ensure highest joint strength. In addition, for certain cold-setting adhesives applied pressure will not be of great impact, in the analysed scope, on the strength of adhesive joints. Therefore, with a view to increasing joint strength, while analysing technological factors attention must be paid to other factors as well.

REFERENCES

1. Godzimirski J.: The strength of constructional bonding joints. Printed by WNT, Warszawa 2002.
2. Godzimirski J., Kozakiewicz J., Łunarski J., Zielecki W.: The constructional bonding joint of metal elements in machine building. Printed by Rzeszow Technical University, Rzeszow 1997.
3. Arenas J.M., Narbón J.J., Alía C.: Optimum adhesive thickness in the structural adhesives joint using statistical techniques based on Weibull distribution. International Journal of Adhesion and Adhesives, 30, 2010, 160–165.
4. Critchlow G.W., Yendall K.A., Bahrani D., Quinn A., Andrews F.: Strategies for the replacement of chromic acid anodizing for the structural bonding of aluminium alloys. International Journal of Adhesion and Adhesives, 26, 2006, 419–453.
5. de Morais A.B., Pereira A.B., Teixeira J.P., Cavaleiro N.C.: Strength of epoxy adhesive-bonded stainless-steel joints. Journal of Adhesion Science and Technology, 27, 2007, 679–686.
6. Rudawska A., Kuczmaszewski J.: The bonding of zinc coated sheets. Printed by Lublin Technical University, Lublin 2005.
7. Rudawska A., Kowalska B., Garbacz T.: The influence of the type of adhesives on low-density polyethylene bonded joints strength. Conference materials. 27th Polymer Processing Society, Annual Meeting. 10-14.05.2011 Maroko, Marrakesz.
8. Moussa O., Vassilopoulos P., de Castro J., Keller T.: Early-age tensile properties of structural epoxy adhesives subjected to low-temperature curing. International Journal of Adhesion and Adhesives 35, 2012, 9–16.
9. Borsalino C., Calabrese L., Di Bella G., Valenza A.: Comparisons of processing and strength properties of two adhesive systems for composite joints. International Journal of Adhesion and Adhesives, 27, 2007, 446–457.
10. Adams R.D., Coppendale J., Mallick V., Al.-Hamdan H.: The effect of temperature on the strenght of adhesive joint. International Journal of Adhesion and Adhesives, 12, 1992, 2185–190.
11. Stewart I., Chambers A., Gordon T.: The cohesive mechanical properties of a toughened epoxy adhesive as a function of cure level. International Journal of Adhesion and Adhesives, 27, 2007, 277–287.
12. da Silva L.F.M., Adams R.D.: Adhesive joint at high and low temperatures using similar and dissimilar adherends and dual adhesives. International Journal of Adhesion and Adhesives, 27, 2007, 216–226.
13. http://www.zch.sarzyna.pl (June 2012).
14. http://tds.loctite.com (June 2012).
15. http://www.akd-tools.gr (June 2012).

26

OPTIMIZATION OF PROCESS PARAMETERS FOR ENHANCED MECHANICAL PROPERTIES OF POLYPROPYLENE TERNARY NANOCOMPOSITES

Oladipupo Olaosebikan Ogunleye[1], Taofeeq Olalekan Salawudeen[1], Muyibi A. Suleyman[2], Yusof Faridah[2]

[1] Department of Chemical Engineering, Faculty of Engineering and Technology, Ladoke Akintola University of Technology, P.M.B. 4000, Ogbomoso, Nigeria, e-mail: ooogunleye@yahoo.com

[2] Department of Biotechnology Engineering, Kulliyyah of Engineering, International Islamic University Malaysia, Jalan Gombak, Kuala Lumpur, Malaysia

ABSTRACT

Preparation of Polypropylene ternary nanocomposites (PPTN) was accomplished by blending multiwall carbon nanotube (MWCNT) in polypropylene/clay binary system using a melt intercalation method. The effects of MWCNT loadings (A), melting temperature (B) and mixing speed (C) were investigated and optimized using central composite design. The analysis of the fitted cubic model clearly indicated that A and B were the main factors influencing the tensile properties at a fixed value of C. However, the analysis of variance showed that the interactions between the process parameters, such as; AB, AC, AB^2, A^2B and ABC, were highly significant on both tensile strength and Young's modulus enhancement, while no interaction is significant in all models considered for elongation. The established optimal conditions gave 0.17%, 165 °C, and 120 rpm for A, B and C, respectively. These conditions yielded a percentage increase of 57 and 63% for tensile strength and Young's modulus respectively compared to the virgin Polypropylene used.

Keywords: MWCNT, PPTN, tensile strength and Young's modulus.

INTRODUCTION

Polypropylene is an important plastic, which offers probably the best quality among polyolefins at low cost [1, 2]. It has a wider area of applications, such as packaging, home utensils, casings and automobile accessories compared than any other thermoplastic. In addition, polypropylene has a lower density between 900 and 920 kg/m^3, in comparison to other engineering materials, allowing for potential weight reductions, very good heat resistance and due to its higher crystallinity, it is an excellent moisture barrier and has good optical properties [3, 4].

Polymers in their pure and natural state are either reinforced with organic fillers, such as sisal, flax, jute and wood fibres [5, 6], or particulate fillers, such as tack, $CaCO_3$ and mica. Such polymer composites are credited with better properties relative to their parent matrix and have found a wide array of applications in the civil construction industries [7, 8]. They are term microcomposite simply because the fillers are dispersed in micro scale. However, effort in the field of polymer microcomposite has reached the highest level of optimization because most times higher filler loading is usually required and consequently affect the final material.

Research on polymer nanocomposites became subject of interest since the successful preparation and application of PA/montmorillonite composites by Toyota over a decade ago. Such nanocomposites prepared from PA 6 possessed good mechanical and thermal properties at low filler loading and, therefore, provided a better economical solution in various fields of application [9]. Various studies reported improvement in mechanical [10, 11], thermal and flammability

[12, 13], and barrier [14, 15] properties of thermoplastic by addition of organically modified layered silicates to polymer matrices. This has attracted further studies in the field of polymeric composite.

Polypropylene, among thermoplastic has a wider area of industrial applications. It can be modified in many ways and its fibre reinforced grades compete even with most engineering polymers. Consequently, many attempts have been made to prepare polypropylene-layered silicate nanocomposite (PP/MNC) in order to have a material of better properties compared to the conventional polymer micro-composite. However, practically all composites prepared from organophilic clay and polypropylene do not show improvements in an extent that satisfies the requirements of most applications in terms of mechanical properties because complete dispersion of the silicate is never achieved in the PP matrix [16]. In addition, the achieved properties usually differ from the intended trade-off what affects the quality of the final composite. As a result, effort is ongoing to produce composite that will combine more than one filler in a polymer matrix. Such that there will be shared properties and shortcomings from one filler can be addressed by the other called a secondary filler [14, 17]. Carbon nanotube is suggested in this study as a secondary filler due to its special intrinsic properties and high aspect ratio combined with surface area.

Due to the complexity in the behavior of Polypropylene blend with nanomaterials cum their reaction towards process parameters such as melting temperature, mixing speed and filler loadings, optimization of the process parameters is rarely reported. The present work is therefore aimed at optimizing these process parameters to enhance the properties of PPTN. Such ternary combination will address the most common shortcomings, such as low mechanical properties and trade-off problems peculiar to the binary precursor (PP/MNC) and hence possess better properties.

METHODOLOGY

Materials

Multiwall carbon nanotubes with outer diameter range between 10 nm and 20 nm, length ~30 μm and 95% purity was manufactured by Zyvex Instrument, Germany and supplied by Cahaya BHD SDN, Malaysia. Bentonite clay manufactured by Across Organics was modified in the laboratory using octadecylamine salt. Polypropylene homopolymer with an MFR > 3 g/10 min manufactured by Petronas Malaysia was used as matrix, while maleic anhydride grafted polypropylene (PPgMA oligoma) (Polybond 3200) with 1.2 wt % MA was used as compatibilizer.

Selection of range of process parameters

Range of the selected process parameters was determined by using design expert version 6.0.8, Stat-ease Inc, USA with full factorial, three factors with low and high levels: The selected parameters were varied between 170 and 250 °C for melting temperature, 0.1 to 1% of binary precursor for MWCNT loading and 100 to 200 rpm for mixing speed. These were used in the preparation of ternary nanocomposite according to Table 1 below, the final PPTN were tested for their tensile properties using ASTM D638, and the results were analyzed.

Table 1. Experimental design for determination of range for process parameters

Experimental run	Carbon nanotube (%)	Mixing speed (rpm)	Melting temperature (°C)
1	0.1	200	250
2	1	50	170
3	1	200	170
4	1	50	250
5	0.1	200	170
6	0.1	50	250
7	1	200	250
8	0.1	50	170

Preparation of PP-Clay/MWCNTs nanocomposites

Ternary nanocomposite production was carried out using melt-mixing method in a hakee mixer (Type Rheomix 600P) by employing a two-batch system. First batch involved the production of Polypropylene/Clay binary precursor using 3% layered silicate in the presence of MAgPP as compatibilizer. The second batch system involved the introduction of MWCNT according to experimental design (Table 2). The final nanocomposite was pelletized, hot pressed and molded into standard shape according to ASTM D638. The molded plaque was then subjected to mechanical test using universal tensile machine (LYLOYD Instruments, Type 1025).

Experimental design

Optimization using central composite design (CCD) of the response surface (RSM) is a suitable tool for determination of optimum conditions when many factors affect a desired response [18]. In this work, seventeen experiments were carried out according to the conditions indicated in Table 2. The response values (tensile strength, Young modulus and elongation) are reported in the last three columns of the table.

RESULTS AND DISCUSSION

Range of process parameters using factorial design

The tensile properties of the selected process parameters based on high and low level of values are shown on Table 3. This consideration shows the effect of low and high value of the selected parameters on the tensile properties and hence gave the required range of values for the optimization purpose. The results were analyzed by grouping the experimental blocks into a pair of similar parameters (for example mixing speed and MWCNT loading) and the third parameter (melting temperature) varied. Similarly, melting temperature and MWCNT were fixed and the mixing speed was varied. The analysis of the results showed that the tensile strength of the composite produced at 250 °C melting temperature was lower than that at 170 °C for a fixed MWCNT loading and mixing speed. The tensile properties at 170 °C were in the range of 30 to 32 MPa, for tensile strength and between 1200 and 1700 MPa for Young's modulus while at 250 °C, the properties lied between the range of 20 and 25 MPa tensile strength and 1000 and 1300 MPa Young's modulus respectively. The loss in the tensile properties at higher temperature could be traced to the degeneration in the plastic properties of PP, which resulted in low viscosity, and hence loss of binding effect on the fillers within the matrix. This agrees with the Eisten postulate, equation (1) which states that the effect of filler on modulus is proportional to that on viscosity.

$$\eta = \eta_s (1 + 2.5V_r) \quad (1)$$

Where η represents the viscosity of the composite, η_s represents the viscosity of the matrix and V_r the filler volume fraction. When the viscosity terms in equation (1) is replaced by the modulus term E, the equation gives:

$$E = E_s (1 + 2.5V_r) \quad (2)$$

E and E_s represent the modulus for composite and matrix respectively. Therefore, at 250 °C melting temperature, an irreversible drop in viscosity of the nanocomposites can be explained in terms of the drop in the stiffness measured as a function of Young's modulus.

Table 2. Central composite design for PP/Clay/CNT and responses

Run	Factor A CNT (%)	Factor B Temperature (°C)	Factor C Mixing speed (rpm)	Response 1 Tensile strength (MPa)	Response 2 Young's modulus (MPa)	Response 3 Elongation (%)
1	0.45	190	125	35.60	1847	8.30
2	0.8	170	125	50.26	1797	13.76
3	0.8	190	100	38.42	1735	9.70
4	0.1	190	100	34.53	1821	9.13
5	0.45	170	150	48.72	1634	11.36
6	0.1	150	150	52.30	1641	16.08
7	0.45	170	125	49.37	1725	13.35
8	0.1	170	125	51.37	1821	13.85
9	0.8	150	100	45.42	1849	11.07
10	0.45	150	125	47.78	1677	6.00
11	0.1	190	150	30.14	1725	7.43
12	0.45	170	125	49.37	1725	13.35
13	0.8	150	150	50.96	1663	14.94
14	0.45	170	100	47.11	1600	12.52
15	0.1	150	100	50.79	1780	14.77
16	0.45	170	125	49.37	1725	13.35
17	0.8	190	150	50.26	1797	13.76

The effect of shear was measured in terms of mixing speed. Therefore, at a fixed melting temperature and percentage MWCNT, the results show that tensile properties increase with shear but the rate of increase was not so significant. For example in the case of composites produced at a fixed temperature of 170 °C and 0.1% CNT loading (EH); the tensile strength was 31.27 MPa and 31.58 MPa at 100 and 200 rpm respectively and 1259.12 & 1209.06 MPa for Young's modulus respectively. The same explanation holds for other combinations such as; BC, AF and DG. It is therefore concluded that melting temperature and CNTs loadings have more significant effect on the preparation of PPTN compared to mixing speed.

Also when the filler loading was varied while the melting temperature and mixing speed kept constant (such as: AG, BH, CE and DE). Considering a pair AG, Table 3 shows that at a fixed melting temperature of 250 °C and mixing speed of 200 rpm, the tensile strengths at 0.1 and 1% MWCNT were 25.1 and 22.73 MPa while the Young's modulus 1513.48 and 1280.40 MPa respectively. This shows that at 1% MWCNT loading, the tensile properties are affected traceable to the formation of agglomerates in the polymer matrices. Specifically on the PPTN; increase in CNT may saturate the PP/Clay binary precursor due to the existing filler (layered silicate) in the matrix, thereby increased the rate of formation of agglomerates and hence reducing the tensile properties. Sometimes, Young's modulus may be favored by increasing filler loading but usually at the expense of other properties, such situation was observed in BH where the Young's modulus of the composite sample was 1259.12 MPa at 0.1% CNT and 1621.64 MPa at 1% CNT. Further increase in CNT might lead to a complete loss in both the tensile strength and the Young's modulus.

Test of significance and accuracy of the model

The level of significance and accuracy of the fitted models were tested through the analysis of variance using p-value. As shown in Table 4, it is evident from ANOVA that the third order polynomial model for both the tensile strength and Young's modulus are highly significant with p-value of 0.0018 and 0.0002 respectively knowing fully well that for p-value < 0.05, the model terms are significant and for p-value < 0.01, the model terms are highly significant [19]. However, the model terms for the elongation are not significant due to its high p-value (0.2598). This result is also justified from the experimental data, which showed an inconsistency pattern for the elongation.

Following the above arguments, it is evident from Table 4 that the variables that are highly significant are B, B^2, AB and AC while A^2 AB^2 and ABC are significant factors in the case of tensile strength where A represents the MWCNTs loading (%), B represents the temperature (°C) and C the mixing speed (rpm). In the case of Young's modulus (Table 6), the highly significant factors are; B, A^2, B^2, C^2, AB, AC, BC, A^2B, A^2C, AB^2 and ABC while only A is significant. From the above analysis both linear and polynomial effects of parameters were significant, meaning that they can act as limiting factor and little variation in their values can alter the properties (Tensile strength and Young's modulus) of the composite. Considering the third model (Elongation), no significant factor was recorded either in the linear or polynomial model (Table 4). This implies that the addition of CNT though enhanced other properties, but did not have any significant effect on elongation. It can therefore be inferred from the analysis of results that Table 4 are true a representation of the model and can be used to determine the responses of the composite to changes in the

Table 3. Tensile properties of PPTN for determination of range for process parameters

Sample PP/Clay/CNTs	Melting temperature (°C)	Mixing speed (rpm)	CNTs loading (%)	Tensile strength (MPa)	Young's modulus (MPa)
A	250	200	0.1	25.1	1513.48
E	170	200	0.1	31.58	1209.06
F	250	100	0.1	26.11	1121.80
H	170	100	0.1	31.27	1259.12
B	170	100	1.0	26.45	1621.64
D	250	100	1.0	23.53	1264.18
C	170	200	1.0	31.15	1375.24
G	250	200	1.0	22.73	1280.40

Table 4. Summary of statistical analysis using ANOVA

Statistical analysis	p-value		
	Tensile strength	Young's modulus	Elongation
Model	0.0018	0.0002	0.2598
A	0.4097	0.0324	0.9782
B	0.0018	0.0001	0.4741
C	0.2584	0.0127	0.7093
A^2	0.0171	< 0.0001	0.1055
B^2	0.0009	0.0006	0.0499
C^2	0.3937	< 0.0001	0.5025
AB	0.0009	0.0037	0.1283
AC	0.0031	0.0032	0.2356
BC	0.8682	0.0002	0.6511
A^2B	0.6451	0.0003	0.1304
A^2C	0.2164	0.0004	0.4054
AB^2	0.0246	0.0089	0.861
ABC	0.0133	0.0005	0.6069

$p < 0.05$ indicate the model terms are significant.
$p < 0.01$ indicate the model terms are highly significant

process parameters. However, if elongation is of prime importance in the application of nanocomposite, the plasticizer, mostly phthalate esters, can be added in a predetermined quantity.

To further reaffirm the significance of the fitted model using regression analysis, the coefficients of determinations (actual R^2 and adjusted R^2) were computed and compared. This is because R^2 only measures the amount of reduction in the variability obtained by using the regressor variables [20]. However, a large value of R^2 does not necessarily imply that the regression model is a good one and as a result, adjusted R^2 were computed and compared with actual R^2 to justify the significance of the model. Table 5 shows that the coefficient of determination R^2 and adjusted R^2 for both the tensile strength and Young's modulus are near to 1 which implies that the regression models are highly significant. In comparison with elongation, though R^2 of 91% is large enough but when compared with adjusted R^2 (52%), there is a wide gap. This also justifies the anomalous behavior of elongation. Adequate precision measures the signal to noise ratio (S/N). A ratio greater than 4 is desirable [19]. Hence adequate precission for both tensile strength and Young modulus ($\approx$ 30 and 60) are large enough for the significancy of the model. However, the adequate precission for elongation of 4.8 is just adequate but less significant or practically insignificant. This also is in agreement with other observation made so far in this research.

Table 5. Measure of significance using R^2 and adequate precision

Statistical tool	Tensile strength	Young's modulus	Elongation
R-squared	0.9973	0.9994	0.9112
Adj R-squared	0.9857	0.9966	0.5265
Adq. precision	29.868	60.624	4.892

Model equations

In order to predict the optimal regions for the production of ternary nanocomposite considering the three responses [tensile strength (T/S), Young's modulus (Y/M) and elongation (E)] within the experimental constraints: CNT loadings (A), melting temperature (B) and mixing speed (C). A third order polynomial model was fitted to the experimental results using the Design Expert software and the following models were developed for each of the responses.

Model equations in terms of actual factors:

$$T/S = -627 + 430A + 7.23B + 1.79C + 57.12A^2 - 0.02B^2 - 4.35AB - 2.49AC - 0.01BC - 0.25A^2C + 6.72E - 003AB^2 + 0.02ABC \quad (3)$$

$$Y/Mod = 531 + 1776A + 3B + 24C + 8487A^2 - 0.06B^2 - 0.18C^2 - 76.27AB + 8AC + 0.11BC - 36.83^2B - 12.57A^2C + 0.32AB^2 + 0.01ABC \quad (4)$$

$$Elong = -249 - 122.23A + 3.33B + 0.09C + 179.42A^2 - 9.47E - 003B^2 + 1.65E - 003C^2 + 0.72AB - 0.87AC - 3.01E - 003C - 1.21A^2B + 0.40A^2C + 1.50E - 004AB^2 + 3.895E - 003ABC \quad (5)$$

Following the model equations above, it can be inferred that equation 3 to 5 are better combinations and can be used for the determination of the required process parameters for the production of PP-ternary nanocomposites of a known tensile properties.

Condition for optimum responses

The relationship between the responses and the experimental variables are illustrated graphically by plotting three-dimensional (3D) response surface graphs (Figures 1 to 3). The *y*-axis represents any of the responses (T/S, Y/M or E) while *x* and *z*-axes represent any two of the three independent variables. In this case, *x* and *y* are melting temperature and MWCNT percentage respectively, while mixing speed was fixed at 125 rpm. In Figure 1 examination of three-dimensional plot showed that the tensile strength is in the optimum

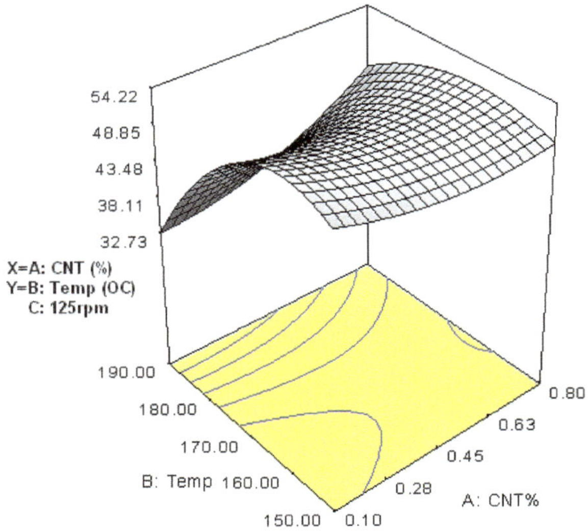

Fig. 1. Response surface plots showing the effect of temperature and MWCNT loadings on the tensile strength of PPTN

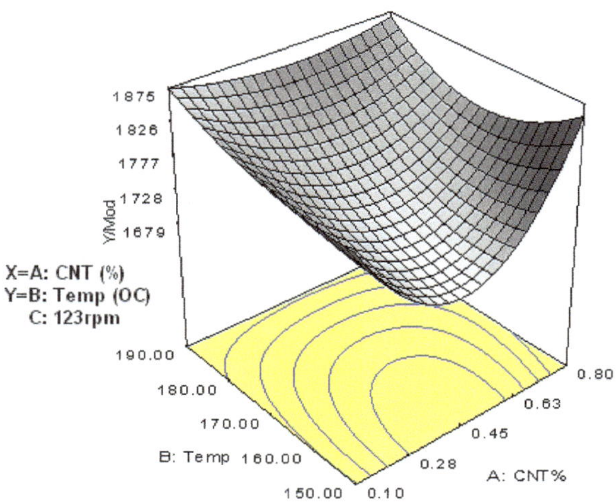

Fig. 2. Response surface plots showing the effect of temperature and MWCNT loadings on the Young's modulus of PPTN

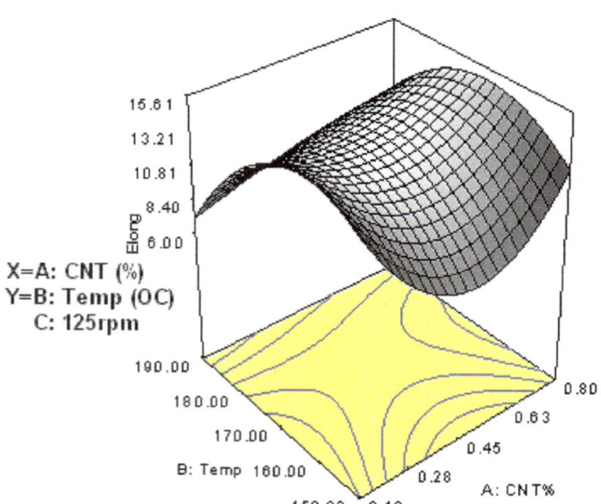

Fig. 3. Response surface plots showing the effect of temperature and MWCNT loadings on the elongation of PPTN

region when the temperature lies between 150 and 170 °C and MWCNT loadings is between 0.1 and 0.28%.

However, if the targeted property is Young's modulus, Figure 2 showed that the optima region also lies between 150 and 170 °C while the CNT loadings lies between 0.28 and 0.63%. Generally, this implies that for the production of PP-ternary nanocomposite with high tensile strength and Young's modulus as the targeted properties, the favorable temperature is between 150 °C and 170 °C and the MWCNT loading is between 0.1 and 0.6% considering Figure 1 and 2.

Regarding the elongation, no consistent pattern could be established statistically for the effect of the process parameters due to low predicted R^2, high p-value and low adequate precision however, some general trends which agreed with the predicted optimal regions such as melting temperature of 150 to 170°C and 0.28 to 0.63 MWCNT loading are indicated in the response surface curve as shown in Figure 3.

Validation of the model

In order to justify the optimization results obtained in Figure 1, 2 and 3, and to determine the exact optimum point a set of supplementary experiments were carried out based on the suggested optima conditions. The results are as shown in Table 6.

The highest tensile properties were obtained at optimum conditions, which are when the MWCNT content was 0.17%, melting temperature of 165.45 °C and mixing speed of 119.11 rpm. This is also justified from the surface response plots (Figures 1 and 2) where the optimum region lies between 0.1 and 0.28% MWCNT for the tensile strength and between 0.28 and 0.63% MWCNT for the Young's modulus. The optimum temperature lies generally between 150 and 170 °C.

CONCLUSIONS

In this study, ternary nanocomposites of Polypropylene, which comprises of modified nanoclay and multiwall carbon nanotubes have been prepared, characterized and the operating parameters optimized. The three key process parameters considered are MWCNT loadings, melting temperature and the mixing speed and the major responses were tensile strength, Young's modulus and elongation. This study clearly developed

Table 6. Validation of the developed model

NO	Process Parameters			Predicted		Experimental	
	CNT (%)	Temp. (OC)	M/S (rpm)	T/S	Y/M	T/S	Y/M
1	0.17	165.45	119.11	51.98	1757	52.30	1759.32
2	0.45	151.31	121.16	47.61	1677	48.19	1679.26
3	0.61	153.03	121.47	47.96	1718	47.43	1726.14

three statistical models that can be used to predict the operating parameters once the tensile properties are fixed for the production of PP-ternary nanocomposite. Hence, novel plastic materials of specific properties can be developed for various industrial applications.

REFERENCES

1. Mittal V. Polypropylene–layered silicate nanocomposites: Filler matrix interactions and mechanical properties. Journal of Thermoplastic Composite Materials, 20, 2007, 575–599.
2. Nello P. Polypropylene handbook. Second edition, Hanser Publishers, Munich, Hanser Gadner Publications, Cincinnati 2005, 326–334,
3. Karian H.G. Handbook of polypropylene and polypropylene composites. Marcel Dekker, NY. 2003, 155–158.
4. Valentini L., Biagiott I.K., Kenny J.M. Lopez M., Manchando M.A. Physical and mechanical behaviour of single wall carbon nanotube/polypropylene/ethylene-polypropylene-diene rubber nanocomposite. Journal of Applied Polymer Science, 89(10), 2003, 2657–2663.
5. Hu N., Masuda Z., Yamamoto G., Fukunaga H., Hashida T., Qin J. Effect of fabrication process on electrical properties of polymer/ multi-wall carbon nanotube nanocomposites composites. Part A. Applied Science and Manufacturing, 39(5), 2008, 893–903.
6. Xun X., Krishnan J., Caroline M.,, Nicholas S. Life Cycle assessment of wood-fibre reinforced polypropylene composite. Journal of Material Processing Technology, 198, 2008, 168–177.
7. Ma J., Qi Z., Hu Y. Synthesis and characterization of polypropylene-clay nanocomposites. Journal of Applied Polymer Science, 82, 2001, 3611–3617.
8. Thostenson E.J., Ren Z., Chou T.W. Advances in science and technology of carbon nanotubes and their composites. A review. Composite Science and Technology, 61, 2001, 1899–1912.
9. Bafna A., Beaucage G., Mirabella F., Mehta S. 3D hierarchical orientation in polymer-clay nanocomposite film. Polymer, 44(4), 2003, 1103–1115.
10. Fornes T.D., Yoon P.J., Keskkula H., Paul D.R. Nylon 6 nanocomposites: The effect of matrix molecular weight. Polymer, 42(25), 2001, 9929–9940.
11. Oya A., Kurokawa Y., Yasuda H. Factors controlling mechanical properties of clay mineral/ polypropylene nanocomposites. Journal of Material Science, 35(5), 2000, 1045–1050.
12. Gilman J.W. Flammability and thermal stability studies of polymer layered-silicate (clay) nanocomposites. Applied Clay Science 15, 1999, 31–49.
13. Isam Y.Q. The potential of carbon nanotubes production as a filler element in polymer nanocomposite: A review article. Minia Journal of Engineering and Technology, 30(1), 2011, 1–21.
14. Chen T.K., Tien Y.I., Wei K.H. Synthesis and characterization of novel segmented polyurethane/clay nanocomposite. Polymer, 41(4), 2000, 1345–1353.
15. Sudip R., Siew Y.Q., Allan E., Xiao D.C. The potential use of polymer-clay nanocomposites in food packaging. International Journal of Food Engineering 2(4), 2006, 4–13.
16. Laszlo S., Agnes A., Pukanszky B.Jr., Julius G.V., Punkaszky B. Morphological characterization of PP/Clay nanocomposites across the length scales of the structural architecture. Micromolecular Materials and Engineering, 291(7), 2006, 858–868.
17. Salawudeen T.O., Suleyman A.M., Qasim H.S., Ma'an F.A., Faridah Y., Isam Y.Q. Improving the polypropylene-clay composite using carbon nanotube as secondary filler. Energy Research Journal 1 (2), 2010, 68–72.
18. Manel M, Souhail B., Moncef C., Christelle R., Michel P., Christophe B., Hamadi A. Optimization of pecting extraction from lemon by-product with acidified date juice using response surface methodology. Carbohydrate Polymers, 74, 2008, 185–192.
19. Stat-Ease, Design Expert, Version 6.0.8. Available on wwwstatease.com(2002).
20. Douglas C.M. Design and analysis of experiments. 5th edition. Wiley, NY, 2000, 388–424.

27

EVALUATION OF THE EFFECTIVENESS OF THE SHOT PEENING PROCESS FOR THIN-WALLED PARTS BASED ON THE DIAMETER OF IMPRESSIONS PRODUCED BY THE IMPACT OF SHOT MEDIA

Kazimierz Zaleski[1], Stanisław Bławucki[1]

[1] Faculty of Mechanical Engineering, Lublin University of Technology, 36 Nadbystrzycka Str., 20-618 Lublin, Poland, e-mail: k.zaleski@pollub.pl; s.blawucki@pollub.pl

ABSTRACT

The paper presents the results of studies on the effectiveness of the shot peening process for thin-walled parts based on the diameter of impressions produced by the impact of shot media. It is also described the current applications of this treatment for the thin-walled parts. The authors described an innovative measuring device which was used in the research. The material used in the studies was non-alloy steel C45. The relationships between beads diameter and thickness of workpieces on diameter of impression were presented.

Keywords: shot peening, thin walled parts, metal treatment.

INTRODUCTION

Shot peening has been widely used in the engineering industry for almost one hundred years. The process has been partially normalized and is fully automated. Current research on the process involves examining the influence of various technological parameters on stress and deformation as well as surface layer properties of a workpiece. The aim of further research is to determine the relationships between conditions of the shot peening process and the material, geometry and rigidity of parts treated by shot peening [1].

The shot peening process is basically characterized by three categories of variables [2]. The first one is the intensity of treatment. This parameter determines the amount of energy of shot media (e.g. beads or pellets). The second parameter is saturation, or the increase in compressive stresses in the surface layer as a function of time of treatment. The final variable is the degree of coverage, that is, the percentage of surface which underwent the actual elastic-plastic deformation during processing [2].

An interesting group of structures are thin-walled workpieces, where the relationships between treatment-induced stresses and strains and the parameters of this treatment are non-linear. This results from lower weight of produced parts, which entails their reduced wall thickness. Although construction elements produced thereby are lighter and cheaper, their rigidity and fatigue strength are lower. This problem can be solved by the application of shot peening as finishing, which will improve some properties of the surface layer of such parts [3].

To better illustrate the problem of rigidity and strength, it is worth mentioning cases where the application of shot peening leads to a significant improvement of fatigue life of parts. A classic example is that of a low-pressure steam turbine rotor (Fig. 1a). The rotor is rotating with a velocity of about 3000 rev/min, the paddle extends due to the action of the centrifugal force and high temperature. This leads to the occurrence of tensile stresses and vibrations in the region of blade-rotor connection as well as hydrogen embrittlement on the face of the blade. A negative consequence of this are frequent and very costly breakdowns of steam turbines in thermal power plants. By treating the blades and rotor disc surfaces with shot peening, stream turbine failures can be eliminated [4].

a)
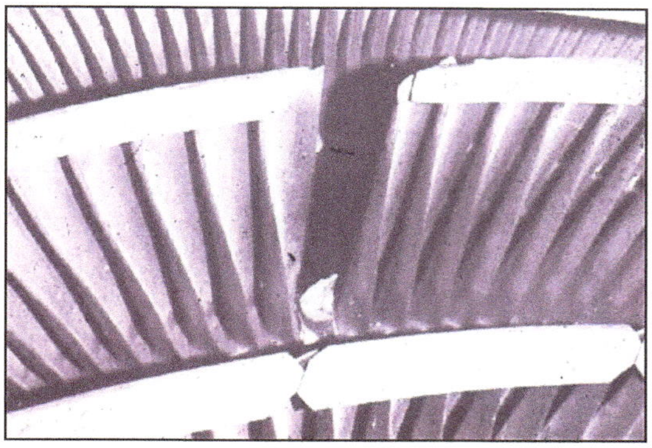
b)

Fig. 1. Blade failure in: a) stream turbine in a steam power plant [4], b) axial compressor of an aircraft jet engine [5]

Compressor rotors in aircraft jet engines are critical for both engine efficiency and flight safety. Many tragic accidents in aviation history resulted from the damage of a single blade of the axial compressor rotor. Given that, the rotational speed of the compressor reaches up to 20 thousand rev/min, even the slightest rotor dynamics imbalance leads to higher vibrations and rapid damage of the jet engine. The blades can be damaged due to their additional extension and friction against the housing or due to the impact of a foreign body, e.g. chunks of ice from the wing. The shot peening technique is widely used to improve tensile strength and flexural strength of rotating components of an axial compressor, which, in turn, improves the fatigue life of blades and their resistance to catastrophic damage [6].

Over the past two decades the German company KSA GmbH developed and implemented a large-scale automated shot peening technology at the Airbus factory. The method describes upon a phenomenon of controlled plastic deformation of material due to a concentrated impact of a steel bead shot. Low intensity treatment leads to production of convex surfaces, while high-intensity shot peening produces concave surfaces (Fig. 2a) [7]. The process can also be run on both sides simultaneously to produce cylindrical cones from a metal plate disc. This is done by the application of 6 mm diameter beads to treat the inner surface of the work piece, 4mm diameter beads to its outer surface, and a lower intensity shot flow. The above forming method can be used to produce properly corrugated plating for aircraft and space shuttles, which previously required manual adjustment [7].

In the pneumatic conveying of polymer pellets in the chemical industry, pipe conveyors would often fail, as their temperature is greatly increased due to prolonged friction of the medium being conveyed against smooth walls of the pipeline. The increase in the pipe wall temperature caused local plasticization of the pellets, which, in turn, led to conveyor congestion and halt of the conveying process. The application of intensive shot peening to the internal surfaces of pipes caused their roughness and surface layer hardness to increase significantly. The increase in their surface roughness led to a decrease in the temperature of the pipeline walls, while the significant increase in hardness contributed to reduced consumption of the conveyor surface [9].

a)

b)
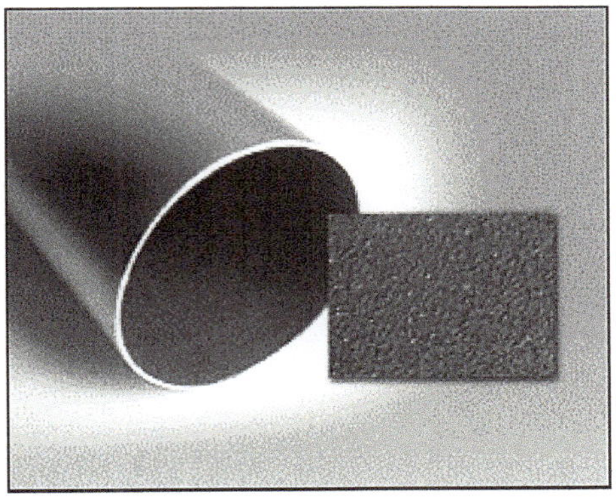

Fig. 2. Shot peening of thin metal sheets: a) aircraft plating [8], b) cylindrical inner surface of a pipeline / transporter [9]

An analysis of the literature reveals that shot peening has a variety of applications in many sectors of industry. An important issue regarding the selection of machining parameters is the assessment of shot peening effectiveness.

Previous studies on shot peening effectiveness did not take into account the fact that thin-walled materials undergo elastic deformation due to the impact of shot media. This phenomenon has a significant influence on effectiveness of the shot peening technique and should, therefore, be taken into consideration at the stage of selecting machining parameters.

The purpose of this study was to determine the effect of workpiece thickness on the diameter of an impression formed by the shot media. The observed variation in the diameter impression is regarded as a measure of shot peening effectiveness.

RESEARCH METHODOLOGY

The aim of the study involved determining the effect of workpiece thickness, shot media diameter and the distance from the point of workpiece fixing on the effectiveness of the shot peening process, reflected in the diameter of the impression formed on the workpiece surface due to the impact of shot media (Fig. 3). The preliminary examination involved determination of a range of the tested parameters, i.a. three diameters D of the bearing beads applied, five thicknesses g of the test samples with the dimensions of 90 mm × 15 mm × *g*, and five distances *L* from the point of workpiece fixing.

The fixed parameters of the experiment included the type of workpiece material (C45 non-alloy steel), surface roughness (after grinding and manual lapping) in the range of $Ra = 0.32 \div 0.63$ µm, shot working pressure of 0.3 MPa and bead impact path set to 260 mm. At constant pressure, the shot medium velocity varied from about 4 m/s for 4.00 mm diameter beads to about 6 m/s for 5.50 mm diameter beads. For the beads with a diameter of 4.00 mm, a loss of compressed air occurred at the walls of the measuring tunnel due to a relatively considerable difference between the cross-sectional areas of the beads and the channel. For beads with a diameter of 5.50 mm, this difference was substantially lower. Hence, the velocity of the shot medium increased with an increase in its diameter.

The experiment was planned using a monoselective test method to investigate the influence of kinetic energy of beads with a diameter *D* (4.00 mm, 5.00 mm and 5.50 mm, respectively), for samples with a thickness *g* (0.50 mm, 1.00 mm, 1.50 mm, 2.00 mm, 3.00 mm) and a distance *L* from the point of workpiece fixing (10 mm, 20 mm, 30 mm, 40 mm, 50 mm) on the diameter of produced impressions. Using the combination D×*g*×*L*, we obtained 75 measuring points. The measurements were repeated five times at each point. Mean values and standard deviations were calculated. The results were put to statistical and multivariate linear regression analysis. The validation test demonstrated that the test stand had appropriate measurement accuracy.

The measurements were performed on the test stand for making impressions (Fig. 4). It consisted of a measuring system (2), a compressor (1) and an electronic control system (5). The components were fixed to the base (2) of the stand. Inside the rack (14) there was a square section hollow tunnel and a decompression slot. The upper part of the rack was provided with two sensors (13) for measuring the time of bead travel between the sensors. The upper part of the test stand comprised a stage (12) with a special clamp (10). The clamp made the workpiece (11) rigidly secured, so as to allow the beads to impact the workpiece surface at varying distances from the point of its fixing. Between the workpiece and the stage there was a small air space which enabled free deflection of the workpiece due to the shot medium impact.

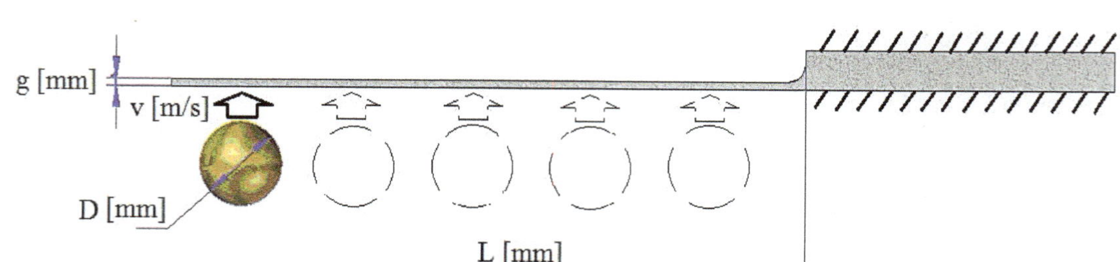

Fig. 3. Test parameters for a workpiece with thickness *g*

A ball bearing bead (4), also called shot medium, gains a set airspeed in the wind tunnel of the rack (14) by a dose of compressed air. Compressed air with a pressure of 0.3 MPa is located in the cylinder (3). The compressed air is produced by a compressor (1) equipped with an indicator gauge (15). When pressed, the trigger (8) opens the solenoid valve (9) which conveys a dose of compressed air from the cylinder (3) into the wind tunnel of the rack (14), setting the bead (4) to the initial velocity *v*. The digital signal from the sensors (13) is transmitted over the wires (7) directly to the electronic control system (5) where two travel times of the shot medium (4) are conditioned and then displayed on the display (6). The design of the experimental test stand is claimed in patent application no. P.408630, which relates to assessing efficiency of the shot peening process for thin-walled parts based on the diameter of the produced impressions.

The diameters of the impressions on the workpiece surface were measured using a toolroom microscope. Figure 5 shows the relationship between impression diameter and workpiece thickness, with the shot medium diameter and distance from the point of fixing maintained constant. The image is magnified ten times.

RESULTS AND ANALYSIS

It has been found that there is a strong correlation between the thickness *g* of the workpiece and the diameter *d* of the impression produced by the shot medium with the known diameter *D* and the velocity *v* (Fig. 5). The shot peening efficiency increases with an increase in the transverse dimension of the workpiece; for thicker samples, however, the relationship $d = f(g)$ is less evident (Fig. 6). This can result from insufficient rigidity

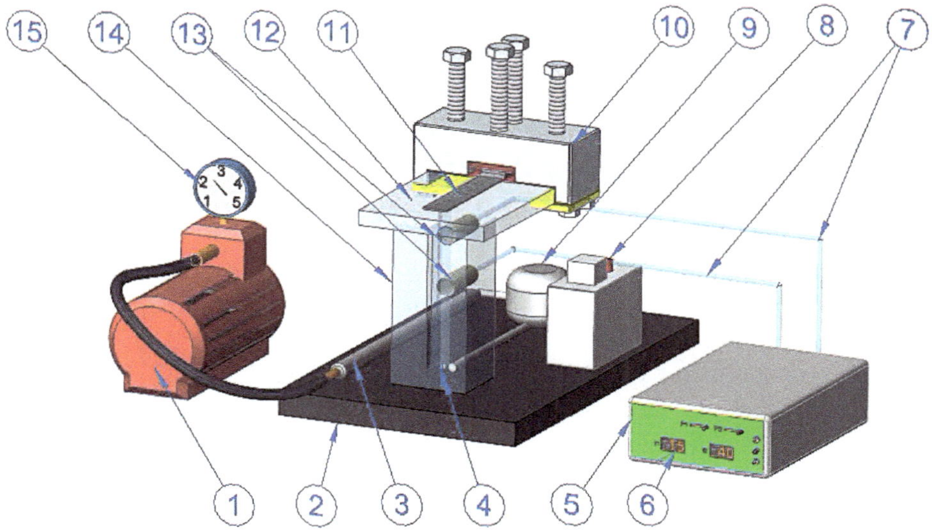

Fig. 4. Test stand for assessing effectiveness of the shot peening process (described in the text)

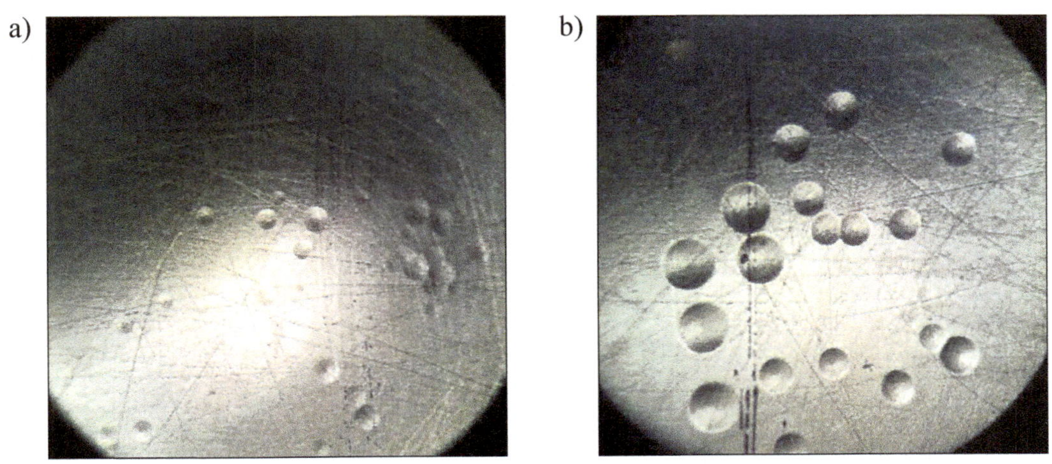

Fig. 5. Impressions on the workpiece surface: a) workpiece thickness $g = 0.50$ mm, b) workpiece thickness $g = 3.00$ mm, magn. 10×

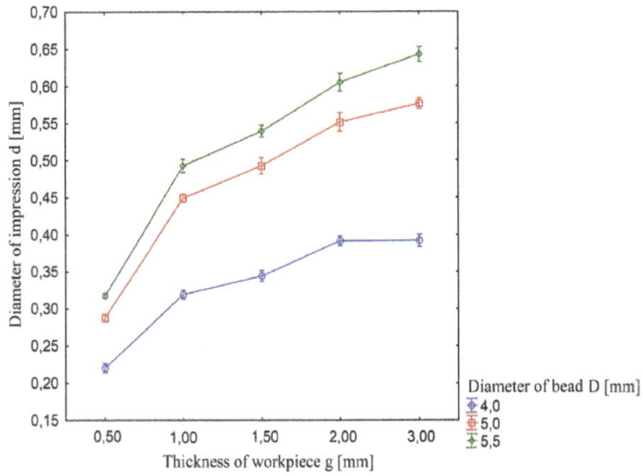

Fig. 6. Diameter of the resulting impression as a function of workpiece thickness for different bead diameters

of thin samples. Probably, due to the impact of the shot medium, a thin-walled sample (e.g. with a thickness $g = 0.50$ mm) undergoes a temporary deflection in the bead-workpiece contact area, which results in dispersion of the impact kinetic energy of the elastic deformation work. Consequently, the shot peening efficiency for thin samples can be more than half lower than for thicker samples. To produce strain on the surface of a thin-walled workpiece that is comparable to that of thicker samples, shot peening velocity must be significantly increased.

The results do not reveal the effect of the distance L from the point of workpiece fixing on the diameter d of the workpiece impression (Fig. 7). The higher the distance between the point of fixing and the impact point is (Fig. 3), there is no observable reduction in the diameter of the impression which would indicate a decrease in the efficiency of the shot peening process. What can be observed, however, are small deviations from

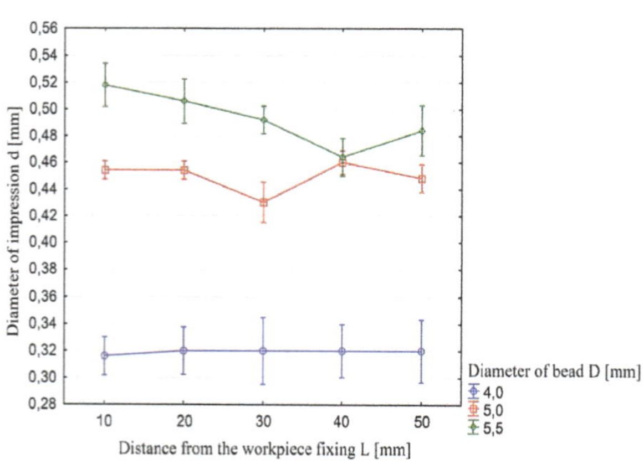

Fig. 7. Diameter of the resulting impression as a function of distance from the workpiece fixing for different bead diameters

the mean diameter of the impressions at the distance $L = 30$ mm from the point of workpiece fixing; these deviations can be observed for different bead diameters. This can be attributed to the occurrence of mechanical vibrations which differ from those in other regions of the workpiece. It is recommended that additional tests be conducted to investigate this region.

As expected, there is a relationship between the diameter D of the shot medium (ball bearing beads) and the diameter d of the produced impression (Fig. 8). An increase in the diameter of the beads leads to an increase in the diameter of the impression. One can also observe that the relationship $d = f(D)$ no longer holds when higher diameter beads are used, despite the described increase in velocity. It can be claimed that with the increase in the bead diameter, the contact surface at the point of impact increases, but the impact depth decreases.

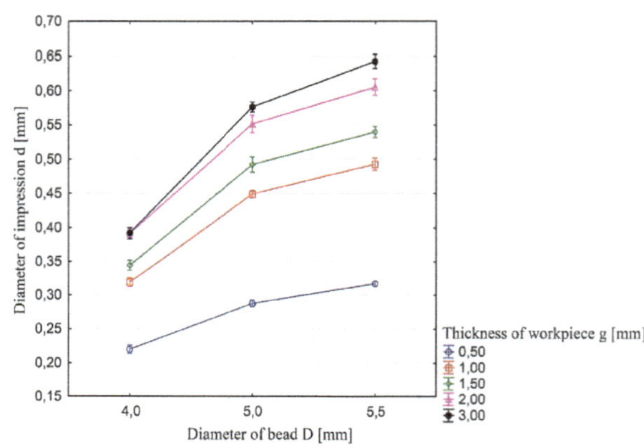

Fig. 8. Diameter of the resulting impression as a function of bead diameter for different workpiece thicknesses

Figure 9 illustrates the influence of individual parameters on the efficiency of processing variables. The chart shows the interaction between the parameters used in the experiment and their varying effects on the diameter of produced impressions. The input parameters can be set in such a manner to produce the smallest (approx. 0.20 mm) diameter impression (e.g. when $D = 4.00$ mm, $g = 0.50$ mm, $L = 30$ mm). It is also possible to determine the local extreme of the maximum impression diameter, e.g. when $D = 5.50$ mm, $g = 3.00$ mm and $L = 30$ mm. When the parameters are set in this way, the impression diameter is about 0.63 mm, so it is three times higher than the one in the previously discussed case.

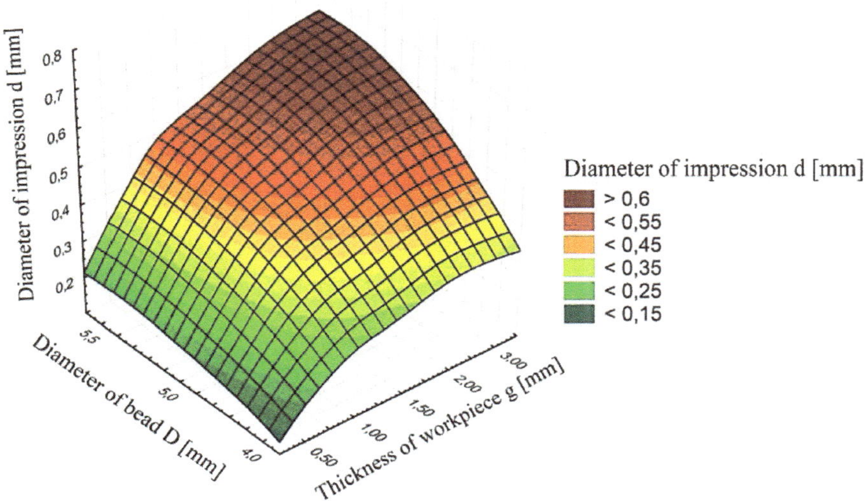

Fig. 9. Diameter of the resulting impression as a function of workpiece thickness and bead diameter

CONCLUSION

The experiment allowed us to determine the effect of selected geometric parameters on the effectiveness of the shot peening process for thin-walled parts. The process effectiveness was determined based on the diameter of produced impressions. As a result, the following conclusions have been drawn:

1. Workpiece thickness has the greatest effect on shot peening effectiveness. The smallest impressions were observed for the lowest thickness workpiece (g = 0.50 mm). It can be assumed that such parts should be subjected to more intense treatment to produce the effect obtained for thicker samples.
2. The distance from the point of workpiece fixing does not affect the effectiveness of the shot peening process as measured by the diameter of impressions produced by the shot medium.
3. The diameter of the shot medium affects the size of the resulting impression. The higher the diameter of the shot medium (beads) is, the higher is the diameter of produced impressions, and thus the higher is the effectiveness of the shot peening process.
4. Further studies should focus on identifying the form of workpiece mechanical vibrations induced by the shot medium. The use of a time-lapse camera in the impact area should provide interesting results in this regard.
5. The range of shot medium diameters applied in investigations should be higher: the diameters should be smaller than 4.5 mm and greater than 5.50 mm.

REFERENCES

1. Bhuvaraghan B., Sivakumar M.S., Prakash O., Overview of the effects of shot peening on plastic strain. Work Hardening and Residual Stresses, Computational Materials, Editor: Wilhelm U. Oster. Nova Science Publishers, 2009, 49–117.
2. Kyriacou S., Shot-Peening mechanics a theoretical study. ICSP6, San Francisco, 1996, 505–516.
3. Bozdana A. T., On the mechanical surface enhancement techniques in aerospace industry – a review of technology. Aircraft Engineering and Aerospace Technology, 77 (4), 2005, 279–292.
4. Ralph J., Ortolano P.E., Kleppe R. L., Shot peening in stream turbines. Technical Services, Metal Improvement Company, Southern California Edison Company, 2005, 1–17.
5. Yoon J., Effect of rain & snow on jet engines: //www.aerospaceweb.org/question/propulsion/q0293.shtml. February 2007.
6. Nowell D., Duo´ P., Stewart I.F., Prediction of fatigue performance in gas turbine blades after foreign object damage. International Journal of Fatigue, 25, 2003, 963–969.
7. Friese A., One of the World's Largest Shot peening machines installed at airbus, Metal Finishing News, Vol. 5, 2004, 18–20.
8. Kenny C., Shaping parts with shot peen forming. The Shot Peener, 3, 2014: 7–8.
9. Campaigne J., The S.P. Staff, Airblast AFC is on the right track. The Shot Peener, 4, 2013, 44–46.

THE VERIFICATION OF THE INFLUENCE OF HEATING STEEL CHARGE PARAMETERS ON THE THICKNESS OF SCALE LAYER

Agnieszka Benduch[1], Jarosław Boryca[1]

[1] Department of Industrial Furnaces and Environmental Protection, Faculty of Process & Material Engineering and Applied Physics, Technical University of Częstochowa, Al. Armii Krajowej 19, 42-200 Częstochowa, Poland, e-mail: abenduch@wip.pcz.pl

ABSTRACT

Parameters of heating are understood as the conditions in which heating process is conducted to obtain the required steel properties. During the heating process, on the surface of heated elements, layer of scale is formed as a by-product,. The selection of appropriate heating parameters allows for achieving the required layer of scale, and thus the predictability of the process effects . The article presents sample calculations of heating parameters for the charge of steel and the analysis of the results of conducted experiments.

Key words: heating of steel, oxidation, scale.

List of designations

A – the surface area, m^2,
t – temperature, °C,
m_0 – mass of the specimen before heating, kg,
m_3 – mass of the specimen after complete cleaning, kg,
α – value of combustion air excess ratio,
x_{Fe} – part of iron in the scale (0.74),
s – thickness of scale layer, m,
z – surface steel loss for scale, kg/m^2,
τ – heating time, h,
ρ – density of scale, (3900 kg/m^3),

INTRODUCTION

In the process of heating steel before plastic working, in addition to heat exchange-related processes, the processes of chemical interaction also occur between the furnace atmosphere and the surface of the steel being heated. As a result of oxidation, scale forms, which causes not only the losses of the heated material,slows down the heating process and increases the consumption of fuel, but also adversely affects the service file of heating equipment [1–3]. The most important external factors influencing the formation of scale are:

- heating time,
- temperature of steel charge,
- temperature of furnace chamber,
- furnace atmosphere [4, 5].

The selection of appropriate heating parameters can also help to obtain the specified thickness of the scale layer, thus to enable the predictability of the results of a given heating process. The "appropriate" heating process parameters are understood as the adjustment of the heating duration and temperature and such a composition of gaseous atmosphere that gives the obtained scale layer the assumed thickness [6]. A scale layer, even a thin one, makes an isolating layer for the heat transfer due to its low thermal conductivity. Within the porous charge heating region, the contact heat conduction plays a key role, therefore, the oxidation phenomenon is crucial throughout the heat treatment process. The research on the scale layer thickness, which is described in the paper, constitutes an introduction to a doctoral thesis that deals with the heating of porous charge covered with an oxide film.

The literature concerned with heat exchange problems lacks information on the influence of the scale layer on the heat exchange process during heating charge of a porous nature. Examples of this type of charge can include:

- bundles of sections, flats, bars and tubes,
- coils of sheet, strip and wire rod,
- wire wound on a reel,
- small elements in bulk quantities, e.g. screws, nuts.

The conducted research is aimed at defining how the scale layer influences the heat exchange process in a bundle of steel rods and flats. For this purpose, it is necessary to analyze the scale formation process and to determine the heating parameters that assure a specific scale layer thickness to be achieved. When the thicknesses of scale layers forming on heating conducted according to calculated and assumed parameters are known, it will be possible to commence investigation to examine the heat exchange in the bed of steel elements covered with a scale layer.

THE HEATING PARAMETERS

The determination of the heating parameters is understood as the selection of the proper conditions under which the heating process is to be conducted. Due to the fact that heat treatment operations run at a high level of thermal energy consumption, the proper selection of heating parameters becomes a crucial issue. Efforts should, therefore, be made to optimize heat treatment processes by carefully selecting the heating parameters [7, 8]. The effect of the correct heating is a uniform temperature field on the entire cross-section of the treated element [9].

The most frequently applied methods of heating are:

- slow heating charge together from furnace,
- accelerated heating; charge places in furnace heating to suitable temperature,
- fast heating; charge places in furnace about initial higher temperature from the temperature required in the heating process.

The calculations of heating time for obtaining definite thickness of scale layer in work were executed, accepting following principles:

- thickness of scale layer $s = 1.5$ mm $= 0.0015$ m,
- temperature $t = 1100$ °C,
- value of combustion air excess ratio $\alpha = 1.1$.

The thickness of scale layer is calculated using the relationship [1, 5]:

$$z = 356 \cdot \tau^{0,75} \cdot \alpha^{0,85} \cdot \exp\left[\frac{7500}{T}\right] \qquad (1)$$

From Equation (1), the following can be obtained from the heating time τ:

$$\tau = \sqrt[0,75]{\frac{z}{356 \cdot \alpha^{0,85} \cdot \exp\left[\frac{7500}{T}\right]}} \qquad (2)$$

The thickness of scale layer is calculated using the relationship [2,3]:

$$s = \frac{z}{\rho \cdot x_{Fe}} \qquad (3)$$

From Equation (3), we obtain:

$$z = s \cdot \rho \cdot x_{Fe} \qquad (4)$$

where: z – surface steel loss for scale, kg/m^2,
ρ – density of scale $\rho = 3900$ kg/m^3,
x_{Fe} – part of iron in the scale $x_{Fe} = 0.74$.

From here:

$z = 0.0015 \cdot 3900 \cdot 0.74 = 4.329$ kg/m^2

By substituting the data to the equation (2) the heating time τ is calculated using the relationship:

$$\tau = \sqrt[0,75]{\frac{4.329}{356 \cdot 1.1^{0.85} \cdot \exp\left[\frac{7500}{1373}\right]}} =$$

$= 3.648$ h $= 218.9$ min

THE TEST BENCH AND TESTING METHODOLOGY

The steel loss was determined using carbon steel St3S samples sized 30×30×50 mm, which were heated up in the furnace. Figure 1 presents the photos of samples. General view of heating stand is presented in Figure 2.

The basic element of the test bench is an electric furnace, type KS 520, with heating chamber dimensions of 220×220×520 mm. A combustion chamber with a gas burner are integrated with

Fig. 1. Carbon steel samples St3S, used for research

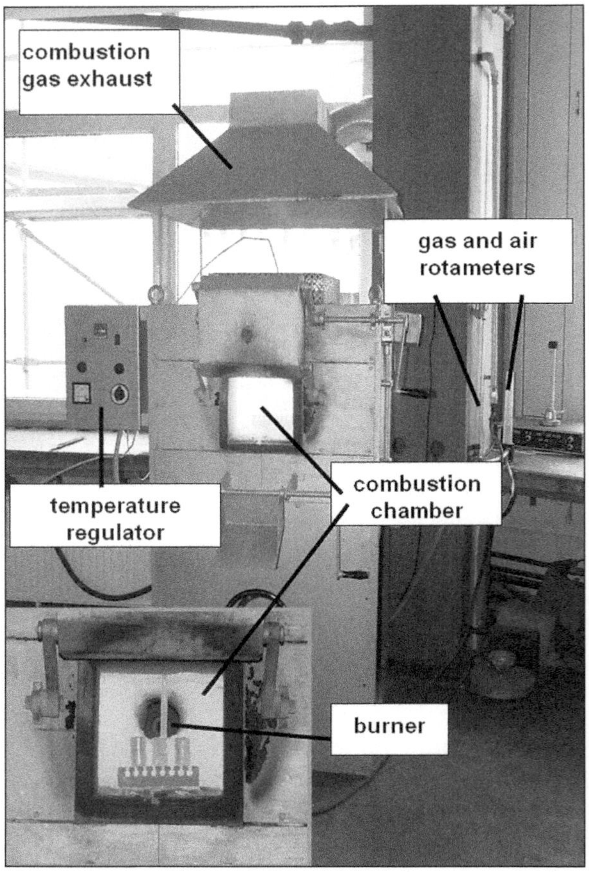

Fig. 2. General view of heating stand for steel samples, electric-gaseous furnace KS 520

the furnace. The burner performs the role of a gaseous atmosphere generator. The value of excess air ratio, and thus the chemical composition of combustion gas, is regulated in the burner. The remaining thermal power of the furnace is supplied in a form of electric power, which allows the temperature being precisely controlled at any location of the heating chamber of the furnace.

The temperature in the furnace is controlled with TROL – 9090 regulator. The accuracy of temperature control is ±1.0 K. The furnace chamber temperature is measured with PtRh-Pt control thermocouple and a three NiCr-Ni measuring thermocouple.

The supervisory sample was placed in the destination of temperature verification in the axis of steel charge, in the furnace chamber with thermocouple NiCr–Ni. The value of excess air ratio is controlled by continuous measurement of gas flow intensity and the air supplied to the burner. The intensity of gas and air flow is measured with rotameters. The flow of gas and air carried out suitably $V_g = 0.488$ m³/h, $V_p = 5.0$ m³/h. The analysis of combustion gas composition was conducted with the TESTO 350 analyzer. Result of measurements of combustion gas composition are summarized in Table 1.

Table 1. The composition of combustion gas in a research

Component of combustion gas	Value	Individual
CO_2	10.8	%
O_2	2.0	%
NO	65.0	ppm
NO_x	68.0	ppm
CO	0	ppm

The investigation on three different methods was conducted:

- I method: The sample was placed in the furnace chamber and which was then heated to the temperature 1100 °C. Then it was heated for 3 hour 39 min.
- II method: The studied sample was inserted into the furnace chamber heated to the temperature 1100 °C. Then it was heating by 3 hour 39 min.
- III method: The sample was placed in the furnace chamber heated to obtain temperature 1100 °C in the axis of the sample. Then it was heating by 3 hour 39 min.

ANALYSIS OF RESULTS

The heating process was carried out for the assumed and calculated parameters. The aim of the undertaken tests was to verify the obtained scale layer thickness for the analytically determined heating parameters.

After leaving the heating unit, the heated samples were allowed to cool down. Scale that formed during the heating process for all of the

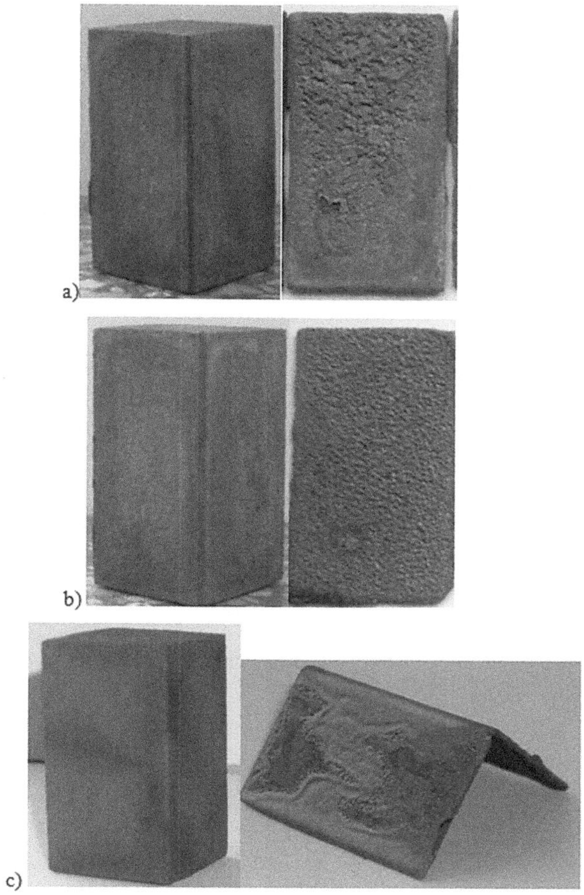

Fig. 3. Photographs illustrating the sample after heating and a layer of scale formed: a) I method, b) II method, c) III method

three methods came off from the steel substrate very easily, as shown in Figure 3.

The small adhesion caused ideal separation of scale layer from the sample surface. This permitted exact qualification of scale thickness. Measurements of the thickness of scale layer dimension were done with a micrometer with accuracy ± 0,01 mm.

The results of measurements of scale layer thickness, measured at three points on each side of the sample are summarized in Table 2.

By examining the final effect of the tests it could be concluded that the thickness of the scale layer formed in the process of heating according to method III attained the assumed value. For heating according to method I, a deviation of 0.15 mm from the assumed scale layer thickness was noted, while for heating by method II, a deviation of 0.23 was found.

These slight deviations from the assumed scale layer thickness are due to, among other things, the arrangement of the sample in the furnace, since the sample base that contacted the furnace bottom was distinguished by a thinner scale layer. This happens because this is the location where the access of the furnace atmosphere is more difficult. Moreover, it should be noted that the lowest scale layer thickness was obtained for heating method II. This is due to

Table 2. The values of the scale layer thickness

Number of the method	Place of measurement	Values of the thickness of scale layer, mm			
		Point 1	Point 2	Point 3	Value average
Method I	Side of sample 1	1.40	1.43	1.38	1.40
	Side of sample 2	1.41	1.32	1.29	1.34
	Side of sample 3	1.50	1.42	1.40	1.44
	Side of sample 4	1.41	1.32	1.30	1.34
	Basic of sample 1	1.50	1.40	1.20	1.36
	Basic of sample 2	1.25	1.18	1.20	1.21
	Value average				1.35
Method II	Side of sample 1	1.37	1.30	1.27	1.31
	Side of sample 2	1.35	1.31	1.30	1.32
	Side of sample 3	1.30	1.27	1.25	1.27
	Side of sample 4	1.31	1.26	1.27	1.28
	Basic of sample 1	1.36	1.25	1.22	1.27
	Basic of sample 2	1.20	1.10	1.15	1.15
	Value average				1,27
Method III	Side of sample 1	1.50	1.47	1.45	1.47
	Side of sample 2	1.50	1.50	1.51	1.50
	Side of sample 3	1.51	1.50	1.50	1.50
	Side of sample 4	1.50	1.49	1.48	1.49
	Basic of sample 1	1.50	1.49	1.47	1.48
	Basic of sample 2	1.48	1.41	1.40	1.43
	Value average				1,48

the fact that during heating according to method II the sample was held in the scale formation promoting temperature (above 900 °C) for much shorter time.

It should also be noted that in the case of heating according to methods I and II the furnace working temperature was 1100 °C; however, the temperature on the charge cross-section reached the value of 1064 °C. This slight difference could have also contributed to the deviations in the scale layer thickness values.

Heating conducted according to the assumptions of method III was aimed at eliminating the cross-section temperature difference. Therefore, the sample was heated until temperature of 1100 °C was obtained in the axis. With the use of this mode of heating it was found that the obtained scale layer thickness conformed the assumption.

SUMMARY

The results of the conducted tests indicate the possibility of using analytical relationships for modelling the thickness of the scale layer forming in the charge heating process. However, the sample heating method should be considered. The scale forming phenomenon might have a significant influence on the results of porous charge testing; therefore, allowing the charge oxidation process should be considered in the context of both experimental and model studies concerning this problem. This possibility will be exploited in the investigation of the heating of porous charge covered with a scale layer.

REFERENCES

1. Boryca J.: Przyczepność warstwy zgorzeliny powstałej w procesie nagrzewania wsadu stalowego, Praca doktorska, Politechnika Częstochowska, Częstochowa 2005.
2. Boryca J., Kieloch M., Piechowicz Ł.: Wpływ grubości warstwy zgorzeliny na przyczepność, Nowe technologie i osiągnięcia w metalurgii i inżynierii materiałowej, Seria Metalurgia nr 48, Częstochowa 2005, 88-91.
3. Kieloch M.: Energooszczędne i małozgorzelinowe technologie nagrzewania wsadu stalowego, Wydawnictwo Politechniki Częstochowskiej, Częstochowa 2002.
4. Kieloch M., Boryca J., Piechowicz Ł.: Przyczepność zgorzeliny w procesie nagrzewania wsadu stalowego, Hutnik, 1, 2005, 7-12.
5. Kieloch M.: Racjonalne nagrzewanie wsadu, Wydawnictwo Politechniki Częstochowskiej, Częstochowa 2002.
6. Benduch A., Halusiak B., Poskart A.: Nagrzewanie elementów wsadu stalowego dla uzyskania określonej grubości warstwy zgorzeliny, In: Dyja H., Kawałek A. (Eds.) New Technologies and Achievements in Metallurgy and Materials Engineering. A Collective Monograph No 24, Chapt. 1. Częstochowa 2012, 11-14.
7. Luty W. (Ed.): Poradnik inżyniera. Obróbka cieplna stopów żelaza, WNT, Warszawa 1997.
8. Kieloch M.: Technologia i zasady obliczeń nagrzewania wsadu. Wydawnictwo Politechniki Częstochowskiej, Częstochowa 1995.
9. Wyczółkowski R., Szecówka L.: Wybrane zagadnienia związane z nagrzewaniem wsadu stalowego o strukturze porowatej. Nowe technologie i osiągnięcia, Częstochowa 2009, 337-359.

THE SHEAR STRENGTH OF THE ROUND CLINCHING JOINTS FORMED BY USING EXTENSIBLE DIES

Waldemar Witkowski[1]

[1] Rzeszow University of Technology, Powstańców Warszawy 12, 35-959 Rzeszów, Poland, e-mail: wwitkowski@prz.edu.pl

ABSTRACT

The clinching joining technology is one of the most popular joining technologies by redrawing sheet material. The joining process parameters, the sheet material and its arrangements influences joints' strength. Cylindrical axial-symmetrical joints formed by using rigid die and punch are still developed. The change of the rigid die on the die with movable segments affect the forming process and joint strength parameters. So there is a need to do experimental researches of the possibilities of joint formation by unchanged punch geometry and different die shape. In this article the forming process parameters (forming force, process energy consumption and its standard deviations) and joints strength parameters (maximum shearing force, total dissipated energy and dissipated energy up to 0.3 maximum force), according to the ISO 12996 standard, for the joints formed with using die with 2, 3 and 4 movable segments were presented. The punch geometry was unchanged and the minimum thickness of the embossment was also unchanged. For the die with 2 and 4 segments the load force direction influence on the joints strength was also presented.

Keywords: sheet metal joining by redrawing, ISO 12996, joints strength,

INTRODUCTION

Development of new material used in the engineering design, especially in automotive industry, in household equipment industry or in aircraft industry, cause the right choice of the appropriate joining methods. Classical joining methods such as welding, soldering or temporary fastening cannot be always implemented to join modern materials. Therefore, alternative joining technologies are still being developed. One of such method is joining by pressing the materials. This technology, depending on the tools shape, use of additional fastener has many variations. The most common type of variations is joining by redrawing sheet without any additional fastener (clinching – CL) formed by using solid tools with axis-symmetrical shape. This shape of tools provides isotropic properties of the joint [1, 2, 3].

The basic geometry of the forming tools includes: punch diameter, diameter and depth of the die, the radius of the punch and die and the angle of inclination of the punch and die. These dimensions affect the proper formation and the joints strength. For different joined materials, its arrangement and thicknesses the tools geometry should be selected separately. The basic parameter of the joint is the minimum thickness of the joints embossment (X), which determines the properly joint formation. Too low value of the parameter X causes that the material of the upper sheet lose its coherence in the bottom sheet. Too large value of X causes the interlock is formed not properly. The influence of the parameter X and punch diameter on the joints strength was presented in [4].

Numerical researches involving the forming process and the strength of the joints made by pressing [5, 6, 7] represent a large contribution to the joints technology development. The determination of the tools geometry for the different material their thickness and its arrangement cause an increase in the use of computer simulation of forming process in order to reduce experimental studies [8, 9, 10]. The experimental research are usually related to the joints strength in shearing and tension test [11, 12]. Some publications present results of the joints strength in multi-axis load state [4, 13], because the joint in real constructions need to carry the load different than shear or tear force.

The change of the shape and geometry of the die, with the punch unchanged, affecting the joint forming process and the joints strength. Influence the sheet thickness and forming force on the interlock shape and size of clinching and hybrid (clinching and glue) joints formed by using the die with three movable segments was presented in [14, 15]. However, there is no publication about the influence of the number of movable segments of the die and the load direction related to the die segments on the joints strength in shearing test.

In this publication an experimental analysis of the clinching joints strength was presented. The joints were made by using the unchanged punch and different dies with movable segments (2, 3 and 4 segments). The influence of the force load direction related to the segments direction was presented. To compare the joints made by using different number of movable segments of the die the joints parameters, according to the ISO 12996 standard [16], were used.

EXPERIMENTAL RESEARCH

For the experimental research the DX51D+Z/275 (mat. number 1.0226) sheet material was used. The thickness of the samples was *1*±0.05 mm. Basic mechanical properties of the sheet material are:

- elastic modulus (Young's mod.) $E = 210000$ MPa,
- the yield strength $R_{p0.2} = 175$ MPa,
- the tensile strength $R_m = 375$ MPa,
- elongation of 80 mm $A_{80} = 32\%$,
- the coefficient of lateral constrictions (Poisson's ratio) $v = 0.3$.

The chemical composition of the steel and maximum content in % is: C (0.1), Mn (0.7) P (0.1), Ti (0.06), S (0.035), Al (0.025), N (0.01), Fe (other).

The lap joints were performed in the Pressed Joint Laboratory of the Department of Mechanical Engineering at Rzeszow University of Technology. The press frame, with electric drive type EMPK, manufactured by ToxPressotechnik company was used [17]. The accuracy of the punch positioning in relation to the die was 0.01 mm, and maximum forming force was set to 100 kN. The measuring and recording of the forming force as function of the punch displacement , with accuracy of 0.5% of the forming force, was enabled by force control systems of the press. The technological parameters of the forming process were chosen in accordance to the recommendation of the press manufacturer [18]. Measuring and recording of the forming force was carried out only in the forming step with punch speed $V = 3$ mm/s, until the parameter X was reached. The minimal thickness of the embossment (X) was set up to a value 0.5 mm (25% of total thickness of joined sheets). The real value of parameter X was measured by using Mitutoyo digital calipers with measurement accuracy of 0.01 mm in range of 20 mm. The punch diameter was 5.6 mm.

The dies with movable segments were designed and manufactured to ensure the interlock forming without losing coherence of sheet material. The die depth was 1.4 mm, and the outer joint diameter of the embossment was 8 mm (Fig. 1).

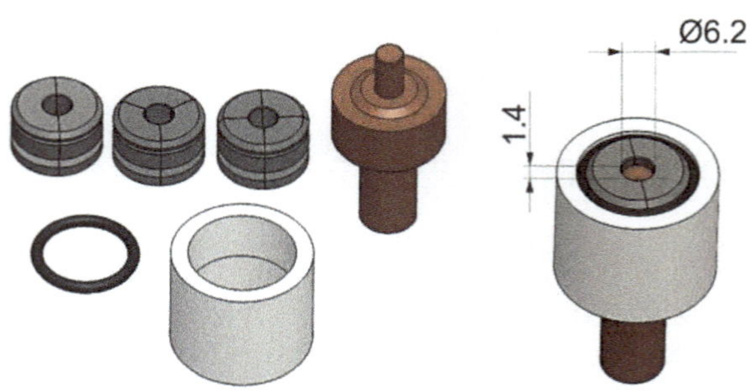

Fig. 1. CAD models of forming tools

The sheets samples with the length of 110 mm, width 40 mm and 1 mm thickness were joined in five samples for each type of die and force load direction. The length of the overlapping elements was 40 mm. In Figure 2 the load force directions were presented for each type of dies.

Static strength test for lap joints were carried out in accordance with guideline set in ISO 12996 standard. Shear tests were performed on UTS 100 testing machine equipped with the extensometer system (the measuring calipers were placed 20 mm from the joint, Figure 3).

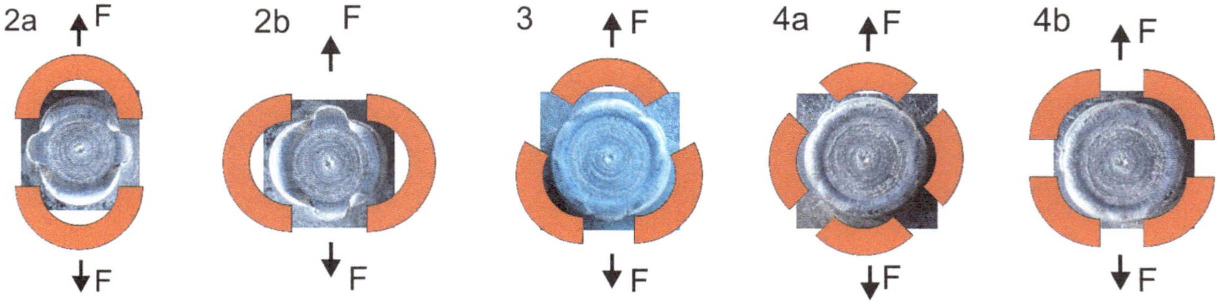

Fig. 2. Directions of load force in accordance to the die segments and samples nomenclature

Fig. 3. UTS 100 testing machine with extensometer system

RESULTS AND ANALYSIS

During the joint forming the force-displacement curves of the punch were recorded. The average of the forming energy and the forming force with their standard deviations were calculated (Fig. 4). For all dies there was no difference in forming process. There was no crack in the cylindrical part of the embossment. The number of movable segments does not significantly affect the forming force and forming energy.

The lap joint samples joined with parameter X=0.5 mm, were tested in shearing test. The joints were destructed by losing sheet material coherence in cylindrical part of the interlock. There were no samples with partial coherence loosing and pulling out form the bottom sheet.

The main parameters of the joint strength is the maximum load force (F_{max}). The value of this force was different for all die variants. In ISO 12996 standard other parameters characterizing the joint strength were given, such as: total dissipated energy (E_t), dissipated energy up to 0.3 maximum force ($E_{0.3Fmax}$), dissipated energy up to maximum force (E_{Fmax}), displacement up to fracture (s_t), displacement up to 0.3 maximum force ($s_{0.3Fmax}$) and displacement up to maximum force (s_{Fmax}), stiffness in elastic range, elastic load limit and others (Fig. 5). The values of some of the joint parameters were presented in Table 1.

The lap joint separation was not observed at maximum value of the shear force. After achiev-

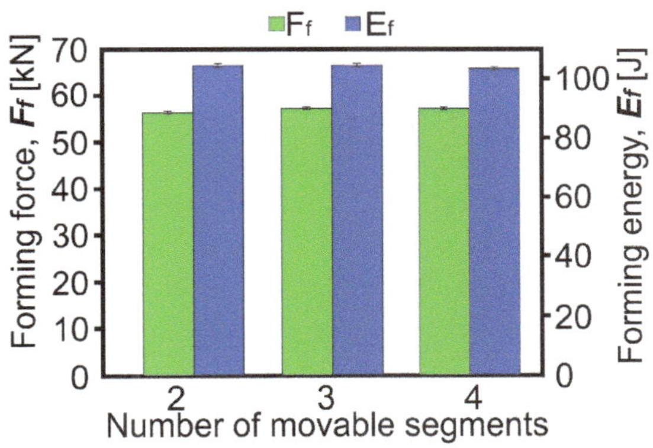

Fig. 4. Forming force and process energy consumption of clinching joining technology with using extensible die

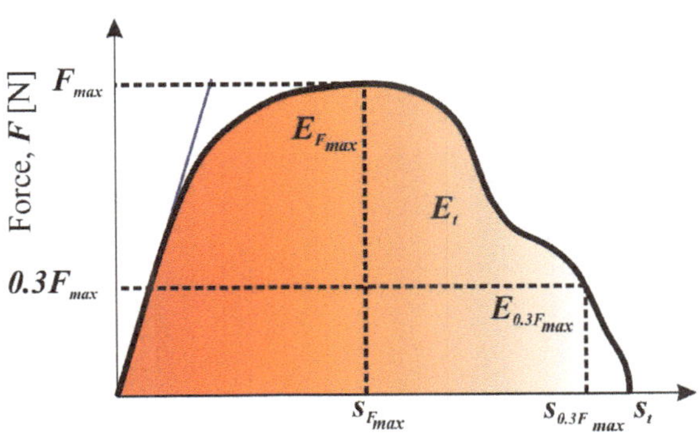

Fig. 5. Force-displacement curve and the joint parameters according to the ISO 12996 standard

Table 1. Joints parameters according to the ISO 12996 standard

Movable segments die configuration	Maximum shearing force F_{max} [N]	Dissipated energy up to 0.3 maximum force ($F=0.3F_{max}$) $E_{0.3Fmax}$ [J]	Total dissipated energy E_{max} [J]	$E_{0.3Fmax}/E_{max}$ (%)
2a	2745	1.354	3.371	0.402
2b	2731	1.328	3.294	0.403
3	3112	2.019	4.705	0.429
4a	3228	2.031	4.411	0.440
4b	3186	1.907	4.351	0.438

ing maximum shear force the shear test was continued until completely lap separation. The displacement for maximum shearing force ratio to the displacement until completely separation was 78, 77%, 83%, 65%, 68% (for the die arrangements: 2a, 2b, 3, 4a, 4b). The changing of the load force direction for the die with 2 and 4 movable segments did not affect the joint strength and the joint strength parameter.

The shape and the dimension of the interlock determines whether the joint was formed properly (without any crack and sheet material coherence loosing). The changing of the number of the movable segments of die, in relation to unchanged punch geometry and value of parameter *X*, was to demonstrate that the interlock can be formed for each used die. The correct shape of interlock resembles the letter *S*. During the joint formation with using the die with movable segments these segments move radially and diverge. The sheet material fill the space between segments and cylindrical part of the die and also between each segment. So the interlock has a different shape and dimensions (t_s – dimension of the interlock in bottom sheet and t_n – dimension of the interlock in upper sheet) depending on the joint cross-section. In Figure 6 on the left side is shape of interlock formed between the movable segment and cylindrical part of the die, and on right side the interlock is formed between two movable segments and cylindrical part of the die.

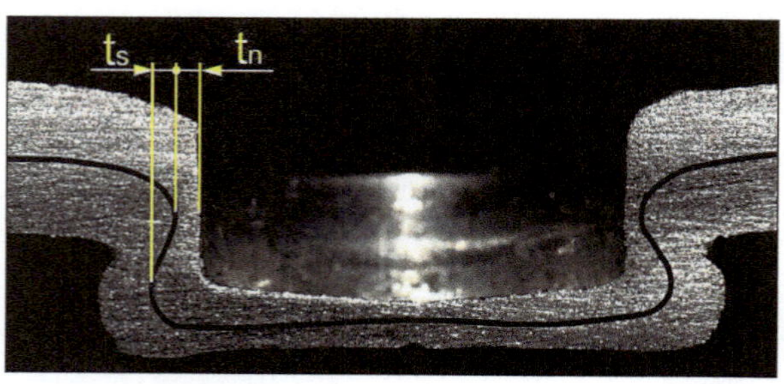

Fig. 6. The interlock cross-section of the joint formed by using the die with 3 movable segments

For the die with 3 movable segments the loading force was directed through the interlock with two different shapes and dimensions. For other die segments configuration of the force was directed by interlock with the same shape (between segment and die or between two segments and die).

CONCLUSION

The experimental researches were carried out to present the possibilities of the clinching joint formation with using unchanged punch geometry, the value of parameter *X* and different number of die movable segments. For all die the interlock was formed properly without any crack und material failure. The joint forming parameters and the joint strength were presented. The forming process energy consumption and forming force had a high repeatability for all dies. The number of the movable segments did not significantly affect on the forming process and its parameters (energy consumption and forming force). The load force direction in the shear test did not also affect the joint strength. The interlock shape was not axially-symmetrical because the sheet material flow, not only between the segment and the cylindrical part of the die, but also between the movable segments.

REFERENCES

1. Mori K., Bay N., Fratini L., Micari F., Tekkaya E.: Joining by plastic deformation. CIRP Annals - Manufacturing Technology, 62, 2013, 673–694.
2. Mucha J.: Współczesne techniki łączenia cienkich blach – zaciskanie przez wytłaczanie (Clinching). Mechanik, 80, 2007, 932–939.
3. Mucha J., Bartczak B., Analiza przetłoczeniowego łączenia blach. Archiwum Technologii Maszyn i Automatyzacji, 3/31, 2011, 59–68.
4. Mucha J., Witkowski W.: The clinching joints strength analysis in the aspects of changes in the

forming technology and load conditions. Thin-Walled Structures 82, 2014, 55–66.

5. Cacko R., Czyżewski P.: Numeryczne modelowanie kształtowania i obciążania połączeń nitowanych bezotworowo. Przegląd mechaniczny, 7-8, 2005, 50–53.
6. He X., Zhao L., Yang H., Xing B., Wang Y., Deng Ch., Gu F., Ball A.: Investigations of strength and energy absorption of clinched joints. Computational Materials Science, 94, 2014, 58–65.
7. Mucha J.: The analysis of lock forming mechanism in the clinching joint. Materials and Design, 32, 2011, 4943–4954.
8. Coppieters S., Lava P., Baes S., Sol H., Van Houtte P., Debruyne D.: Analytical method to predict the pull-out strength of clinched connections. Thin-Walled Structures, 52, 2012, 42–52.
9. Oudjene M., Ben-Ayed L., Delamézière A., Batoz J.-L.: Shape optimization of clinching tools using the response surface methodology with Moving Least-Square approximation. Journal of Materials Processing Technology, 209, 2009, 289–296.
10. Kaščák Ľ., Spišák E.: Clinching as a non-standard method for joining materials of dissimilar properties. Zeszyty Naukowe Politechniki Rzeszowskiej, Mechanika, RUTMech, 3/84 (2012), 31–41.
11. Witkowski W.: Wytrzymałość okrągłych połączeń przetłoczeniowych typu clinching podczas próby ścinania i rozrywania. Czasopismo Naukowe Mechanik, 12, 2014, 990–992.
12. Markowski T., Mucha J. Witkowski W.: Eksperymentalna analiza wytrzymałości okrągłych połączeń przetłoczeniowych formowanych jednolitą matrycą. Wybrane zagadnienia i problemy z zakresu budowy maszyn – cz.1. Oficyna Wydawnicza Politechniki Rzeszowskiej 2014, 99–110.
13. Mucha J., Witkowski W: Eksperymentalna analiza wytrzymałości okrągłych połączeń przetłoczeniowych blach ze stopu aluminium EN AW-5754 w stanie O/H111 poddanych złożonemu stanowi obciążenia. Przegląd Mechaniczny 2, 2015, 21–24.
14. Lambiase F., Di Ilio A.: Finite element analysis of material flow in mechanical clinching with extensible dies. Journal of Materials Engineering and Performance, 2(6), 2013, 1629–1636.
15. He X., Liu F., Xing B., Yang H., Wang Y., Gu F., Ball A.: Numerical and experimental investigations of extensible die clinching. The International Journal of Advanced Manufacturing Technology, 74(9-12), 2014, 1229–1236.
16. ISO 12996-2013 Mechanical joining – Destructive testing of joints – Specimen dimensions and test procedure for tensile shear testing of single joints.
17. http://www.tox-pl.com (01.01.2015).

THE INFLUENCE OF LAP SIZE ON SHEAR STRENGTH OF ADHESIVE JOINTS

Andrzej Kubit[1], Barbara Ciecińska[1], Kamil Drozd[1]

[1] Department of Manufacturing Processes and Production Engineering, The Faculty of Mechanical Engineering and Aeronautics, Rzeszow University of Technology, Powstanców Warszawy 12, 35-959 Rzeszow, Poland, e-mail: akubit@prz.edu.pl; bcktmiop@prz.edu.pl; drozduwp@wp.pl

ABSTRACT

Adhesive joints may be used instead of forced-in joints, welded and riveted joints, soldered and twisted connections. They are characterized by a lot of advantages, thanks to them the adhesive joints are willingly used in manufacturing processes. These advantages include low price of joint, reduction of weight of the final construction, improvement of durability, reliability and quality and improving visual appearance. From this point of view it should be concluded that experimental tests of adhesive joints are grounded. In the paper the results of experimental tests of static shear strength of single-lap joints are presented. These joints are made with two constructive glues. The specimens made from AISI 4130 steel were glued. Their surfaces were prepared in two versions before gluing: grinded with abrasive paper by hand and cleaned by sand-blasting. The results were analyzed from variable size of the lap point of view.

Keywords: adhesive joints, lap size, shear strength.

INTRODUCTION

Adhesive bonding consists of placing a thin layer of glue between bonded surfaces. This process allows to joint two objects with adhesive forces. Adhesive bonding also gives possibility to joint elements made from different materials – metallic with metals, ceramic, plastics or even glass. In practice it may be the only way to joint materials. Gluing is good manner to joint elements with different thickness or for such of them where jointing by other methods is impossible. Bonding a is simple and economic process, joints gives possibility to maintain high durability of construction, they do not provoke negative internal stresses because of small hardening shrinkage. The glue may function as a seal because of its chemical resistance. The modification of glue composition by fillers may lead to obtaining substances with unique characteristics. This special treatment may lead to significant changes of glue properties: they may be resistant to high temperature, conduct electricity or be dielectrics in reverse, have different flexibility, fatigue or impact strength, hardness, heat resistance or life. These modifications are sometimes used in joints whose temperature of exploitation changes in great degree [1].

The bonding process entails several basic operations. The steps of the creation of each joint are almost the same and it is important to follow them in a precisely defined order. There are:

- preparing surface before glue covering,
- preparing and spreading the glue
- jointing the surfaces
- glue hardening – the joint making
- the finishing and quality control of joint [2–4].

In adhesive bonding the correct pretreatment of the surface is important – it is deciding factor to achieve durable joint. Thanks to these operations the elimination of every kind of pollution on the surface is possible, e.g. dirt, fat, corrosion products, dust. The surface may also expand and may achieve its proper energy state [2, 5, 6].

At present there are different ways of achieving technological effects described above. However, sometimes there are also some difficulties, e.g.: problematic inspection manners, the requirement of retain short time between the following activities, the necessity of keeping the tools and equipment clean [2].

Typical actions performed during the pretreatment of surfaces are:

- Degreasing – in order to remove fats, lubricants, dust, organic and non-organic substances. This step may be realized in the steams and liquids of solvents. Typical substances to this purpose are: acetone, ethylic alcohol or other alkaline chemicals emulsified the lubricants and grease [2, 6, 7];
- Cleaning in acid or alkaline bath – it is removing oxides and scale from surfaces of the part; these layers are unnecessary and may lead to some problems in the next stage of surface pretreatment. Cleaning is conducting in the adjusted to the sort of material bath [3, 6, 8];
- Rinsing – in acid or alkaline bath, sometimes it is carried out also after special treatment. It is aimed at removing remains of galvanic or cleaning bath, dust after special treatment. Cold or hot, demineralized or distilled water is use. Next, the surface should be cleaned in ethanol. Even small parts of etching bath determine corrosion or other negative reactions in glues and prime coats [2, 6, 7].
- Drying – removing of remains of water from previously rinsed elements [2].
- Special treatment of surface – for preparing thin layers on the surfaces, they consist of salts (phosphates, fluorides, sulfates, nitrides, chromates, and other similar compounds), whose presence on the surfaces before gluing increases adhesive joints strength [5, 8].
- Coating with primer or glue – in time which lasts as short as possible. Parts with primer or other cover should be protected with fat-free paper and storing without the air access in the temperature 20 °C [2, 7].

All components should be mixed in correct proportions to receive gluing substance. The amount of glue and hardener, process of mixing and carefulness are important. In mixing of glues different kinds of mixers are used, this is a guarantee of precise in jointing different substances. Excess or lack of hardener and absence of thoroughness are the cause of low quality of joint. Then the joints may be soft, clammy and ductile or have low strength [3, 6, 9]. The consistency of adhesives and their homogeneity are the cause of quality of joint. Excess of glue gives thick joint which dries a long time and has poor strength. Lack of glue makes problems with wettability of surface, gives thin and poor joint [4].

During the activity with the preparing of gluing mass it is absolutely necessary not to making accidental bubbles. The heating and mechanical mixing gives good results in this. The bubbles in the joint reduce the strength about 5–15% [3]. Another important problem is time of glue life. It starts while the mass is prepared and from this moment the adhesive properties depend in the joint. It is important also during the realization of technological process, while the sequence and length of operations are planning. After an opening of container with glue its properties are modified with time. What is more, length of life of glue is not synonyms to the time of storage [3, 6].

After the preparation of the mass the covering of jointed surfaces takes place. It is very important operation and its manner depends on the form of glue, its viscosity, shape of jointed elements or even on type of production [6].

During the surface covering by glue the same thickness, regular field and correct wettability are required. Elements may be heated for less viscosity and high wettability. Heating is required in the case of glue in a form of bars, solids or powder [8].

After the covering the surfaces by the glue the jointed parts should be connected to each other in the shortest time and fixed not yet hardened joint [10].

THE METHODOLOGY OF EXPERIMENT

Because of many problems in gluing the experimental research was conducted. The influence of size of lap on the shear strength of a single-lap joints was investigated. To the preparation of test pieces the constructional steel AISI 4130 consisted of C (0,22–0,29%), Mn (0,40–0,70%), Si (0,17–0,37%), Cr (0,80–1,10%), Ni ($\leq$ 0,30%), Mo (0,15–0,25%), V ($\leq$ 0,05%), W ($\leq$ 0,20%), Ti ($\leq$ 0,05%), Cu ($\leq$ 0,25), As ($\leq$ 0,08%) was used [11]. This steel is applied in the manufacturing processes of loaded parts, exposed to hitting, torsion and vibration for example in aeronautics [12].

The test pieces were joined by Araldite 2011 and 2014-1 glues. These glues are characterized by long life time, high shear strength and tear off strength, static and dynamic load resistance. They are different in time of bonding and the proportion of epoxy resin to the hardener (Table 1) [11].

Table 1. The properties of epoxy resins Araldite 2011 and Araldite 2014-1 [11]

Epoxy resin	Araldite 2011	Araldite 2014-1
Proportion of components	1:1	2:1
Life time (min)	100	60
Color (after mixing)	Wan-yellow	Grey
Time of bonding 20 °C (h)	7	3
Time of hardening w 20 °C (h)	36	36

For gluing the test pieces of dimensions $d \times l \times g$: 25×100×1 mm with 5-repetition was made. Variable value was the length of lap z and it was: $z_1 = 8$ mm, $z_2 = 12{,}5$ mm, $z_3 = 16$ mm, $z_4 = 24$ mm (Fig. 1).

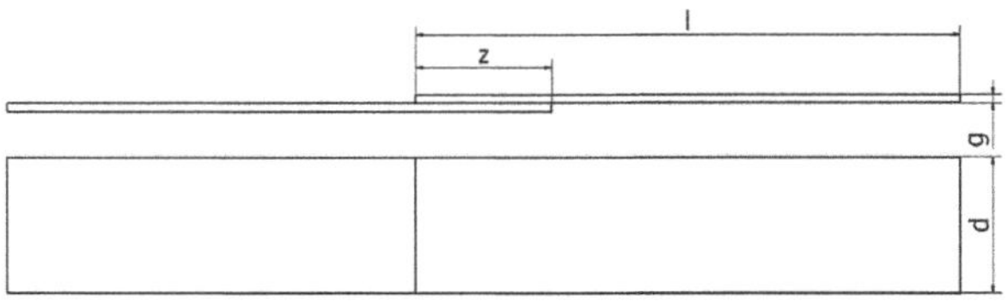

Fig. 1. The scheme of adhesive joint in the experiment

The surfaces before gluing were made in two options: treated by abrasive paper P80 and mechanically by sand-blast cleaning (corundum A90). Then the pieces were cleaned in acetone in ultrasonic washer and dried in room temperature. Next, the bonded pieces were placed into the special bonding holder for hardening the joints. After the jointing the test pieces were broken in the Zwick/Roell Z100 tensile testing machine.

RESULTS

In the experimental research the value of shear force P_t was obtained. The mean value of P_{t-m} and static shear strength R_t from the relation (1) were calculated:

$$R_t = P_{t-m} / A \qquad (1)$$

where: A – the field of lap.

The results of tests and calculations are presented in the Tables 2–5. Mentioned below graphs

Table 2. The results for z_1

The length of lap: $z_1 = 8$ mm; $A_1 = 200$ mm^2				
Test piece number	Force P_t [N]			
	Araldite 2011		Araldite 2014-1	
	handmade surface	sand-blast cleaning	handmade surface	sand-blast cleaning
1	3230.243	4580.718	3500.823	5498.408
2	3021.059	4501.421	4038.446	5059.642
3	3399.728	4553.418	3031.017	4697.395
4	3182.038	4711.192	2827.701	4849.223
5	2752.909	4898.991	2281.864	4763.577
P_{t-m}	3117.200	4619.150	3135.970	4973.650
R_t	15.585	23.246	15.68	24.868

Table 3. The results for z_2

The length of lap: $z_2 = 12.5$ mm; $A_2 = 312.5$ mm^2				
Test piece number	Force P_t [N]			
	Araldite 2011		Araldite 2014-1	
	handmade surface	sand-blast cleaning	handmade surface	sand-blast cleaning
1	3478.465	5398.833	3742.796	5549.746
2	4284.306	5959.551	4442.110	6056.149
3	3904.567	5591.105	4807.640	5909.373
4	3884.153	5452.010	4362.317	5689.107
5	4407.530	5703.114	4823.385	5878.906
P_{t-m}	3991.800	5620.920	4435.650	5816.660
R_t	12.773	17.987	14.194	18.613

Table 4. The results for z_3

The length of lap: z_2 = 16 mm; A_3 = 400 mm^2				
Test piece number	Force P_t [N]			
	Araldite 2011		Araldite 2014-1	
	handmade surface	sand-blast cleaning	handmade surface	sand-blast cleaning
1	4193.315	6061.828	4843.896	6256.972
2	3897.565	5792.773	4361.675	6021.921
3	4143.484	5813.575	4353.719	6302.023
4	4130.527	6163.991	3762.965	6220.482
5	4070.587	5780.942	3500.831	6656.479
$P_{t\text{-}m}$	4087.100	5992.620	4164.620	6291.580
R_t	10.218	14.807	10.412	15.729

Table 5. The results for z_4

The length of lap: z_4 = 24 mm; A_4 = 600 mm^2				
Test piece number	Force P_t [N]			
	Araldite 2011		Araldite 2014-1	
	handmade surface	sand-blast cleaning	handmade surface	sand-blast cleaning
1	5693.161	7842.041	4364.676	8202.019
2	6573.163	7748.961	4734.915	8205.555
3	4785.892	7546.047	4221.458	8259.362
4	5127.682	8442.654	4402.663	7965.750
5	5397.028	7969.401	3983.050	8317.146
$P_{t\text{-}m}$	5515.380	7909.820	4341.350	8189.700
R_t	9.192	13.183	7.236	13.650

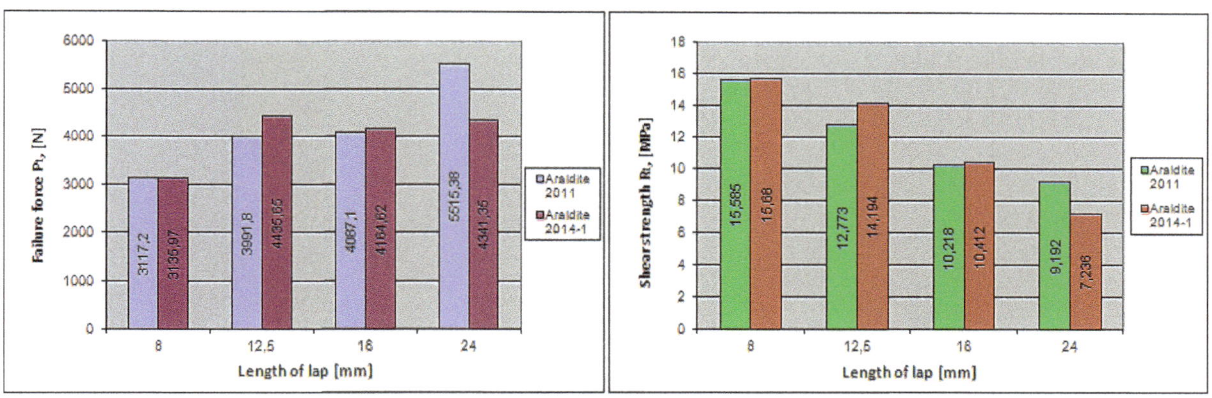

Fig. 2. Influence of length of lap on failure force (a) and static shear strength (b) of single lap adhesive joints which surfaces were prepare by treated by abrasive paper P80

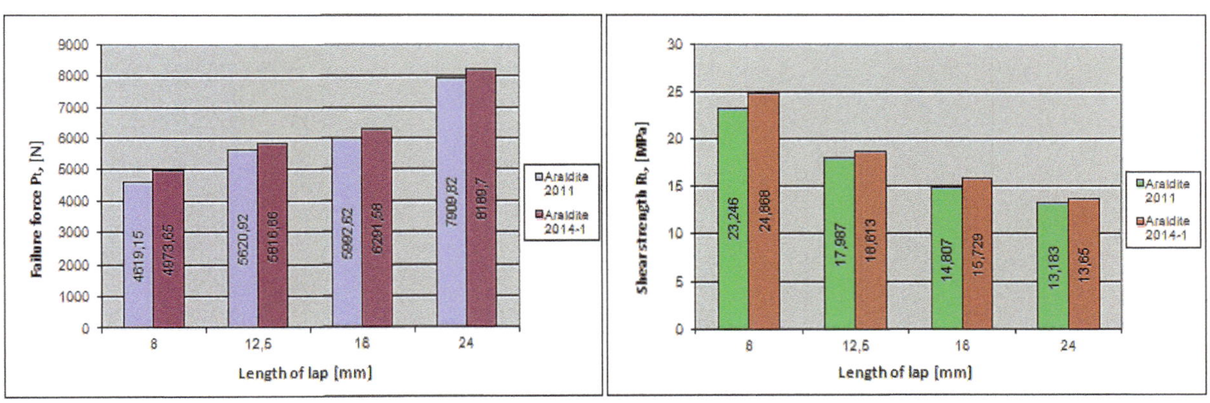

Fig. 3. Influence of length of lap on failure force (a) and static shear strength (b) of single lap adhesive joints which surfaces were prepare mechanically by sand-blast cleaning (corundum A90)

show graphic interpretation of influence of length of lap on failure force and static shear strength for both used adhesives and surface preparation methods (Figures 2 and 3).

CONCLUSIONS

The length of lap has influence on the value of shear force and shear strength of single-lap joints. With the growth of lap size the value of shear force also increases. It has a direct relation with the growth of field of the lap. The shear strength decreases with the increase of length of the lap.

From the experiment it follows that with the same valued parameters of gluing by the epoxy resins Araldite 2011 and 2014-1 the shear strength has not great differences between values.

Significant influence on achieved static shear strength with the same length of the laps has the manner of preparing the surfaces before gluing. In the case of sand-blasting the value of force was several times higher than in the case of handmade cleaning with the abrasive paper use. From this point of view the surface pretreatment has influence on the achieved results (not only joint construction).

REFERENCES

1. Mirski Z., Piwowarczyk T.: Podstawy klejenia, kleje i ich właściwości. Przegląd Spawalnictwa 8, 2008.
2. Czaplicki J., Ćwikliński J., Godzimirski J., Konar P.: Klejenie tworzyw sztucznych. WKŁ, Warszawa 1987.
3. Godzimirski J., Kozakiewicz J., Łunarski J., Zielecki W.: Konstrukcyjne połączenia klejowe elementów metalowych w budowie maszyn. OWPP, Rzeszów 1997.
4. Porębska M., Skorupa A.: Połączenia spójnościowe. PWN, Warszawa 1993.
5. Kłonica M., Kuczmaszewski J.: Ozonowanie stali 316L na potrzeby klejenia. TAiM, 2, 2014.
6. Kuczmaszewski J., Rudawska A.: Klejenie blach ocynkowanych, WUPL. Lublin 2005.
7. Bastian H.W.: Technika zamocowań: kołki, śruby, nity, zszywki, klejenie. Arkady, Warszawa 1996.
8. Cagle Ch.V.: Kleje i klejenie. WNT, Warszawa 1977.
9. Godzimirski J.: Wytrzymałość doraźna konstrukcyjnych połączeń klejowych. WNT, Warszawa 2002.
10. Godzimirski J., Komorek A., Rośkowicz M., Smal T., Tkaczuk S.: Tworzywa adhezyjne: zastosowanie w naprawach sprzętu technicznego. WNT, Warszawa 2010.
11. www.huntsman.com (access 14.01.2015)
12. www.saarstahl.com (access 14.01.2015)

31

THE STUDY OF CUTTING FORCES AND THEIR AMPLITUDES DURING HIGH-SPEED DRY MILLING MAGNESIUM ALLOYS

Ireneusz Zagórski[1], Józef Kuczmaszewski[1]

[1] Department of Production Engineering, Mechanical Engineering Faculty, Lublin University of Technology, 36 Nadbystrzycka Str., 20-618 Lublin, Poland, e-mail: i.zagorski@pollub.pl; j.kuczmaszewski@pollub.pl

ABSTRACT

This paper presents the state of knowledge on cutting forces values during milling of different magnesium alloy types. Additionally, the results of own research on cutting forces and their amplitudes after milling magnesium alloys used for manufacturing elements applied in the aerospace industry are included. The subject of the analysis was the influence of technological parameters and the tool type on cutting forces and their amplitudes. The analysis is important due to the correlation between cutting forces and temperature in the cutting area, and, consequently, the safety of machining and strain of the workpiece.

Keywords: high-speed dry milling, magnesium alloys, aviation elements, machinability, cutting forces.

INTRODUCTION

Aircraft structure elements made from magnesium alloys are commonly subjected to milling, in both roughing and finishing machining operations. Maintaining accuracy in shape and dimension of the workpiece is as significant as e.g. achieving the required surface finish quality, as it is related to such machinability indicators as temperature in the cutting area or cutting forces. Moreover, the aim is to reduce the processing time, which might be attainable with the application of appropriate methods of efficient high speed machining. The reason for applying efficient machining is the prospect of reducing processing time, increasing process efficiency and, as a result, reducing production costs.

High Speed Machining (HSM) is a term meaning machining with high speeds. HSM is defined as a high efficiency machining method which allows achieving high quality surface. Additionally, this type of processing is characterised by high parameters (feed per tooth, cutting speed) and low depth of cut. Consequently, a small chip cross-section and an enhanced chip removal are achieved. Increased cutting speed leads to an increase in feed rate per tooth, which, in turn, induces a decrease in main time, which implies higher production efficiency of the process [1, 13].

What seems vital is the fact that during HSM the increase in cutting speed (with constant process efficiency) results in a reduction of cutting forces. Consequently, milling thin-walled elements (e.g. up to 0.1 mm) becomes possible [13]. A division into conventional machining and HSM is commonly assumed. According to that division, HSM occurs when an increase in cutting speed v_c leads to reduction in cutting forces, which may be expressed as follows [1]:

- $\partial F/\partial v_c < 0$ – HSM,
- $\partial F/\partial v_c > 0$ – conventional machining.

High efficiency machining (high productivity) includes HPM (high performance machining) or HPC (high performance cutting). HPC is a process which strives for optimal usage of the spindle for removal of large volumes of material per unit of time. Higher cutting speed v_c ensures increased process efficiency, but with the simultaneous occurrence of higher feed per tooth. It is argued that HPC is suitable for preliminary machining and profiling, while HSC excels in finishing [4].

For instance, the application of HSM allows to reduce (4 times) the processing time with no effects on the surface finish quality or shape accuracy [6]. Achieving the desired shape accuracy is possible due to reduced cutting forces during high speed machining.

CURRENT STATE OF KNOWLEDGE

Great emphasis is placed on safety during machining of magnesium alloys. Improving material removal processing of magnesium alloys as well as widening magnesium alloys application range seems to be of great significance [9]. Common problems occurring during dry milling of magnesium alloys are as follows: [2, 7, 8, 10] risk of chip ignition, inaccurate shape and dimensions resulting from adhesive processes and flank built up (FBU), low wear resistance of the tool during milling of strengthened alloys as well as impeded chip removal outside the cutting area.

Cutting forces cause strain of the processed machine components. Additionally, so called deformations of chip thickness were found to increase with the increase of cutting forces. The shear angle decreases with decreasing undeformed chip thickness and, as a result, the shear plane temperature changes (increases). This phenomenon can be explained using the term appropriate cutting forces (appropriate cutting force is the correlation between the cutting force and chip cross-section). The cutting force increases with decreasing undeformed chip thickness. Shearing energy per unit volume increases the volume of removed metal when cutting forces increase causing an increase in cutting temperature in the sear plane [3, 5].

Both adhesion and flank built-up can induce high fluctuations of cutting forces as well as reduction of surface finish quality and accuracy in shape and dimension. When milling magnesium alloy AZ91HP using a tool with a PCD cutting edge with the application of oil cooling, cutting forces are low and they increase linearly with increasing feed rate [7]. PCD tools allow to obtain superior quality surface finish, lower cutting forces and are less prone to overheating due to low coefficient of friction in the tool – workpiece contact [11]. Due to lower cutting forces occurring during processing using tools with polycrystalline diamond cutting edges, especially with reduced cross-section of machined layers, considerably low temperatures are observed [2, 4, 7, 10, 11].

During processing with a Kordell design tool within technological parameters of 'conventional' machining, change in cutting speed v_c, induced higher values of cutting force components and their amplitudes in the case of casting magnesium alloy AZ91HP. However, with change in feed per tooth f_z, the components F_x, F_y and their amplitudes increased along with an increase in feed. Higher values of cutting forces and their amplitudes, similarly as in the case of change in cutting speed v_c, were observed for the AZ91HP alloy [5, 12].

When machining with 'classical geometry' tools, cutting forces and their amplitudes are influenced by changes in feed per tooth f_z, rather than change in cutting forces v_c. The highest cutting force value was observed for AZ91HP alloy machined with a tool with a PCD cutting edge. Changes in feed per tooth induce an approximately linear increase in cutting force components and their amplitudes. Highest values of cutting force components F_y and F_z occurred with v_c=300 m/min. What seems intriguing is that cutting force components decrease with cutting speed increasing up to v_c=400 m/min. No considerable differences in cutting force values were observed for both AZ31 and AZ91HP alloys within the entire range of feed and changes in cutting forces [5].

METHODOLOGY AND SCOPE OF RESEARCH

The main aim of the study was to analyse the effect of such parameters as cutting speed v_c and feed per tooth f_z on cutting forces and their amplitudes during machining AZ31 and AZ91HP magnesium alloys. Processing was performed on vertical machining centre Avia VMC800HS. The tools applied in the study were 16 mm end mills: carbide TiAlN coated, carbide Kordell design and one with a PCD cutting edge. A constant milling width a_e = 14 mm and the following technological parameter ranges were used: a_p = (0.5–6) mm, f_z = (0.05–0.3) mm/tooth, v_c = (400–1200) m/min.

Forces were measured using a Kistler 9257B dynamometer and a 5017B amplifier. The dynamometer allows to measure forces from -5 kN to +% kN. Cutting forces are defined as 'functional' indicators of machinability.

Figure 1 shows a plan for a research on the influence of multiple factors on cutting forces and their amplitudes. The changing parameters are as follows: cutting speed v_c, feed rate per tooth f_z,

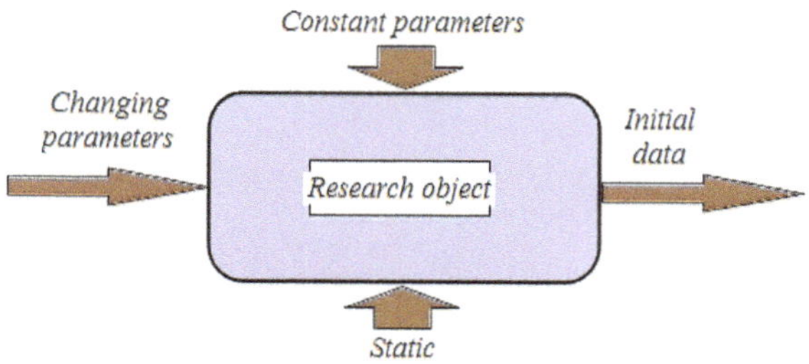

Fig. 1. Research plan for the analysis of cutting forces and their amplitudes in high-speed machining

type of tool material and type of workpiece material. Depth of cut a_p and cutting width a_e values were constant. Cutting forces (F_x, F_y, F_z) as well as cutting force amplitudes (AF_x, AF_y, AF_z) served as initial data.

Kordell design is characterised by three wave-shaped cutting edges. This type of tools is dedicated for processing aluminium alloys (so-called first choice) as well as magnesium alloys. Additionally, the characteristic features of Kordell geometry are helix angle λ = -40°, maximal depth of cut a_p = (7–45) mm, no corner radius r_ε, a positive rake angle γ = (9–12)° for reducing cutting forces. Characteristic features of such a carbide Kordell design tool are presented in Figure 2.

The obtained results allowed finding the following correlations:

- maximum values of cutting force components F_x, F_y, F_z as a function of technological parameters of processing;
- maximum values of cutting force amplitudes AF_x, AF_y, AF_z as a function of technological parameters of processing.

TEST RESULTS AND THEIR ANALYSIS

Figute 3 shows the time history of cutting force components F_x, F_y and F_z during milling of magnesium alloy AZ31 with a carbide TiAlN coated tool.

Figure 4 presents the effect of change in cutting speed v_c on cutting force component F_x and its amplitude AF_x. The influence is ambiguous. The decrease in cutting force and its amplitude most probably is a consequence of transition to HSM (increase in v_c – decrease in cutting forces). This correlation is observed for tools with so called 'classical geometry' of the cutting edge.

Figure 5 shows the effect of change in feed per tooth f_z on cutting force F_x and its amplitude AF_x. The cutting force component and its amplitude were found to increase with increasing feed per tooth. The increase in cutting force F_x and its amplitude AF_x is approximately linear.

Figure 6 presents the influence of change in cutting speed v_c on cutting force F_y and its amplitude AF_y. The influence is ambiguous. In the case of 'classical geometry' tools, the force and its amplitude initially increase, but then begin to decrease. This process indicates transition to HSM. In the case of a Kordell design tool, the influence of change in cutting speed v_c is reverse.

Figure 7 shows the influence of change in feed per tooth on cutting force F_y and its amplitude AF_y. With the application of a tool with a PCD cutting edge as well as a Kordell design tool cutting speed and its amplitude increase along with the increase in feed per tooth. The increase in force F_y and its amplitude AF_y is approximately linear. A different influence of f_z is observed when a carbide TiAlN coated tool is applied. In this case, the force and its amplitude increase until feed per tooth reaches f_z = 0.15 mm/tooth, for higher feed per tooth values, the force and its amplitude decrease.

Figure 8 shows the influence of change in cutting speed v_c on cutting force F_z and its amplitude AF_z. In the case of 'classical geometry' tool, the force and its amplitude initially increase, but then begin to decrease, which indicates transition to HSM. In the case of a Kordell design tool, the influence of change in cutting speed v_c is reverse.

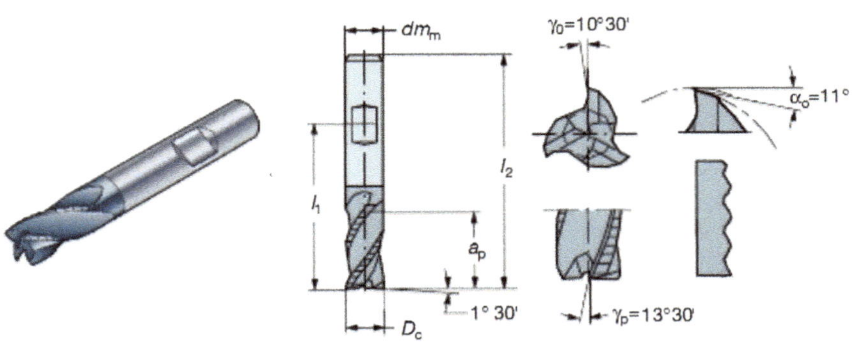

Fig. 2. Catalogue image of a Kordell end mill [14]

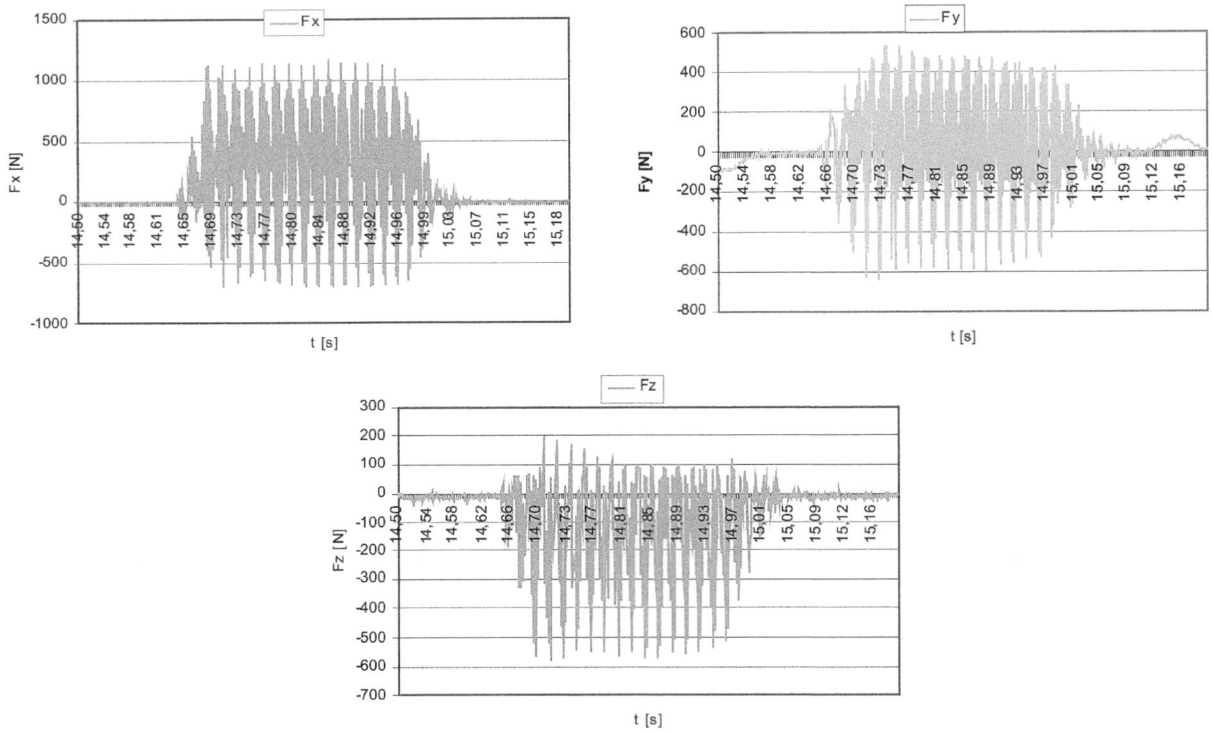

Fig. 3. Time history of cutting force components during milling of magnesium alloy AZ31 using a TiAlN coated carbide tool: a) component F_x, b) component F_y, c) component F_z; $v_c = 800$ m/min, $f_z = 0.3$ mm/tooth, $a_p = 6$ mm

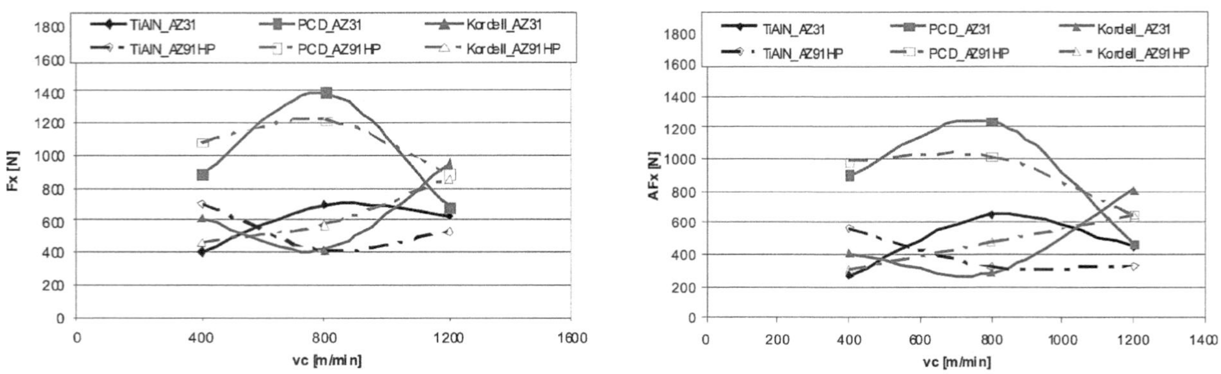

Fig. 4. The effect of cutting speed v_c on the value of a) cutting force component F_x and b) its amplitude AF_x during milling of magnesium alloys AZ31 and AZ91HP using different tools; $f_z = 0.15$ mm/tooth, $a_p = 6$ mm

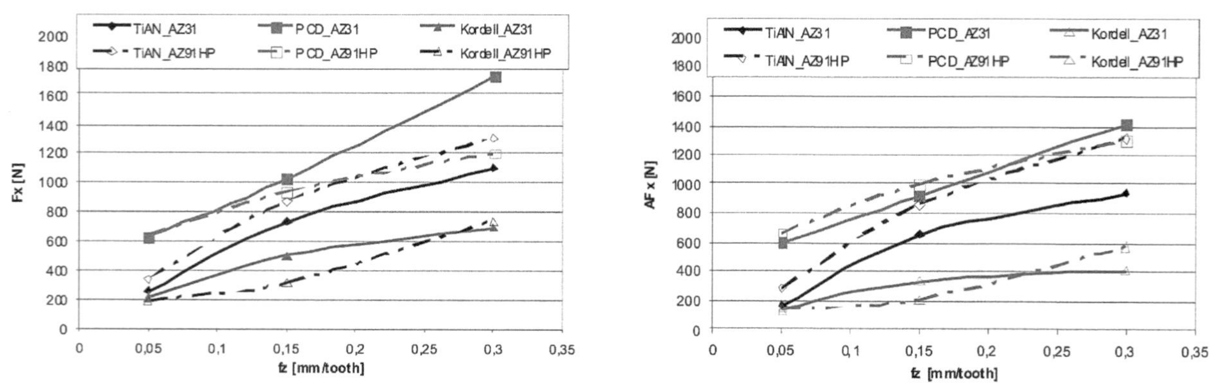

Fig. 5. The effect of change in feed per tooth f_z on the value of a) cutting force component F_x and b) its amplitude AF_x during milling of magnesium alloys AZ31 and AZ91HP using different tools; $v_c = 800$ m/min, $a_p = 6$ mm

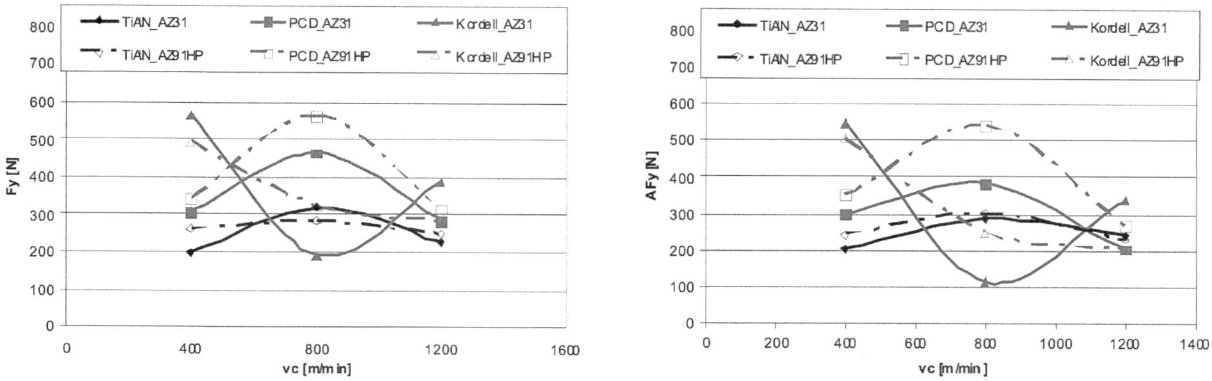

Fig. 6. The influence of change in cutting speed v_c on the value of a) cutting forces component F_y and b) its amplitude AF_y during milling of magnesium alloys AZ31 and AZ91HP using different tools; f_z=0.15 mm/tooth, a_p=6 mm

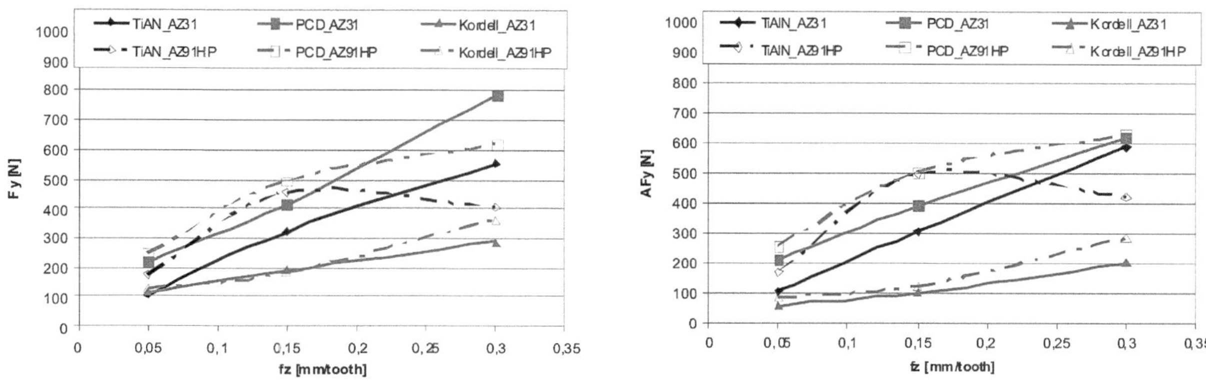

Fig. 7. The influence of feed per tooth f_z on the value of a) cutting force component F_y and b) its amplitude AF_y during milling of magnesium alloys AZ31 and AZ91HP using different tools; v_c= 800 m/min, a_p= 6 mm

Figure 9 shows the effect of change in feed per tooth f_z on cutting force F_z and its amplitude AF_z. Cutting force component and its amplitude was found to increase with increasing feed per tooth. The increase in cutting force F_z and its amplitude AF_z is approximately linear.

CONCLUSIONS

The conducted analyses, study works as well as own research allow to formulate the following conclusions:

1. In high-speed machining it is feed per tooth f_z rather than cutting speed v_c which is of greater influence on cutting forces and their amplitudes.
2. In the majority of instances cutting forces were higher when the tool applied had a PCD cutting edge.
3. In the case of change in cutting speed v_c, cutting force components F_x, F_y and F_z reached highest values (in the majority of instances) for cutting speed equalling v_c= 800 m/min. This correlation applies when tools with 'classical geometry' cutting edges are used.
4. What might be found interesting is the fact that cutting force components decreased with cutting speed increasing to the value of v_c=1200 m/min, in the case of tools with 'classical geometry' cutting edges.
5. An increase in feed rate per tooth causes approximately linear increase in cutting force components and their amplitudes.
6. No significant changes in cutting forces were observed with changing feed per tooth f_z in the case of magnesium alloys AZ31 and AZ91HP. The change occurred with changing cutting speed v_c.

Acknowledgment

Financial support of Structural Funds in the Operational Programme - Innovative Economy (IE OP) financed from the European Regional Development Fund, No POIG.01.01.02-00-015/08-00 is gratefully acknowledged.

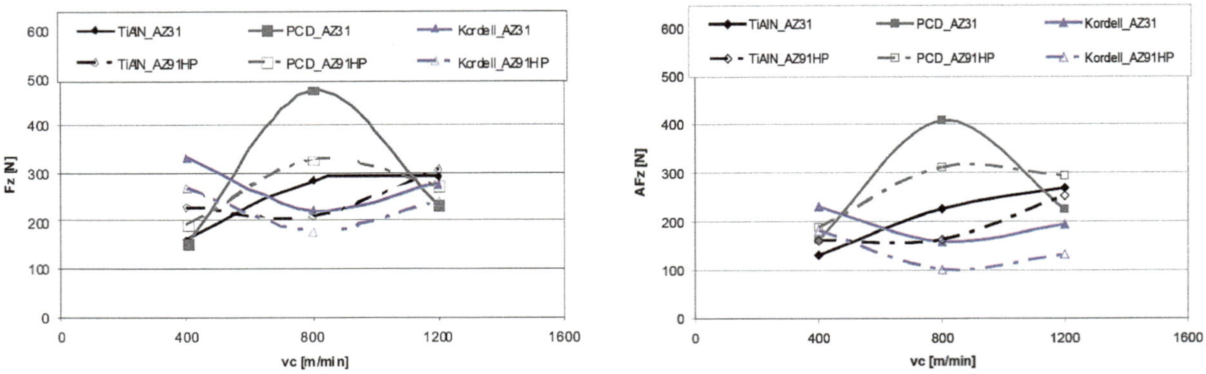

Fig. 8. The influence of cutting speed v_c on the value of a) cutting force component F_z and b) its amplitude AF_z during milling of magnesium alloys AZ31 and AZ91HP using different tools; $f_z = 0.15$ mm/tooth, $a_p = 6$ mm

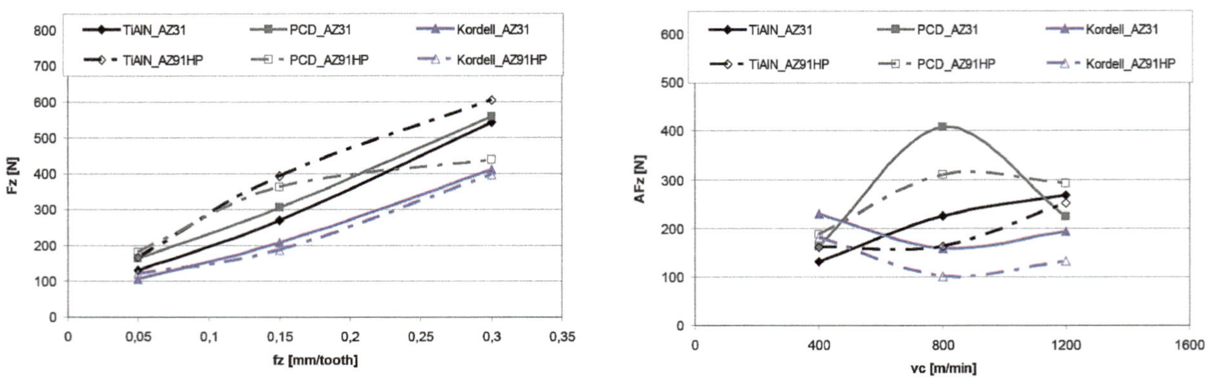

Fig. 9. The influence of feed per tooth f_z on the value of a) cutting force component F_z and b) its amplitude AF_z during milling of magnesium alloys AZ31 and AZ91HP using different tools; $v_c = 800$ m/min, $a_p = 6$ mm

REFERENCES

1. Adamski W. Manufacturing development strategies in aviation industry. Advances in Manufacturing Science And Technology, 2010, 34(3): 73-84
2. Akyuz B. Machinability of magnesium and its alloys. TOJSAT: The Online Journal of Science and Technology, July 2011, 1(3): 31-38.
3. Fang F.Z., Lee L.C., Liu X.D. Mean Flank Temperature Measurement in High Speed Dry Cutting. Journal of Materials Processing Technology, 2005, 167: 119–123.
4. Grzesik W. Fundamentals of the machining of constructional materials. WNT, Warsaw 2010.
5. Kuczmaszewski J., Zagórski I. The study of cutting forces and amplitudes during the milling magnesium alloys. Postępy Nauki i Techniki, 2012, 15: 228-237.
6. Leigh E., Tlusty J., Schueller J. Applying high-speed machining techniques on rotor grip configurations, Presented at the American Helicopter Society 55th Annual Forum, Montréal, Quebec, Canada, May 25–27, 1999, p. 1-16.
7. Oczoś K.E. Efficient machining of magnesium alloys. Mechanik, 2000, 7: 467-474.
8. Oczoś K.E. Extension of the magnesium alloys application range. Mechanik, 2009, 5-6: 386-400.
9. Oczoś K.E., Kawalec A. Light metals forming. PWN, Warsaw 2012.
10. Tönshoff H.K., Denkena B., Winkler J., Podolsky Ch. Technology of Mg and Mg alloys: Machining. Edited by Friedrich H.E., Mordike B.L., Springer 2006, p. 398-418.
11. Tönshoff H.K., Winkler J. The influence of tool coatings in machining of magnesium. Surface and Coatings Technology, 1997, 94-95: 610-616.
12. Zagórski I., Kuczmaszewski J. Selected problems of effectiveness of processing of magnesium alloys with a Kordell design tool. Management and Production Engineering Review, 2011, 4: 90-96.
13. Zalewski A. Optimal HSM machining strategy for improving production efficiency. Design and Engineering Constructions, Dec. 2007, 3: 23-26.
14. http://www.sandvik.coromant.com/, Sandvik Coromant Catalogue, Metalcutting Technical Guide: Milling, 15.03.2011.

BROADBAND CONCEPT OF ENERGY HARVESTING IN BEAM VIBRATING SYSTEMS FOR POWERING SENSORS

Andrzej Rysak[1], Michael Müller[2], Marek Borowiec[1], Jarosław Zubrzycki[1], Grzegorz Litak[1], Anna Godlewska-Lach[1], Volker Wittstock[2]

[1] Faculty of Mechanical Engineering, Lublin University of Technology, 20-618 Lublin, Poland; e-mail: a.rysak@pollub.pl; m.borowiec@pollub.pl; j.zubrzycki@pollub.pl; g.litak@pollub.pl; anna.godlewska@pollub.eu.pl

[2] Technische Universität Chemnitz, Institut für Werkzeugmaschinen und Produktionsprozesse Reichenhainer, 09126 Chemnitz, Germany, e-mail: michael.mueller@mb.tu-chemnitz.de; volker.wittstock@mb.tu-chemnitz.de

ABSTRACT

Recent demand for powering small sensors for wireless health monitoring triggered activities in the field of small size efficient energy harvesting devices. We examine energy harvesting in an aluminium beam with a piezoceramic patch subjected to kinematic harmonic excitation and impacts. Due to a mechanical stopper applied, inducing a hardening effect in the spring characteristic of the beam resonator, we observed a broader frequency range for the fairly large power output. Impact nonlinearities caused sensitivity to initial conditions and appearance of multiple solutions. The occurrence of resonant solution associated with impacts increased efficiency of the energy harvesting process.

Keywords: piezoceramic, energy harvesting, vibration, broadband effect, impacts.

INTRODUCTION

The existence of non-linearities in mechanical systems usually shows complex dynamical responses by a non-trivial dependence of amplitude and frequency, vibration localization, and also by occurrence of multiple solution. These nonlinear effects can improve the effectiveness of energy harvesting systems in capturing the vibration kinetic energy, through the so called broadband frequency effect [1–7]. Corresponding devices of energy harvesting contain mechanical resonators and additional energy transducers for transforming ambient mechanical energy into the electric form. This direction of reseach was continued by Blystad and Halvorsen [8]. On the other hand, the papers [9–11] studied extensively micro-electromechanical systems proposing applications of the electrostatic devices for energy harvesting. Additionally, Gu and Livermore [9] presented experimental results of models of energy harvesting devices in which a low frequency resonator impacts a high frequency harvesting resonator. Such experiments prove that the efficiency of the electrical power transfers can be significantly improved. The paper continues studies on non-linear impacting piezocaramic harvesters with stoppers, drawing attention to broadening the energy harvesting efficiency characteristic demonstrated by authors of Gu and Livermore [9] and Soliman et al. [12,13] and Borowiec et al. [14]. Finally, Rysak et al. [15] performed modal analysis of the broadband vertical beam energy harvester with the moving mass including additional impacts and friction phenomena. In the analyzed real experimental model, the influence of an applied mechanical stopper on the efficiency of the piezoceramic layers in the vibrating structure is investigated. Note that the impacts of the cantilever into the stopper not only limit vertical displacements but do also change the spring characteristics of the vibrating beam which increases the resonant resonant frequency.

PZT material included in the piezoelectric, initially polarized fibres was prepared in Chemnitz TU and is characterized by a unique electrical

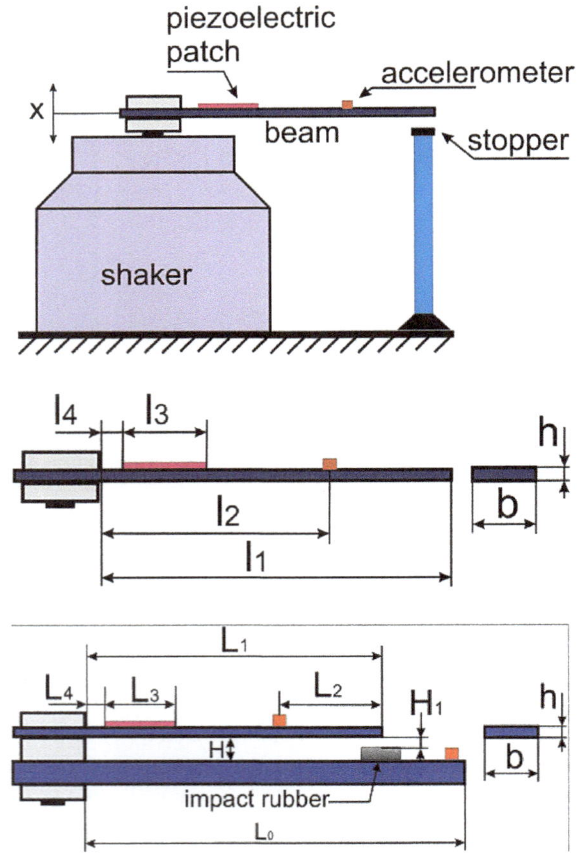

Fig. 1. Experimental stand (a), note the location of the piezoceramic patch, and accelerometer (sensor). Characteristic dimensions of the excited aluminum beam (b), and the two impacting beam system (c). To minimize the risk of fracture of the piezoceramic element we placed the thin layer of rubber on the stopper in both configurations (see Fig. a-b and c). The system geometrical and material parameters are included in Table 1. The lower beam is Fig. 1c is much more rigid having the resonance frequency fairly outside the examined region

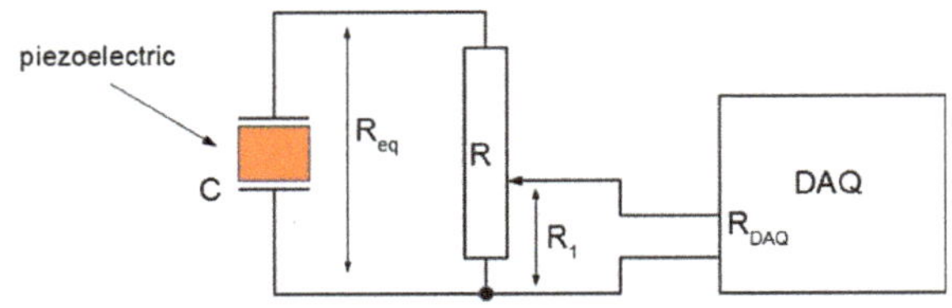

Fig. 2. Electrical circuit set with the electro-motive force on the piezoceramic element; the circuit of the voltage divider was used in order to adjust the measured voltage to the range available for the DAQ card

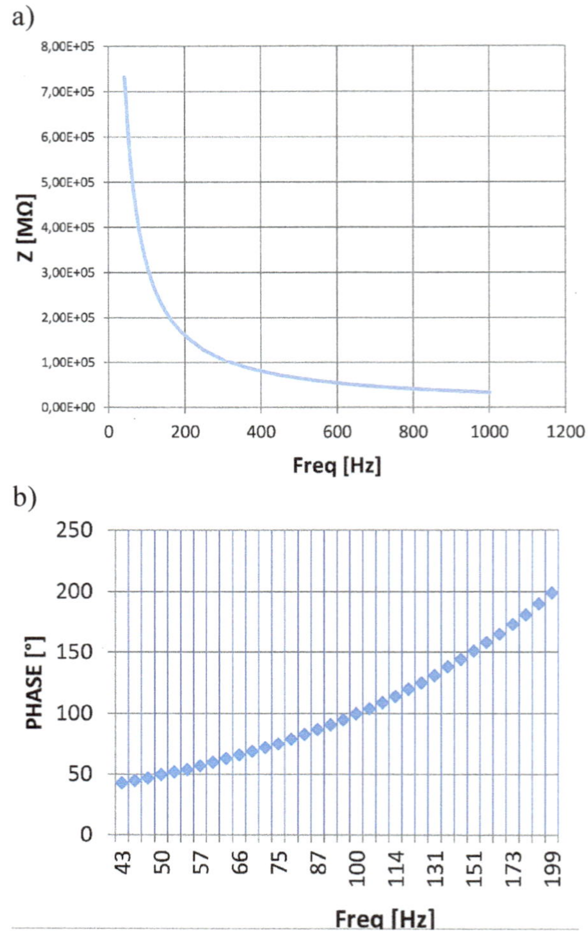

Fig. 3. Frequency dependence of the impedance (3a), frequency dependence of the angle θ (3b); Research on the meter Hioki 3532-50

impedance. Figure 3 shows the measured impedance amplitude and related angle provided by the HIOKI 3532-50 impedance meter. Impedance is inversely proportional to frequency and the angle changes are due to the influence of small ohmic resistivity caused by electrical lead connections.

EXPERIMENTAL MODEL

The measurements have been performed on a vibration generator TIRAvib TV 50101, which operated in a sine mode under controlled conditions. The schematic setup is presented in Fig. 1a. The beam with attached piezoceramic layers was excited in vertical direction in the vicinity of the principal resonance. The piezoceramic patch consisted of ten piezoceramic fibres (PZT) utilizing d_{31}-effect. The fibers (PZT) were bounded with Elecolit isotropic conductive adhesive and insulated by globe top (epoxy resin). We worked in two configurations; with the fixed (Figs. 1a-b) and moving frame (Fig. 1c) stoppers.

During the beam vibration, the electric power output is generated by the piezoelectric elements while the electrical circuit is loaded by corresponding resistors (Fig. 2). The electrical output was measured by means of data acquisition set (DAQ) (see Fig. 2) as:

$$U_{DAQ} = U_{out}\frac{R_1}{R} \quad (1)$$

where: R_1/R is the voltage divider presented in Figure 2, applied to avoid exceeding the measurement voltage range. In the experiments we used piezoceramic fibre patch.

Note that the two discussed configurations (Figs. 1 a-b and c) differ in the excitation the first of them (Figs. 1a-b) is characterized in additional parametric excitation as the effective distance between the beam and stopper is modulated with the excitation frequency:

$$h_1' = h_1 - \frac{A}{\omega^2}\sin(\omega t) \quad (2)$$

On the other hand, in the second case (Fig. 1c), the distance H^1 is fixed.

ANALYSIS OF EXPERIMENTAL RESULTS

The measurements have been done using two different distances between the fixed and moving stoppers and vibrating beam (Fig. 1). The results for two sets are shown in Figures 3a and b, by means of resonance curves of gravitation force captured by the accelerometer mounted on the beam. Figure 4a presents behaviour in case of stopper gap (measured for a motionless beam) being 6.5 mm, and in Figure 4b gap being 11 mm. Finally, Figure 3c represents the double beams with the distance between them of 14.3 mm.

Firstly, it is easy to notice the difference for increasing gap (see Figs. 4a and b), when the stopper is less engaged. Moreover, vibrating the beam with impacts leads to increasing the energy harvesting efficiency through the resonant solution appearing for higher frequencies. Such behaviour manifests in hardening the beam stiffness characteristic and next broadening the reached level of the measured amplitude. The beam was subjected to kinematic periodic excitation with the increasing and decreasing frequency sweeps (in Figs. 4a, b, blue and red lines respectively). On the right hand side of the resonance centre we observe at least two solutions indicating resonant (impacting) and non-resonant solutions due to the stopper induced hardening of the effective beam stiffness. Interestingly, the two solutions (see blue and red lines) are also observed on the left hand side of the resonance center. This looks like stiffness softening originating in the contact between the beam and stopper. Note that this contact was realized through a thin layer of rubber.

The broadening in the non-resonant solution branch is presumably caused by air circulations between macroscopic beam and stopper. Note that it decreases with the growing distance between the beam and stopper. Secondly, to clarify

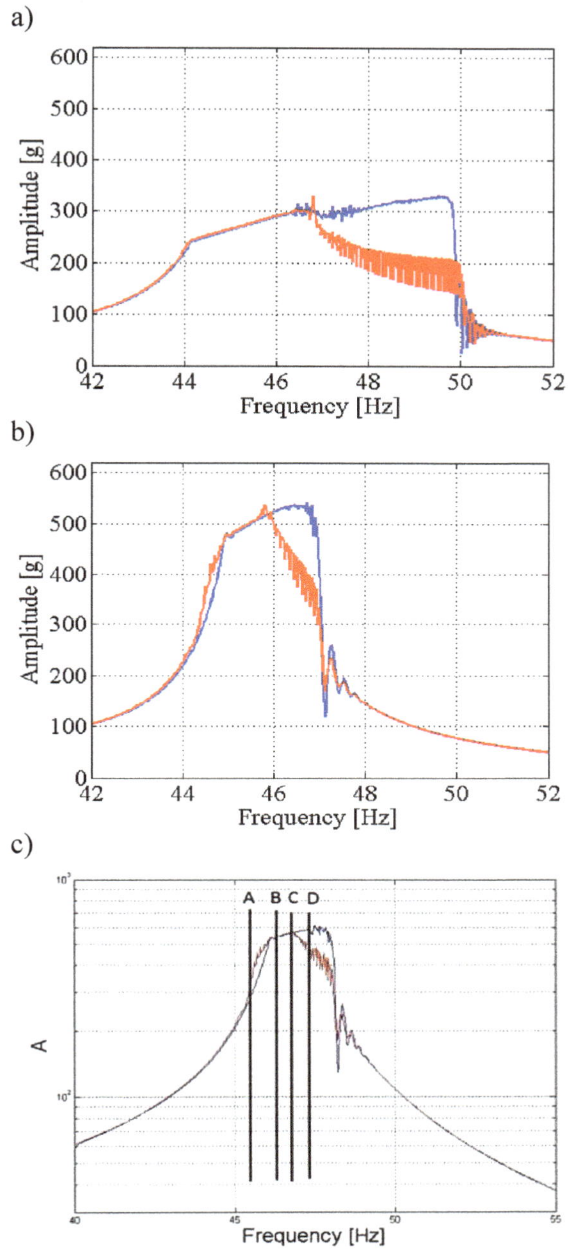

Fig. 4. Amplitude of inertial acceleration output versus frequency at different gap sizes between beam and independent stopper: h^1 =6.5 mm (a) and 11 mm (b). Figure (c) presents the same dependency of the amplitude A as a function of frequency for the system of double beams with the distance between beams equals to H_1 =14.3 mm. The blue and red lines correspond to up and down frequency sweep, respectively. Note that we limited the discussion to the first mode shape of beam vibration

Table 1. Parameters of the mechanical resonator and electrical circuit

Symbol and value	Description
b = 20 mm	beam width
h = 1.5 mm	beam thickness
h^1 , h_1'	distances between the beam and stopper for the stationary and moving set, respectively (Fig. 1a)
$L_1 = l_1$ = 163 mm	beam length
l_2 = 98 mm, L_2 = 65 mm	positions of the accelerometer
$L_3 = l_3$ = 20 mm	length of the piezoceramic layer
$L_4 = l_4$ = 12 mm	position of the piezoceramic layer
H_1 = 14.3 mm	distance between beams in double beam system
R_{DAQ} = 127 kΩ	internal impedance of DAQ
R_{eq} = 30.5 kΩ	equivalent load for the attached circuit
E = 69 GPa	Young modulus of the beam
ρ = 2.7 g/cm³	density the beam material
A = 2 g	acceleration amplitude
C = 4.9 nF	capacity of the piezoceramic layer
11.5 mm × 0.24 mm	dimensions of the piezoceramic
× 0.26 mm	fibers (PZT)
N = 10	number of parallel fibres in the piezoceramic layer
ω	excitation frequency

the dynamical response (of Figs. 4a-b cases) we present Figure 5 reflecting the evolution of the voltage time series along the resonance curve for the gap between the beam and stopper of 11 mm (Fig. 4b). The characteristic impact influences are noticeable in Figures 4b and c. On the other hand, Figures 4d-f correspond to non-resonant solutions. Finally, in Fig. 6 we show the corresponding time series for the last case (of Fig. 4c). Here the impacts are also clearly visible in the contact regions = 46.46 Hz (b) and 46.82 Hz (c), while the outer cases Figure 6a and 6d indicate the motion without impacts.

CONCLUSIONS

We studied the effect of stopper on vibration energy scavenging in two configurations (fixed and moving frame stoppers). Decreasing the distance between stopper and mechanical beam resonator introduced hardening to the system stiffness. Consequently, it also influenced the process of energy harvesting significantly, leading to increasing the frequency range and finally

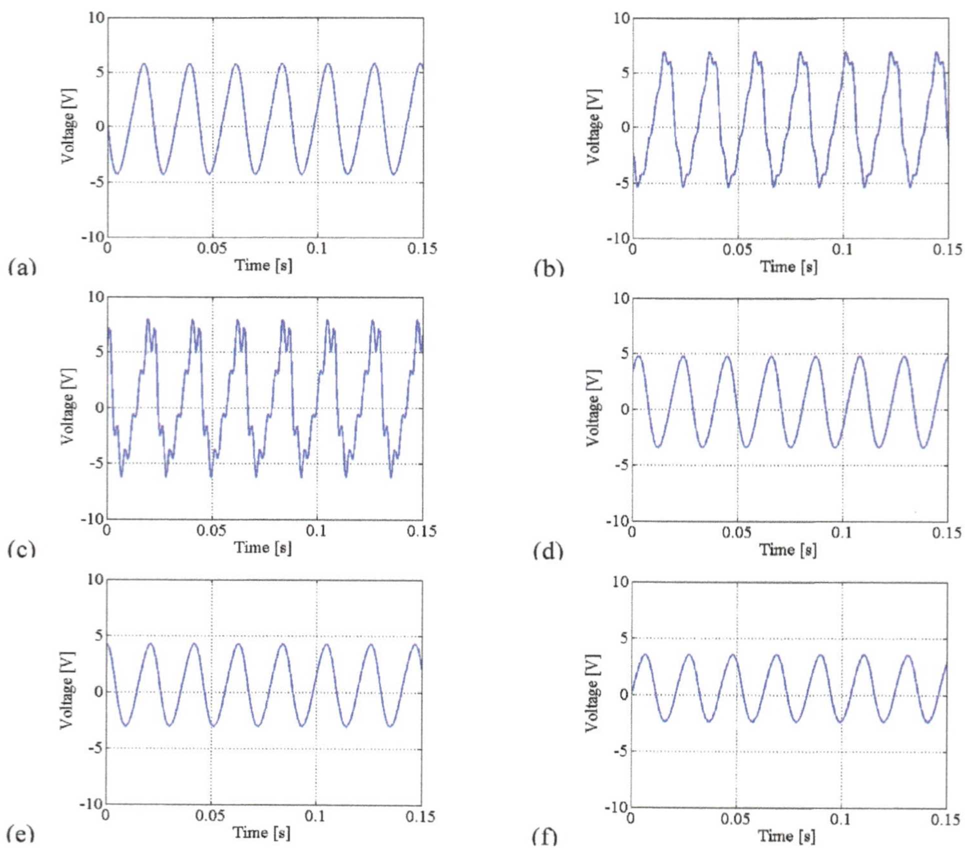

Fig. 5. Time series of voltage output measured at the gap size 11mm (case presented in Fig. 3 b). The excitation frequencies for consecutive time series were: (a) f_{exc} = 44.70 Hz, (b) f_{exc} = 45.40 Hz, (c) f_{exc} = 45.86 Hz, (d) f_{exc} = 46.56 Hz, (e) f_{exc} = 46.72 Hz, (f) f_{exc} = 47.06 Hz respectively. The voltage variance (in V² units) proportional to the power output for the consecutive cases are following: 12.32 (a), 16.5 (b), 18.33 (c), 8.2 (d), 6.61 (e), and 4.36 (f)

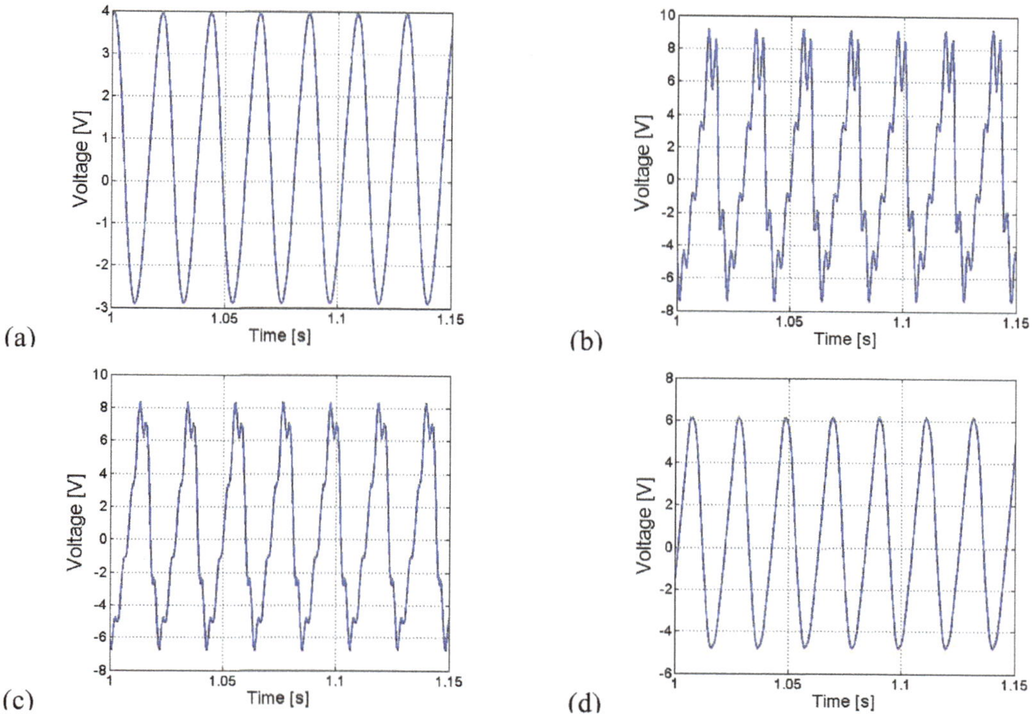

Fig. 6. Time series of voltage output measured at the gap size 14.3 mm (case presented in Fig. 3 c). The excitation frequencies for consecutive time series were: (a) f_{exc} = 45.40Hz, (b) f_{exc} = 46.46 Hz, (c) f_{exc} = 46.82 Hz, (d) f_{exc} = 47.32 Hz, respectively. Values of frequencies were marked on Figure 3c with subsequent names: A, B, C, and D The voltage variance (in V^2 units) proportional to the power output for the consecutive cases are following: 5.81 (a), 22.23 (b), 24.13 (c), 14.60 (d)

to the broadband effect. The evolution of the acceleration and voltage outputs measured along the amplitude-frequency diagram confirms the increase of the vibration amplitude, and consequently, the improvement of the power output in the frequency broad band form. In that region of frequency we noticed multiple solutions by doing up and down frequency sweeps. Obviously the improvement energy harvesting is associated with increasing range of impacting solution. It is worth to notice that in cases of the resonant (impacting) and non-resonant (non-impacting) solutions the percentage of the corresponding basins of attraction area are changing with the frequency. According to the findings of Borowiec et al. [14] where the ratio is changing from 1 to 0 for growing frequency.

Acknowledgments

This research was supported by the Polish National Science Center (A.R., M.B., J.Z., and G.L.) under the grant agreement No. 2012/05/B/ST8/00080 and by the Deutsche Forschungsgemeinschaft (DFG) in context of the Collaborative Research Centre/Transregio 39 PT-PIESA (M.M. and V.W.), subproject A2.

REFERENCES

1. Beeby S.P., Tudor M.J., White N.M., Energy harvesting vibration sources for microsystems applications. Measurement Science and Technology 17, 2006, 175–195.
2. Mitcheson P.D., Yeatman E.M., Rao G.K., Holmes A.S., Green T.C., Energy harvesting from human and machine motion for wireless electronic devices. Proc. IEEE 96, 2008, 1457–1486.
3. Erturk A., Inman D., Piezoelectric Energy Harvesting. Chichester: John Wiley & Sons Ltd. 2011.
4. Harne R.L., Wang K.W., A review of the recent research on vibration energy harvesting via bistable systems, Smart Mater. Struct. 22, 2012, 023001.
5. Friswell M.I., Ali S.F., Adhikari S., Lees A.W., Bilgen O., Litak G., Nonlinear piezoelectric vibration energy harvesting from a vertical cantilever beam with tip mass J. Intellig. Mat. Syst. Struct. 23, 2012, 1505–1521.
6. Litak G., Friswell M.I., Kwuimy C.A.K., Adhikari S., Borowiec M., Energy harvesting by two magnetopiezoelastic oscillators with mistuning, Theoret. Appl. Mech. Lett. 2, 2012, 043009.
7. Kwuimy C.A.K., Litak G., Borowiec M., Nataraj C., Performance of a piezoelectric energy harvester dri-ven by air flow, Appl. Phys. Lett. 100, 2012, 024103.

8. Blystad L-C. J., Halvorsen E., A piezoelectric energy harvester with a mechanical end stop on one side, Microsyst. Technol. 17, 2011, 505–511.
9. Gu L., Livermore C., Impact-driven, frequency up-converting coupled vibration energy harvesting device for low frequency operation Smart Mater. Struct. 20, 2011, 045004.
10. Le C.P., Halvorsen E., Sorasen O., Yeatman E.M., Microscale electrostatic energy harvester using internal impacts J. Intellig. Mat. Syst. Struct. 23, 2012, 1409–1421.
11. Le C.P., Halvorsen E., MEMS electrostatic energy harvesters with end-stop effects, J. Micromech. Microeng. 22, 2012, 074006.
12. Soliman M.S.M., Abdel-Rahman E.M., El-Saadany E.F., Mansour R.R., A wideband vibration-based energy harvester, J. Micromech. Microeng. 18, 2008, 115021.
13. Soliman M.S.M., Abdel-Rahman E.M., El-Saadany E.F., Mansour R.R., Design procedure for wideband micropower generators, J. Mircoelectromech. Systems 18, 2009, 1288–1299.
14. Borowiec M., Litak G., Lenci S., Basins of attraction in a simple harvesting system with a stopper. In: J.T.A Machado, D. Baleanu, A. Luo (Eds.), Discontinuity and Complexity in Nonlinear Physical System, Vol. 6, 315, 2014, 315–321.
15. Rysak A., Scheffler M., Gier J., Borowiec M., Litak G., Broadband vertical beam energy harvester with the moving mass. Latin American Journal of Solids and Structures, submitted 2013.

COMPARATIVE STUDY IN THE PASSIVE FORCE AND CUTTING TORQUE IN THE MILLING PROCESS OF POLYMER MATRIX COMPOSITES AND ALUMINUM ALLOYS

Krzysztof Ciecieląg[1], Kazimierz Zaleski[2]

[1] Department of Mechanical Engineering, Lublin University of Technology, ul. Nadbystrzycka 36, 20-618 Lublin, Poland, e-mail: krzysztofciecielag@wp.pl

[2] Department of Mechanical Engineering, Lublin University of Technology, ul. Nadbystrzycka 36, 20-618 Lublin, Poland, e-mail: k.zaleski@pollub.pl

ABSTRACT

This paper presents the results of a study undertaken to investigate the passive force and cutting torque during the milling of carbon fiber reinforced plastics saturated with epoxy resin and two aluminum alloys: AlSi21CuNi (AK 20) and 7075 (PA 9). The milling process was conducted using end mills with diamond inserts. The machining parameters were changed equally for each material as a result of which the passive force and cutting torque during the milling of these materials could be compared.

Keywords: polymer composites, silumin, duralumin, passive force, cutting torque.

INTRODUCTION

Polymer matrix composites have been of interest to engineers and scientists for decades. They are used in aviation, shipbuilding, automotive industry, sports and medicine. Their machining, that is cutting, turning, milling and drilling, is conducted using tools which have appropriate geometry and are made of wear-resistant materials. For each machining type, it is therefore necessary to know the "phenomena" that occur in the processing area. The relationships between variable machining parameters and values of cutting forces and cutting torques are of vital importance here.

There are numerous studies devoted to polymer composites, a new type of construction materials. The studies mainly focus on the description of their surface quality after machining operations such as turning, drilling or milling. Also, the studies discuss cutting forces and cutting torques.

One of the works on polymer matrix composites which analyzes the forces in milling is [1], written by D. Kalla, J. Sheikh-Ahmad and J. Twomey, in which cutting force values in the end milling of unidirectional and multidirectional composites are predicted. The predicted force values and the measured force results are then compared. Finally, the authors demonstrate the possibility to predict forces in the milling of polymer composites using the method employed. The authors of [2] describe the research undertaken to develop a new analytical method based on the energy method to predict forces that occur during the processing of polymer matrix composites with different fiber orientation (90° to 180°). The authors of [3] investigate cutting forces during milling grooves in a carbon composite, using two types of diamond cutters, in order to find the most favorable fiber orientation depending on the cutting force values. It has been found that the optimum conditions for machining composites occur with the fibers oriented at an angle of 45°/135°, which requires the lowest cutting force values. In [4], R. Teti offers a general description of polymer matrix composites, metal matrix composites and ceramic matrix composites. The turning, cutting, drilling and milling of each of these materials are described. The effect of tool wear on cutting forces in the machining of polymer composites with glass fibers is described in [5]. It has been found that the tool wear increases the cutting force values. In [6] and [7], methods for producing composites as well as the way of forming them are discussed.

The aim of this study is to investigate the effect of the cutting parameters, such as cutting speed and feed rate, on the passive forces and cutting torques during the milling of carbon fiber reinforced plastics (CFRP). Additionally, the results of milling silumin and duralumin are presented for comparative purposes. These materials have a low specific gravity and relatively high strength. Aluminum alloys exhibit a good resistance to corrosion, while the composites are characterized by a high corrosion resistance.

TEST STAND AND TOOLS

The carbon fiber reinforced plastics and aluminum alloy machining was conducted on a vertical machining center, Avila - VMC 800 HS. The samples were mounted on a milling table with dimensions of 1000 mm by 540 mm and displacement in three axes: X (800 mm), Y (540 mm), and Z (620 mm). A vice is mounted on the table, and to the vice the samples of the composite and aluminum alloys are attached in the working space during the milling operation. In the machine spindle, a Kistler dynamometer, model 9125A11A2, is mounted. It is used to measure the cutting torque and passive (vertical) force. A general view of the test stand is shown in Figure 1.

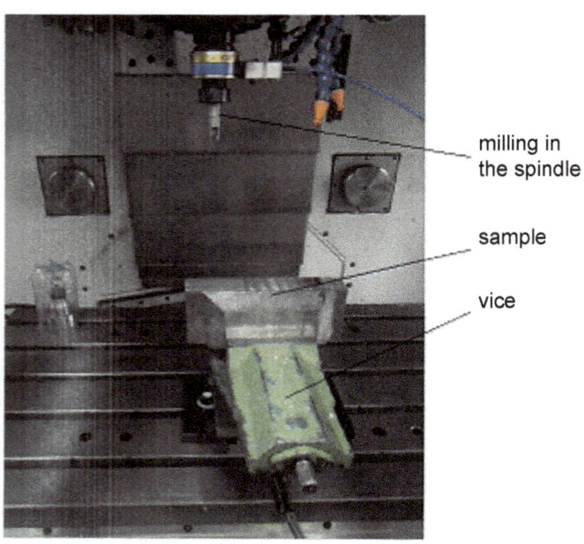

Fig. 1. Test stand

The testing tools (shown in Figure 2) were Iscar-manufactured cutters with diamond inserts (PCD), marketed under the symbol APKW 100304 PDR ID5 with radiused corners of 0.4 mm.

Fig. 2. End mill with diamond inserts

PREPARATION OF SAMPLES

The creation of the composite samples with dimensions of 200 by 200 by 10 mm consisted of the supersaturation of carbon fiber mats with epoxy resin, traded under name GR / EP 985-GF-3070, and positioning them in a 0–90 configuration. For a sample with a thickness of 10 mm there were about 50 pieces of mat. The temperature, humidity, and the amount of impurities in the controlled room were maintained at a preset level, namely at a temperature range of 18–30 °C, humidity up to 60%, and the amount of impurities not exceeding 10 000 particles per 1 m^3.

After positioning each fabric, the samples were put into a furnace and soaked for 120 minutes at the temperature of 177 °C and under the pressure of 0.3 MPa. After their removal from the furnace, the samples were left to cool and then the resin was cut.

The second material to be tested was silumin AlSi21CuNi, a casting aluminum alloy. The chemical composition of the alloy is listed in Table 1.

Table 1. The chemical composition of the alloy AlSi21CuNi

Chemical element	Si	Fe	Cu	Zn	Ti	Mn	P	Ni	Pb	Cr	Mg	Al
Percentage	21.92	0.40	1.27	0.02	0.004	0.14	0.007	0.91	0.004	0.006	0.47	rest

Table 2. The chemical composition of the alloy 7075

Chemical element	Si	Fe	Cu	Zn	Ti	Mn	Cr	Mg	Al
Percentage	0.40	0.50	2.00	6.10	0.20	0.30	0.28	2.90	rest

The third tested material was duralumin 7075. The chemical composition of the alloy is listed in Table 2.

THE EXPERIMENT

In the experiments, the effect of the feed rate and cutting speed on the passive force and cutting torque in milling carbon fiber reinforced plastics as well as silumin AlSi21CuNi and duralumin 7075 were investigated. The machining was performed using a 20 mm diameter end mill with diamond inserts (PCD ID5). The feed varied in the range from 0.04 mm/tooth to 0.28 mm/tooth. The cutting speed ranged from 100 m/min to 340 m/min. Both parameters were changed at a constant depth of cut equal to 1 mm. Each experiment was repeated 6 times.

The passive force and cutting torque were measured using the dynamometer mounted in the spindle of the machining center. The force and torque values were measured over the entire passage length. More than 10 000 measurements were obtained. 10 highest and 10 lowest results for the passive force and cutting torque were filtered out. Mean values of the maximum and minimum cutting force and cutting torque were calculated. The amplitude of the passive force and cutting torque was determined by subtracting the mean maximum value from the minimum value.

An example of the passive force distribution over time is shown in Figure 3. Each measurement was made of six samples, which allowed for determining a standard deviation.

RESULTS

Investigating the passive forces, it can be observed that the maximum forces increase with increasing the feed at a constant cutting speed. This dependence applies to the composite as well as to silumin and duralumin (Figure 4). In the case of duralumin, the passive force values are the highest and they vary from 400 N to 900 N. In case of polymer composite, the force values are approximately three times lower. The standard deviation for the maximum forces is in the range from 25 N to 80 N.

The dependence of the passive force amplitude on the feed is shown in Figure 5. Both the force and its amplitude increase with increasing the feed.

In case of duralumin, a change in the cutting speed at a constant feed value results in a decrease in the passive forces. In case of silumin and the composite, the effect of the cutting speed on the passive force is not as considerable it as can be seen in Figure 6.

Figure 7 shows the dependence of the passive cutting force amplitude on the cutting speed. The

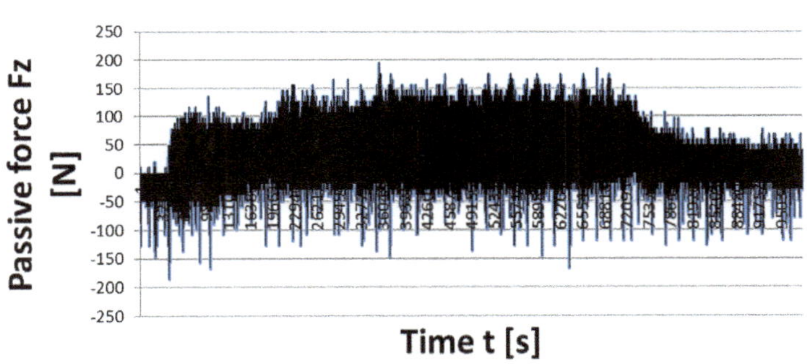

Fig. 3. Distribution of passive force during milling

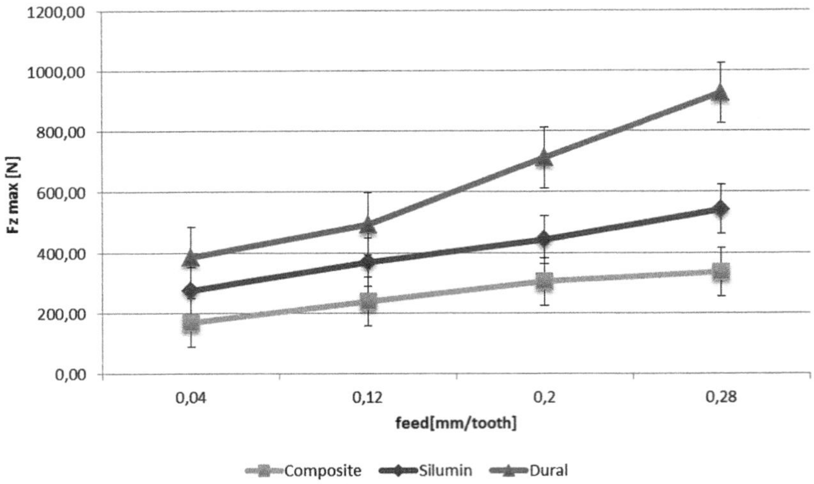

Fig. 4. Dependence of passive forces on feed

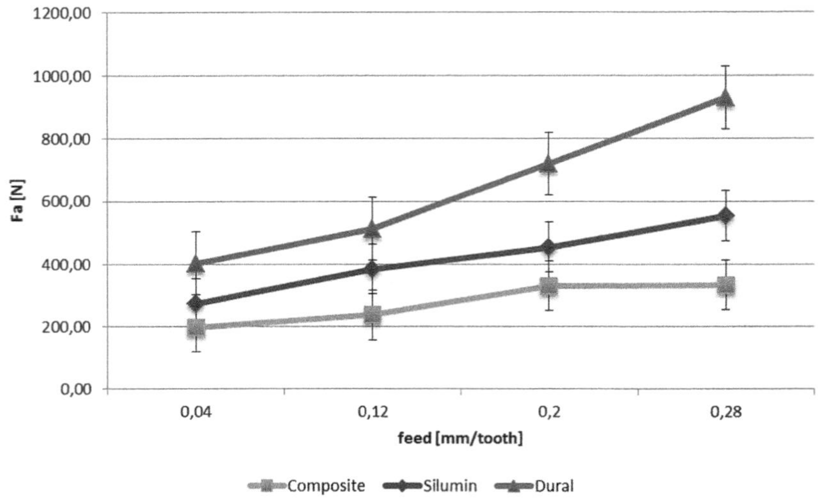

Fig. 5. Dependence of passive force amplitude on feed

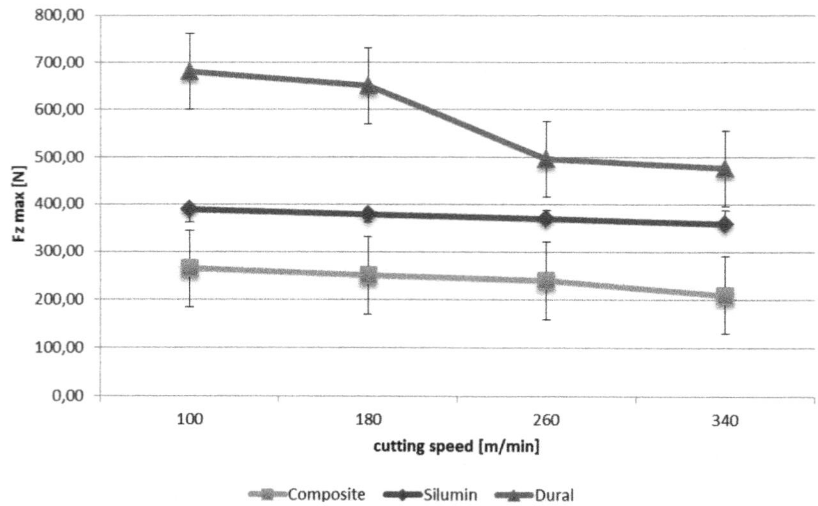

Fig. 6. Dependence of passive forces on cutting speed

graph clearly illustrates a decrease in the forces for duralumin. In case of the composite and siluminthis effect is smaller.

The investigation of the cutting torque has shown that the highest values is recorded when milling the composite (they are about 2 – 3 times

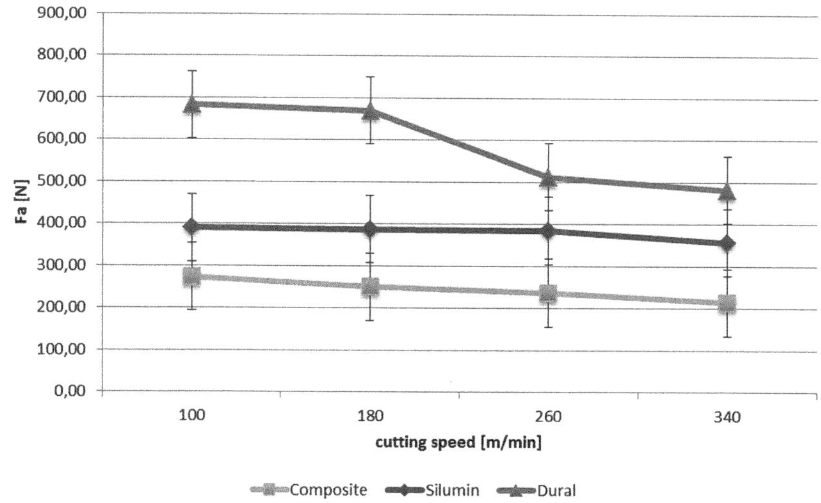

Fig. 7. Dependence of passive force amplitude on cutting speed

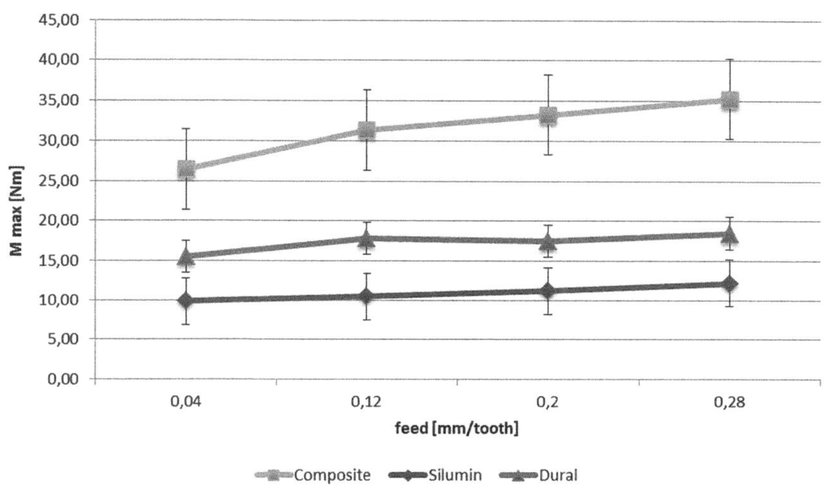

Fig. 8. Dependence of cutting torque on feed

higher compared to aluminum alloys). Increasing the feed causes a small increase in the maximum cutting torque (Figure 8). The maximum values of the cutting torque in the composite milling process vary in the range from 25 to 35 Nm. The standard deviation for the maximum cutting torque is from 2 Nm to 5 Nm. Figure 9 shows the dependence of the cutting torque amplitude on the feed. An increase in the cutting torque amplitude, especially for the polymer composite, is visible.

Investigating the relationship between the maximum cutting torque and the cutting speed, an interesting observation can be made. In case of silumin and duralumin, the cutting torque decreases with increasing the cutting speed, but the reverse is true for the polymer composite (Figure 10). Figure 11 shows the dependence of the cutting torque amplitude on the cutting speed. Analyzing this figure, another interesting observation can be made: the cutting torque amplitude increases for the composite, simultaneously decreasing for silumin and duralumin.

CONCLUSIONS

Summing up the results, it can be observed that increasing the feed rate at a constant cutting speed causes the increase in the passive forces and their amplitude. This dependence applies to all the investigated materials. With an increase in the cutting speed, the passive forces and their amplitude decrease, especially in case of duralumin. The maximum cutting torque increases with the increase in the feed for all the investigated materials. However, the increase in the cutting speed causes a decrease in the maximum cutting torque in case of aluminum alloys and an increase in the cutting torque when milling the composite. This may prove that the cutting edge has a different effect on such a complex material as the examined composite.

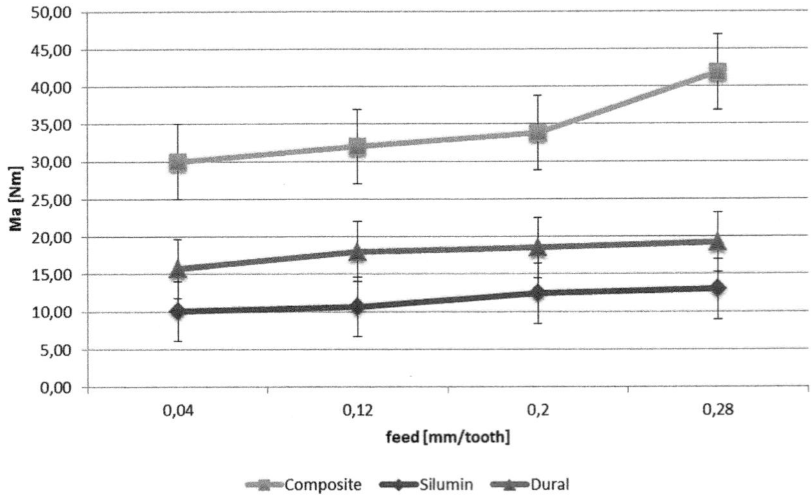

Fig. 9. Dependence of cutting torque amplitude on feed

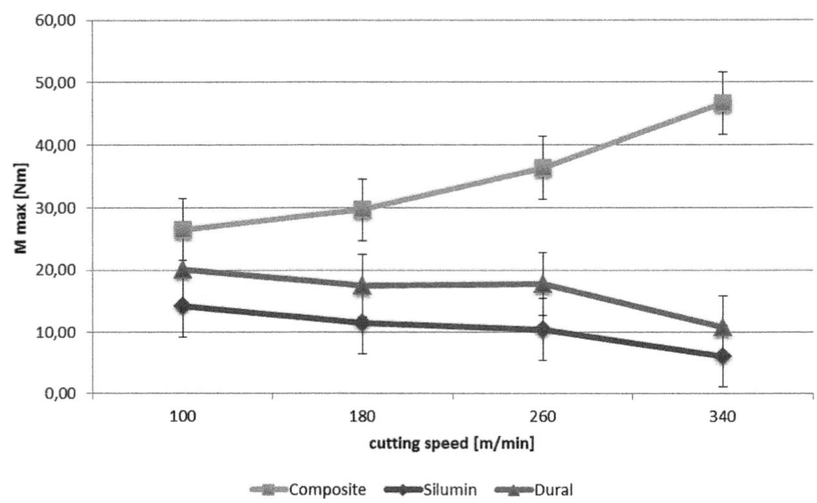

Fig. 10. Dependence of cutting torque on cutting speed

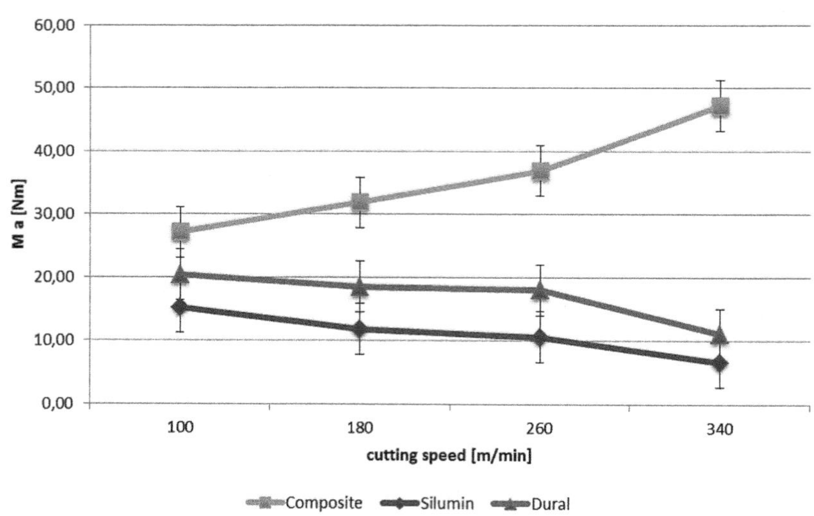

Fig. 11. Dependence of cutting torque amplitude on cutting speed

Acknowledgement

Financial support of Structural Funds in the Operational Programme – Innovative Economy (IE OP) financed from the European Regional Development Fund – Project "Modern material technologies in aerospace industry", No POIG.01.01.02-00-015/08-00 is gratefully acknowledged.

REFERENCES

1. Kalla D., Sheikh-Ahmad J., Twomey J.: Prediction of cutting forces in helical end milling fiber reinforced polymers. Int. J. Mach. Tools and Mfg., 50, 2010, 882–891.
2. Sahraie Jahromi A., Bahr B.: An analytical method for predicting cutting forces in orthogonal machining of unidirectional composites. Composite Science and Technology, 70, 2010, 2290–2297.
3. Karpata Y., Bahtiyarb O., Değerb B.: Mechanistic force modeling for milling of unidirectional carbon fiber reinforced polymer laminates, International Journal of Machine Tools and Manufacture, 56, 2012, 79–93.
4. Teti R.: Machning of Composite Materials. Annales of the CIRP, 51(2), 2002, 611-634.
5. Caprino G., De lorio I., Nele L., Santo L.: Effect of tool wear on cutting forces in the orthogonal cutting forces in the orthogonal cutting of unidirectional glass fibre-reinforces plastics. Composites, Part A, 27A, 1996, 409-415.
6. Boczkowska A., Kapuściński K., Lindemann Z., Witemberg-Perzyk D., Wojciechowski S.: Kompozyty, Oficyna Wydawnicza Politechniki Warszawskiej, Warszawa 2003.
7. Ochelski S.: Metody doświadczalne mechaniki kompozytów polimerowych, WNT, Warszawa 2004.

DETERMINATION OF TEMPERATURE IMPACT ON CMM RESIDUAL KINEMATIC ERRORS DISTRIBUTION

Wiktor Harmatys[1], Piotr Gąska[1], Adam Gąska[1], Maciej Gruza[1], Jerzy Sładek[1]

[1] Laboratory for Coordinate Metrology (M-10), Faculty of Mechanical Engineering, Cracow University of Technology, Jana Pawła II 37, 31-864 Kraków, Poland, e-mail: wiktorharmatys@gmail.com; pjgaska@gmail.com; agaska@mech.pk.edu.pl; gruzam@interia.pl; sladek@mech.pk.edu.pl

ABSTRACT

In case of coordinate machines that use CAA correction matrix, the issue of kinematic errors analysis may be based on the determination of residual error distribution. Temperature changes have an impact on CMM kinematic structure, which may cause the differences in the map of residual errors. As for today, the residual errors were analysed only for the reference temperature. No research was undertaken on the residual errors changes depending on the temperature variations. This paper presents the experiment aimed at residual errors analysis and resulting errors distributions for different temperatures.

Keywords: residual errors, CMM correction.

INTRODUCTION

Today, in the era of continuous development of technology and industry, it becomes even more important to ensure adequate quality of manufactured products. This task is entrusted to Coordinate Metrology which is an essential tool for quality control in modern companies. The biggest advantage of Coordinate Measuring Technique (CMT) is probably its versatility. Since information about the measured object geometrical features are obtained by determining coordinates of points on the object's surface, one device can be used for many measurement tasks. The main tool of CMT is Coordinate Measuring Machine (CMM). In the CMMs the kinematic pairs can shift in mutually perpendicular directions, which determine the Cartesian axes x, y, z and defining the machine basic coordinate system. Thanks to utilization of different types of probing systems, measurements with CMMs may be automated. The unquestionable advantage of CMMs is their high accuracy and repeatability. The most precise machines offer accuracy up to tenths of a micron, and the works on machines that would be capable of measuring in nanometres are in progress. However, the problem of CMMs accuracy is still important, especially evaluation of measurement uncertainty. Existing methods of uncertainty estimation recommended for CMMs are costly, especially due to necessity of multiple repetitions of measurements and the cost of required standards. Moreover, these methods require the personnel with certain knowledge and skills of percipient analysis of the results. Simulation methods of uncertainty estimation are different approach to the problem. Their functioning is most often connected with the development of the so-called virtual machines, which are responsible for multiple simulation of measuring task. Simulation methods have proved to be an interesting solution, but they are used almost exclusively in the laboratories of the best metrological research centres [1, 2]. This fact is caused by a number of factors, however, the most important ones are: duration of the tests which proceed the preparation of virtual machine for a specific CMM model, and the variability of the conditions in which the machine is working. The models developed so far assume machine functioning at reference conditions (temperature of 20°C). Deviations from this temperature may cause changes in the geometric parameters, especially those expressing the length which in turn leads to the variability of the CMM kinematic errors. Currently, at the Laboratory of Coordinate Metrology (LCM) the research are undertaken on the virtual machine which would take into consideration both of mentioned arguments. Developed at LCM virtual machine consists of two main modules: one responsible for machines kinematics modelling and one devoted to probing systems functioning. Distribution of residual errors is a basis for module responsible for kinematic errors modelling, it can be also used in determination of the optimal position of measured object in the machine measuring volume [3]. The paper describes the experiment that involved determination of residual errors distribution in different temperatures, in order to check the influence of temperature variability on kinematic errors. The obtained results as well as conclusions are also presented in the paper.

THE ERRORS RELATED TO CMM KINEMATICS

Machine's kinematic errors are one of the main factors affecting the accuracy of any CMM. CMM kinematic errors were the subject of a number of studies [4, 5] which resulted in the development of models of machine errors, for example full rigid body or reduced rigid body models. The most commonly used description of CMM geometrical errors include 21 components:

- positioning errors for each machine axis – xtx, yty, ztz;
- errors of mutual perpendicularity – xwy, xwz, ywz;
- translation errors – xty, xtz, ytx, ytz, ztx, zty;
- rotation errors – xry, xrz, yrx, yrz, zrx, zry, xrx, yry, zrz.

The geometrical errors defined in such a manner are a basis for operation of CAA (Computed Aided Accuracy) Matrix. The matrix itself as well as the methods of its determination are further described in [6]. The usage of this matrix has become a common practice, because it allows to correct most of the errors components associated with the machine kinematic. The Figure 1 shows the values of individual components before (red bars) and after (green bars) the CAA matrix was used.

As it can be seen, some errors were almost completely corrected. The remaining errors include both random errors and systematic errors that cannot be corrected or their correction would be unviable, namely residual errors. Since most of the currently used machines, utilize CAA correction system, the issue of CMM kinematic error analysis can be transferred to the analysis of residual errors. However, because the residual errors take into account the random errors components, they should be expressed rather by the probability distribution than certain values of error.

The residual errors distribution can be expressed by assigning the standard deviations of points coordinate reproduction to chosen reference points in measuring volume of machine. The standard deviations represent the probability of reference point reproduction. The exemplary distribution of residual points is presented in Figure 2. The red arrows represent the standard deviation in scale 100000 : 1. All values are given in [mm].

EXPERIMENT AND RESULTS

The distribution of residual errors is a basis for operation of one of the modules of virtual machine which is responsible for the kinematic errors modelling. Special methodology for residual errors determination was prepared at LCM for

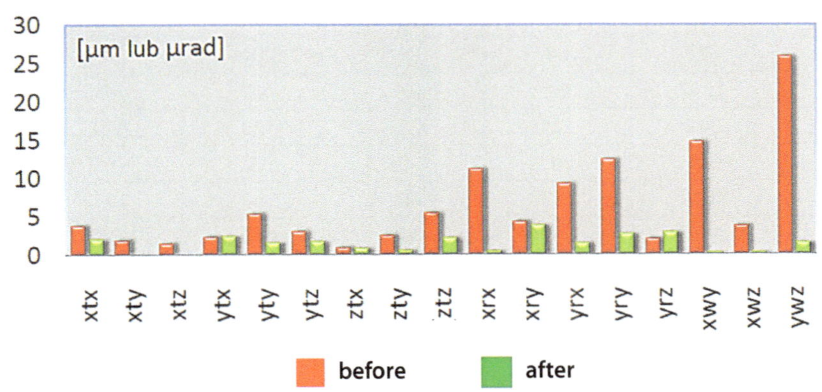

Fig. 1. The geometrical errors before and after applying the CAA correction

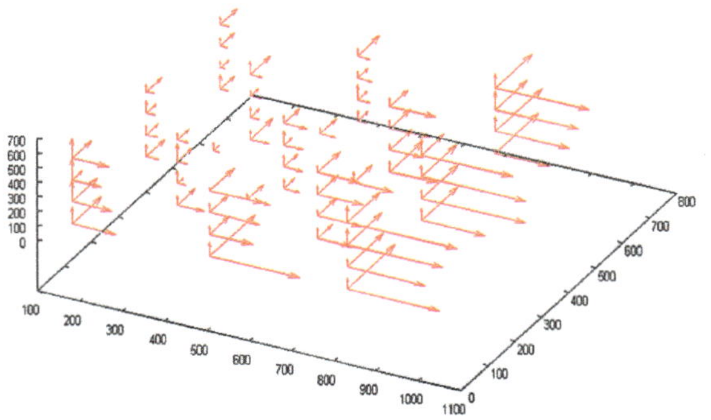

Fig. 2. The example of distribution of residual errors distribution

the purpose of the developed virtual machine [7]. Residual errors are determined in chosen points which defines the reference grid. The required measurements are taken using LaserTracer system which uncertainty for spatial measurements can be expressed by the following equation: $U = 0{,}2 \pm 0{,}3 \cdot L/1000$ µm. LaserTracer is a part of laser tracking devices family. It measures the distance to the target (retroreflctor) whose movements it follows. During measurements retroreflector is mounted on the machine in the place of probe head. LaserTracer as a more accurate device is used in order to check if machine indicate a correct coordinates of a reference point. The machine is programmed in such a way that retroreflector approach to each node of reference grid from several directions (14 approach direction for each point) and then moves to the another grid node. As the LaserTracer is able to measure only length, the multilateration method need to be used to determine the coordinate values. That means that the whole measuring sequence should be repeated at least four times, each time placing the Laser Tracer in different position in CMM measuring volume. During all required measurements the CAA matrix installed on the machine have to be active. The described methodology was used to determine the residual errors distribution in different temperatures.

Measurements described in this paper were performed on the Zeiss WMM850S machine, located in the LCM at the Cracow University of Technology. This machine has a moving bridge, and it is characterized by measuring volume of 800\1200\700 mm. It is placed in the air-conditioned room, which allows temperature control. The three following temperatures for residual errors determination were chosen: 20±0,2 °C; 21±0,2 °C and 22±0,2 °C. The ambient conditions during all measurements were monitored using temperature sensors which were mounted in measuring volume of the machine and close to the scales corresponding to each axis. The readings from sensors indicate that the temperature during the test was within the specified range. The residual errors were determined in 36 reference points distributed evenly in measuring volume of the machine. The grid nodes nominal coordinates are presented in Table 1.

Table 1. The grid of reference points

Point number	x [mm]	y [mm]	z [mm]	Point number	x [mm]	y [mm]	z [mm]
1	180	242	64	19	180	242	338
2	180	627	64	20	180	627	338
3	180	1012	64	21	180	1012	338
4	440	242	64	22	440	242	338
5	440	627	64	23	440	627	338
6	440	1012	64	24	440	1012	338
7	700	242	64	25	700	242	338
8	700	627	64	26	700	627	338
9	700	1012	64	27	700	1012	338
10	180	242	201	28	180	242	475
11	180	627	201	29	180	627	475
12	180	1012	201	30	180	1012	475
13	440	242	201	31	440	242	475
14	440	627	201	32	440	627	475
15	440	1012	201	33	440	1012	475
16	700	242	201	34	700	242	475
17	700	627	201	35	700	627	475
18	700	1012	201	36	700	1012	475

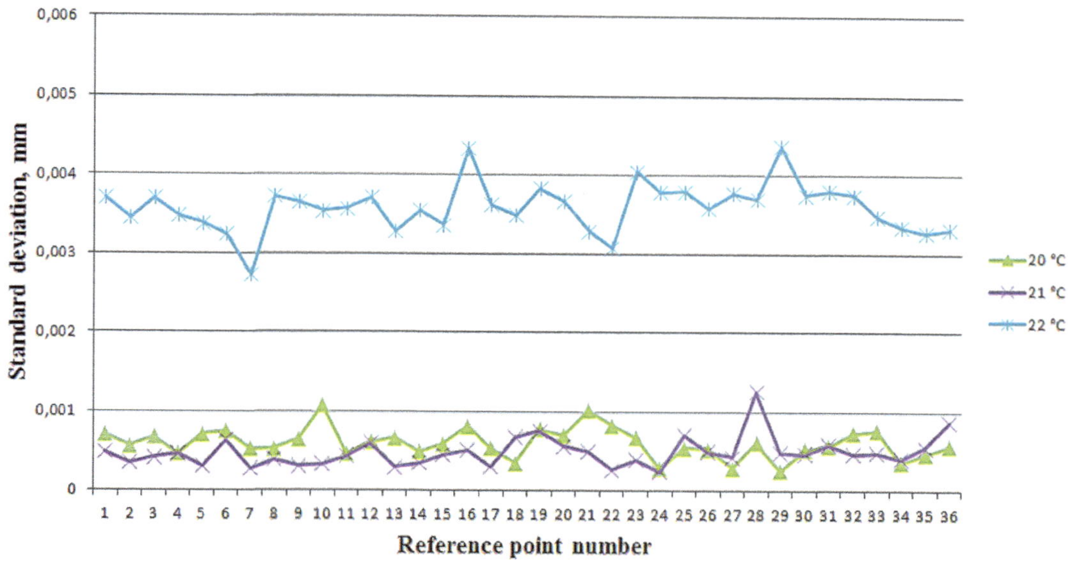

Fig. 3. Standard deviation of point reproduction in *x*-axis, for different temperatures

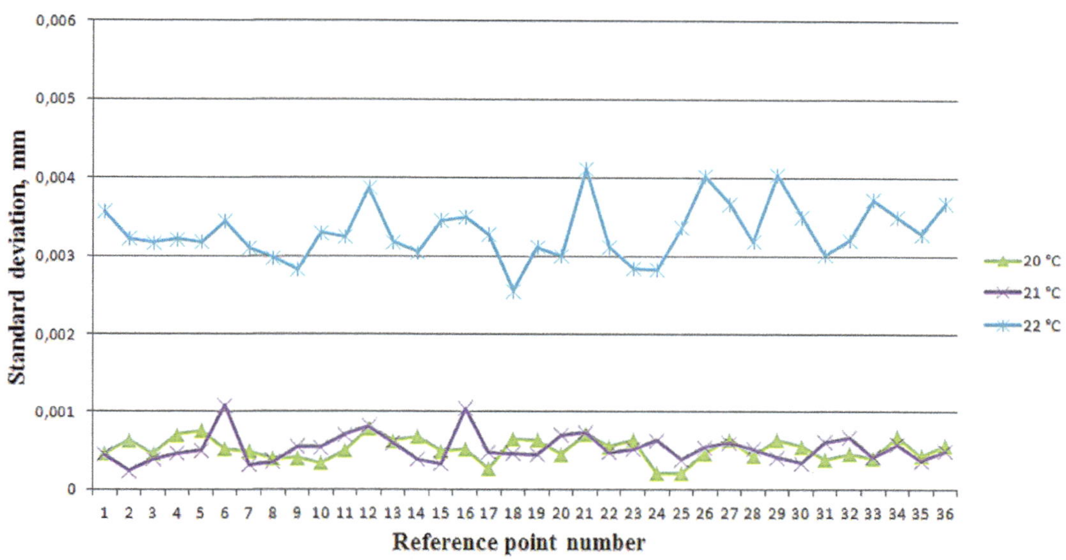

Fig. 4. Standard deviation of point reproduction in *y*-axis, for different temperatures

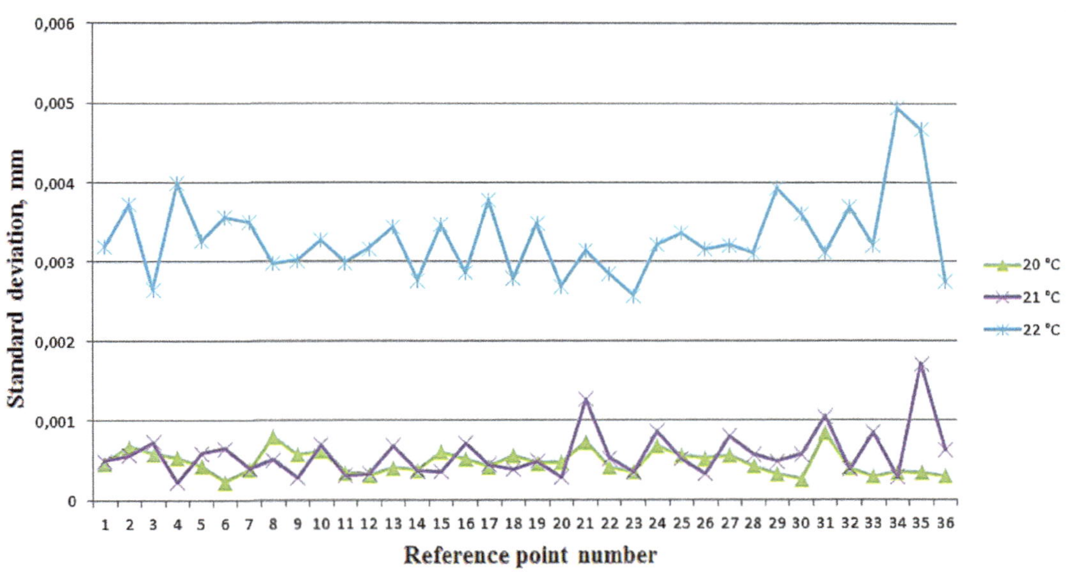

Fig. 5. Standard deviation of point reproduction in *z*-axis, for different temperatures

The obtained results are presented in Figures 3–5 which show the standard deviation of each point reproduction respectively in x, y and z axis for all three examined temperatures. Point numbers correspond to the numeration from Table 1.

CONCLUSIONS

The obtained results indicate that changes in the temperature noticeably affects the residual errors derived from machine kinematics. The results obtained for the 22± 0,2 °C are characterized by not only the biggest values, but also the highest variability, which reaches two micrometers. Also the results obtained for 21± 0,2 °C, show the considerable volatility. The temperatures bigger than 22 °C are rarely met in measuring rooms even in industrial conditions so studies at higher temperatures are unnecessary. The further research need to be undertaken in order to verify if the similar results would be obtained also for the temperatures lower than reference temperature of 20 °C, what appears to be likely. If these suppositions are confirmed, it will mean that Virtual machines that could be successfully implemented in the industry should take into consideration thermal influences. In case of virtual machine developed at LCM that means that residual errors distribution (which is a base for a module that simulates the CMM kinematic errors) should be determined for different temperatures. The number of temperatures in which tests ought to be performed should be chosen on the basis of experience of measuring personnel which work with a specified CMM model.

Acknowledgements

Reported research was realized within confines of project financed by Polish National Center for Research and Development No: LIDER/06/117/L-3/11/NCBR/2012.

REFERENCES

1. Trapet E., Franke M., Hartig F., Schwenke H., Wadele F., Cox M., Forbers A., Delbressine F., Schnellkens P., Trenk M., Meyer H., Morltz G., Guth Th., Wanner N.: Traceability of coordinate measuring machines according to the method of the Virtual Measuring Machines, PTB-F-35, Germany, Braunschweig, 1999.
2. Sładek J.: Modelowanie i ocena dokładności maszyn oraz pomiarów współrzędnościowych. Praca habilitacyjna, Wydział Mechaniczny, Politechnika Krakowska, 2001.
3. Gąska A., Krawczyk M., Kupiec R., Ostrowska K., Gąska P., Sładek J.: Modeling of the residual kinematic errors of coordinate measuring machines using LaserTracer system, The International Journal of Advanced Manufacturing Technology, 73, 1-4, 2014, 497–507.
4. Trapet E. et al.: Traceability of coordinate measurements according to the method of the virtual measuring machine. Final report, European Commission SMT- EDG XII/C/5 Project No. Mat. 1 – CT94-0076.
5. Schwenke H., Knapp W., Haitjema H., Weckenmann A., Schmitt R., Delbressine F.: Geometric error measurement and compensation of machines – An update. CIRP Annals – Manufacturing Technology, 57, 2008, 660–675.
6. Gąska A., Gruza M., Gąska P., Karpiuk M., Sładek J.: Identification and correction of coordinate measuring machine geometrical errors using LaserTracer systems, Advances in Science and Technology Research Journal, 7, 20, 2013, 17–22.
7. Gąska A.: Modeling the accuracy of coordinate measurement with use of Monte Carlo method. Ph.D. Dissertation, Cracow University of Technology, 2011.

IMPACT OF DEPTH OF CUT ON CHIP FORMATION IN AZ91HP MAGNESIUM ALLOY MILLING WITH TOOLS OF VARYING CUTTING EDGE GEOMETRY

Olga Gziut[1], Józef Kuczmaszewski[2], Ireneusz Zagórski[2]

[1] Student of Mechanical Engineering Faculty, Lublin University of Technology, 36 Nadbystrzycka Str., 20-618 Lublin, Poland

[2] Department of Production Engineering, Mechanical Engineering Faculty, Lublin University of Technology, 36 Nadbystrzycka Str., 20-618 Lublin, Poland, e-mail: j.kuczmaszewski@pollub.pl; i.zagorski@pollub.pl

ABSTRACT

Safety of Mg milling processes can be expressed by means of the form and the number of fractions of chips formed during milling. This paper presents the state of the art of magnesium alloys milling technology in the aspect of chip fragmentation. Furthermore, the impact of the depth of cut a_p and the rake angle γ on the number of chip fractions was analysed in the study. These were conducted on AZ91HP magnesium cast alloy and milling was performed with carbide tools of varying rake angle values ($\gamma = 5°$ and $\gamma = 30°$). It was observed that less intense chip fragmentation occurs with decreasing depth of cut a_p. The number of chip fractions was lower at the tool rake angle of $\gamma = 30°$. The test results were formulated as technological recommendations according to the number of generated chip fractions.

Keywords: high-speed dry milling, magnesium alloys, machinability, magnesium chips, end-mill geometry.

INTRODUCTION

The industry applications set a number of requirements to any technology in use, and increasing manufacturing efficiency is among the most vital. In milling this could not be achieved without advances in technological parameters of machining (predominantly v_c and f_z). Subtractive machining, including milling, bases heavily on the changes of these parameters. Although the depth of cut a_p is adjusted for particular applications, its impact is frequently disregarded. In magnesium alloy milling it is of utmost importance to limit the formation of dust fraction, which is highly susceptible to ignition. It is known that there is a correlation between the number of chip fractions and the increase or decrease of certain technological parameters and other analysed factors. Therefore, it appears necessary to investigate into the most efficient solutions, in terms of both effectiveness and safety of the realised process.

Chip formation is an inseparable aspect of milling process. However, the chips formed in magnesium alloy machining might pose a serious threat to the health of milling machine operators, what is more, the machine life can be shortened, as fine chips can lead to incorrect work of machine elements, such as: bearings, cables or guideways. The danger of uninitiated chip and dust ignition is a further reason behind the development of milling technology towards chip formation monitoring during milling.

Modern machining technologies are characterised by increasing process efficiency without sacrificing the quality of surface. This could be referred to as the optimal process performance. Modern technologies of manufacturing include: HSM machining (offering high efficiency and quality, machined surface roughness $Ra \leq 1$ μm) [17], HPM, HPC (offering optimal use of spindle performance with simultaneous high material removal rate, technique resembling roughing or shaping machining) or HSC (finishing method) [7, 14, 15]. The application of these methods of machining makes the process much more time-efficient. In addition, milling (high-speed in particular) typically serves as a roughing and finishing method [7, 17].

Another key factor describing safety of the realised machining process is temperature in the cutting zone [3, 11], which has a direct impact on the number of chip fractions formed in milling process [20]. There is a certain correlation between chip fragmentation and all technological parameters of milling. The type of formed chips (ergo chip ignition) is determined by, *e.g.* cutting velocity, the depth of cut or the type (chemical composition) of magnesium alloy [1]. Chip morphology and shape play an equally important role, along with their mass (of both leading and intermediate fractions) [8, 9], ignition initiation analysis and partial melting on the surface of Mg alloys [2, 6, 18]. Magnesium alloys have distinct morphology, on the one hand, they are of lamellar plate structure, on the other hand, their surface is smooth and shiny [4].

Technological considerations designate surface quality (expressed by surface roughness parameters) and cutting forces (affecting the dimensional and shape accuracy) as interesting indices of machinability [10, 16, 17, 19]. The analysis of cutting forces during milling with tools of varying cutting edge geometry shows that lower force values and amplitudes were observed at rake angle $\gamma = 30°$ [5]. Moreover, the increase of cutting forces results in deformed chip thickness, hence the increase of temperature in the cutting plane [3, 12, 13].

METHODOLOGY

The present study was conducted to analyse the impact of changing a_p and rake angle on the number of intermediate fractions of chips. Figure 1 shows the plan for tests conducted with tools of varying cutting edge geometry. The primary objective of the study was to determine technological parameters which would guarantee safety conditions of milling process. The process itself was performed on a vertical machining centre, Avia VMC800HS. The following cutting conditions were accepted: width of engagement a_e = 14 mm, cutting velocity v_c = 800 m/min, varying rake angle $\gamma = (5 \div 30)°$, and the technological parameters of milling: $a_p = (0.5 \div 3)$ mm, at $f_z = 0.05$ mm/tooth and $f_z = 0.15$ mm/tooth. The tests were carried out on AZ91HP magnesium cast alloy.

The description of chips formed in milling tests necessitated the introduction of the following terms:

- fraction – a population of elements of a particular size or dimensions,
- fragmentation – during machining, formation of an intermediate chip fraction of different shape and mass, alongside the leading fraction (prevailing and characteristic).

The division into the leading (main) and intermediate fractions was based on the shape and size of chips, fraction A being the biggest, fraction D the smallest chips. The term 'leading fraction' was ascribed to fraction A, and smaller fractions were referred to as 'intermediate fractions'. The reference for the chip fraction division was derived from the Polish Norm PN-ISO 3685 1996, which was, however, a classification referring to turning steel.

RESULTS

Table 1 shows the correlation between a_p at f_z = 0.05 mm/tooth and the number of recorded chip fractions and their shape. The highest number of fractions was observed at the lowest depth

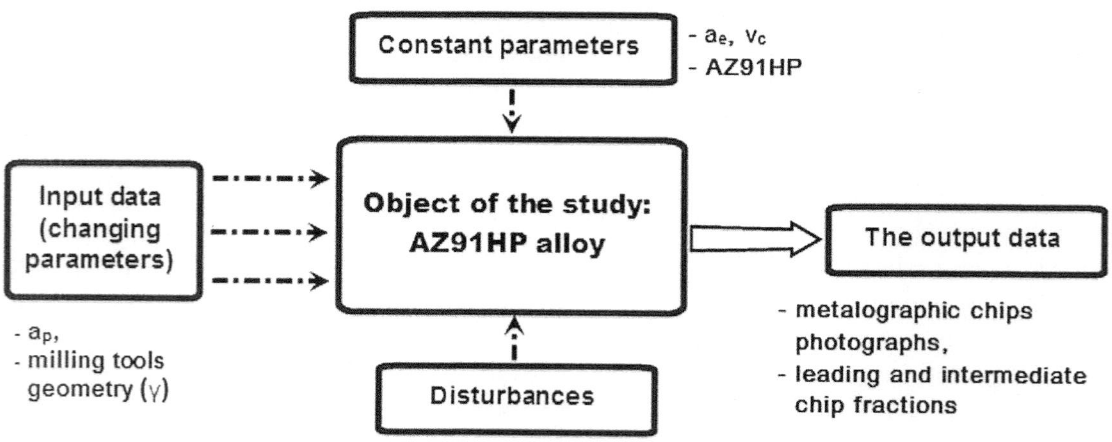

Fig. 1. Test plan for assessment of a_p and γ impact on chip form

of cut, fraction A was recognised as washer-type helical, fraction B washer-type helical short, fraction C elemental and D – dust. Fractions A and B obtained at a_p = 1.5 mm are described as tubular compressed because their width is higher than in the case of chips formed at a_p = 0.5 mm. Fraction C was described as elemental. At the depth of cut a_p = 3 mm the shape of chips in the recognised fractions was tubular compressed short (fractions A and B). Noted differences between fractions concerned chip dimensions. Chips found in fraction C are recognised as elemental.

Table 1 presents the correlation between the cutting speed and the shape of formed chips and the number of fractions separated in milling at the rake angle of $\gamma = 5°$. Figure 2 shows examples of chip shapes observed in milling at the change of depth of cut (feed f_z = 0.05 mm/tooth).

Figure 2 shows a selection of fractions determined at the depth of cut of a_p = 0.5 mm and the biggest fraction at a_p = 1.5 mm. Fraction D chips (Fig. 2a) are classified as dust, due to their small width and delicate structure, contributing to their easy braking. According to size, the second small-

Table 1. Impact of depth of cut a_p on chip shape and number of fractions (AZ91HP magnesium cast alloy): a) a_p = 0.5 mm, b) a_p = 1.5 mm, c) a_p = 3 mm, at v_c = 800m/min, f_z = 0.05 mm/tooth and $\gamma = 5°$

Parameter \ Fraction	All fractions	Leading fraction A	Intermediate fraction B	Intermediate fraction C	Intermediate fraction D
a) a_p = 0.5 mm f_z = 0.05 mm/tooth	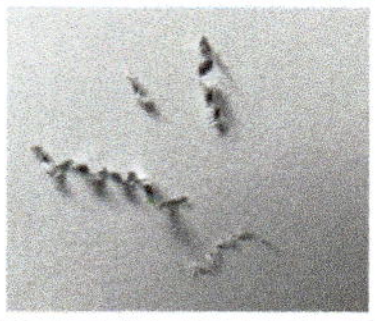				
b) a_p = 1.5 mm f_z = 0.05 mm/tooth	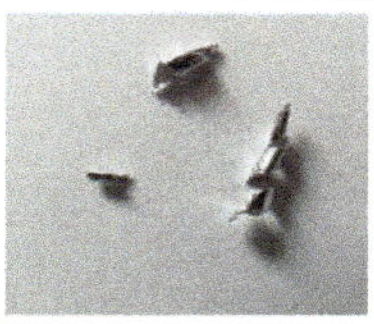	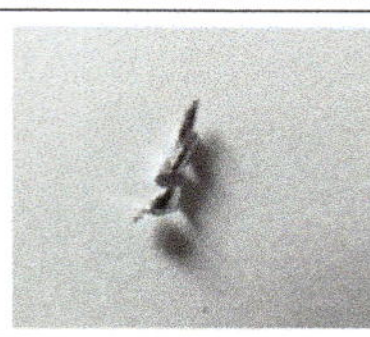	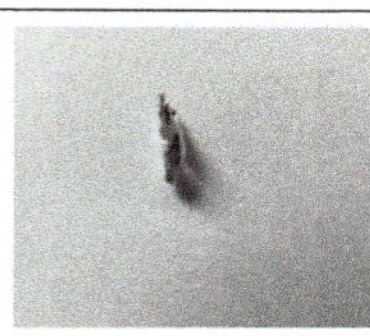	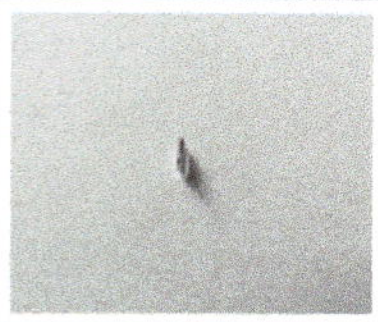	no fraction detected
c) a_p = 3 mm f_z = 0.05 mm/tooth	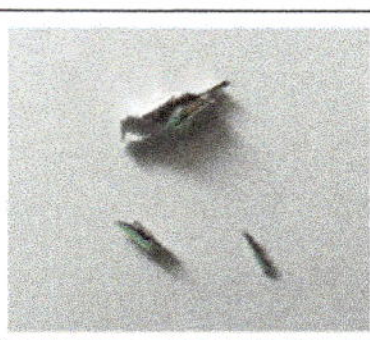	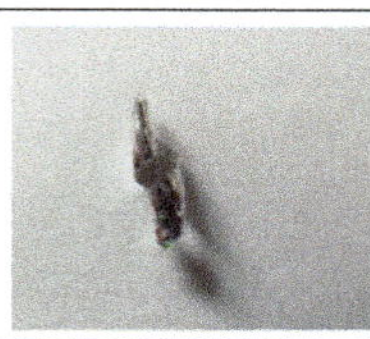	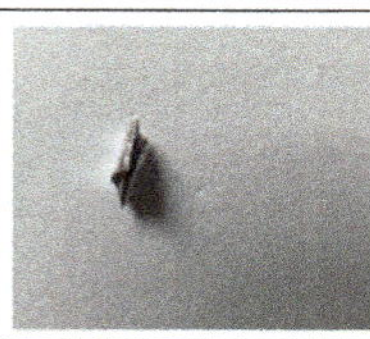		no fraction detected

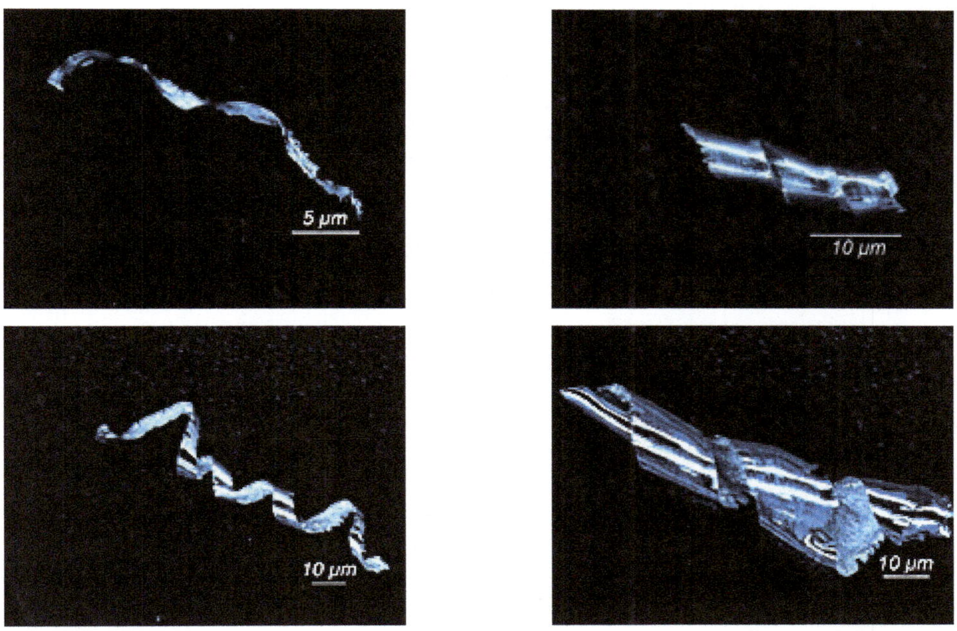

Fig. 2. Chips formed in AZ91HP magnesium cast alloy milling for selected chip fractions and depths of cut a_p (f_z = 0.05 mm/tooth): a) a_p = 0.5 mm, fraction D, b) a_p = 0.5 mm, fraction C, c) a_p = 0.5 mm, fraction A, d) a_p = 1.5 mm, fraction A

est fraction was C (Fig. 2b), which consisted of very fine chips of a tightly wound helix. Fraction A was represented by the longest, helical chips of smooth external surface and lamellar internal surface. The most noticeable difference between the leading fractions at a_p = 0.5 mm (Fig. 2c) and a_p = 1.5 mm (Fig. 2d) is the width of chips. Chips formed at the tested milling parameters are less convenient in terms of storage or transport, as due to their coiled shape they require more storage space.

Table 2 shows chip fractions formed with a changed depth of cut a_p at the feed rate f_z = 0.15 mm/tooth. Table 2 indicates that when the depth of cut was the lowest, 3 chip fractions were formed. Due to their small size chips in fraction C are considered as dust, and chips belonging to fractions A and B as washer-type helical. Chips of fraction B are notably shorter than those of fraction A, therefore could be regarded as elemental chips, however, their shape resembles the washer-type helical type. Fractions A and B contain tubular compressed chips at the depth of cut a_p = 1.5 mm. Fraction C includes arc chips, whereas fraction D, elemental. At the depth of cut a_p = 3 mm tubular compressed chips were formed (fraction A and B) and loose arc chips (fraction C and D). It ought to be mentioned that fraction A chips are significantly longer than in fraction B, and chips of fraction D are more coiled than chips representing fraction C.

Figure 3 presents chips classified as the smallest and the biggest fractions at different depths of cut. At the depth of cut a_p = 0.5 mm fraction A (Fig. 3a) contained the longest chips, similarly to fraction A at a_p = 3 mm (Fig. 3c). Considerable differences were observed not only in terms of width but the tightness of the helix as well. Fraction A chips at a_p = 0.5 mm had a tighter helix and their structure was quite uniform, whereas the leading fraction at the depth of cut of 3 mm had a notably less uniform chip, with distinct bends and cuts on the edges. Chips belonging to fraction C at the depth of cut a_p = 0.5 mm were distinctly different from other chips in terms of size and shape. These were small elements of highly frayed edges, very well seen in magnified picture. At the depth a_p = 0.5 mm three fractions were determined. Fractions A and B were classified as washer-type helical, however, fraction A chips were over three times as long as chips of fraction B and C, the latter were classified as dust. At the depth of cut of a_p = 1.5 mm fractions A and B were determined as tubular compressed, and chips of fraction C as elemental. Both fractions formed at a_p = 3 mm were classified as tubular compressed.

Table 3 presents the impact of cutting velocity on the shape of formed chips and the number of fractions obtained in milling conducted at the tool rake angle γ = 30°. The analysis of table 3 indicates that the change of the depth of cut a_p, at f_z = 0.05 mm/tooth, does impact the shape of chips and the degree of their fragmentation. Figure 4 shows chip shapes formed during milling of AZ91HP magnesium alloy with a changed depth of cut a_p at the feed rate f_z = 0.05 mm/tooth.

Table 2. Impact of depth of cut a_p on chip shape and number of fractions (AZ91HP magnesium cast alloy): a) a_p = 0.5 mm, b) a_p = 1.5 mm, c) a_p = 3 mm, at v_c = 800m/min, f_z = 0.15 mm/tooth and γ = 5°

Fraction / Parameter	All fractions	Leading fraction A	Intermediate fraction B	Intermediate fraction C	Intermediate fraction D
a) a_p = 0.5 mm f_z = 0.15 mm/tooth	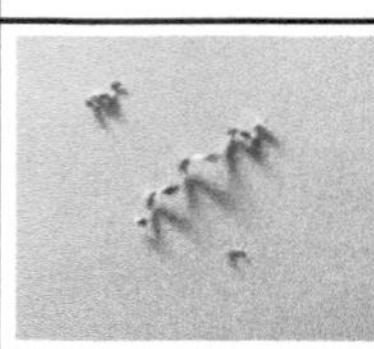	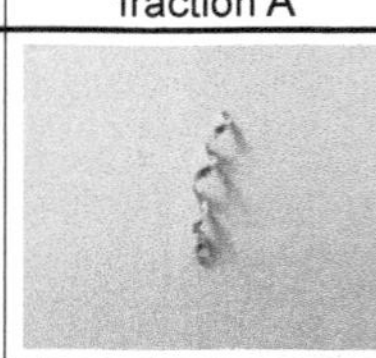	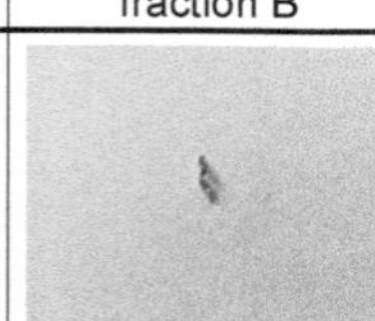	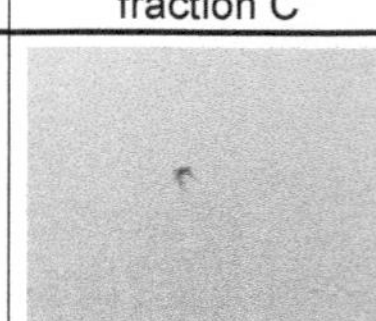	no fraction detected
b) a_p = 1.5 mm f_z = 0.15 mm/tooth	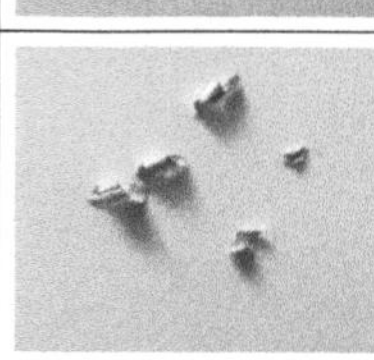	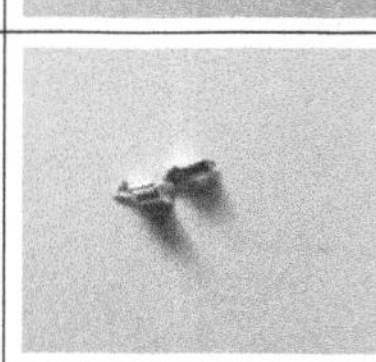			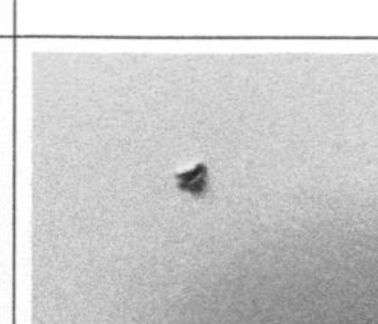
c) a_p = 3 mm f_z = 0.15 mm/tooth	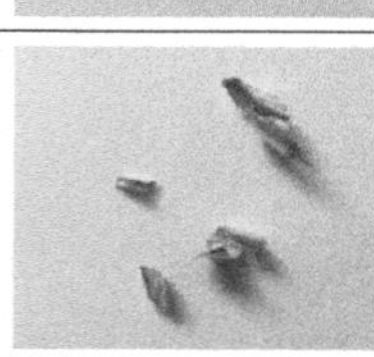	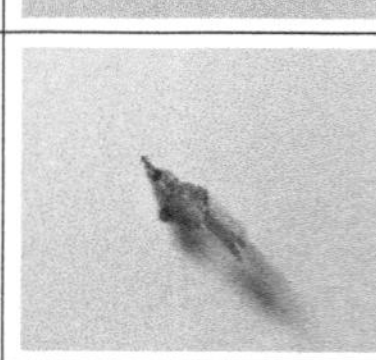		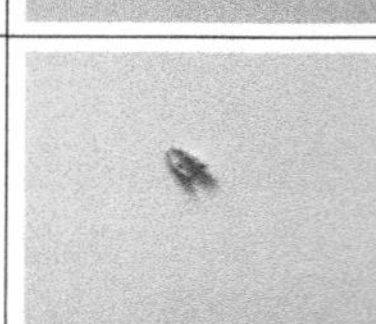	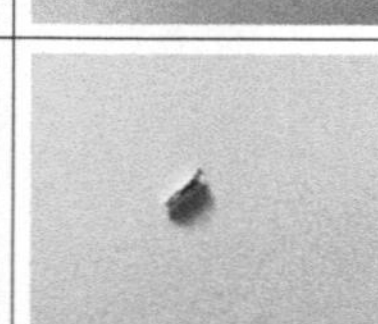

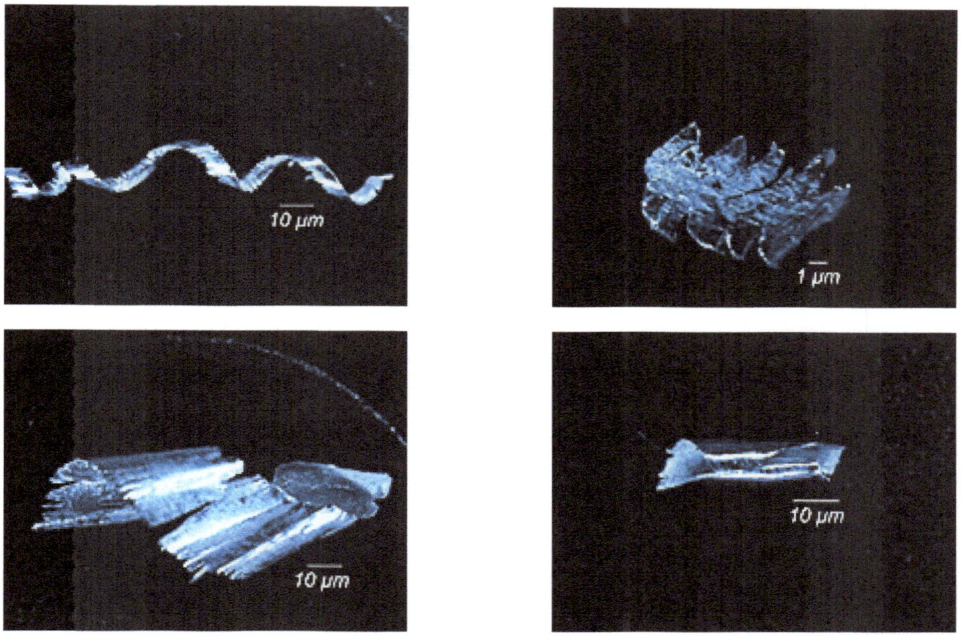

Fig. 3. Chips formed in AZ91HP magnesium cast alloy milling for selected chip fractions and depths of cut a_p (f_z = 0.15 mm/tooth): a) a_p = 0.5 mm, fraction A, b) a_p = 0.5 mm, fraction C, c) a_p = 3 mm, fraction A, d) a_p = 3 mm, fraction D

Chips of fraction C (Fig. 4b) were of small size and therefore classified as dust. Fraction A (Fig. 4a) consisted of long coiled chips, which due to their shape could be difficult in transport and storing. The outer surface of chips formed at the highest depth of cut was shiny and deformations were visible only at the edges. The tightness of the helix at γ = 30° was higher than in the case of γ = 5°, which is most clearly visible in the leading fraction of the depth of cut a_p = 3 mm.

Table 4 presents chip photographs illustrating the correlation between the depth of cut a_p (at f_z = 0.15 mm/tooth) and the shape and number of fractions classified. It can be observed that the highest number of fractions was observed at the depth of cut a_p = 1.5 mm, while at the other depths there were 3 fractions. When the depth of cut was the lowest, fractions A and B (washer-type helical chips) and C (dust) were distinguished. The width of chip increased with the increase of the depth of

Table 3. Impact of depth of cut a_p on chip shape and number of fractions (AZ91HP magnesium cast alloy): a) a_p = 0.5 mm, b) a_p = 1.5 mm, c) a_p = 3 mm, at v_c = 800m/min, f_z = 0.05 mm/tooth and γ = 30°

Fraction / Parameter	All fractions	Leading fraction A	Intermediate fraction B	Intermediate fraction C	Intermediate fraction D
a) a_p = 0.5 mm f_z = 0.05 mm/tooth	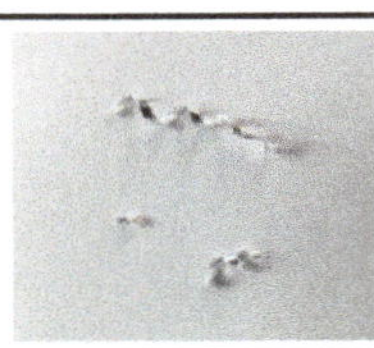	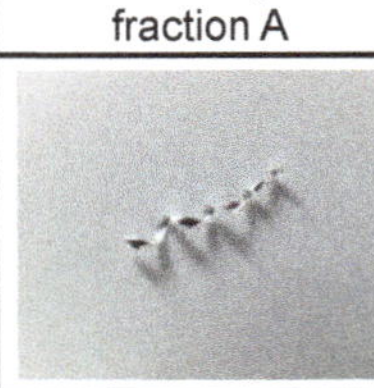	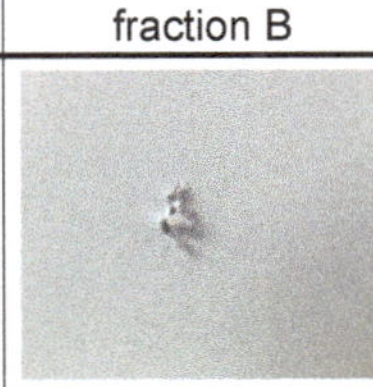	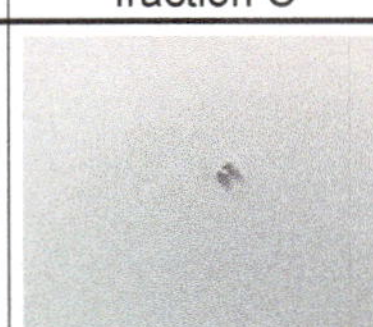	no fraction detected
b) a_p = 1.5 mm f_z = 0.05 mm/tooth	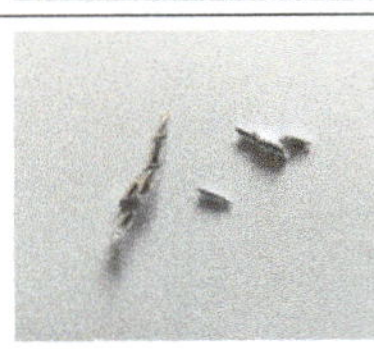		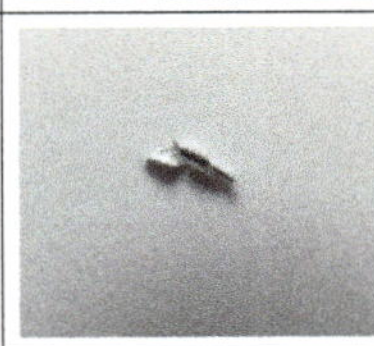		no fraction detected
c) a_p = 3 mm f_z = 0.05 mm/tooth	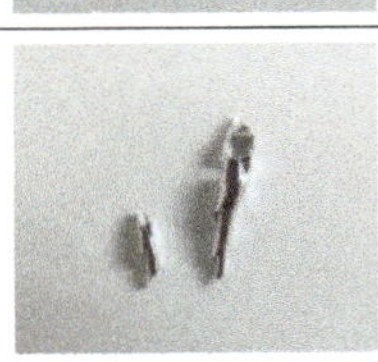	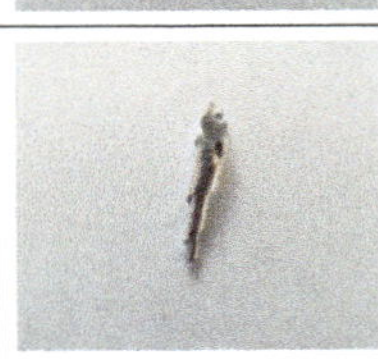	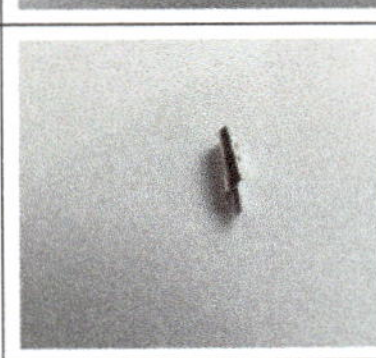	no fraction detected	no fraction detected

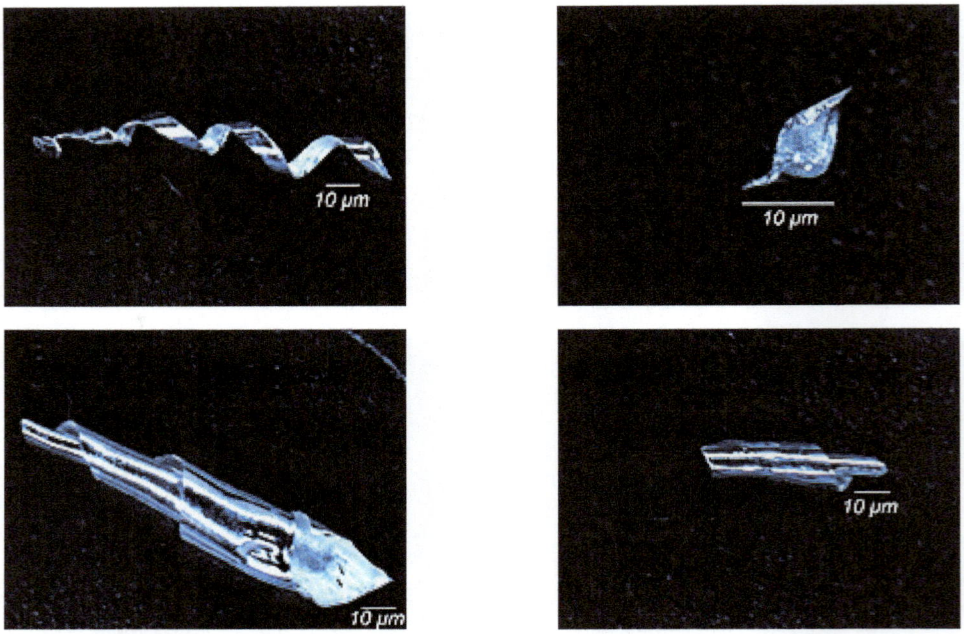

Fig. 4. Chips formed in AZ91HP magnesium cast alloy milling for selected chip fractions and depths of cut a_p (f_z = 0.05 mm/tooth): a) a_p = 0.5 mm, fraction A, b) a_p = 0.5 mm, fraction C, c) a_p = 3 mm, fraction A, d) a_p = 3 mm, fraction B

cut, therefore, fractions A and B (a_p = 1.5 mm) were recognised as tubular compressed chips. Fraction D represented dust and fraction C contained elemental chips. At the depth of cut a_p = 3 mm the formed chips were of tubular compressed shape. Figure 5 shows chip shapes formed during AZ91HP magnesium alloy milling, performed with a changed depth of cut a_p at the feed rate f_z = 0.15 mm/tooth.

Figure 5 shows the smallest fractions distinguished at the depths of cut a_p = 0.5 mm, a_p = 1.5 mm and a_p = 3 mm. In the case of a_p = 0.5 mm and a_p = 1.5 mm the smallest fractions represented small chips of tight helical shape and smooth edges, classified as dust. While the length of chips was roughly similar for fractions C and D (a_p = 1.5 mm), the main difference concerned the shape and size of chips, which could be described as arc. Chips of fraction C (a_p = 3 mm) had a characteristically tighter helix in comparison to other smallest chip fractions.

Table 4. Impact of depth of cut a_p on chip shape and number of fractions (AZ91HP magnesium cast alloy): a) a_p = 0.5 mm, b) a_p = 1.5 mm, c) a_p = 3 mm, at v_c = 800m/min, f_z = 0.15 mm/tooth and $\gamma = 30^o$

Parameter \ Fraction	All fractions	Leading fraction A	Intermediate fraction B	Intermediate fraction C	Intermediate fraction D
a) a_p = 0.5 mm f_z = 0.15 mm/tooth	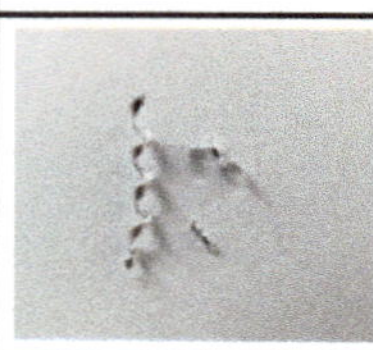	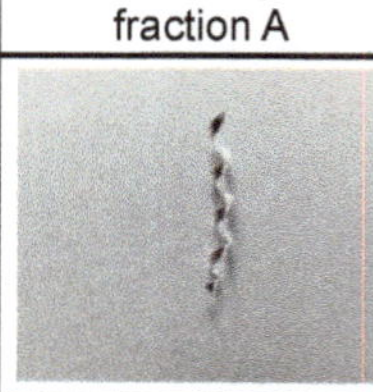	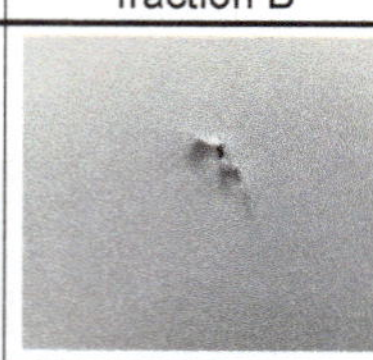	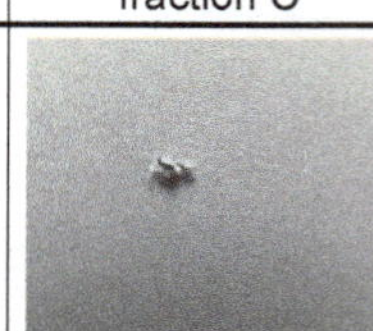	no fraction detected
b) a_p = 1.5 mm f_z = 0.15 mm/tooth	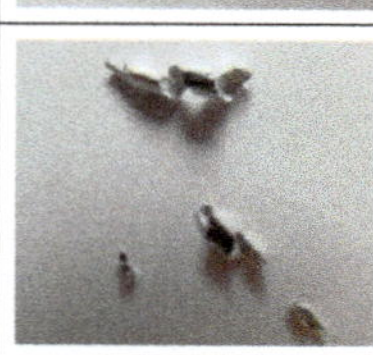		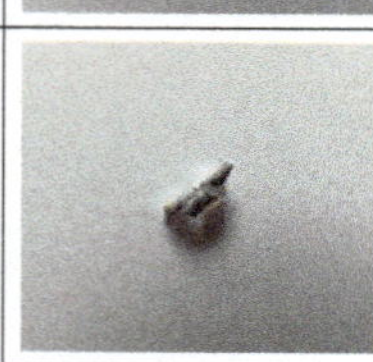	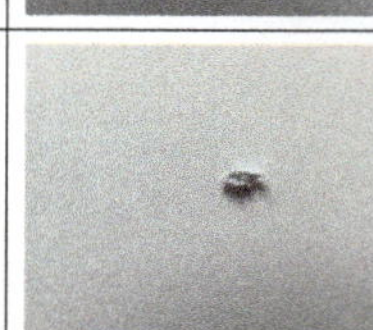	
c) a_p = 3 mm f_z = 0.15 mm/tooth	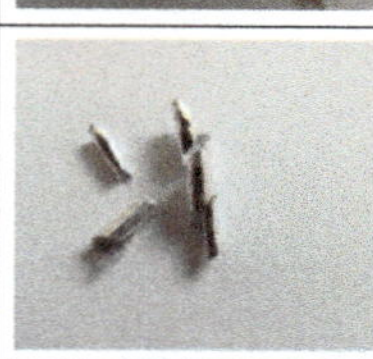		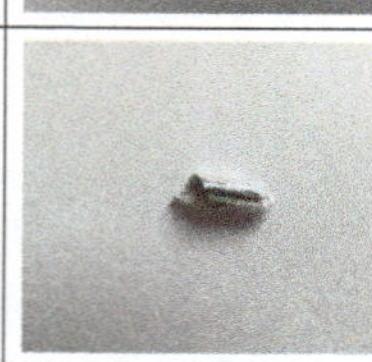	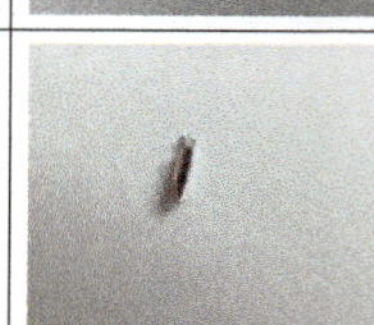	no fraction detected

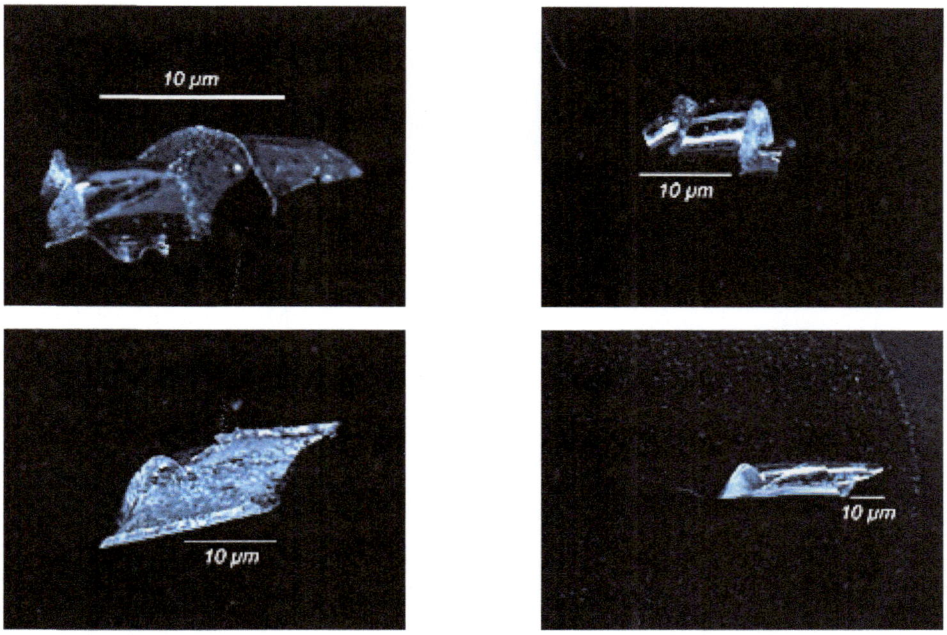

Fig. 5. Chips formed in AZ91HP magnesium cast alloy milling for selected chip fractions and depths of cut a_p (f_z = 0.15 mm/tooth): a) a_p = 0.5 mm, fraction C, b) a_p = 1.5 mm, fraction D, c) a_p = 1.5 mm, fraction C, d) a_p = 3 mm, fraction C

CONCLUSIONS

The conducted study points at the type of changes regarding technological parameters of milling (mainly a_p). The tests conducted within the analysed area of variables focused on the depth of cut a_p. The tests assessed the impact of both the depth of cut a_p and the rake angle of the tool γ on the number of chip fractions formed in Mg alloy milling. The presented study may serve as an indicator of parameters for effective and safe AZ91HP magnesium alloy milling.

A potentially hazardous area, accepting the formation of a visible dust fraction as a criterion of safety, is when the depth of cut was the lowest: a_p = 0.5 mm and a_p = 1.5 mm. These parameters contributed to increased levels of dust fraction formation in milling. Moreover, other chips formed in these conditions were either washer-type helical or tubular compressed, which makes their transport and storage highly inconvenient.

The presented results clearly show that the fewest chip fractions were formed at the tool rake angle γ = 30°. It appears that the parameter which is specifically responsible for the number of formed chip fractions is the depth of cut a_p. Moreover, at the lowest depth of cut the highest amount of dust fractions was observed: intermediate fraction D (a_p = 0.5 mm, f_z = 0.05 mm/tooth, γ = 5°), intermediate fraction C (a_p = 0.5 mm, f_z = 0.15 mm/tooth, γ = 5°), intermediate fraction C (a_p = 0.5 mm, f_z = 0.05 mm/tooth, γ = 30°), intermediate fraction C ($\mathbf{a_p}$ = 0.5 mm, f_z = 0.15 mm/tooth, γ = 30°) and intermediate fraction D (a_p = 1.5 mm, f_z = 0.15 mm/tooth, γ = 30°). The form of chip obtained in the depths of cut between 0.5÷3 mm is undesired, as its thin and coiled shape is unfavourable in terms of chip evacuation, transport and storage.

The conclusions drawn from the results of the study are as follows:

1. Fewer fractions, hence a lesser degree of chip fragmentation, are formed at the tool rake angle γ = 30°.
2. The largest number of fractions was obtained with the change of the depth of cut a_p (when compared with conducted as a part of this study, however not described in this analysis, other technological parameters).
3. A hazardous area with regards to dust fraction formation is at the lowest depths of cut: a_p = 0.5 mm and $\mathrm{a_p}$ = 1.5 mm.
4. The most advantageous chips, easily evacuated from the milling machine's workspace and not requiring additional storage space, are formed when the depth of cut is equal or higher than a_p = 3 mm.
5. The present study concludes that AZ91HP magnesium alloy should be conducted at considerably high values of a_p and 'sharp' tool geometry.

Acknowledgements

Financial support of Structural Funds in the Operational Programme – Innovative Economy (IE OP) financed from the European Regional Development Fund, Nr POIG.01.01.02-00-015/08-00 is gratefully acknowledged.

REFERENCES

1. Akyuz B.: Machinability of magnesium and its alloys. The Online Journal of Science and Technology, 1(3), 2011, 31–38.
2. Arai M., Sato S., Ogawa M., Shikata H.I.: Chip control in finish cutting of magnesium alloy. Journal of Materials Processing Technology 62, 1996, 341–344.
3. Fang F.Z., Lee L.C., Liu X.D.: Mean flank temperature measurement in high speed dry cutting, Journal of Materials Processing Technology, 167, 2005, 119–123.
4. Guo Y.B., Salahshoor M.: Process mechanics and surface integrity by high-speed dry milling of biodegradable magnesium–calcium implant alloys. CIRP Annals – Manufacturing Technology, 59, 2010, 151–154.
5. Gziut O., Kuczmaszewski J., Zagórski I.: The influence of technological parameters and geometric features of a cutting edge on cutting forces during AZ91HP alloy milling. Advances in Manufacturing Science and Technology, 38(2), 2014, 83–92.
6. Hou J.Z., Zhou W., Zhao N.: Methods for prevention of ignition during machining of magnesium alloys. Key Engineering Materials, Vol. 447–448, 2010, 150–154.
7. Kuczmaszewski J.: Efektywność wytwarzania elementów lotniczych ze stopów aluminium i magnezu. Komputerowo Zintegrowane Zarządzanie, Oficyna Wyd. Polskiego Towarzystwa Zarządzania Produkcją, Opole 2011, 7–18.
8. Kuczmaszewski J., Zagórski I.: Badania fragmentowania wiórów w procesie frezowania stopów magnezu (in Polish). Mechanik, Vol. 8–9, 2014, 321–328.
9. Kuczmaszewski J., Zagórski I.: Control of the weight, ignition temperature and actual temperature of chips produced when machining some selected magnesium alloy grade (in Polish). Mechanik, Vol. 10, 2012, 824–828.
10. Kuczmaszewski J., Zagórski I.: The study of cutting forces and amplitudes during the milling magnesium alloys (in Polish). Advances in Science and Technology Research Journal, Vol. 15, 2012, 228–237.
11. Le Coz G., Marinescu M., Devillez A., Dudzinski D., Velnom L.: Measuring temperature of rotating cutting tools: Application to MQL drilling and dry milling of aerospace alloys. Applied Thermal Engineering, Vol. 36, 2012, 434–441.
12. Li, H.Z., Liu, K., Li, X.P.: A new method for determining the undeformed chip thickness in milling.Journal of Materials Processing Technology, 113(1–3), 2001, 378–384.
13. Lotfi Sai, Bouzid W., Zghal A.: Chip thickness analysis for different tool motions for adaptive feed rate. Journal of Materials Processing Technology, 204(1–3), 2008, 213–220.
14. Oczoś K.E.: Extension of the magnesium alloys application range (in Polish), Mechanik, Vol. 5–6, 2009, 386–400.
15. Oczoś K.E., Kawalec A.: Light metals forming (in Polish). PWN, Warsaw 2012.
16. Zagórski I., Kuczmaszewski J.: The study of cutting forces and their amplitudes during high-speed dry milling magnesium alloys. Advances in Science and Technology Research Journal, 7(20), 2013, 61–66.
17. Zagórski I., Pieśko P.: Badania porównawcze chropowatości powierzchni wybranych stopów magnezu po frezowaniu narzędziem pełno-węglikowym oraz PKD. Postępy Nauki i Techniki, No. 8, 2011, 53–58.
18. Zhao N., Hou J., Zhu S.: Chip ignition in research on high-speed face milling AM50A magnesium alloy. Second International Conference on Mechanic Automation and Control Engineering, 15–17 July 2011, Inner Mongolia, China.
19. http://www.nanomag.us/Hydro_Mg_Brochure_Machining_Mg.pdf Machining magnesium, z dnia 19.01.2012.
20. Polska Norma: Badanie trwałości noży tokarskich punktowych, PN–ISO 3685:1996.

COMPARISON OF GEOMETRIC PRECISION OF PLASTIC COMPONENTS MADE BY SUBTRACTIVE AND ADDITIVE METHODS

Paweł Fudali[1], Waldemar Witkowski[1], Dawid Wydrzyński[1]

[1] The Faculty of Mechanical Engineering and Aeronautics, Rzeszow University of Technology, al. Powstańców Warszawy 8, 35-959 Rzeszów, e-mail: pfudali@prz.edu.pl; wwitkowski@prz.edu.pl; dwktmiop@prz.edu.pl

ABSTRACT

The paper presents information on manufacturing processes of plastic components. Basic subtractive and additive methods are described. There were also manufactured elements of fan housing by using this two types of methods. Then, the elements were measured using a touch probe. The obtained results were analyzed, on which a comparison of components' geometric accuracy was performed.

Keywords: rapid prototyping, fused deposition modelling, machining, touch probe measurement, points cloud.

INTRODUCTION

The manufacture of plastic components is currently dominated by injection moulding. The method is suitable for mass production because the time of a single piece is very short. However, the preparation process, the production of the punch, die and tooling is very time-consuming and costly. In many cases, we need to make individual elements, in this case, it is reasonable to apply the additive and subtractive methods. The article focuses on two methods of preparation of plastic components. Two identical parts were made, the first one by fused deposition modelling (FDM) method and the second by milling. Both elements were compared from the point of complexity and the results obtained.

SUBTRACTIVE AND ADDITIVE MANUFACTURING

Subtractive manufacturing is a process in which a piece of material is cut into a final shape and size by a material-removal process.

Subtractive manufacturing can be divided into:

- machining,
- chip machining,
- abrasive machining,
- erosion machining.

Depending on the desired shape and roughness of the surface there are used different types and methods of machining. They differ between themselves in used tools and the nature of the movements which perform the tool and the workpiece [1, 2].

The principal machining processes are turning, milling, boring, drilling, sawing, shaping, planing, and grinding. Except those mentioned machining methods, others such as reaming, countersinking, broaching, honing and lapping are known. Depending on the accuracy, shape and size of the work surface, there are the following types of machining: coarse, fine and very fine, called finishing.

The term additive manufacturing refers to the production technologies that create objects through a sequential layering process. Elements that are manufactured additively can be used anywhere during the product life cycle, from preproduction (rapid prototyping – RP) to full-scale production (rapid manufacturing – RM), in addition to tooling applications.

The most popular methods of RP are: fused deposition modelling (FDM), Stereolithography

(SLA), 3D printing (3DP), Electron Beam Freeform Fabrication (EBF), Direct metal laser sintering (DMLS), electron beam melting (EBM), selective heat sintering (SHS), selective laser sintering (SLS), Laminated object manufacturing (LOM), Digital Light Processing (DLP) etc.

Rapid Prototyping is an additive technology which is capable of providing a physical model of the designed object from a 3D model in CAD system. Regardless of which method will be used to make an actual element, CAD model is converted to STL format, which is an informal language of this technology. Then, a transverse division of model into layers is conducted, the thickness of which, in the case of FDM technology, depends on the size of the nozzle used for part construction (0.127–0.33). Real-time execution of Rapid Prototyping methods is disproportionately shorter than standard machining. This methods create both illustrative models and functional elements, used directly in the construction of machines or equipment and those that are used for example in a foundry [6].

Models are made of thermoplastic materials. Most materials used in model manufaturing are PC (Polycarbonate), ABS (acrylonitrile – butadiene – styrene), and their modification example ABSI (methyl methacrylate ABS) and mixtures.

FDM METHOD

One of the method of rapid prototyping is FDM (Fused Deposition Modeling). FDM is a technique involving the extruded deposition of plasticized model and supporting material with two-nozzle head. The material is deposited on a working platform, which decreases gradually with a coating of plasticized material (in a form of filaments) forming the model in accordance with the arrangement of successive layers of a virtual STL model. Nozzle during imposing plasticized material moves over the platform in the XY plane, and the platform moves along the Z axis [6]. The scheme of the method is shown in Figure 1.

As a part of this work a fan housing element was made in FDM technology. Figure 2a illustrates a CAD model of the considered element prepared in SolidWorks program, and in Figure 2b the same model with superimposed grid saved in STL format is presented.

Then the STL file was exported to the device in which it was manufactured. Figure 3 shows the final element.

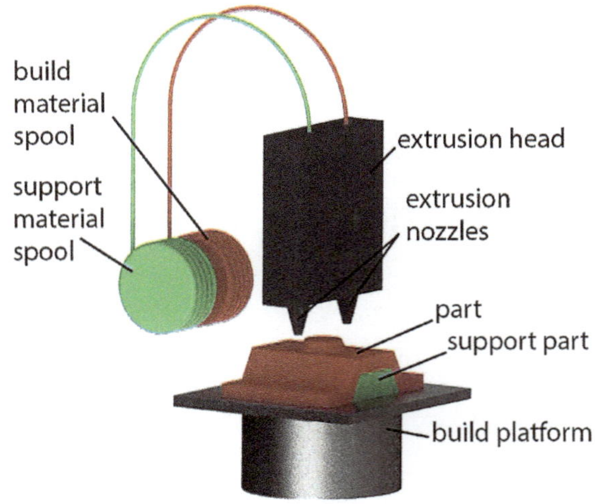

Fig. 1. The scheme of FDM method

a)

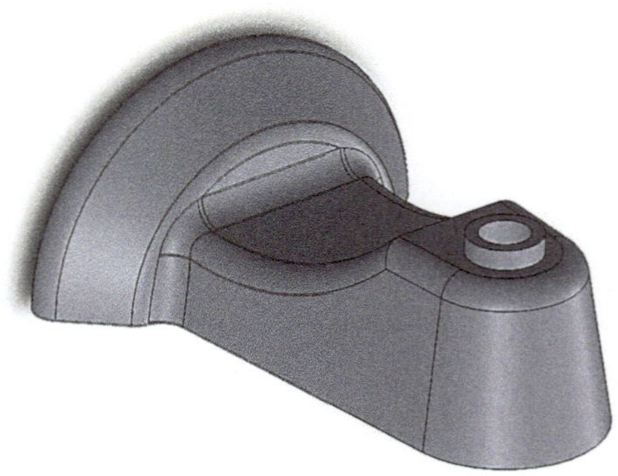

b)

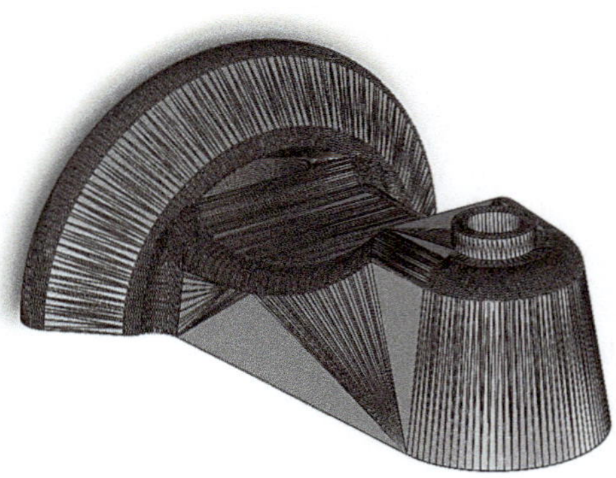

Fig. 2. A fan housing element, a) CAD model, b) STL model

Fig. 3. Element made in FDM method

MILLING

Virtual model has been saved in data exchange format (.STL). An integrated CAD/CAM software (Inventor/HyperMILL) was used for preparing the machining simulation. The advantage of the integration of CAD-CAM systems is the ability to make changes of the detail design features with automatic updates designed machining paths (no need to re-define them).

Subsequent machining operations, the parameters and the used tools are shown in Table 1. Operations from 1 to 2 are roughing operations, while others are finishing.

Table 1. Machining operations, parameters and tools

No	Operation (HyperMill)	Tools	The machining parameters
1.	Rough	End mill Ø16	S = 3000 [rpm] f = 5000 [mm/min] a_p = 2 [mm] machining allowance 0.15 [mm]
2.	Rough	End mill Ø8	
3.	Flat		S = 3000 [rpm] f = 5000 [mm/min] a_e = 0.1 [mm] (distance between paths)
4.	Naba sides		
5.	Steep	Shell end mill Ø16	
6.	Foot	Ball end mill Ø8	
7.	Ridge		
8.	Rounded	Ball end mill Ø4	
9.	Steep	Ball end mill Ø4	

Examples of roughing path (machining operation number 1) are shown in Figure 4a. For this operation, Figure 4b shows the appearance of the blank after the machining. Prepared motion tools path were used to create NC code.

The part was machined on the CNC HAAS VF2 milling machine. Modeling disc was used as a blank. The mechanical and physical properties of the semi-finished product, are given in Table 2.

CAD model after the virtual machining simulation and real part after machining are shown in Figure 5.

ANALYSIS OF THE COMPONENT ACCURACY

Both parts (machined and FDM model) were mounted on the CNC machine (V2 HAAS). Using the touch probe (RENISHAW OMP40-2) points on the objects were measured. While measuring flat surfaces perpendicular or parallel to the direction of measuring NC driver software automatically compensates the contact coordinates of the measuring ball radius [2].

The problem during the measurement of the surface at an angle is that the machine controller, depending on the direction of approach (perpendicular or parallel to the Z-axis) compensates the coordinates of the measuring tip along the feed axis (Fig. 6) [2]. The solution of this problem would be:

- rotation of the model in A, B or C axis (if machine allows to control this axis) by an angle of slope of the wall (which must be known) so that the measuring ball in contact with the object along the normal to the surface being measured [3, 5];
- use the correction methods (method of cross-product, cross-product method using weighting factors, the method uses a distance crite-

Fig. 4. Roughing model a) after path generation, b) after machining simulation, c) after milling

Table 2. Mechanical and physical properties of the modelling disc material

Material name	Color	Shore hardness	Compressive strength [MPa]	Bending strength [MPa]	Density [kg/dm^3]	Heat deflection temperature [^{0}C]
Ebazell 80	beige	approx. 30	approx. 8.6	approx. 12	0.08	approx. 80

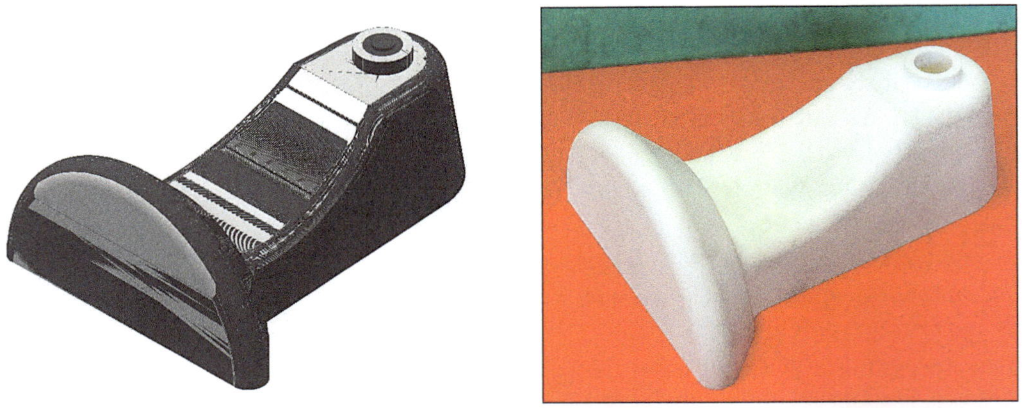

Fig. 5. The model after virtual machining simulation in program HyperMILL and after the milling

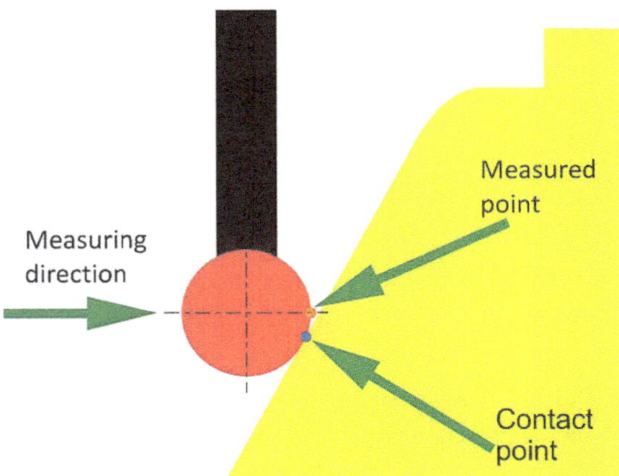

Fig. 6. Measured point on slope walls

rion to select neighboring points) when free surfaces are measured [7].

The points cloud from the measured model were imported into CATIA. In this environment, they were changed into surfaces representing the model [4]. The side surfaces of the object were shifted by the radius of the ball in the direction of the measuring probe directions to the model.

A comparative analysis of the real geometry of the model and the reference model was made. Referenced model was a CAD model in STL format, used to create a machining program and the implementation of a rapid prototyping method.

The results of the analysis are shown in Figure 7. In Table 3 the maximum absolute values of the differences between the model and the actual model are shown.

On flat surfaces the measuring area was defined (Fig. 7). During measuring on the rounded surfaces the measuring head can be easily damaged because of its bending. Measurement of the hole consisted of determining the diameter at different depths and the position of its axis.

Table 3. Accuracy of the performed models

Specification	Ebazell 80 [mm]	FDM [mm]
Surface parallel to the milling machine table	<0.016	>0.13
Surface perpendicular to the milling machine table	<0.016	>0.13
Internal hole	<0.032	>0.13
Slope surface	<0.016	>0.13

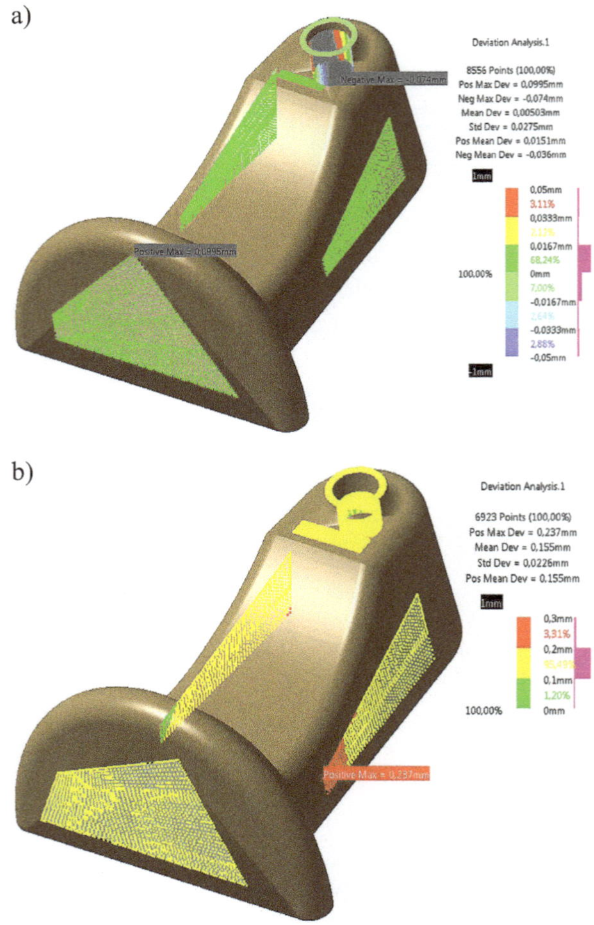

Fig. 7. Analysis of the accuracy of geometric object made in a) subtractive method, b) additive method

CONCLUSION

Basing on our studies we can conclude the following:

- model made in machining (milling) have smaller errors than the object made in an additive method,
- model geometry errors made in additive method are largely due to the thickness of the material layer,
- curved surface measurement using the probe is not possible due to the nature of measurement (pressure of measuring ball causes deflection of the probe and its further movement would cause measuring head damage),
- performance of individual item made by machining is much more labour-intensive (CNC machine and machining program are required) than in FDM method (FDM ~126 min, machining ~150 min),
- accuracy measurements directly at the milling machining (in one fixing) allow to avoid an error in determining the base and allow dimensional control during processing,
- both methods are ideal for the production of a single prototype.

REFERENCES

1. Chen J., Ling C.: Improving the machine accuracy through machine tool metrology and error correction, Advanced Manufacturing Technology, 11, 1996, 198–205.
2. Dornfeld D., Dae-Eun L.: Machine design for precision manufacturing. In: Precision Manufacturing. Springer Verlag, USA 2008.
3. Hsu Y.Y, Wang S.S.: Mapping geometry errors of five-axis machine tools using decouple method. Int. Precis. Technol., 1, 2007, 123–132.
4. Hylewski D.: Próba weryfikacji metody doboru parametrów digitalizacji powierzchni swobodnej dla celów inżynierii odwrotnej w systemie CATIA V5. Wyd. Politechniki Poznańskiej, Poznań 2011.
5. Uddin M., Ibaraki S., Matsubara A., Matsushita T.: Prediction and compensation of machining geometric errors of five-axis machining centers with kinematic errors. Precision Engineering, 33, 2009, 194–201.
6. Budzik G.: Odwzorowanie powierzchni krzywoliniowej łopatek części gorącej silników lotniczych w procesie szybkiego prototypowania. Oficyna Wydawnicza Politechniki Rzeszowskiej, Rzeszów 2009.
7. Karbowski K.: Podstawy rekonstrukcji elementów maszyn i innych obiektów w procesach wytwarzania. Wyd. Politechniki Krakowskiej, Kraków 2008.

SIMULATION STUDY OF LONGITUDINAL FORCES IN THE COUPLING DEVICE OF HEAVY FREIGHT TRAINS

Józef Stokłosa[1], Marek Jaśkiewicz[2]

[1] Faculty of Transport and Computer Science, University of Economics and Innovations, Projektowa 4, 20-209 Lublin, Poland, e-mail: jozef.stoklosa@wsei.lublin.pl

[2] Faculty of Mechatronics and Machine Building, Kielce University of Technology, 1000-lecia Państwa Polskiego 7, 25-314 Kielce, Poland, e-mail: m.jaskiewicz@tu.kielce.pl

ABSTRACT

On the LHS line (Broad-gauge Metallurgical Line), far out West of the railway line with a gauge of 1520 mm, heavy goods trains for a gross weight 5500 tons and a length of 850 m are operated. The article presents the results of a simulation study of the forces that occur in the automatic coupling device of SA-3 type of Russian production train consisting of 60 coal wagons of Russian construction of gross mass 91 tons each. The train moves on the 1520 mm gauge tracks curve S type (the radius of curvature of curves 300 m). Simulation studies were conducted using the Train Module of program to dynamic study multi-elements systems of Universal Mechanism UM 6.0.

Keywords: heavy freight trains, the longitudinal forces in the coupling device, Nadal criterion, modeling, simulation, the theory of Kalker's contact.

INTRODUCTION

The introduction of freight train operations with increased gross weight and increased length is one of the ways to improve railways capacity , especially one-track railways. Permissible length of freight trains shall not exceed the length of the usable main track and at least one master track additional on railway stations. In practice, the length of freight trains in Poland does not exceed 600 m, a weight of 3200 tons gross and depends on factors such as a profile of the railway line, power traction vehicle (or total power of all traction vehicles), the length of station track on intermediate stations [11,14].

On separated railways, both in Europe and on other continents, freight trains of mass gross much excess of 3200 tones are exploited. In Russia and China the weight of heavy freight trains reaches 9000 and even 12 000 tones. In the USA, Canada trains weighing 15 000 gross tons operate. In Europe heavy freight trains with a gross weight exceeding 3200 tons are exploited among others on Betuwe line of a length of 160 km, which leads from the port of Rotterdam to the German border. The line is double track and electrified at 25 kV. Maximum gross weight for trains on the Betuwe line is 5000 tons.

From the moment of the entry into operation of the upgraded by Newag diesel locomotives of 311Da (modernized locomotive ST44) the company PKP LHS tried to conduct on the broad-gauge railway line no 65 (Polish/Ukrainian border – South Sławków station) heavy trains of mass gross 5500 tons and a length of 850 m. Train of the mass is drawn by two diesel locomotives 311Da.

SIMULATION MODEL

The experience of the countries in which heavy trains operate show that transitional processes have particular importance in the process of operation of such trains: acceleration and braking. Studies have shown that at that time significantly increased thrust forces in coupling devices. Their amplitude is increased to a value that may result in rupture of the coupling [1, 4, 10, 12].

In the volume of cargo carried out by LHS line the largest share has iron ore which is imported from Ukraine to Dąbrowa Górnicza (Steelworks Katowice). LHS line is non-electrified monorail broad-gauge line (1520 mm), which allows the transport of cargo from Ukraine without reloading of the goods at the border from broad-gauge cars to normal cars.

Ore Transport takes place in coal wagons of Russian construction (Figure 1) equipped with double-axle bogies type 18-100. Mass of unloaded wagons is 22 000 kg, capacity 69 000 kg [3].

Simulation studies were conducted using the Train and Train3D modules as part of a program to simulate multi-elements systems called Universal Mechanism v. 6.0 [13].

Universal Mechanism is the program oriented to practical engineers, students and teachers, in other words it is oriented to all people who are involved in problems of dynamics of machines and mechanisms. Mechanical systems are described by means of representing them as systems of rigid bodies connected by various kinematical pairs and force elements, so-called multibody systems.

UM has advanced postprocessor, which includes: linear analysis, statistics, multivariant calculation and optimization and export of results. UM is a useful tool for the computer-aided modeling of multibody and hybrid systems of various types: complex aerospace structures, robots, railway vehicles, automobiles, cable systems, etc. It gives an opportunity to solve both direct and inverse kinematic, dynamic, and control problems.

The program package includes module UM Loco, which is intended for simulation of dynamics of railway vehicles of different types (diesel and electric locomotives, passenger and freight wagons) in both straight and curve railroad tracks. The simulation is performed in time domain by means of numeric integration of differential or differential-algebraic equations of motion. UM Loco allows the user to create fully parameterized models of vehicles.

Model of train created with 60 coal wagons of gross weight 91 tones each was tested (Figure 2).

Wagons are connected by automatic couplers type SA-3 embedded in dampers Sh-2-T type of non-continuous draw gear (Figure 3).

This is a common construction used in freight wagons used by the railways of the Russian, Ukrainian, Belarusian, Kazakhstan. Automatic coupler SA-3 takes on longitudinal forces extending during the drive of train, and compressive forces occurring during braking.

The schema of coupling device with damper was modeled as part of the elastic-frictional element placed in rigid frame-stepping of wagon according to schema: (spring + frictional element) + spring (Figure 4).

The research was conducted for the rail vehicle-track with the rails type R65 presenting a

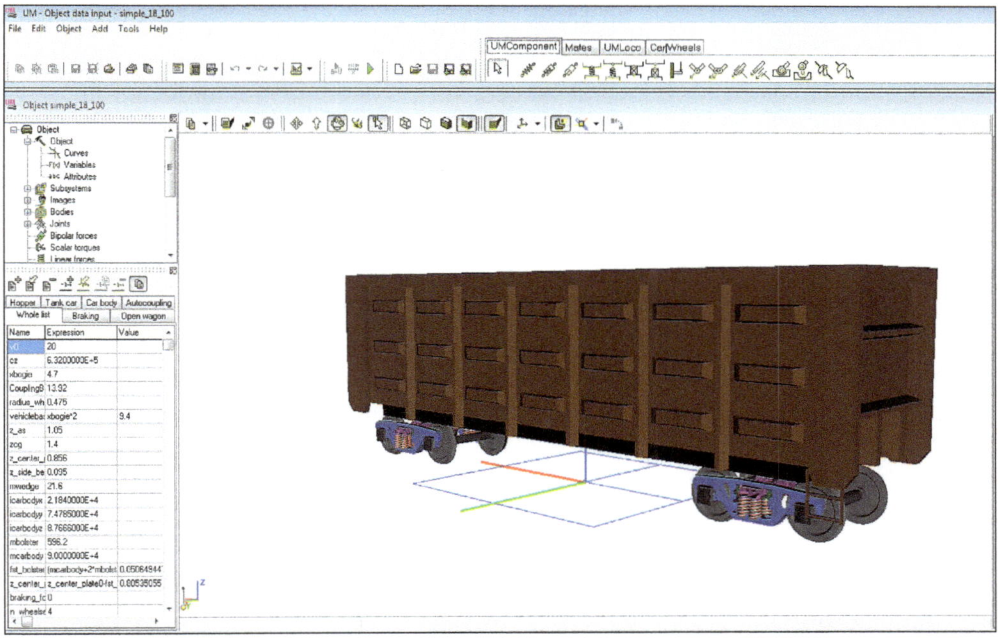

Fig. 1. Model of coal wagon type 12-757 of the Russian production with bogies 18-100 made in Universal Mechanism loco program

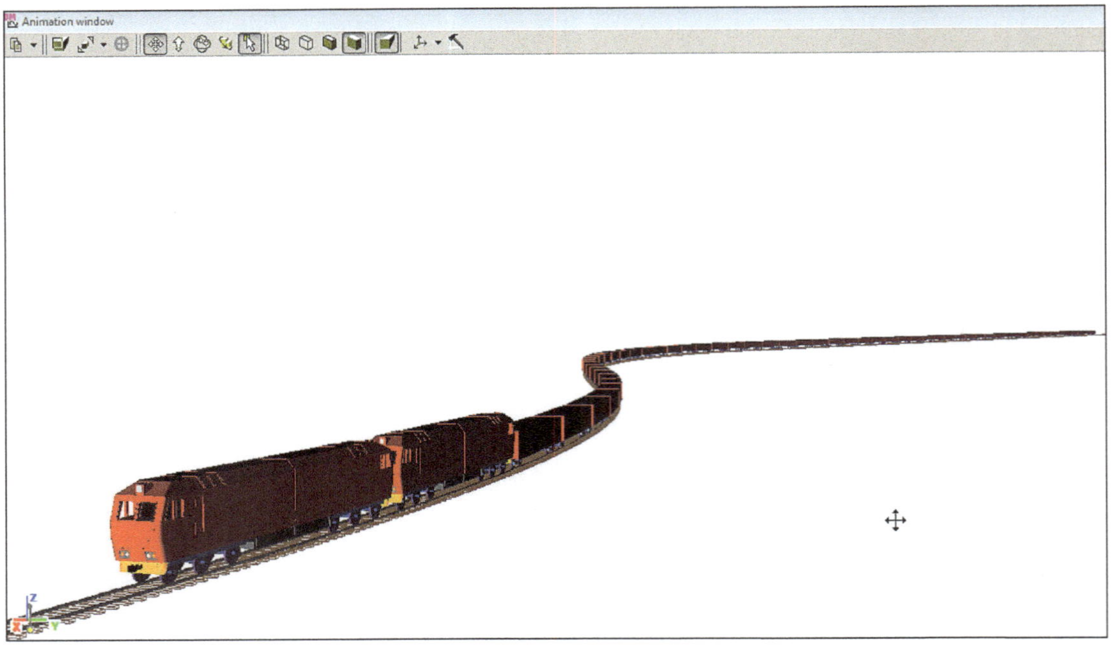

Fig. 2. Model of freight train made in UM program.

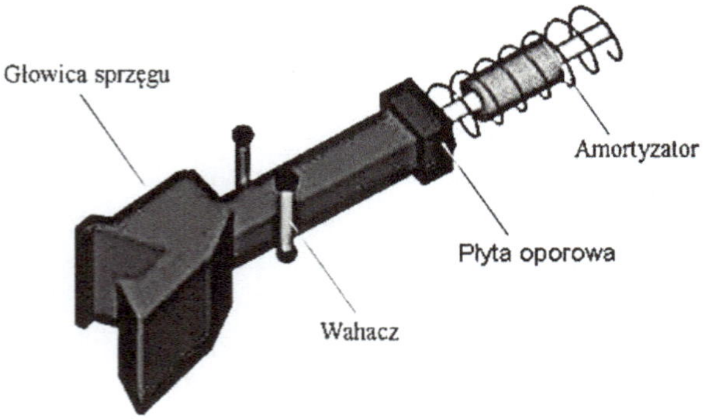

Fig. 3. Automatic coupler SA-3 along with frictional damper Sh-2-T [13]

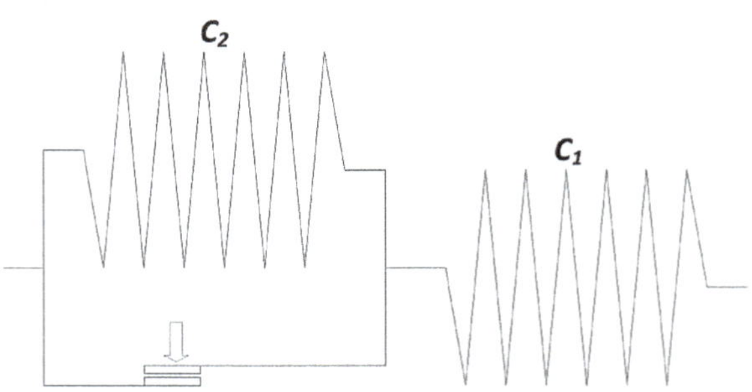

Fig. 4. Rheological model of SA-3 coupler C1, C2 - the stiffness of the springs

basic type of rails on the LHS line (the standard type of rails on busy rail lines in Russia, and in the former Soviet republics) and the wheels of PU1Sz-950 profile and the diameter of the new wheel 950mm.

The composition is driven by two diesel locomotives type 311Da (PKP ST40) operated from the driver's cab of the first locomotive. The locomotive ST40 is a heavy traction vehicle based on two three-axle trucks in a Co - Co system, develops the power of 2133 kW (2900 KM) with total weight of 126,9 tons [6]. The driving force, followed by six traction motors type ED 118A GE, was asked in the form of the traction characteristics of the locomotive (Figure 5).

The train was moving on the track curve S. On the Figure 6 described the components of a geometric system of the track and line profile is shown in Figure 7.

In addition, it is assumed that on the straight sections rails have not used profile, while in the curves and spirals external rail course has worn profile unilaterally on the inner side of the ongoing rail friction wheel flanges on rolling stock (Figure 8).

The model includes a basic resistance and resistance to motion in the curvature of the track. Due to the fact that in the test train composed of Russian construction wagons and moves along a wide track (1520 mm) to describe the resistance forces train traffic is based on fundamental patterns used on Russian railways [9].

The essential resistance to movement of diesel locomotives and electric [9]:

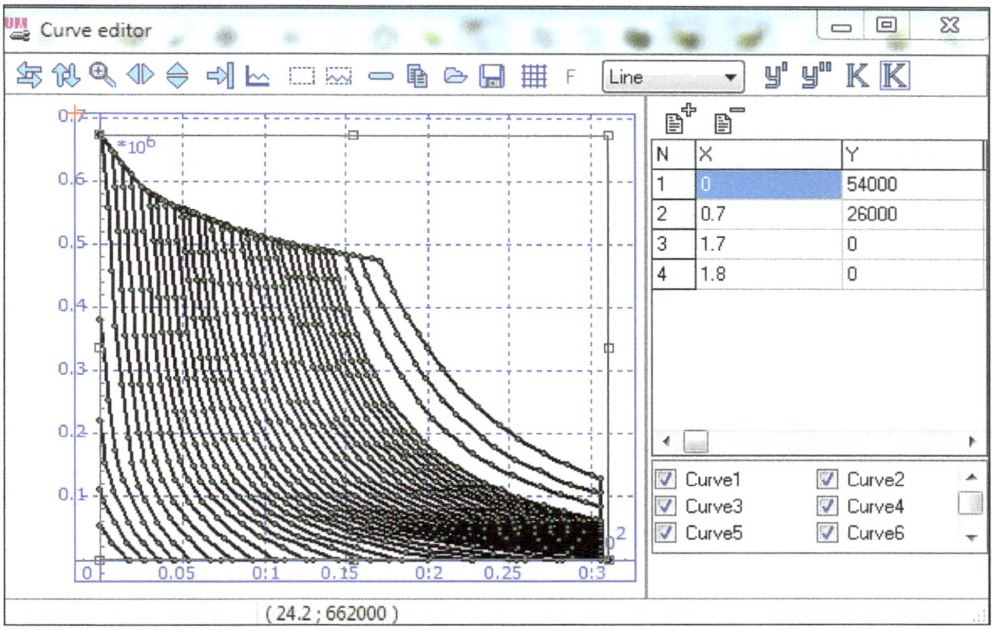

Fig. 5. Sample traction characteristics of electric locomotive

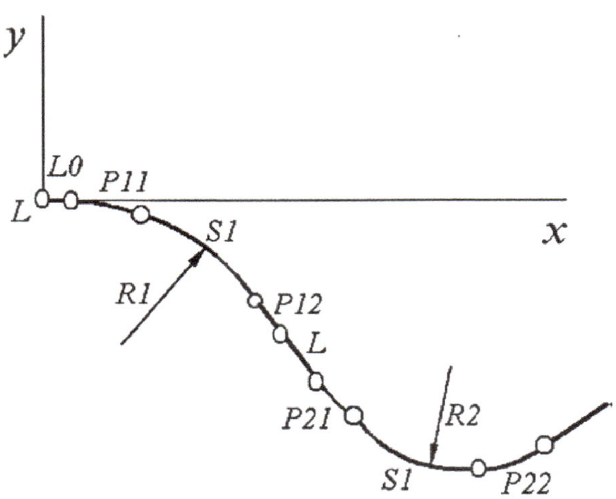

Fig. 6. Macro-geometry of railway line: L0, L – straight segment, P11, P12, P21, P22 – temporary curves, S1 – an arc of radius R1, H1 – superelevation of track, dy – widening of the track in the arc

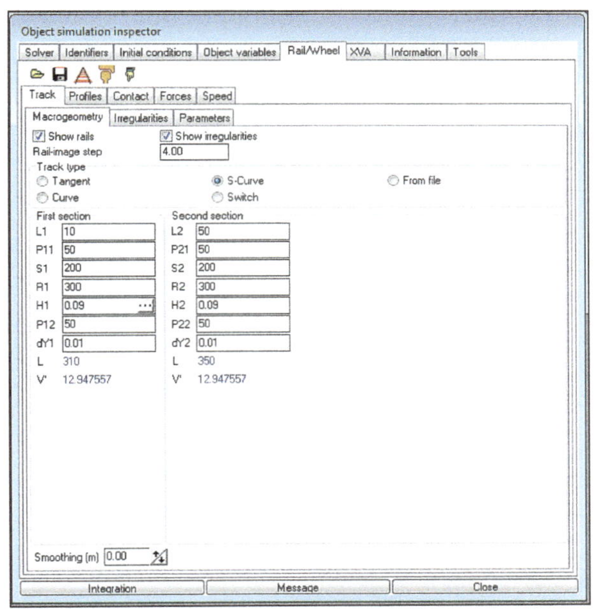

Fig 7. Profile line specified in UM program

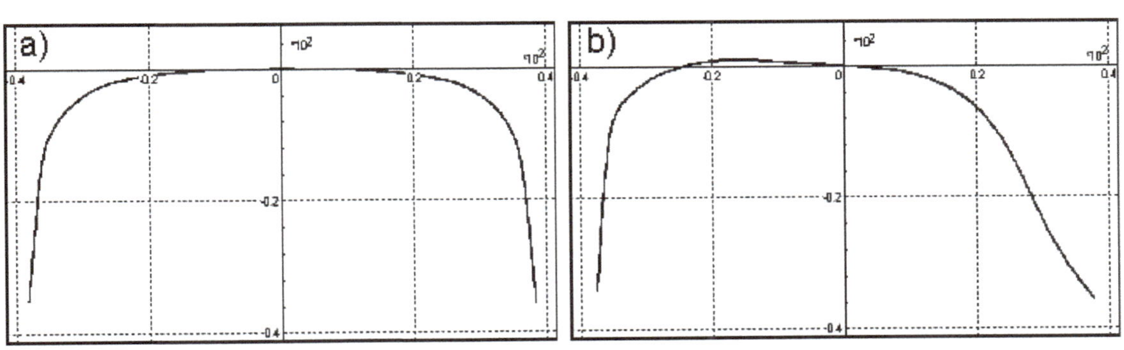

Fig. 8. Profile rail head type R65: a) new rail, b) splint worn unilaterally (external line of the arch rail road)

$$W_{oL} = \frac{9{,}81(2{,}4+0{,}009v\cdot 3{,}6+0{,}00035(v\cdot 3{,}6)^2)Q_L}{1000}\ [N] \tag{1}$$

The essential movement resistance four-axles wagons [9]:

$$W_W = 9{,}81\left(\frac{0{,}7\ Q_W}{1000} + (3 + 0{,}09\ v \cdot 3{,}6 + 0{,}002(v \cdot 3{,}6)^2)4\right)\ [N] \tag{2}$$

where:
W_{oL} – the essential resistance to movement of the locomotives in [N],
Q_L – the mass of the locomotive in [t],
Q_W – the weight of load wagon in [t],
v – the speed of train in [m/s].

The additional traffic resistance – resistance to motion in the curvature of the track [9]:

$$W_Ł = \frac{a}{R} \cdot \frac{S_k}{L}\ [N] \tag{3}$$

where:
$W_Ł$ – resistance to motion in [N],
R – the radius of the arc in [m],
S_k – the length of the curvature of the track in [m],
L – the length of train in [m],
a – the empirical factor

The relationship of forces in the wheel/rail contact from micro-skids is evaluated by using a simplified non-linear contact Kalker's theory [5].

The UM Program allows to insert to a simplified model of train wagon made in 3D technology anywhere in the train set (Figure 9). Inserting 3D wagon allow sets of forces that occur in the wheel/rail contact point and the designation of the value of the coefficient of Y/Q – criteria for safety against derailment based on designation of the transverse forces Y to the vertical Q at the point of wheel/rail contact.

RESULTS OF SIMULATION STUDIES

The train was moving at a speed of v = 10 m/sec. At the start of simulation the face on the first locomotive was at the beginning of the straight section of the track (the beginning of the line L0 in Figure 6). Simulation time has been chosen in such a way that the whole composition has travelled by the curved section of the track.

Braking process begins in the 5th second of simulation. The train moves in braking mode. Let us assume that the speed of propagation of air in the brake line (the so-called wave speed braking) is 250 m/s. As a result the start or end of the braking process is not simultaneous in all wagons.

Figure 10 shows a graph of longitudinal forces in the coupling in the middle wagon. significant compression force jumping to a value of 500 kN are visible.

In the last wagon of train there are also momentary compressive forces of considerable values up to 500 kN (Figure 11). As a result, the operation of such large compressive forces can lead to derailment of wagons.

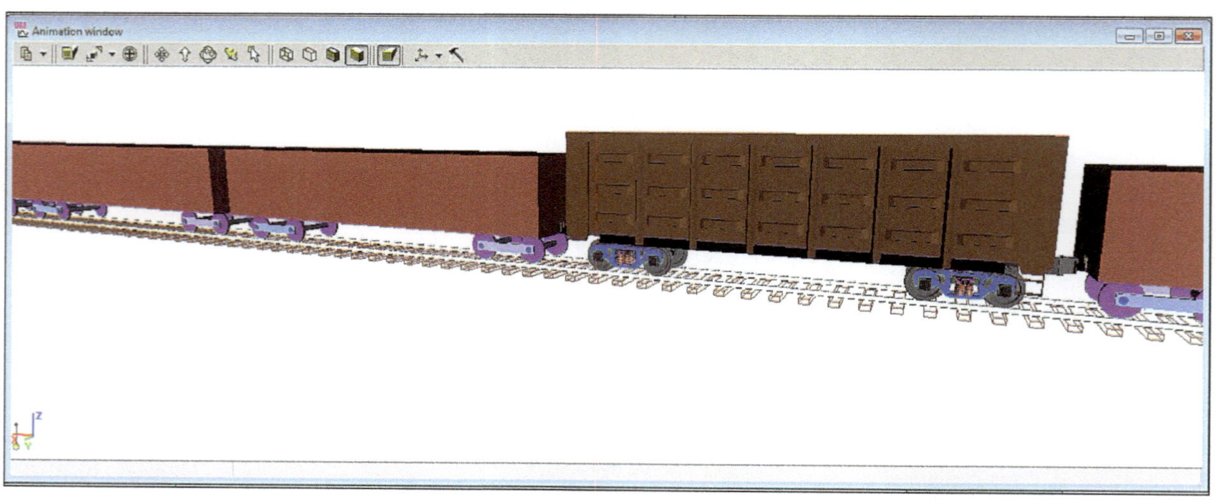

Fig. 9. View of portion of the model train with inserted wagon model 3D to determine the forces in the wheel/rail contact

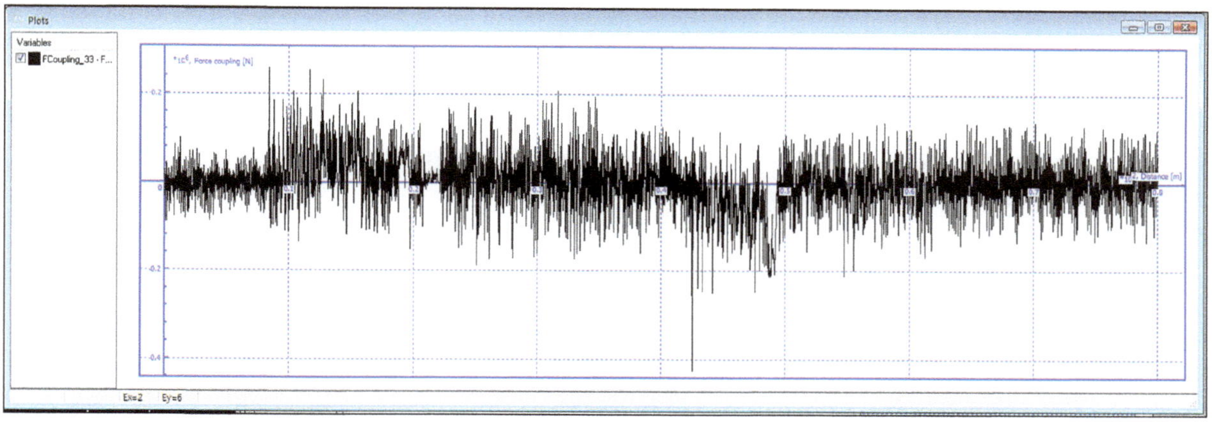

Fig. 10. The longitudinal force in compression coupling SA-3 in wagon No 33 (middle wagon)

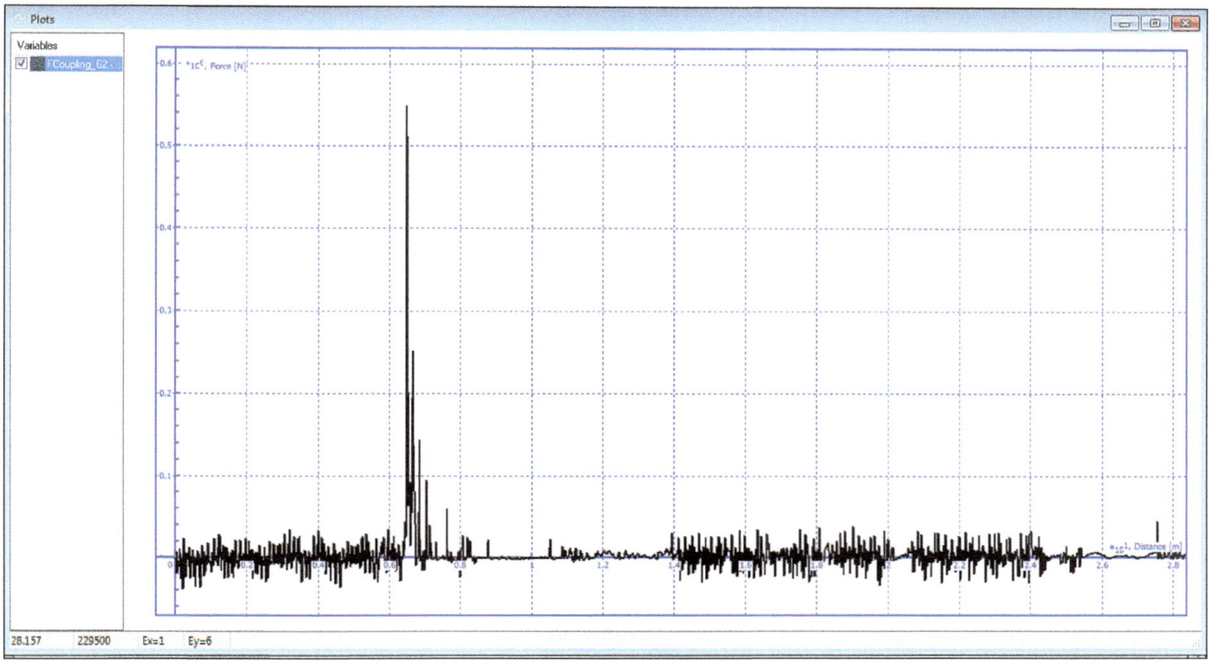

Fig. 11. A longitudinal forces in the coupling in the last wagon

CONCLUSIONS

The article gives the part carried out by the authors, research the dynamics of heavy freight train. In further research, according to the authors, should be taken into account the exact parameters of the LHS line - the status of the track, the degree of wear of the rails, the inclination of the line. In addition, research should be subject to the status of the braking system. The authors are aware that the company PKP LHS is not the owner of the moving wagons on the track wide from Hrubieszów to Slavkov, however, research on the dynamics of long trains serve to get to know the process of braking such heavy formations.

REFERENCES

1. Delooz F. Heavy haul trains in Europe. 9th International Heavy Haul Conference IHHA, Shanghai 2009.
2. EN 14363: Railway applications – Testing for the acceptance of running characteristics of railway vehicles. Testing of running behavior and stationary tests. European Committee For Standardization, 2005.
3. Gruzowyje wagony kolei 1520 mm. Albom sprawochnik. Wydawnictwo Transport, Moskwa 1989.
4. Hongyan Zh. Dynamic Interact Heavy Haul Railways on Track Structures. Chinese Railways, No. 1, 2005: 38-42.
5. Kalker J.J., Piotrowski J. Some New Results in Rolling Contact. Vehicle System Dynamics, 18, 1989.

6. Marciniak Z., Durzyński Z. Projekt modernizacji lokomotyw spalinowych serii ST44. TTS 9, 2005.
7. Pogorelov D.Yu. Computer Modeling of Rail Vehicles Dynamics. Mechanics and Tribology of Transport Systems, Rostov-on-Don 2003: 226-231.
8. Pokropiński B. Lokomotywy spalinowe produkcji polskiej. WKiŁ, Warszawa 2009.
9. Pravila tiagovyh raschiotov iz pojezdnoy raboty. M. Transport. 1995.
10. Pugi L, Fioravanti D, Rindi A. Modelling the longitudinal dynamics of long freight trains during the braking phase. 12th IFToMM World Congress, Besançon (France) 2007.
11. Rozporządzenie Ministra Infrastruktury z dnia 18.07. 2005 r. w sprawie ogólnych warunków prowadzenia ruchu kolejowego i sygnalizacji, Dz. U. Nr 172, poz. 1444.
12. Schmidt S. et al. ZEVrail, 9, 2009: 358-364.
13. Universal Mechanism 6.0, Simulation of Longitudinal Train Dynamics. User's manual, 2010.
14. Założenia techniczno-eksploatacyjne kierowania i sterowania ruchem (KSR) dla PKP, uzgodnione z PKP DG KA – pismem nr KA2b-5458-05/97 z dnia 23.05.1997 r.

RESEARCH ON THE CONTENT AND FILLER TYPE ON INJECTION SHRINKAGE

Tomasz Jachowicz[1], Ivan Gajdoš[2], Volodymyr Krasinskyi[3]

[1] Lublin University of Technology, Faculty of Mechanical Engineering, Department of Polymer Processing; Nadbystrzycka 36, 20-618 Lublin, Poland, e-mail: t.jachowicz@pollub.pl

[2] Technical University of Kosice, Faculty of Mechanical Engineering, Department of Technologies and Materials, Masiarska 74, 04001 Kosice, Slovakia, e-mail: ivan.gajdos@tuke.sk

[3] Lviv Polytechnic National University, Department of Chemical Technology of Plastics, S. Bandera str. 12, 79013 Lviv, Ukraine, e-mail: vkrasinsky82@gmail.com

ABSTRACT

The paper presents the phenomenon of injection shrinkage. The definition of shrinkage and shrinkage types are presented. The main factors affecting shrinkage value of injection-molded part are discussed. Based on the experimental tests conducted using injection-molded parts made from polypropylene filled with glass fiber and talc, the dependence of injection-molded part shrinkage on filler content and selected parameters of the injection molding process has been determined.

Keywords: polymer processing, injection molding, shrinkage, filler, composite injection-molded part.

INTRODUCTION

The injection molding is one of the most important manufacturing processes for producing parts made of polymer materials. There are many different variants of this process, all of them characterized by cyclical nature. Injection molding can be used to process practically all kinds of polymer materials, particularly thermoplastic and thermoset materials, but also rubber compounds, liquid silicones and composite materials. As a result of injection molding, an injection-molded part is produced; depending on both the injection molding process variant employed and the polymer used, the injection-molded part is characterized by a very wide range of functional and strength properties as well as variations in shape, structure, dimensions and production accuracy [2, 4, 9, 19].

Injection-molded part design is predominantly affected by the following factors: polymer type, processing method and molding conditions, tool design and the processing capacity of injection molding machine . All these factors directly or indirectly affect the final dimensions, structure and shape of the injection-molded part, its consistency with the tolerance range adopted during the design process and molding anomalies (material and surface defects) that are significant from a user's point of view [6, 12, 18].

SHRINKAGE AND ITS TYPES

Shrinkage is defined as the decrease in the volume or dimensions of a plastic part relative to the volume or corresponding dimensions of the mold cavity, with the decrease occurring both during the final stage of the process and within a specified time period after the process [4, 6, 12].

Shrinkage of plastic materials can be classified according to several criteria. With regard to time and place of occurrence, shrinkage can be divided into primary shrinkage and secondary shrinkage. Primary shrinkage means the decrease in part dimensions during cooling (for thermoplastics) or hardening (for thermoset plastics) in the mold cavity and shortly after demolding. It is conventionally assumed that primary shrinkage lasts 16 hours [4, 9, 12]. Secondary shrinkage is defined as the change in the volume and dimensions of a molded part as a function of time that occurs even after primary shrinkage and approaches, at the same time, a constant value. The sum of primary shrinkage and secondary shrinkage is called overall shrinkage [4, 12, 16].

With regard to changes in the geometric dimensions of a polymer product, shrinkage can be divided into the following categories: linear, volumetric, mean, longitudinal and transverse. Linear shrinkage is the decrease in a given linear dimension of an item relative to the corresponding linear dimension of the mold cavity, usually expressed as a percentage. Moreover, it should be mentioned that there is also a term "actual shrinkage" which takes into account the thermal expansion of cavity material that occurs during processing [1, 4, 5].

Owing to variations in the orientation of polymer macromolecules, shrinkage values differ depending on the direction measured; consequently, transverse shrinkage and longitudinal shrinkage can be distinguished. Transverse shrinkage refers to the linear dimension transverse to the direction of polymer flow in the mold cavity, while longitudinal shrinkage is the linear shrinkage measured along the flow direction. Transverse shrinkage is lower than longitudinal shrinkage. The difference between the values of transverse shrinkage and longitudinal shrinkage is called anisotropic shrinkage [4, 8, 12, 16].

Another criterion used to classify shrinkage concerns processing methods. For this reason, the term "technological shrinkage" has been coined, which can encompass, among others, injection molding shrinkage, compression shrinkage, extrusion shrinkage and casting shrinkage [4, 9, 12, 15].

The relationship of change in part dimensions to time is called shrinkage characteristic curve. Basic characteristic curve describes one of the processing methods (injection molding, compression or casting), encompassing the time period from the moment the mold cavity is filled with the polymer melt until secondary shrinkage is over. A comprehensive characteristic curve applies to the main processing method for producing a part (e.g. injection molding), taking into account an additional processing method performed at elevated temperature in order to minimize the risk of shrinkage and to ensure the dimensional stability of the part (e.g. heat treatment). This characteristic curve encompasses the time period from the moment the mold cavity is filled with the polymer melt until the additional processing technique is over, which is considered identical to the end of secondary shrinkage. Examples of visual interpretation of shrinkage characteristic curves for an amorphous plastic and a semicrystalline material are shown in Figure 1 and Figure 2 [4, 6, 16].

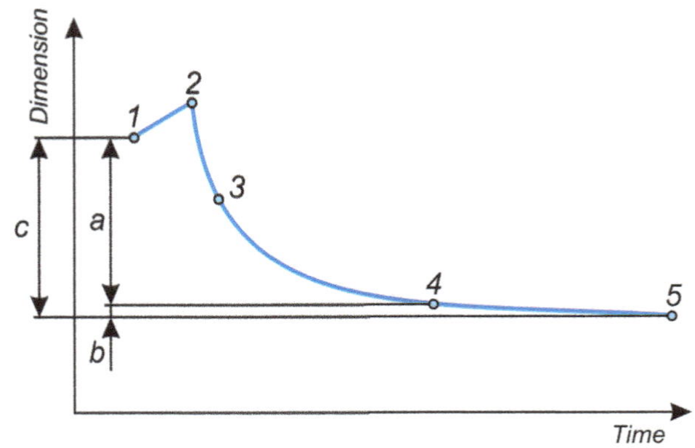

Fig. 1. Basic characteristic curve of shrinkage for amorphous part: a – primary shrinkage, b – secondary shrinkage, c – overall shrinkage, 1 – mold cavity dimension at initial cavity filling temperature, 2 – dimension of mold cavity heated by polymer, 3 – part dimension after demolding, 4 – part dimension after primary shrinkage, 5 – part dimension after secondary shrinkage [9]

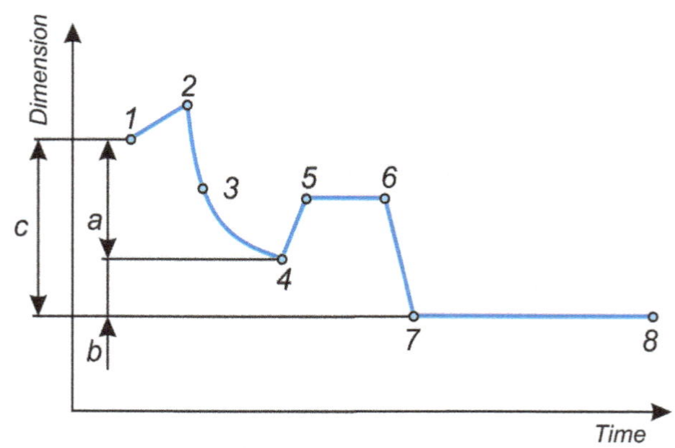

Fig. 2. Comprehensive characteristic curve of shrinkage for semicrystalline part: a - primary shrinkage, b – secondary shrinkage, c – overall shrinkage, 1 – mold cavity dimension at initial cavity filling temperature, 2 – dimension of mold cavity heated by polymer, 3 –part dimension after demolding, 4 – part dimension after primary shrinkage and at the beginning of heat treatment, 5 and 6 – part dimensions during heat treatment, 7 – part dimension after heat treatment, 8 – part dimension after secondary shrinkage [9]

FACTORS AFFECTING SHRINKAGE

With its origin in physiochemical phenomena, shrinkage is affected by many factors that can be grouped according to the following: [1, 4, 9, 12, 20]:

- polymer material (density, molecular weight, crystalline or amorphous structure),
- part (shape, dimensions, dimensional accuracy and tolerance range),
- processing conditions (polymer temperature, mold temperature, product cooling and cooling method applied),
- tool design (flow system design, mold cavity shape),
- physical or chemical treatments applied to the molded part (normalizing, annealing, chlorination).

The effect of processing conditions on mold shrinkage is difficult to determine. When thermoplastics are processed by injection molding technique, shrinkage is mainly caused by the decreasing temperature of the polymer being processed in the final stages of the injection cycle, and it depends on technological parameters of the process (temperature, pressure, time and cooling conditions) as well as on the parameters of subsequently employed thermal and thermo-chemical treatments. In contrast, when thermoset plastics are processed by compression method, compression shrinkage is significantly affected by chemical processes that occur in the polymer being processed at the final stages of the compression cycle [4, 5, 12].

It has been established that crystalline materials are more prone to shrinkage than amorphous materials [1, 8, 11]. The shrinkage phenomenon has been best investigated for the injection molding process, hence the majority of dependences describing the effect of molding conditions on shrinkage value have been formulated with regard to injection-molded parts. Table 1 lists values of linear primary shrinkage for the most important polymer materials [16].

Table 1. Value of linear primary shrinkage [16]

Polymer type	Shrinkage value [%]
Polystyrene	0.5 – 0.7
ABS resins	0.4 – 0.6
Low-density polyethylene	1.0 – 1.5
High-density polyethylene	2.0 – 4.0
Polyamide	1.2 – 3.0
Poly(vinyl chloride)	0.1 – 0.4 hard
	1.0 – 3.5 soft
Polypropylene	1.0- 2.0
Poly(methyl methacrylate)	0.4 – 0.8
Polycarbonates	0.4 – 0.8
Polyoxymethylene	3.0 – 5.0

Polymer materials with higher density exhibit higher shrinkage. It is particularly important in the case of polyolefines, polyamides and polyacetals. Polymers with lower density crystallize to a lower degree due to their stronger chain branching, which leads to lower shrinkage. Shrinkage decreases with the increase in melt flow rate [1, 4, 14].

Injection molding parameters, in particular pressure and holding time as well as mold temperature and cooling time, have a significant effect on shrinkage [1, 4, 11, 13]. Generally, if the injection pressure, holding pressure and injection temperature increase, injection shrinkage of thermoplastics decreases, whereas increasing mold temperature and injection rate leads to increased shrinkage. If the holding time is too short, which is usually the outcome of premature freeze-off, or if the holding pressure is too low, depressions or shrinkage cavities can occur. If the holding pressure is too high, it results in excessive increase in polymer pressure and its packing in the mold cavity, leading to scratches and cracking of the injection-molded part [1, 4, 5, 11]. The temperature of the mold cavity surface and the wall thickness of the molded part determine the cooling rate uniformity. With extending the cooling time and increasing the wall thickness of the molded part, the heat transfer becomes slower, so, as a consequence, the molecule chains have more time to level up the stresses, while semicrystalline materials have more time even for crystallization [3, 4, 10]. Although the increase in the mold temperature leads to the increase in overall shrinkage, the fraction of primary shrinkage is then so low that sufficient dimensional accuracy of the molded part can still be obtained [3, 5, 6, 19].

The injection-molded part design and the way of polymer feed determine shrinkage uniformity in particular fragments of the molded part. The non-uniformity of wall thickness leads to different shrinkage values, which induces internal stresses and warpage [4, 11, 17, 20]. With very thin walls of molded parts, strong macromolecule orientation and anisotropic shrinkage occur. Shrinkage is clearly lower if the molded part does not have any sharp edges, for example in the place where the bottom changes into a side wall. This is due to the fact that rounded shapes facilitate polymer flow, which results in lower pressure losses [6, 9, 12].

The design of the injection mold is critical to shrinkage uniformity in the injection-molded part. The most commonly mentioned design factors that affect shrinkage value are as follows: the number and arrangement of mold cavities, flow path length, type of a gating system as well as the place where the polymer is forced into the mold cavity, cross section and shape of pinpoint gates as well as the arrangement of temperature adjusting channels. Parts molded in multi-cavity molds are characterized by considerable variations in shrinkage values. Shrinkage increases along the polymer flow direction in the mold injection channels with the increase in the distance of the mold cavity from the central injection channel. This is due to the decrease in the pressure along the polymer flow path. Shrinkage is lower in the vicinity of the ingot than at the distant edge of the molded part; for this reason, many injection points are used to level out shrinkage value [6, 12, 20].

Anisotropic shrinkage of polymer materials without fillers is relatively low, particularly in the case of amorphous materials. The anisotropy of filled materials mainly depends on the filler content and shape, the shape and dimensions of the molded part, the way of forcing polymer into the mold cavity (which is connected with mold cavity filling and filler orientation), as well as the dimensions and shape of the pinpoint gate [3, 7, 13, 14].

As additional substances, e.g. fillers, are introduced into the material at different stages of processing, they significantly change both the properties of products obtained and the process itself. As far as the injection molding process is concerned, fillers, both as powders and short fibers, exert a positive impact as they lead to the reduction in injection molding shrinkage and the dimensional stability of the injection-molded part. Depending on injection molding conditions and the structure of the injection-molded part, the application of fibrous fillers can result in distinct fiber orientation, which leads to anisotropic shrinkage. The addition of fillers can, however, deteriorate the operational conditions of the injection molding machine deteriorate, exerting an erosive impact on the barrel and screw surfaces; what is more, some fillers impair the rheological properties of plastics and make polymer flow difficult [4, 7, 13].

A detailed discussion of the effect of individual factors on the shrinkage phenomenon is beyond the scope of the present paper. However, the problem has been thoroughly presented in the literature [4, 8, 9, 11]. To give an example of the effect of selected factors on shrinkage value, the dependence of primary injection shrinkage on injection pressure and side gate dimensions for highly crystalline polyoxymethylene called Duracon M140-44 is shown in Figure 3 [10]. To give another example of the effect of selected factors on shrinkage value, the dependence of difference of primary injection shrinkage on wall thickness and gate size for highly crystalline polyoxymethylene Duracon M270-44 is shown in Figure 4 [10].

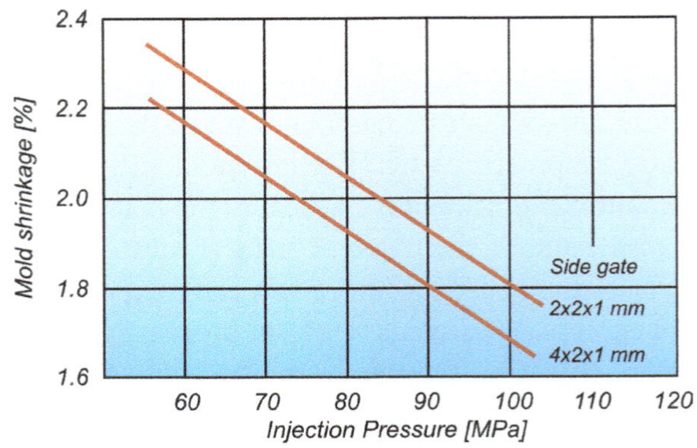

Fig. 3. Dependence of primary injection shrinkage on injection pressure and side gate dimensions for highly crystalline polyoxymethylene Duracon M140-44 [10]

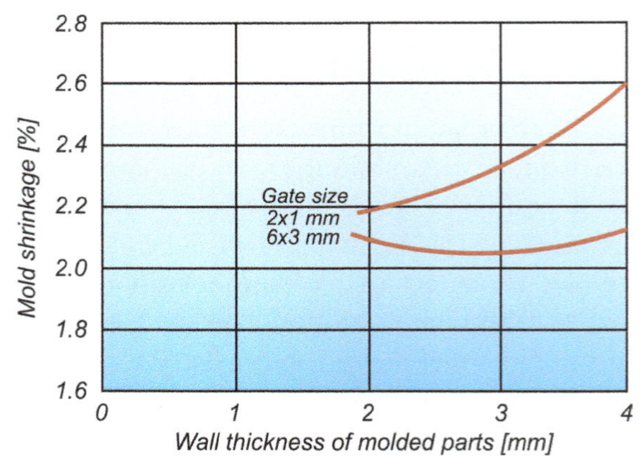

Fig. 4. Dependence of difference of primary injection shrinkage on wall thickness and gate size for highly crystalline polyoxymethylene Duracon M270-44 [10]

EXPERIMENTAL TESTS

To conduct the experimental tests investigating the effect of filler type and content as well as of selected parameters of the injection molding process on linear shrinkage value, injection-molded parts in the shape of tensile test specimens were used. The injection-molded parts were obtained using a laboratory screw injection molding machine, with variable factors being: injection time set to 4 and 6 seconds, cooling time set to 15, 30 and 45 seconds, respectively. The injection-molded parts were made of polypropylene, marketed under the name of Moplen EP440G, manufactured by Basell Orlen Polyolefins. The fillers used included: short glass fiber, as a ready PP/GF concentrate, and talc, also in a form of a ready PP/T concentrate. For injection, polymer batches containing: 10%, 20%, 30%, 40% and 50% of glass fiber, 10%, 20%, 30% and 40% of talc, and pure propylene without filler, were used. The temperature in particular heating zones of the plasticizing system was increased from 220 °C to 240 °C, reaching 245 °C in the injection nozzle. The tool used was a two-cavity injection mold thermostated at a temperature of 50 °C. Necessary for further calculations of linear shrinkage, the dimensions of the mold cavity were as follows: length L = 152.00 mm, width H = 20.15 mm and thickness B = 4.00 mm.

Corresponding to the dimensions of the mold cavity, the actual geometric dimensions of the injection-molded parts produced with particular batches, differing in filler type and content and obtained at different injection parameters, were measured with an accuracy of 0.01 mm by means of an electronic measuring device. The measurements were made three times: immediately after demolding (to analyze primary shrinkage), after twenty-four hours and after one week (to analyze secondary shrinkage). Following the preliminary calculations, the length L of the injection-molded part, as the longest dimension of the examined product, was used for further presentation of the results.

Figure 5 and Figure 6 offer examples of the charts illustrating the change in the dimension L of the injection-molded part after demolding as a

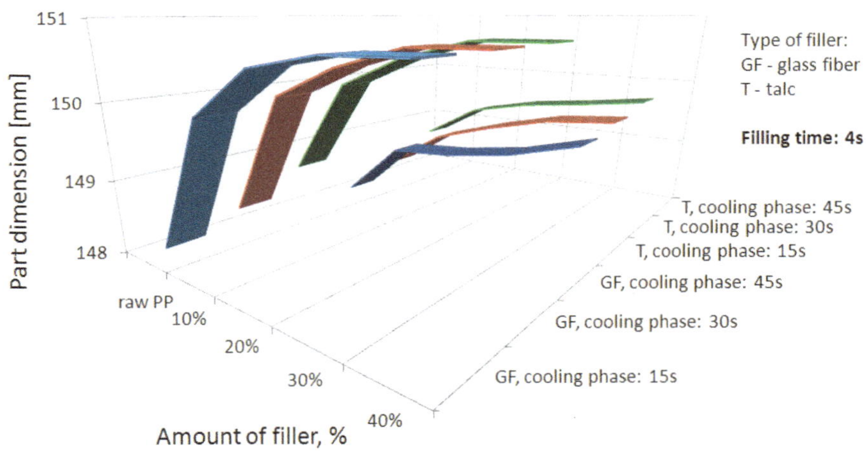

Fig. 5. Change in the dimension L of the glass fiber-filled part and talc filled part after demolding, depending on the filler content and cooling time of the part in the mold. The part was produced at filling phase time of 4s

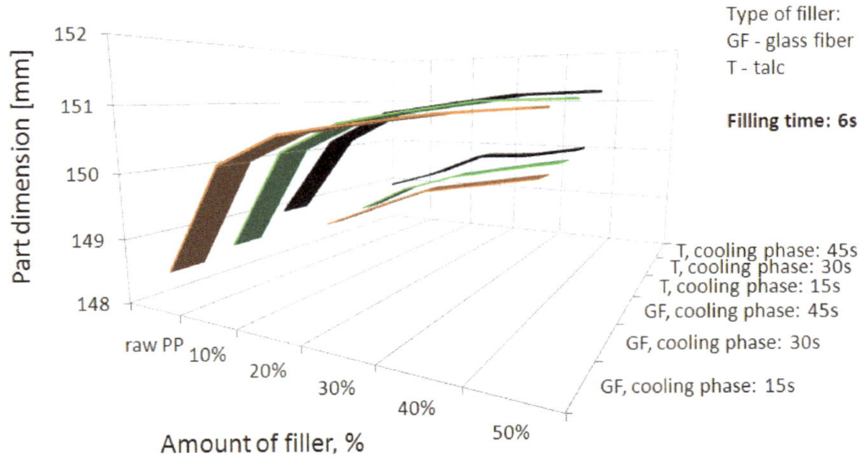

Fig. 6. Change in the dimension L of the glass fiber-filled part and talc filled part after demolding, depending on the filler content and cooling time of the part in the mold. The part was produced at filling phase of 6s

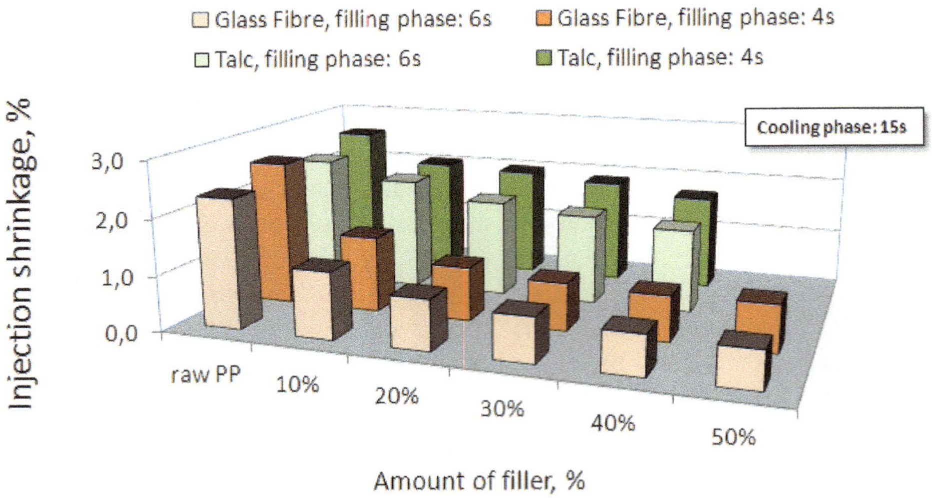

Fig. 7. Effect of the filler type and content on the linear injection shrinkage of polypropylene injection-molded parts produced at filling phase time of 4s and 6s also with a cooling phase time of 15s

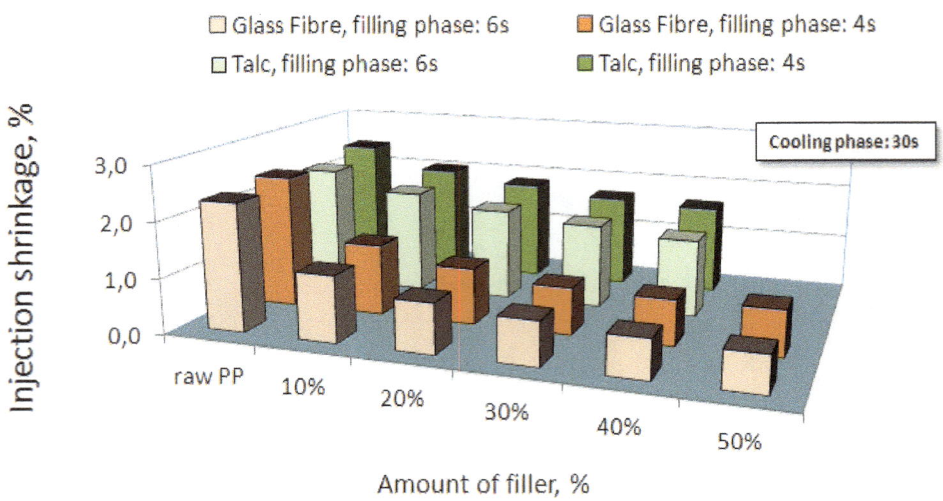

Fig. 8. Effect of the filler type and content on the linear injection shrinkage of polypropylene injection-molded parts produced at filling phase time of 4s and 6s also with a cooling phase time of 30s

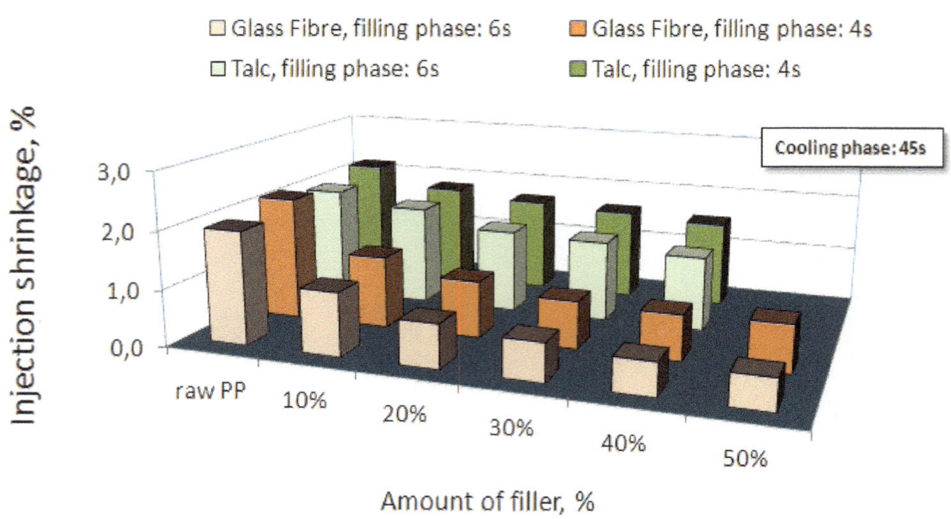

Fig. 9. Effect of the filler type and content on the linear injection shrinkage of polypropylene injection-molded parts produced at filling phase time of 4s and 6s also with a cooling phase time of 45s

function of the injection time and cooling time of this part. Figure 5 offers a comparison between the change in the lengths of the parts filled with glass fiber and talc that were produced at an injection time of 4 seconds, while Figure 6 presents the variations in the dimensions of the parts produced with the injection time set to 6 seconds. Extending the injection time and cooling time led to increasing the values of the investigated linear dimensions, whereas the differences between the corresponding dimensions were the more clear, the lower the filler content. Figure 7, Figure 8 and Figure 9 present the effect of the injection time and cooling time on the values of linear injection shrinkage, depending on the filler type and content.

Increasing the filler content led to decreased shrinkage, this effect being most visible for the glass fiber-filled part whose shrinkage sometimes decreased nearly twice, compared to the shrinkage of the part without any filler. The addition of talc resulted in decreased shrinkage, too, yet the decrease was not as considerable as in the case of glass fiber. With glass fiber, the decrease in shrinkage could be observed when the filler content was changed from 10% to 20%, and then to 30%. A further increase of the glass fiber content in polypropylene did not cause any visible decrease in shrinkage. As for the talc filler, the decrease in shrinkage was not as considerable as the one observed for glass fiber, and it proceeded at a uniform rate along with increasing the talc content in the material.

The extension of the injection time, which enables higher polymer packing in the mold cavity, affected the increase in the part dimensions more than the extension of the cooling time of the part in the mold.

CONCLUSIONS

Given both the complexity of the injection molding process for polymer materials and a relatively high number of factors that affect the process and its results, it is necessary that polymer properties, tool design and machine processing capacity be known in order to determine the optimal conditions of injection molding. The performed experimental tests were undertaken to investigate selected problems of the injection molding process, including factors such as changes in properties of the polymer being processed (filler type and content) and changes in some relatively easily adjustable injection parameters (injection time and cooling time of the part in the mold). The phenomenon of injection molding shrinkage of polymer materials, which changes both the shape and dimensions of the molded part, is one of the fundamental problems to be solved at the stage of both tool design and parameter selection for this processing method. The conducted analysis of injection molding shrinkage has demonstrated that shrinkage value can be effectively affected if injection molding conditions are selected properly, within the processing capacity of the injection molding machine; if necessary, it is also possible to modify the processing properties of polymers in order to produce molded parts with the desired geometric and strength properties. Both increasing the filler content in the polymer to a given value and extending the injection molding time (by increasing the injection time and cooling time) led to a clear decrease in injection molding shrinkage and, as a result, increased the dimensional accuracy of the injection-molded parts obtained, which means better production quality and enhanced applicability of such parts.

Acknowledgments

This paper is the result of an international project tiled "Technological and design aspects of extrusion and injection molding of thermoplastic polymer composites and nanocomposites" (PIRSES-GA-2010-269177) financed by the European Union under the 'People' programme of the Seventh Framework Programme (FP7-PEOPLE) as Marie Curie Actions – International Research Staff Exchange Scheme (IRSES).

REFERENCES

1. Bociąga E., Jaruga T., Sikora R.: Wybrane zagadnienia wtryskiwania precyzyjnego. Cz. II. Czynniki wpływające na jakość wyprasek precyzyjnych. Polimery 2009, 54, 7-8, 522–529.
2. Bociąga E.: Specjalne metody wtryskiwania tworzyw polimerowych. WNT, Warszawa 2008.
3. Dulebova L., Greškovič F., Spišák E., Duleba B., Garbacz T.: The impact of the amount of filler on the properties of polymer materials. Kovárenství 2012, 44, 33–36.
4. Fisher J. M.: Handbook of molded part shrinkage and warpage. Plastic Design Library, William Andrew Inc., Norwich 2003.

5. Gershon A.L, Gyger L.S., Bruck H.A., Gupta S.K.: Thermoplastic polymer shrinkage in emerging molding processes. Experimental Mechanics 48, 6, 2008, 789–798.

6. Kazmer D.O.: Injection mold design engineering. Carl Hanser Verlag, Munich 2007.

7. Kowacs J.G., Solymossy B., Effect of glass bead content and diameter on shrinkage and warpage of injection-molded PA6. Polymer Engineering & Science 49, 11, 2009, 2218–2224.

8. Kwon K., Isayev A.I., Kim K.H., van Sweden C.: Theoretical and experimental studies of anisotropic shrinkage in injection moldings of semicrystalline polymers. Polymer Engineering & Science 46, 6, 2006, 712–728.

9. Malloy R.A.: Plastic Part Design for Injection Molding. An Introduction. Carl Hanser Verlag, Munich 1994.

10. Polyplastics catalogue: Acetal Copolymer Duracon M90-44. Datasheet. Polyplastics Digital Library (www.polyplastics.com).

11. Postawa P.: Skurcz przetwórczy wyprasek a warunki wtryskiwania. Polimery 50, 3, 2005, 201–207.

12. Rosato D.V., Rosato D.V., Rosato M.G.: Injection Molding Handbook. Kluwer Academic Publisher, Norwell 2000.

13. Shelesh-Nezhad K., Taghizadeh A.: Shrinkage behavior and mechanical performances of injection molded polypropylene/talc composites. Polymer Engineering & Science 47, 12, 2007, 2124–2128.

14. Sikora J.W., Greškovič F.: Wpływ rodzaju napełniacza oraz czasu ochładzania na właściwości mechaniczne wyprasek wtryskowych z osnową polipropylenową. Przetwórstwo Tworzyw 16, 5, 2010, 273–279.

15. Sikora R., Jachowicz T.: Wpływ czasu ochładzania na skurcz przetwórczy wytworów otrzymanych metodą wytłaczania z rozdmuchiwaniem. Polimery 45, 10, 2000, 713–719.

16. Sikora R.: Przetwórstwo tworzyw wielkocząsteczkowych. Wydawnictwo Edukacyjne. Warszawa 1993.

17. Slota J., Spišák E., Gajdoš I.: Modern approaches in development of plastic products. Journal for Technology of Plasticity, 34, 1-2, 2009, 97–104.

18. Smorawiński A.: Technologia wtrysku. WNT, Warszawa 1989.

19. Yang S. S., Kwon T. H.: Numerical and experimental investigation of shrinkage behavior of precision injection molded articles. Polymer Engineering & Science 48, 8, 2008, 1569–1583.

20. Zawistowski H., Frenkler D.: Konstrukcja form wtryskowych do tworzyw termoplastycznych. Wydawnictwo Poradników i Książek Technicznych Plastech, Warszawa 2003.

INFLUENCE OF PROCESS PARAMETERS ON DIMENSIONAL ACCURACY OF PARTS MANUFACTURED USING FUSED DEPOSITION MODELLING TECHNOLOGY

Filip Górski[1], Wiesław Kuczko[1], Radosław Wichniarek[1]

[1] Chair of Production Engineering and Management, Faculty of Mechanical Engineering and Management, Poznan University of Technology, 3 Piotrowo Str., 60-965 Poznań, Poland, e-mail: filip.gorski@doctorate.put.poznan.pl; wieslaw.kuczko@doctorate.put.poznan.pl; radoslaw.wichniarek@put.poznan.pl

ABSTRACT

The paper presents the results of experimental study – part of research of additive technology using thermoplastics as a build material, namely Fused Deposition Modelling (FDM). Aim of the study was to identify the relation between basic parameter of the FDM process – model orientation during manufacturing – and a dimensional accuracy and repeatability of obtained products. A set of samples was prepared – they were manufactured with variable process parameters and they were measured using 3D scanner. Significant differences in accuracy of products of the same geometry, but manufactured with different set of process parameters were observed.

Keywords: rapid prototyping, fused deposition modelling, 3D optical scanning, manufacturing accuracy.

INTRODUCTION

Rapid Prototyping (RP) and Additive Manufacturing (AM), also known as Layered Manufacturing is a group of technologies that allow to produce a physical prototype based only on the 3D CAD model, without need to prepare tooling of any kind. RP technologies have found their place among other, traditional manufacturing technologies – they are invaluable when there is a need of quick manufacturing of a physical prototype of a designed part [5].

Usability of prototypes manufactured with RP technologies is directly related to their parameters, among which accuracy is one of the most important, especially in case of functional prototypes. Manufacturing accuracy of any part is a degree of its compatibility with an ideal part [11]. Two types of accuracy in manufacturing can be distinguished. Dimensional accuracy is a degree of compatibility of linear or angular dimensions with dimensions of a perfectly produced part, dimension deviations being a direct measure of this accuracy. Shape accuracy is a degree of compatibility of specific shapes with the perfect part or with other shapes – direct measures of this type of accuracy are deviations from the ideal cylinder, sphere, straight line, plane etc. Just like all other technical and economic indexes of products manufactured in an additive manner, accuracy is strongly affected by parameters of manufacturing process – products of the same nominal geometry will have entirely different properties if they are manufactured using different sets of values of these parameters. The most important parameter is the model orientation – set of angles between basic planes of the object and the manufacturing direction.

Dependency between manufacturing orientation (and other parameters of additive manufacturing process) and technical and economic product indexes are of particular interest of research facilities all over the world dealing with the additive manufacturing technology. There are studies focused on identification and description of these relations [1, 2, 4] and their generalization and formulation of directives for the control of economic and technical indexes of products by optimal process parameter selection [3, 9]. This paper presents research belonging to this type of study – it presents the results of experimental testing of influence of the orientation on the accuracy and dimensional repeatability of products obtained using one of the most widespread AM technologies – Fused Deposition Modelling (FDM). The acquired results are shown together with the results of earlier work by the authors, regarding the relations between orientation and mechanical properties of additively manufactured products [6]. The authors have also worked on a similar problem in the past (influence of process parameters on accuracy of FDM parts), but the layer filling strategy was the parameter investigated instead of the orientation [10].

The research was performed in the Laboratory of Rapid Manufacturing of Chair of Production Engineering and Management, located in the Faculty of Mechanical Engineering of Poznan University of Technology. The Laboratory frequently cooperates with industry and there are often requests for prototypes of parts manufactured with high accuracy and strength. It is impossible to obtain a product with an optimal combination of these two properties without identifying the relations between them and process properties – and this is what this paper is about.

BASIC INFORMATION

Additive manufacturing using thermoplastics –Fused Deposition Modelling

Fused Deposition Modelling is a process consisting of layered deposition of plasticized build and support material supplied in a form of a wire by an extrusion head (see fig. 1 for process schema). Numerically controlled device deposits its build and support material on the model base, with data about head positioning coming from horizontal cross-sections of the part, prepared on the basis of the 3D CAD model. The ABS material is frequently used, other thermoplastics can be used too, depending on the machine and head type. Obtained models are considerably strong and can be subjected to further treatment by machining, gluing or painting, to obtain desired surface quality. The produced part is ready for use immediately after support material removal [8].

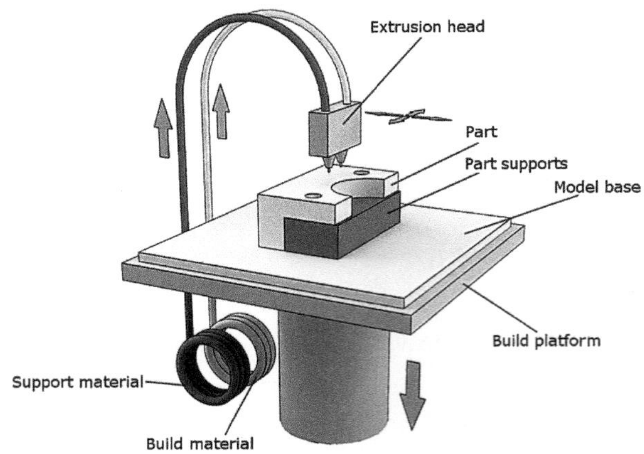

Fig. 1. Schema of the Fused Deposition Modelling process [8]

Influence of the FDM process parameters on features of obtained products

Finished product manufactured using Fused Deposition Modelling technology can be characterized by technical indexes – strength of certain kind (tensile, flexural strength or impact resistance), dimensional and shape accuracy, as well as economic indexes, such as manufacturing time and amount of support and build material used. Many factors have direct influence on these indexes [2]. A phenomenon specific for the described technology is relatively high significance of additive process parameters. Additive technologies make no use of any tooling, it is their most important advantage. The role of the tooling in the aspect of influencing technical and economic indexes is taken by the process parameters (more specifically, the sets of parameters), which may be directly or indirectly controlled by the process engineer (Fig. 2). The orientation of manufactured model during the process is the most important parameter of these which can be changed directly.

Orientation of the model in the working chamber during layered manufacturing process can be intuitively described as an angular difference between plane determining direction of the object division into layers and selected, basic plane of the manufactured object (Fig. 3). Orientation can be unequivocally defined by three angular values. One of them – rotation in the Z axis (around vertical direction) has no importance from the viewpoint of technical and economical indexes, as it has no influence on how the object is divided into layers.

The orientation directly affects the internal structure of the model [6], deciding, among other

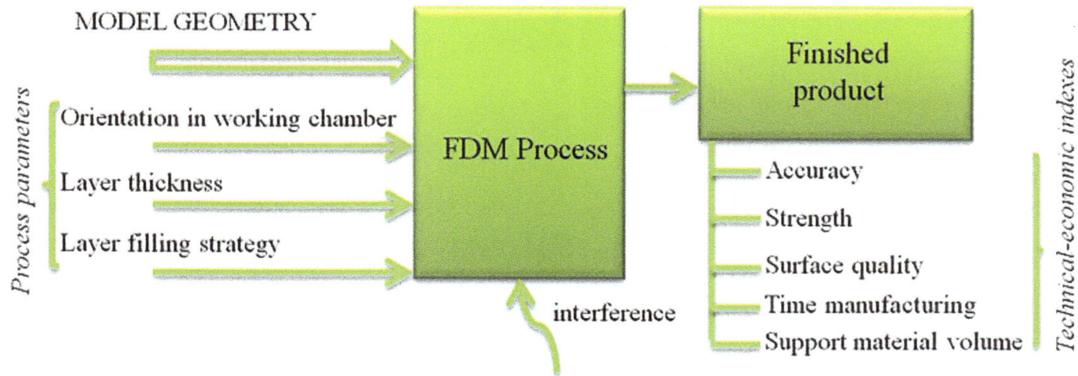

Fig. 2. Parameters of the FDM process and technical and economic indexes of the finished product

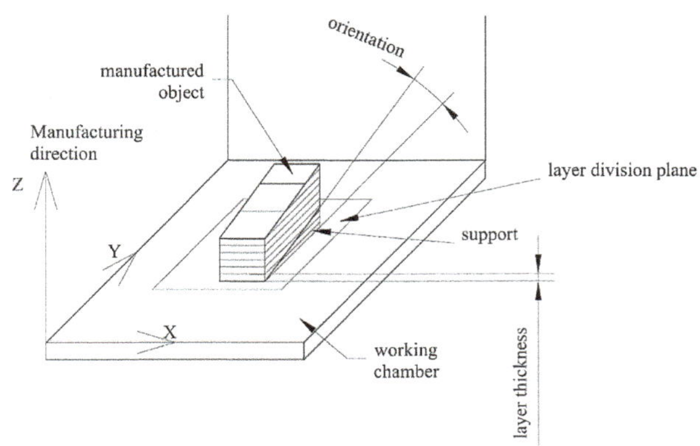

Fig. 3. Orientation of the model in the working chamber [6]

things, about strength, accuracy or surface quality. It is also a relevant parameter in terms of economic aspects of product manufacturing.

CONCEPT AND METHODOLOGY OF RESEARCH

Aim and plan of the study

An exact character of dependencies between the orientation and particular features of products obtained with FDM method has not been fully investigated yet. There were attempts at experimental determination of these relations [1], but it is still an open research problem. This paper is aimed at preliminary identification of the relation between model orientation and two basic characteristics related to accuracy of the obtained products:

- dimensional accuracy, understood as degree of compatibility of basic dimensions of the obtained product with dimensions of the ideal product (nominal dimensions),
- repeatability – degree of dimensional compatibility of two products of the same nominal geometry, manufactured in the same conditions, with identical values of the process parameters.

Obtained results – accuracy and repeatability coefficients for samples of variable orientation – were compared with simultaneously acquired results of tensile tests performed on the same samples. Simplified plan of the research (in chronological order) is shown in Figure 4. Samples were manufactured using FDM technology available in the Laboratory of Rapid Manufacturing. Then they were measured using 3D optical scanner. After measurements, they were subjected to tensile tests.

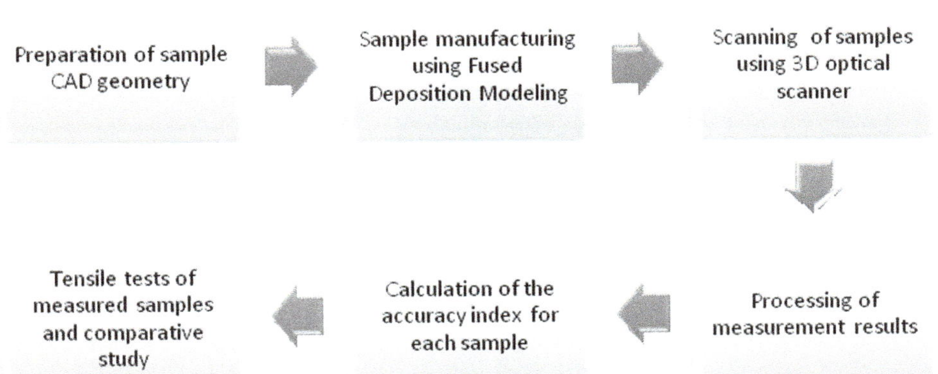

Fig. 4. Plan of the research

Preparation of samples for measurements

For the accuracy study of FDM made parts, tensile test samples were selected as a nominal geometry (they contain both straight and curved profiles, so it is also possible to evaluate shape accuracy). On the basis of the standards describing the strength tests (PN-EN ISO 527), CAD models of samples were prepared (Fig. 5). Then they were transformed to a triangle mesh and loaded into software CatalystEX, used for process control on the available machine – Dimension BST 1200 (Fig. 6).

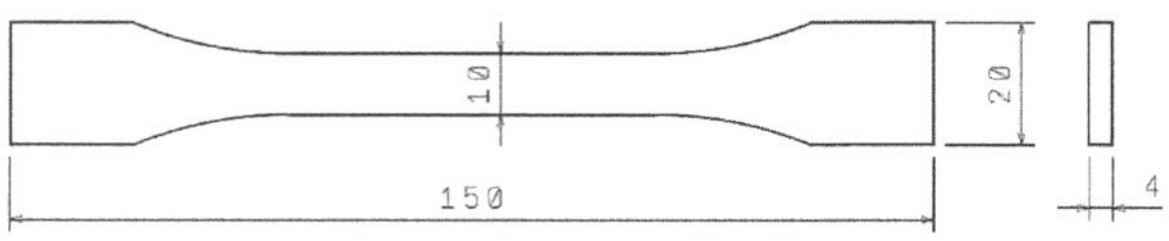

Fig. 5. Sample used in accuracy studies

Using the same geometry, five different samples were prepared by changing the orientation values – two possible values of X orientation were selected (flat and side) and for these, Y orientation was varied by 45 degrees. Samples of the following orientation were manufactured:

- 0° X, 0° Y (flat-orientation sample)
- 90° X, 0° Y (side-orientation sample)
- 0° X, 45° Y
- 90° X, 45° Y
- 90° Y (vertical orientation sample).

Each sample type was manufactured in 3 copies, to perform repeatability study. To refer to each sample type in an easy way, they were assigned IDs, based on their orientation in X and Y axes. Summary of manufactured samples is presented in Table 1. Some of the specimens are shown in Figure 7.

Sample measurement – 3D optical scanning

Three-dimensional scanners use light (mostly white, blue is also used) of a known structural pattern – usually it is a fringe sequence of known, variable width and density. Light is projected on an object and fringe pattern image is registered by cameras. Deformation of the pattern is then analyzed by the software to map each point from the camera matrix with an appropriate point coordinates in space. A single measurement (up to 20 seconds), also named scan, gives as many measured points as the camera matrix has – in case

Table 1. Summary of manufactured samples for measurements

No.	Orient. X [°]	Orient. Y [°]	ID	Manufacturing time [s]	Support material [cm³]	Build material [cm³]
1.	0 (flat)	0	0F	1320	1.96	8.81
2.	90 (side)	0	0S	2760	2.28	8.71
3.	0	45	45F	11520	16.71	8.60
4.	90	45	45S	13080	26.68	8.67
5.	n/a (vertical)	90	90	13320	11.69	8.51

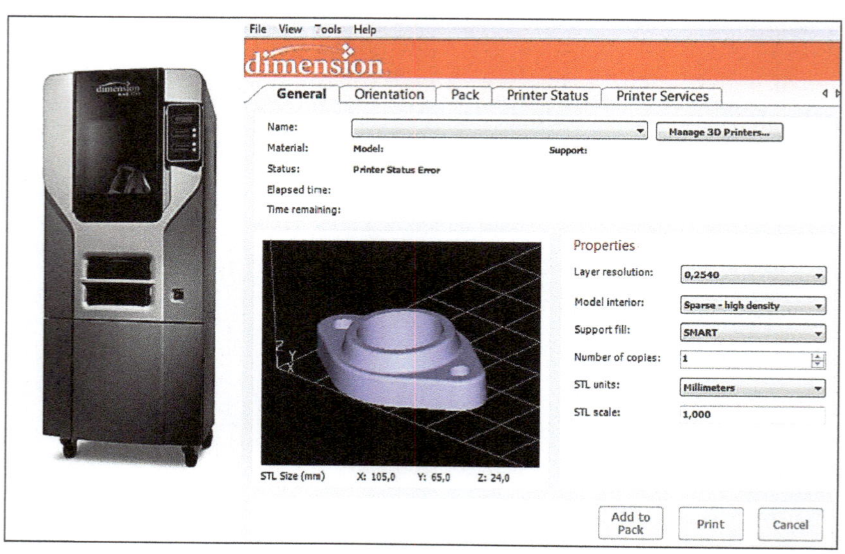

Fig. 6. Dimension BST 1200 machine used for sample manufacturing and main window of the CatalystEX software

Fig. 7. Samples manufactured using FDM technology – same geometry, different orientation

of the scanner used in the research it is a value of 0,8 MPix, so it is 800 000 points, practically representing the whole measured surface. Measurement accuracy can be as high as 0,02 mm [7].

For research discussed in the paper, an ATOS I optical scanner by GOM was used (Fig. 8). To accurately measure small objects – tensile test samples – the smallest possible measurement field available for this scanner was used, with scanning volume of 125 mm per 100 mm per 90 mm. The objects were covered in non-coded targets, used for auto-orientation of the subsequent scans [7], placed in a fixed position (Fig. 9) and measured (Fig. 10).

Processing of the measurement results

After the measurements are finished, a preliminary data processing must be carried out. It consists of the following stages:

- joining the data from all scans together – performed automatically, with some corrections able to be performed manually;
- removal of useless points (representing fixtures or other objects present in the measurement volume), manually or automatically;
- assignment of a coordinate system to the object (basing) by the "plane, line, point" method – definition of Z plane, X axis and point 0 by indication of points forming these geometries (3 points for plane, 2 for line, 1 for coordinate system beginning);
- generation of triangular mesh on the basis of the processed point cloud and export of the generated mesh to an STL format.

Data processing is performed using the software supplied with the 3D scanner by GOM company. Further processing consists of preparation of measuring reports. The following reports have been prepared:

- accuracy report for each sample – comparison with an ideal part geometry (deviations between nominal CAD model and a triangle mesh),
- repeatability report for each pair of the same type of samples – comparison of the two samples (deviations between two triangle meshes).

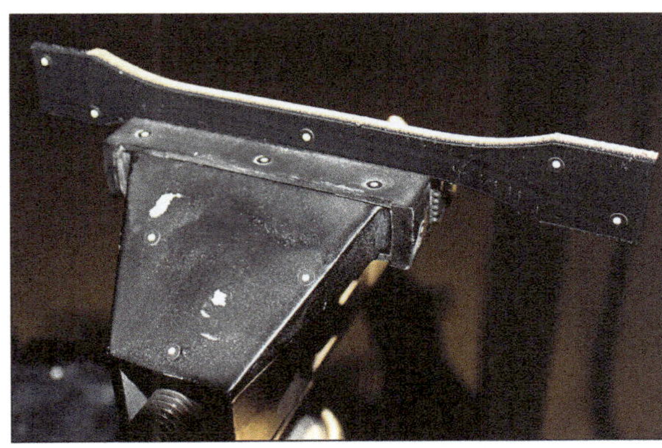

Fig. 9. Non-coded reference targets placed on measured object

Fig. 8. ATOS I scanner used in the research

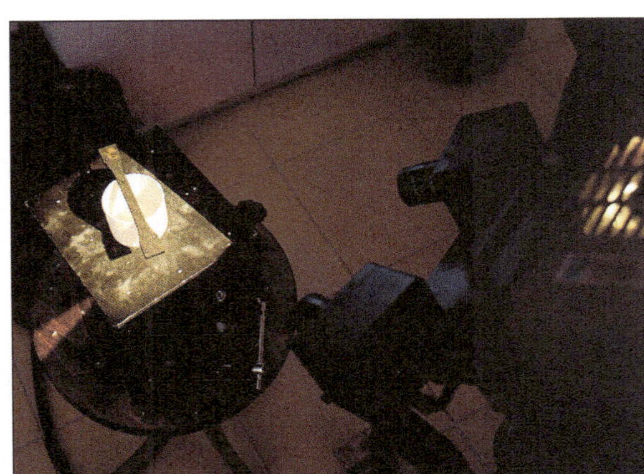

Fig. 10. Measurement of the samples using 3D scanner

The digital data that was compared (namely, triangle mesh with nominal model in 1 and two triangle meshes in 2) was put together using the "best fit" algorithm provided in the GOM software. Reports contained the following information in the visual form:

- coloured deviation map (examples in the Figure 11).
- deviations along the outlines of selected cross-sections,
- deviations from specified dimensions, checked dimensions are shown in the Figure 12a, with exemplary page from the report in the Figure 12b. Each dimension was inspected in several locations – 4 for widths, 2 for length, 7 for thickness.

For repeatability reports, only coloured deviation map and deviations in cross-sections were generated. Each sample type was characterized by two coefficients, a dimensional accuracy coefficient and a dimensional repeatability coefficient. Both were calculated using average deviation values from the reports.

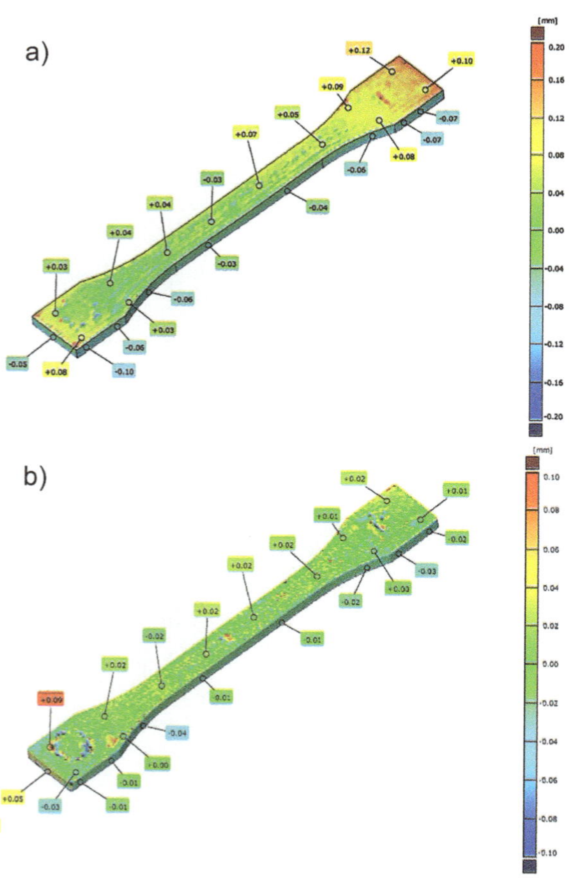

Fig. 11. Examples of deviation maps obtained during the study for sample type 0F: a) deviation from the nominal shape, b) deviation of sample 1 from sample 2

The dimensional accuracy coefficient is basically calculated as an average of deviation from all inspected dimensions, which is why it must be treated only as a general index. The formula for the accuracy coefficient is the following:

$$k_d = \frac{k_{len} + k_{w1} + k_{w2} + k_h}{4} \tag{1}$$

where: k_d – coefficient of dimensional accuracy of the specific sample type,

k_{len}, k_{w1}, k_{w2}, k_{th} – coefficients of accuracy of 4 checked dimensions (length, two widths and thickness), with a single coefficient formula as following:

$$k_x = \frac{\sum_{i=1}^{n} \frac{d_x}{x_{nom}}}{n} \cdot 100 \tag{2}$$

where: x – dimension (x stands for length, widths and thickness, formula (2) was used for all four coefficients),

n – number of samples taken into account (n=3 in this paper),

d_{xi} – average from deviation from the dimension x in sample number i [mm], absolute value

x_{nom} – nominal value of the dimension x (see fig. 5 for all nominal dimension values).

The higher the accuracy coefficient, the lower the accuracy. Ideal part would have an accuracy coefficient of 0. The value ofaccuracy coefficient equal 1 means that average deviation from the nominal dimensions is 1%.

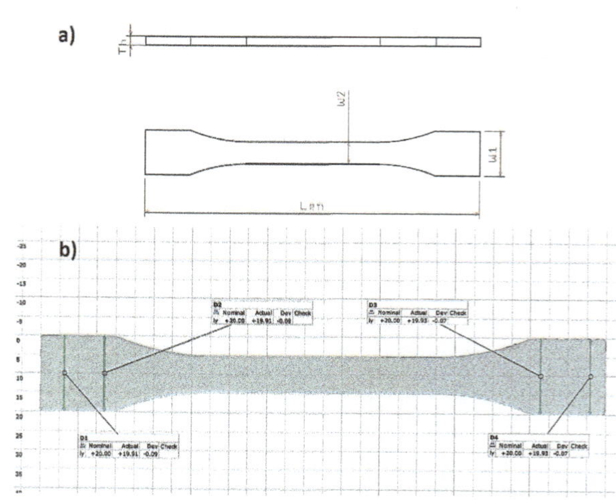

Fig. 12. Dimensions taken into consideration in accuracy analysis, a) indication of dimensions, Len – length, W1 – broad area width, W2 – narrow area width, Th – thickness; b) example of deviation report for W1 dimension, sample ID 0F

The repeatability coefficient is calculated as an average difference in dimensional accuracy coefficients calculated for each sample separately. The formula is as following:

$$k_r = \frac{|k_1 - k_2| + |k_2 - k_3| + |k_1 - k_3|}{3} \quad (3)$$

where: k_r – repeatability coefficient of the specific sample type,

k_1, k_2, k_3 – accuracy coefficients calculated as in formulas (1) and (2), but separately for each of the three samples.

The higher the repeatability coefficient, the lower the repeatability can be achieved for the sample – perfectly repeatable process would result in coefficient equal to 0.

Both coefficients are dimensionless, although they could be treated as a percentage.

RESEARCH RESULTS

Dimensional accuracy and repeatability measurement results

Using formulas (1), (2) and (3), the accuracy and repeatability coefficients were determined for each sample. Table 2 contains the most generalized form of actual research results – coefficients k_d and k_r for each sample type. The coefficients are illustrated in Figure 13. Table 3 contains the components of accuracy coefficient – average coefficients of each checked dimension – their analysis brings some interesting conclusions.

Comparison of accuracy coefficients with tensile strength

To compare the accuracy with the strength of the same FDM made samples, tensile tests were performed, described in greater detail in [6]. Their results, in juxtaposition with accuracy and repeatability coefficients, are contained in Table 4 and the values are presented in the graphical form in Figure 14. The table contains additional data about manufacturing time (from Table 1), for informational purpose. The best values in each category are bolded and worst are shown in a different colour. To make visual comparison easier, k_d and k_r coefficients were inversed – value used for graphs is equal to $100 - k_d/k_r$. Thanks to the inversion, greater height of all the columns in the graph in the Figure 14 represents more beneficial value of a property, so it is easier to compare the values.

CONCLUSIONS

After analysis of the obtained data, the following conclusions can be drawn:

1. It is confirmed that the orientation directly influences both the accuracy and repeatability of FDM parts. Character of the $A = f(O)$ func-

Table 2. Coefficients of accuracy of samples made using Fused Deposition Modelling technology (k_d – accuracy, k_r – repeatability, lower value = better)

Sample ID \ Coefficient	k_d	k_r
0F	0.976	0.090
0S	1.403	0.409
45F	0.665	0.080
45S	0.852	0.603
90	1.046	0.668

Table 3. Coefficients of accuracy of each inspected dimension (lower value = better)

Sample ID \ Coefficient	k_{len}	k_{w1}	k_{w2}	k_{th}
0F	0.021	0.508	0.625	2.750
0S	0.031	0.600	2.817	2.167
45F	0.034	0.354	0.675	1.595
45S	0.038	0.300	0.583	2.488
90	0.193	0.092	0.933	2.964
on average:	0.064	0.370	1.127	2.393

Table 4. Comparison of accuracy with the tensile strength and elongation at break of FDM samples (higher value = better), with manufacturing time included (t_m)

Sample \ Property	$100 - k_d$	$100 - k_r$	σ_m [MPa]	ε_b	t_m [s]
0F	99.024	99.940	19	4.6%	**1320**
0S	98.560	99.727	**22.9**	**7.0%**	2760
45F	**99.335**	**99.947**	12.9	1.5%	11520
45S	99.148	99.598	14.9	1.2%	13080
90	98.954	99.555	11	1.0%	13320

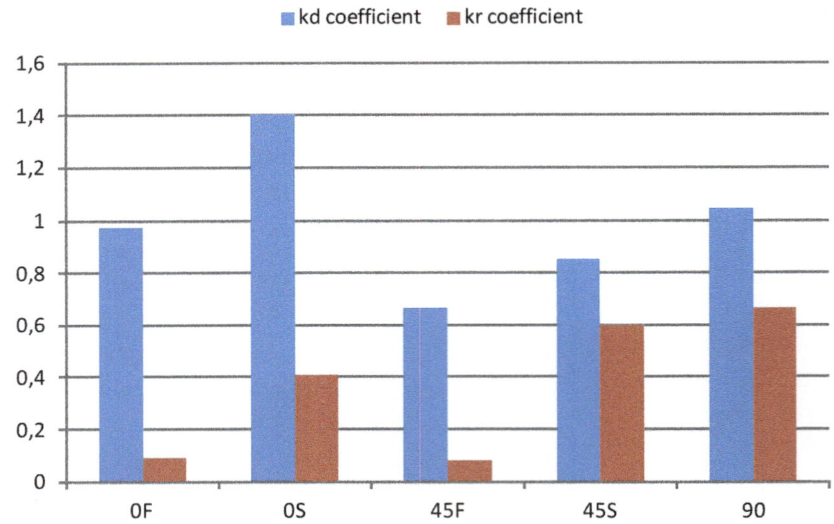

Fig. 13. Coefficients of accuracy and repeatability for FDM samples (lower = better)

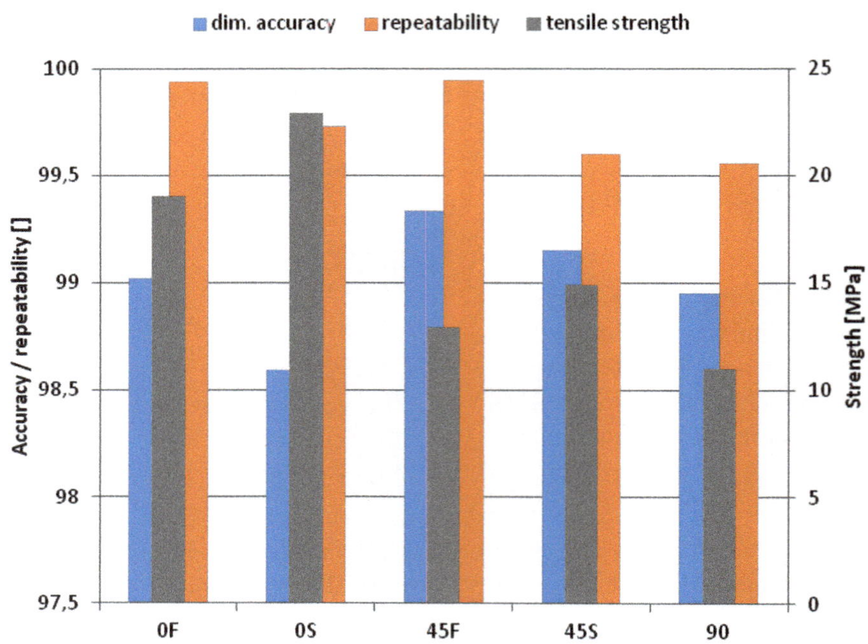

Fig. 14. Juxtaposition of strength and accuracy of FDM samples (higher = better)

tional relation (*A* – accuracy, *O* – orientation) is non-linear and rather hard to describe mathematically with such small number of sample types.

2. Relation between calculated accuracy and repeatability coefficients can also be described as non-linear, at least for coefficients calculated only on the basis of deviations from specific dimensions.
3. Accuracy of FDM made parts should be considered as low in comparison with other plastic forming technologies, with deviations of meaningful dimensions above 2% in some cases.
4. Analysis of deviations of particular dimensions brings an interesting conclusion – deviations are relatively smaller for higher dimensions, i.e. the greater the object size is, the more accurately it will be manufactured. This only applies to relative deviations (i.e. deviation/nominal dimension ratio), as absolute deviation values remain on more or less the same level, regardless of the nominal dimension value.
5. Strength of FDM made parts is not particularly related to the accuracy, although some minor positive coupling can be observed in some cases.

6. The most beneficial sample out of the manufactured 5 types, regarding all the possible criteria (accuracy, repeatability, strength, cost, time, surface quality) with assumption that all criteria are equally significant, is the 0F sample, as a combination of both technical and economic indexes is optimal for this product, although its strength, accuracy and repeatability are not the best.

General conclusion can be drawn that it is not easy to select a proper orientation to meet all the requirements regarding accuracy and strength of the part – there is no perfect combination, therefore, it is fully justified to use artificial intelligence tools to determine optimal orientation, as some authors suggest [3, 9]. For parts with simple geometry, it is usually the best way to focus on the economic effectiveness, because short manufacturing time is usually related to higher strength and in some cases – higher accuracy and better surface quality. It can be explained in a simple way – the more layers the product has, the longer it takes to manufacture it and there are more weak spots which lower the strength in certain tests (bond between layers is based only on adhesion, so each product has lower strength if the load direction is equal to manufacturing direction), so aiming at orientation with the lowest number of layers allows to rise at least two important product parameters, with high chance to rise accuracy and sometimes surface quality as well.

To fully identify the relation between orientation and accuracy, a higher number of sample types is required – for the start, with variation of orientation in Y axis equal to least 15 degrees, which would give 7 samples per one X orientation (14 samples in total for two X orientations) – it would allow to define at least an approximate character of the $A = f(O)$ function. Further experimental studies should be conducted to explore the possibility of controlling all the product technical and economic indexes using orientation value, not only for products of simple geometry, but also for more complex shapes.

Acknowledgments

The studies described in the paper were a part of the pre-doctoral grant financed by the National Science Centre on the basis of decision no. DEC-2011/01/N/ST8/07603.

REFERENCES

1. Ahn S.H. et al.: Anisotropic Tensile Failure Model of Rapid Prototyping Parts - Fused Deposition Modeling (FDM). International Journal of Modern Physics B (IJMPB), 17(8-9), 2003.
2. Bellini A., Guceri S.: Mechanical characterization of parts fabricated using fused deposition modeling. Rapid Prototyping Journal, 9, 2003.
3. Ghorpade A., Karunakaran K.P., Wiwari M.K.: Selection of optimal part orientation in fused deposition modeling using swarm intelligence. Journal of Engineering Manufacture, 221, 2007.
4. Giannatsis J. et al.: Investigating the influence of build parameters on the mechanical properties of FDM parts. Proceedings of the 5th International Conference on Advanced Research in Virtual and Rapid Prototyping, Leiria, Portugal, 28 Sept. – 1 Oct, 2011.
5. Górski F., Kuczko W., Wichniarek R., Dudziak A., Kowalski M., Zawadzki P.: Choosing optimal rapid manufacturing process for thin-walled products using expert algorithm. Journal of Industrial Engineering and Management, 3(2), 2010.
6. Górski F., Wichniarek R., Andrzejewski J.: Wpływ orientacji części na wytrzymałość modeli z ABS wytwarzanych techniką modelowania uplastycznionym tworzywem sztucznym. Przetwórstwo Tworzyw, 9, 2012.
7. Konieczny R., Dudziak A., Grajewski D., Górski F.: Techniki pomiarów optycznych w inżynierii odwrotnej. Comprint, Poznań 2012.
8. Pająk E., Górski F., Wichniarek R., Dudziak A.: Techniki przyrostowe i wirtualna rzeczywistość w procesach przygotowania produkcji. Promocja 21, Poznań 2011.
9. Panda S.K., Padhee S., Sood A.K., Mahapatra S.S.: Optimization of Fused Deposition Modelling (FDM) Process Parameters Using Bacterial Foraging Technique, Intelligent Information Management, 1(2), 2009.
10. Wichniarek R., Górski F., Kuczko W.: Analiza wpływu stopnia wypełnienia modeli FDM na ich dokładność kształtową. Przetwórstwo Tworzyw, 9, 2012.
11. Zawora J.: Podstawy technologii maszyn. WSiP, Warszawa 2007.

IMPACT OF VIBRATORY AND ROTATIONAL SHOT PEENING ONTO SELECTED PROPERTIES OF TITANIUM ALLOY SURFACE LAYER

Kazimierz Zaleski[1], Andrzej Zyśko[2]

[1] Mechanical Engineering Faculty, Lublin University of Technology, 36 Nadbystrzycka Str., 20-618 Lublin, Poland, e-mail: k.zaleski@pollub.pl

[2] Mechanical Engineering Faculty, Lublin University of Technology, 36 Nadbystrzycka Str., 20-618 Lublin, Poland, e-mail: andrzej.zysko@wp.pl

ABSTRACT

This study presents the results of tests on impact of vibratory and rotational shot peening of the Ti6A12Mo2Cr titanium alloy onto the processed object surface roughness and surface layer microhardness. The external surfaces of ring-shaped samples were shot peened. The preceding process consisted of turning with a cubic boron nitride blade knife. Steel beads, having a diameter of 6 mm, were used as a processing medium. The variable parameters of shot peening were vibrator amplitude and shot peening time. The range of recommended technological parameters for vibratory and rotational shot peening was determined. As a result of shot peening, the surface roughness could be reduced by approximately 4 times and the surface layer could be hardened to the depth of approximately 0.4 mm.

Keywords: shot peening, titanium alloy, surface layer, surface roughness, microhardness.

INTRODUCTION

Titanium alloys feature high strength, small density, high corrosion- and resistance to high temperature. These properties make titanium alloys a valuable structural material used in various branches of industry, and particularly in aviation industry. Titanium alloys are used for production of components that are exposed to changing loads during operation. Fatigue strength of such components can be increased by shot peening.

Titanium alloy components are most often shot peened by dynamic jet method, which consists of blow effect of a shot or bead jet ejected mechanically or pneumatically onto the processed object surface. The tests performed by the authors of this study [1] showed that shot peening with a jet of glass beads having the diameter of 55–100 μm within 40 seconds increase the fatigue strength of WT3-1 titanium alloy by 40%. The use of glass beads having the diameter of 40–150 μm in shot peening process proved to be even more effective (fatigue strength increased by 50%) than use of steel beads with diameter of 3 mm (fatigue strength increased by 20%) [2]. The study [3] presents the results of jet shot peening impact onto low-cycle fatigue strength. It was found that jet shot peening of an aircraft engine compressor rotor, made of WT3-1 titanium alloy, increased its durability three times. Shot peening allowed to increase fatigue strength of the components operating not only at room temperature but also at an elevated temperature. It was found that shot peening of the Ti-6A1-4V titanium alloy samples with a cast steel shot jet resulted in almost identical increase of fatigue strength, both during tests at room temperature as well as at 150°C [4].

The fatigue strength of machined elements increased also as a result of vibratory shot peening with free beads. This method consists of attachment of machined objects in a working chamber to which a charge of loose beads is poured, after which the working chamber is set in oscillating motion. The surface strain hardening effect is obtained as a result of loose beads hitting against the machined objects. Application of the controlled shot peening method enabled to evaluate the effect of beads energy, while hitting against the machined surface and density of these impacts onto the Ti-6A1-4V titanium alloy fatigue strength [5].

The increase of fatigue strength of shot peened elements is related to the shape of advantageous properties of the surface layer of such elements during the dynamic shot peening process. The test results of Ti-2.5Cu titanium alloy surface layer properties after surface treatment conducted by various methods are presented in the study [6]. It was found that due to shot peening, the surface layer microhardness in the vicinity of the surface increased by approximately 40% and the own compressive stresses of about 460 MPa were formed. The study [7] presents the test results of impact of shot peening with a jet of 70 μm beads within 8 seconds onto properties of the Ti-6A1-4V titanium alloy surface layer. This process caused a significant increase of surface roughness (value of Ra parameter increased 8 times), increase of microhardness by approximately 18% and forming of compressive residual stresses of about 900 MPa. Increase of surface layer microhardness and shaping of compressive residual stresses were obtained as a result of controlled shot peening of the Ti-6Al-4V alloy [8].

Shot peening by means of free beads may be performed on objects of various shapes. The objects whose shape is approximate to solids of revolution may be shot peened by vibratory and rotational method that consists of joining vibratory shot peening with rotational motion of the object being processed. The existing testing proved a beneficial effect of vibratory and rotational shot peening onto properties of the bearing steel surface layer in a quenched condition (62–64 HRC hardness) [9]. The purpose of this study is to evaluate the effect of vibratory and rotational shot peening onto the titanium alloy surface roughness and strain hardening degree.

TESTING METHOD

Tests were performed on the Ti6Al2Mo2Cr titanium alloy (per WT3-1 Russian standards), whose chemical composition was specified in Table 1. This is an alloy for plastic working, of the a+b structure, used widely for industrial purposes, especially for manufacturing parts for the machines working at changing loads, what justifies the usefulness of shot peening in these parts.

The samples prepared for tests had a ring shape with external diameter D = 56 mm, internal diameter d = 50 mm and length l = 10 mm. The process that preceded shot peening the external surface of samples was turning with the PCLNR2020K12 turning tool with the CNMA120404 inserts with cubic boron nitride blades. The machining parameters were as follows:

- depth of cut a_p = 0.5 mm,
- feed f = 0.07 mm/rotation,
- cutting speed v_c = 140 m/min.

The external surface roughness of samples after turning was Ra = 0.88 to 1.23 mm.

Vibratory and rotational shot peening was carried out on a special stand whose diagram is shown in Figure 1. The tested samples 7 were fixed to the spindle 4, which revolved in a slide bearing 9, seated in side walls of the working chamber 5. The working chamber 5 was set in oscillating motion by means of a vibrator 11. Due to vibration, the beads 6 inside the working chamber were hitting against the external surfaces of samples 7. Uniformity of shot peening on the entire circumference of samples was assured thanks to rotational motion of the spindle 4 with the samples attached. The special nuts 8 and 10 were

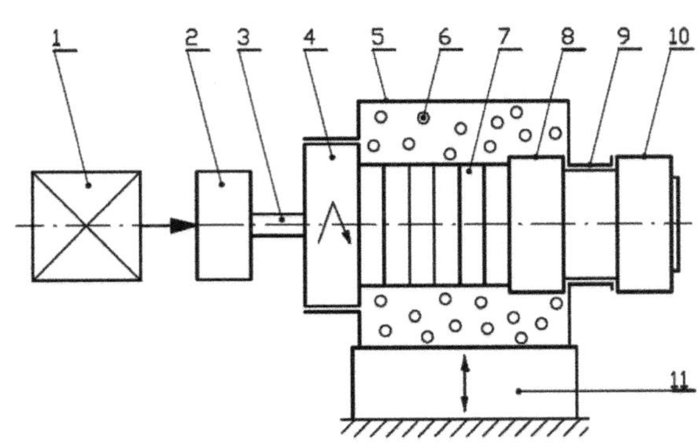

Fig. 1. Schematic illustration of the stand for vibratory and rotational shot peening: 1 – electric motor, 2 – reduction gear, 3 – flexible coupling, 4 – spindle, 5 – working chamber, 6 – bead, 7 – sample, 8, 10 – special nuts, 9 – slide bearing, 11 – vibrator

Table 1. The chemical composition of the Ti6Al2Mo2Cr alloy

Chemical element	Al	Mo	Cr	Fe	Si	C	N	Ti
Percentage	5.8	2.4	1.5	0.4	0.3	0.05	0.04	rest

used to secure the spindle 4 against displacement in axial direction. The spindle 4 was driven from the electric motor 1 via the reduction gear 2 and the flexible coupling 3.

The following constant parameters were applied for vibratory and rotational shot peening:

- vibrator frequency v = 7 Hz,
- bead diameter d_k = 6 mm,
- sample rotational speed n = 2 rpm.

The variable parameters of shot peening were as follows:

- vibrator amplitude,
- shot peening time.

The vibrator amplitude was changed within a = 36–60 mm for the constant shot peening time t = 10 minutes. Also, shot peening time was changed within t = 1–40 minutes at the constant amplitude a = 60 mm.

After shot peening, the tests of selected surface layer properties were performed, which covered:

- surface roughness test,
- surface layer microhardness distribution test.

Roughness of external surface of samples was measured using the Surtronic 3+ instrument of Taylor Hobson production. The measurement direction was parallel to the sample axis.

Microhardness measurements were made using the LM700AT microhardness tester of LECO production, by the Vickers' method, with the load of 981 mN (the 100g weight). The surface layer microhardness distribution was determined on perpendicular microsections. Based on the microhardness distribution, the surface layer strain hardening degree was determined using the following formula:

$$e = \frac{HV_{max} - HV_0}{HV_0} 100\% \tag{1}$$

where: HV_{max} – maximum microhardness,
HV_0 – microhardness of material core.

TEST RESULTS AND THEIR ANALYSIS

The effect of vibratory and rotational shot peening onto the processed surface roughness is presented in Figures 2 and 3. The Ra roughness parameter value is approximately 2–4 times smaller as compared to the surface before shot peening (after turning). Along with the increase of vibrator amplitude within a = 36–52 mm, the Ra parameter value reduced visibly which means that along with the amplitude increase, the energy of shot peening beads hitting against the processed surface increased and, as a result, microunevenness shaped after the process that preceding shot peening was leveled more effectively. In the scope of bigger amplitudes (a = 52–60 mm), it was observed that the amplitude increase was accompanied with the increase of Ra parameter, which probably resulted from the fact that the beads hitting with high energy made dents in the processed surface.

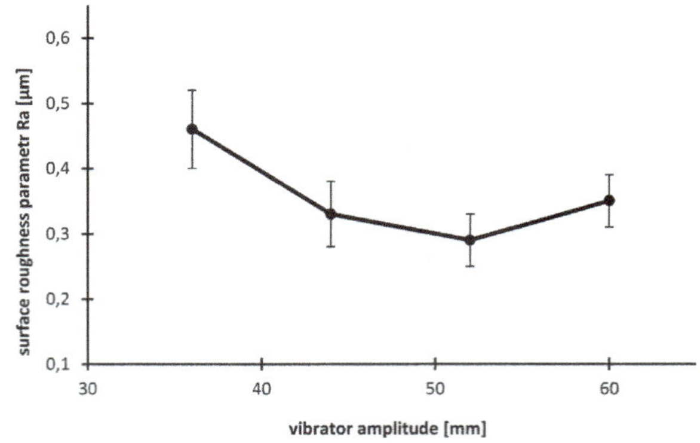

Fig. 2. Effect of vibrator amplitude onto Ra surface roughness parameter

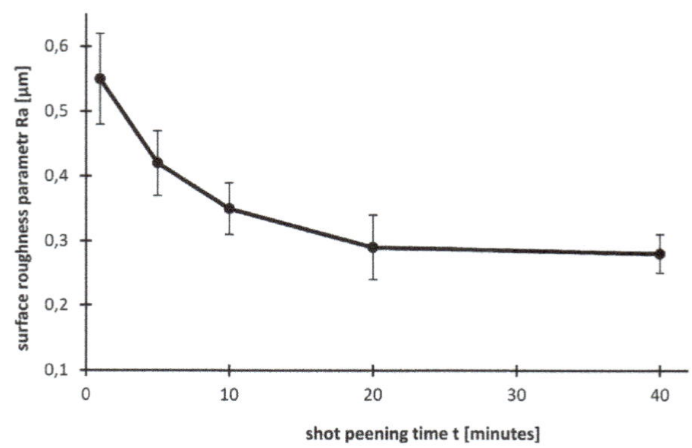

Fig. 3. Effect of shot peening time onto Ra surface roughness parameter

Along with the increase of shot peening time, the Ra roughness parameter value decreased (Figure 3), which was related to the increase of hitting density of burnishing beads onto the surface processed. Besides, along with prolongation of shot peening time, the surface layer of samples got hardened, which resulted in reduction of depth of impression due to beads impacts, and thus reduction of surface roughness. Prolongation of shot peening time above 20 minutes had no effect onto roughness of the surface processed.

The effect of vibratory and rotational shot peening onto microhardness of the processed sample surface layer is presented in Figures 4–8. The surface layer strain hardening degree, calculated per formula (1), and the depth of strain hardening layer g_h (Figure 4) were accepted as the indices that feature the microhardness distribution.

The strain hardening degree was changing along with the change of shot peening parameters and its value was from 6% to 23%. Both vibration amplitude increase and shot peening time had effect onto the surface layer strain hardening degree and such an increase was more visible in the scope of smaller values of shot peening parameters (Figures 5 and 6). The increase of shot peening time above 20 minutes caused only an insignificant increase of strain hardening degree.

Also the depth of hardened layer depends on the burnishing parameters and its value changes within the range from 0.17 mm to 0.42 mm. The curves showing the effect of vibration amplitude and burnishing time onto the hardened layer depth (Figures 7 and 8) have a similar shape as the curves of consolidation degree versus burnishing parameters.

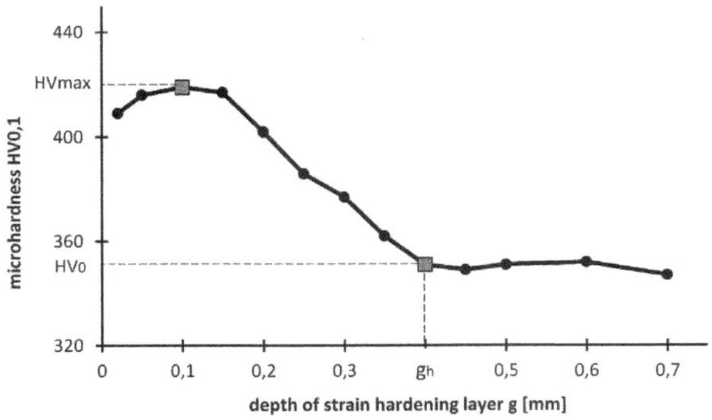

Fig. 4. Typical surface layer microhardness distribution after vibratory and rotational shot peening

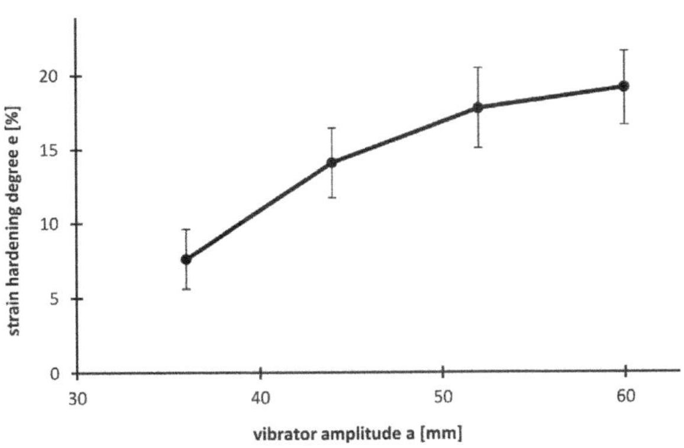

Fig. 5. Effect of vibrator amplitude onto strain hardening degree

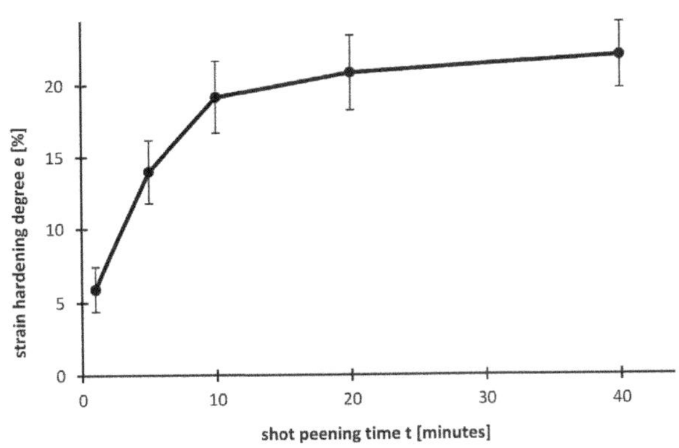

Fig. 6. Effect of shot peening time onto strain hardening degree

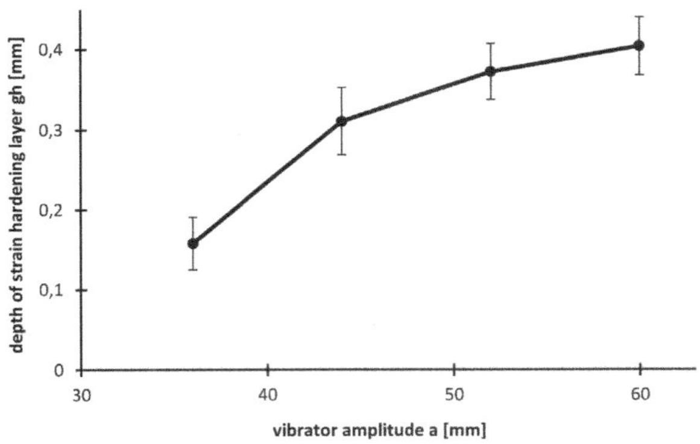

Fig. 7. Effect of vibrator amplitude onto depth of strain hardening layer

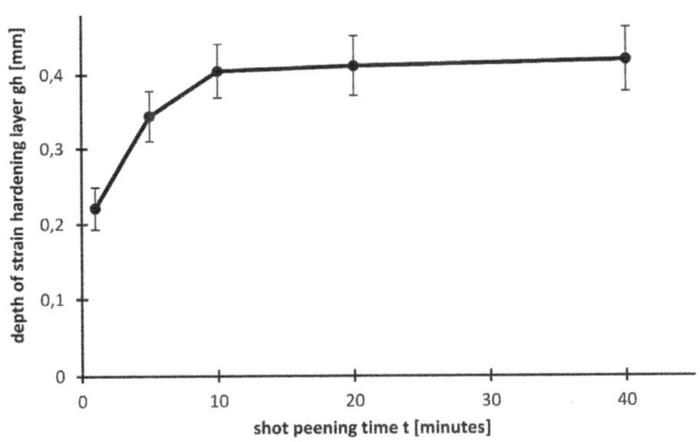

Fig. 8. Effect of shot peening time onto depth of strain hardening layer

CONCLUSIONS

Significant difficulties that occur while grinding titanium alloys (intensive wear of grinding wheel, high roughness of the machined surface) are the reasons for which it is required to search for alternative methods of finishing of machine components made of such alloys. The conducted tests proved that in case of the components of a shape approximate to solids of revolution, such a method may be vibratory and rotational shot peening.

Having analyzed the test results, it can be stated that the vibrator amplitude a = 52–60 mm, and the shot peening time t = 10–20 minutes (it is pointless to increase the shot peening time over 20 minutes due to economic reasons) should be applied. Upon application of the shot peening parameters specified above, it is possible to obtain the surface roughness of approximately Ra = 0.3 μm, the surface layer strain hardening degree of about 20%, and the hardened layer depth of about 0.4 mm.

The changes of processed titanium alloy surface layer properties obtained as a result of vibratory and rotational shot peening are advantageous due to durability of the machined components.

REFERENCES

1. Łunarski J., Zielecki W.: Ocena wytrzymałości zmęczeniowej stopów łopatkowych po procesach implantacji i mikrokulkowania. Zeszyty Naukowe Politechniki Rzeszowskiej Nr 46, Mechanika z. 17, 1988, 109-116.
2. Korzyński M.: Kulowanie strumieniowe kulkami szklanymi i łożyskowymi jako sposób poprawy wytrzymałości zmęczeniowej. Zeszyty Naukowe Politechniki Rzeszowskiej Nr 46, Mechanika z. 17, 1988, 25-31.
3. Sahoo B., Satpathy R.K., Prasad K., Ahmed S., Kumar V.: Effect of shot peening on low cycle fatigue life of compressor disc of a typical fighter class aeroengine. Procedia Engineering 55, 2013, 144-148.
4. Jiang X.P., Man C.-S., Shepard M.J., Zhai T.: Effects of shot-peening and re-shot-peening on four-point bend fatigue behavior of Ti-6Al-4V. Materials Science and Engineering A 468-470, 2007, 137-143.
5. Zaleski K.: The effect of shot peening on the fatigue life of parts made of titanium alloy Ti-6Al-4V. Ekspoatacja i Niezawodnosc – Maintenance and Reliability, 4(44), 2009, 65-71.
6. Maawad E., Brokmeier H.-G., Wagner L., Sano Y., Genzel Ch.: Investigation on the surface and near-surface characteristics of Ti-2.5Cu after various mechanical surface treatments. Surface & Coatings Technology 205, 2011, 3644-3650.
7. Tsuji N., Tanaka S., Takasugi T.: Effect of combined plasma-carburizing and shot-peening on fatigue and wear properties of Ti-6Al-4V. Surface & Coatings Technology 203, 2009, 1400-1405.
8. Zaleski K.: Kształtowanie wybranych właściwości warstwy wierzchniej elementów metalowych w procesie dynamicznego nagniatania rozproszonego. Wydawnictwo Politechniki Lubelskiej. Lublin 2008.
9. Zaleski K.: Wpływ kulkowania wibracyjno-rotacyjnego na stan warstwy wierzchniej stali w stanie zahartowanym. [In:] Świć A.: Projektowanie i automatyzacja procesów produkcyjnych. Wydawnictwo Politechniki Lubelskiej. Lublin 2005, 13-18.

SMART MATERIALS AS MODERN ENGINEERING SUBSTANCES

Adam Ćwikła[1]

[1] Institute of Technical Sciences and Aviation, The State School of Higher Education in Chełm, Pocztowa 54, 22-100 Chełm, Poland, e-mail: acwikla@pwsz.chelm.pl

ABSTRACT

The article presents the type and use of intelligent materials in aviation and medicine. Their basic features and properties have been discussed. Moreover, the authors quote examples of different uses of intelligent materials, both the existing ones and those in design and development stages. Their use is claimed to be increasing the efficiency and reliability of processes and equipment as well as facilitates the development of new mechanisms.

Keywords*:* smart materials, piezo-ceramics, shape memory alloys, aviation, medicine.

INTRODUCTION

The development of civilization is inseparably concerned with the use of materials, hence the names of époques such as The Stone Age, The Bronze Age and the Iron Age. The advancement of science is related to the advancement in construction materials. Researchers keep developing new materials. In 1980's a group of materials called intelligent materials appeared [1]. Certain types of them had been known and used before, yet the discovery of their of specific features generated new interest in them. 1990's and the beginning of the new millennium brought a lot of interest and turbulent advancement in their research. Different names of such materials can be encountered in literature: intelligent materials, smart materials, adaptive materials, or even multifunctional materials.

Presently, so called smart solutions are more and more broadly used. Usually, it is state-of-art electronic or mechatronic systems based on the so called artificial intelligence that are be hidden behind this term. Smart materials are less often associated with intelligent solutions [1, 2].

In spite of rapid development of such materials, no single universal definition of intelligent materials has been developed. Intelligent material is most frequently defined as the material which is capable of reacting to external stimuli by making significant changes in their properties in order to provide a desired and efficient reply to the stimulus [1] (Fig. 1). With the use of mechatronical terminology – intelligent material is the one that should play the role of the sensor, processor and initiating device – an updater that translates the obtained effect. At the same time, the properties should have the features of feed back and feed forward system. It is vital to obtain the effect in real time or in approximate to real time [1]. Figure 1 presents a general idea of smart materials. In these materials such external factors as temperature, humidity, exerted pressure, electric or magnetic field change one or more physical properties of the material. Depending on the type of material and the stimulus, the following properties may change: shape, size, internal structure,

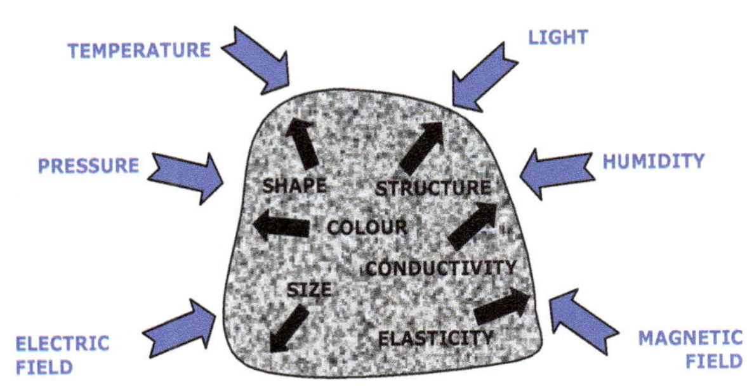

Fig. 1. The general rule of action intelligent materials

colour, electric, magnetic and thermal conductivity as well as resilience [1, 2] (Fig. 2).

Smart type materials are increasingly frequently used as propelling systems, steering elements and sensors. These elements are used to change the properties of dynamic constructions, adapt the shape to the needs and to collect information about the construction.

The branches of technology where such materials are used include medicine, aviation, sport and robotics. The article describes a few uses of smart materials for different applications in medicine and aviation.

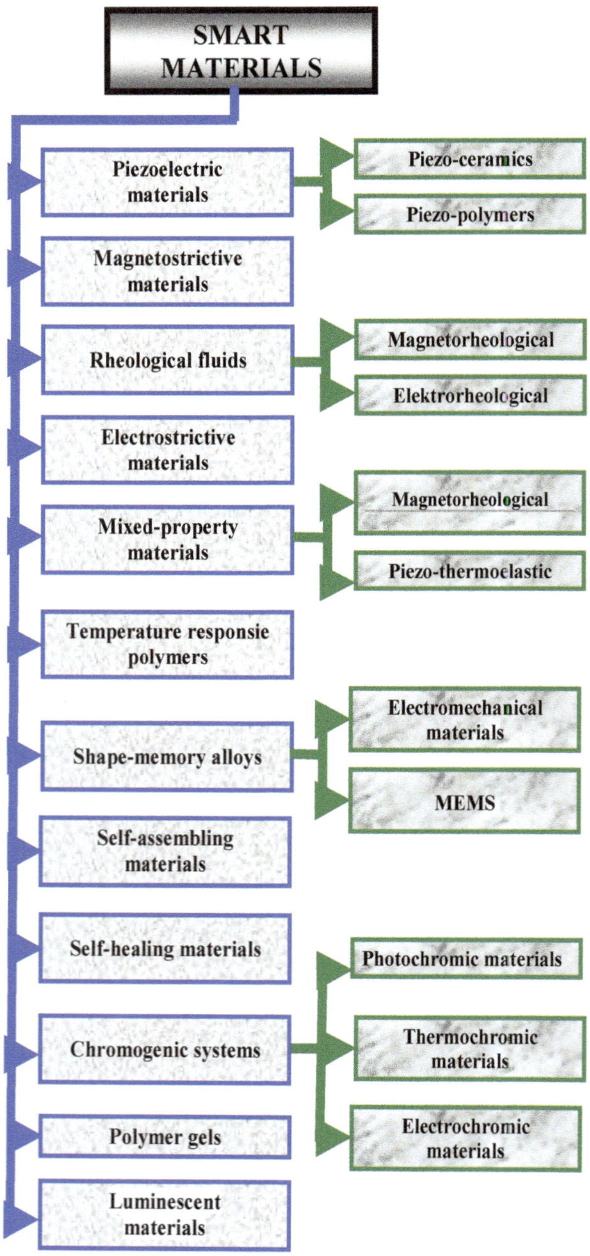

Fig. 2. Types of smart materials

SMART MATERIALS IN AVIATION

Limiting vibration in the tailplane

Modern planes, especially military constructions, reach large speeds and operate in high load factor conditions. In case of some constructions disturbed air stream flowing off the wing edges hits the tailplane causing dangerous vibrations (Fig. 3). The vibrations cause abrupt tailplane construction stability decrease, limit the time allowed in the air and increase failure rate.

Fig. 3. The plane with the "twin" tail

The system for deadening tailplane construction vibrations was ordered by US Air Force and developed by ACX company, in cooperation with Canadian and Australian government agencies [4].

The basic element that deadens vibrations are piezoelectric, formable performing elements. Other elements of the system include acceleration and deformation sensors that provide information for a digital signal processor (DSP).

The driver prepares signals, sent to executing elements in real time via high voltage amplifier that closes the driver circuit. The scheme of the system is presented in Fig. 4. The set belongs to the groups of active vibration reduction.

Piezoelectric performing elements located in the tailplane were designed and ordered for this purpose. The construction, its distribution, isolation of electric connections allow safe use of the system in military purposes, where normal way of providing high voltage required by uncovered piezoceramic materials is hazardous. The technology allows even placing the elements beneath the plane sheathing, dipped in the fuel. Another requirement that every plane element must meet is their weight and size. In

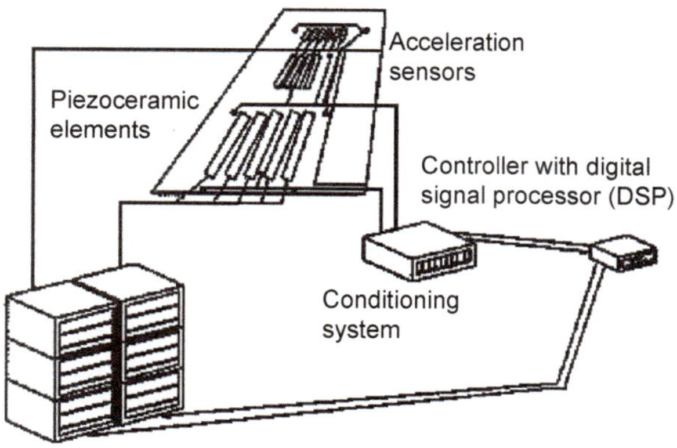

Fig. 4. Vibration Reduction Scheme of the tail of the aircraft

case of actuators, the limiting factor was weight of 10 kg for the tailplane.

With the use of Finite Element Method numerical optimisation of the actuators' thickness and distribution, in order to maximise primary controllability (bending) and secondary (turning) form of tailplane vibration. The conducted tests showed that these forms in experimental constructions are induced at 15 Hz and 45 Hz by the disturbance of air stream running off the wings. As a result of research, piezoelectric plates were located in one of three layers, 0.05 mm thick in the places of actuators on both sides of each tailplane.

The system of sensor acceleration and deformation was designed in such a way that ensures providing information to the feedback system. The data from sensors is used to compute the reply sent to the actuators.

The feedback control system with a double input and output was implemented with a digital computer operating in real time and sending the control impulses at the frequency of 5000 times per second.

Piezoelectric system of vibration damping was tested in two phases. The first stage concerned gathering data in the open (uncontrolled) system of feedback, when the signals from sensors were not sent to the actuators, but were collected for further analysis.

The analysis of signals allowed developing the rules and algorithms which should be followed in the closed system in order to minimise vibrations. The second phase concerned data implementation and testing it during flights. The tests proved over 50% reduction of strains in the core of tail structure.

Active plane manoeuvrability

To large extent plane manoeuvrability depends on the movement of the rear part of wing flaps. It is important that they are reliable and effective. In most planes they are hydraulically driven. The system uses centrally located hydraulic pumps. In order to maintain the flops' correct operations it is necessary that the collective hydraulic line connects each set of flaps. This complex system of pumps and lines is quite difficult and expensive to maintain. What is more its weight is quite significant. Different solutions that allow substituting hydraulic systems were examined. Ono of the possibilities was to use rods with shape-memory alloys (Fig. 5).

The wing flaps' construction is presented in Figure 6. The scheme of the system using rods made of smart materials are presented in Figure 7. The system with smart materials is much smaller and more efficient.

Fig. 5. The aircraft in which the wires used shape memory material

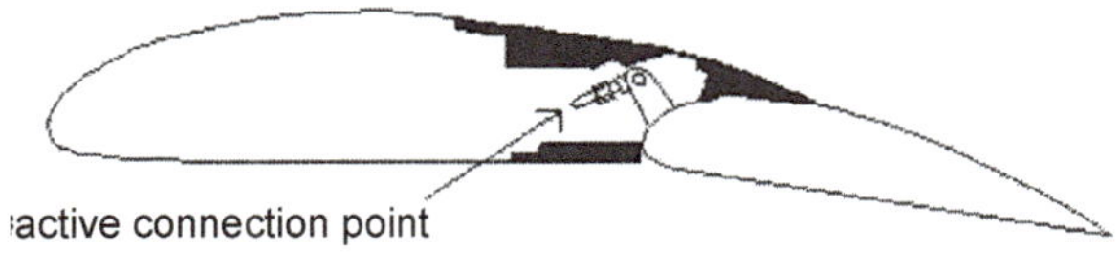

Fig. 6. Schematic of the wing flaps [7]

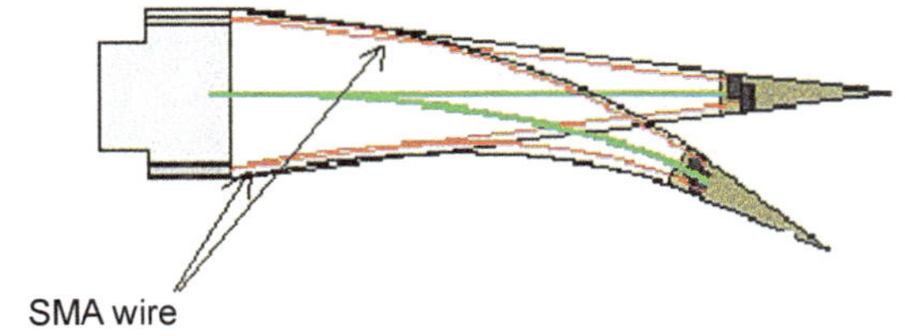

Fig. 7. Schematic of layout of wires comprising a shape memory material [7]

The rods were placed in the upper and lower part of the wing. When the flap falls down, the lower rod is heated, whereas when the flap is raise the upper part rod is heated (Fig. 7) [8].

The rod was shortened under the influence of the electric fields provided by the electric installation eliminates vast hydraulic pipe lines. Without hydraulic system the costs and time of reparations are significantly higher.

Plane constructed with smart materials

In NASA research centre in Virginia, USA scientists construct revolutionary planes with smart materials, which can adapt during the flight (Fig. 8). The planes built in the framework of Morphing project were designed as safe and quick aircrafts which, alike birds, can adapt their shapes to the conditions [5]. NASA is planning to build an unmanned combat aerial vehicle. It is planned to construct an aircraft reaching the speed of 0.5-0.9 mach with wingspan of c.a. 40 feet. The project anticipates the possibility to change the wing geometry index by 200%, the change of wing surface by 50% and the change of wing rotation by 50% and wing angle by 20 degrees. The weight of wing should not exceed the weight of a conventionally built element (Fig. 8) [7].

Additional requirements that the design is to meet is the possibility to modify air inlets into engines, in order to obtain an optimum air stream flow for various speeds and decrease the size of fuel containers after emptying them. Smart materials would also be used to construct "virtual rudders". They could „appear" when needed and „disappear" when they are no longer useful (Fig. 9). NASA mainly relies on piezoelectric materials and shape memory materials. Further plans concern the design of transportation and passenger aircrafts that can fly at the speed significantly exceeding the speed of sound. With smart materials scientists attempt to limit the so called sonic boom.

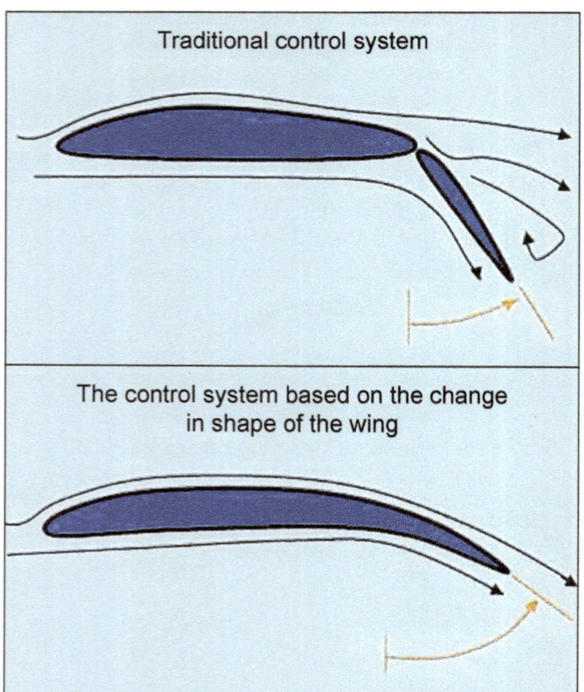

Fig. 8. Comparison of traditional wing and wing made of smart materials

Fig. 9. Models of unmanned aircraft built by NASA with intelligent materials [5]

SMART MATERIALS IN MEDICINE

Staple for osteosynthesis

Staples for osteosynthesis with shape memory are used wherever osteosynthesis is conducted with Blount staples made of traditional implant metals. The advantage in Blount staples made of shape memory alloys over traditional materials concerns the possibility to obtain close and strong connection of the broken bones. This allows quick and controlled treatment. One example is the operation of axial correction of legs (X or O curve) that requires cutting off a wedge in the knee joint (Fig. 10). The use of two, three staples, which are heated with electric current, guarantees stable osteosynthesis [1, 3].

Another example of using the staples is the operation of foot in case of immobilised crurotalar joint (Fig. 11). The use of staples made of shape memory alloys ensures good clamp of resection surfaces. Following the resection and foot correction the staple, alike the traditional one, is adjusted to the bone and heated with two electrodes, until strong connection is obtained. In order t to stabilise the joint 2–3 staples are used.

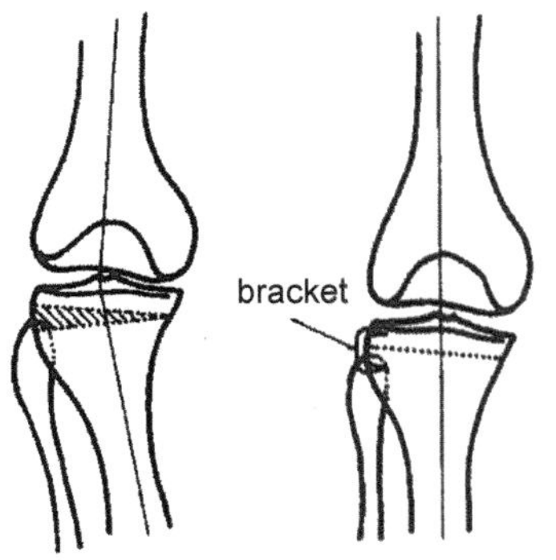

Fig. 10. The use of staples for osteosynthesis after knee surgery

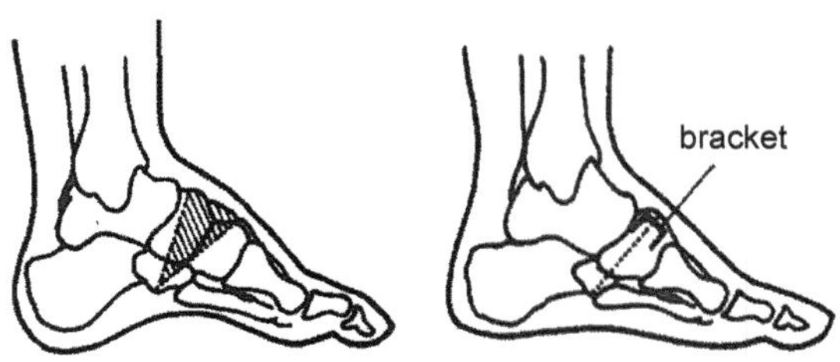

Fig. 11. The use of staples for osteosynthesis for ankle surgery

For clinical purposes staples with transformation temperatures ranging from 42 to 46 °C are used. Prior to delivering to clinics the staples are the subjected to deformations and heating in order to test their behaviour. The staples are delivered in special plastic bags and sterilised with γ radiation.

It is significant to know the strains in broken bones. For this purpose a special device was designed to measure the strength exerted by staples' arms during heating. Due to the construction of the device which does not allow shape changes, strains were induces with the increase of temperature. At 60 °C the strength of 1160 N is induced. Further heating up to 100 °C raises the strength to 2500 N. During the operation the staple is heated up to 55 °C. Its slow cooling down to body temperature causes the decrease of strength. At 37 °C both staples exert the power of 880 N, therefore, single staple exerts the power of 440 N. The value refers to a staples sized 3.5×4.5 mm = 15.75 mm^2. For larger staples with the cross section of 17.5 mm^2 analogical values of strain were obtained.

Staples for curing broken ribs

Another example of the use of shape memory metals in surgery, which makes the operation significantly easier, is the staples made of perforated metal sheets for treating broken ribs (Fig. 12). In subsequent pictures the state after tempering (*a*), deforming (*b*) is presented. In such a state the staple is placed onto a broken rib and heated (*c*). As a consequence, the staple return to the original shape and connects broken elements of the rib (*d*). One advantage of such an operation is not only simplifying the operation but also limiting the springiness of the connection.

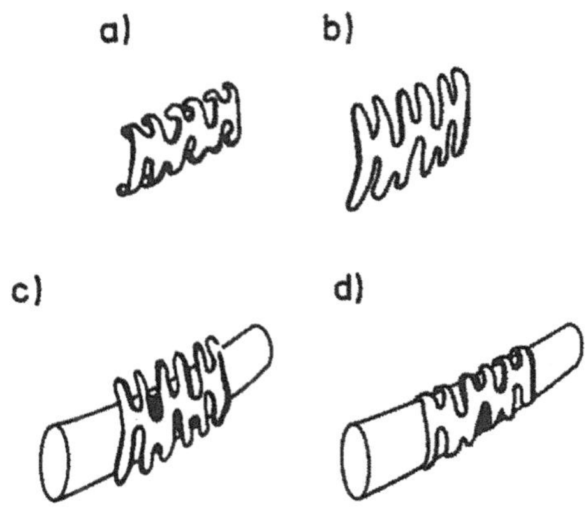

Fig. 12. The use of staples for the treatment of fractures of the ribs [1, 3]

Osteosynthesis plates

In extreme situations of broken bones osteosynthesis plates area used. The plates are attached to both sides of the broken bone with screws. This hastens the fracture treatment [1, 4] and balances the distribution of compressive strains onto the connection surface. In order to obtain even distribution of compressing strains plates with oblique holes or pitches for screws with ball-shaped heads. Osteosynthesis plates made of shape memory alloys have an advantage of tightening large cracks in fractures, what cannot be done with conventional plates. The plate in initial state is lengthened by stretching and attaching to the fracture. While being heated, it shrinks and eliminated the crack sized up to 5 mm. At the moment when both elements connect, the operator cuts of the heat inflow and the position is maintained. In case heat inflow is not cut off, harmful concentration of strains in the fracture from the side of the plate is observed.

Intramedullary rod

In order to treat closed spiral fractures, e.g. in thigh the rods are forced into the medullary cavity. A rod of larger diameter than the cylindrical hole is placed into the bone what stabilises the fracture. Along with numerous advantages, there is one disadvantage. It takes a lot of strength to force the rod into the bone. Removing the rods generates further problems, including leaving the rod as a permanent implant. Such inconveniences disappear when a tube made of NiTi alloy is used instead of the implant. The tube increases its diameter as a result of heating, and decreases when cooled down below the temperature of human body. The rule of using such a tube is presented in Figure 13. 10 mm tube (6), made of metal sheet, narrows down to 8.5 mm (1) and is placed into the hole in the broken bone (1). As a result, it increases its size up to 9.5 mm, what tightens the broken bone (5). After the end of treatment, the place is cooled down in with distilled water or physiological liquid to the temperature below M_s and tube shrinkage down to 9 mm (4). Finally, the tube can be removed without using strength [1, 3].

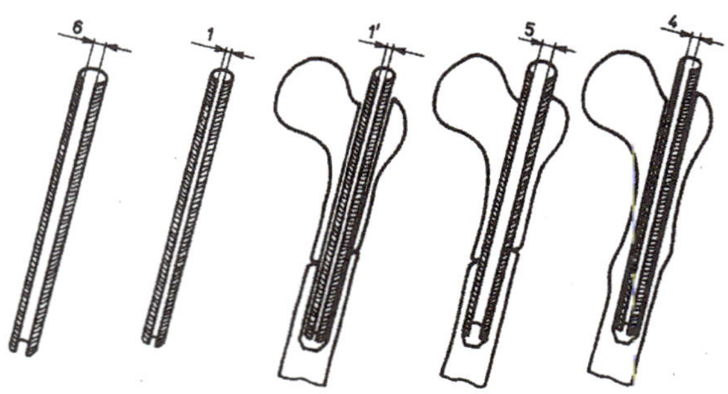

Fig. 13. The use of sleeves for fracture treatment

Blood clot filter

The purpose of this filter, which is implanted into vena cava, is to catch blood clots, caused by various injuries, which can migrate to heart. Filter of different shapes are made of NiTi wire sized 0.5 mm in diameter and 200 mm long. Tempered in icy water, the shaped filter is straightened (deformed) and inserted into the vessel with the use of a catheter. At body temperature a reverse process takes place and the wire return to its original shape – a filter-mesh sized 2 mm. This allows efficient capture of harardous clots. Scientists' experience with tests on dogs is encouraging enough to run clinical tests.

Dental braces

Dental braces are made of NiTi alloys. They are attached to teeth in order to correct the dental displacements (Fig. 14). Such material is replacing stainless steel wires that had been used before. NiTi wires in martensic state are flattened (30% deformation) as a result of their rolling. In this state of pseudo-resiliency the wires are characterised by high resilience.

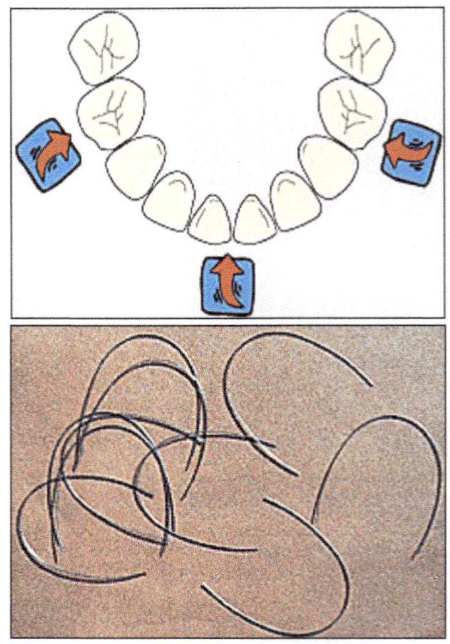

Fig. 14. The use of nitinol as orthodontic wire

Deformed by 90° they return to the original shape, unlike stainless steel wires, which remain curved at 55°. One important advantage of nickel titanium wire, in comparison to steel wires, is the value of Young's module that is 6 times lower, thus for the same bend significantly lower strengths are employed (Fig. 15). From clinical point of view, this means that nickel titanium wire exerts lower and more stable strength on the bite than it is in case of steel.

The use of NiTi wire in orthodontics (Fig. 16) has the following advantages:

- requires smaller bend curve,
- limits the time of operation,
- limits the time of treatment,
- limits tiresomeness for patients,
- lower number of visits at orthdent's,
- higher comfort for patients, due to lower values of extorted strengths onto the teeth.

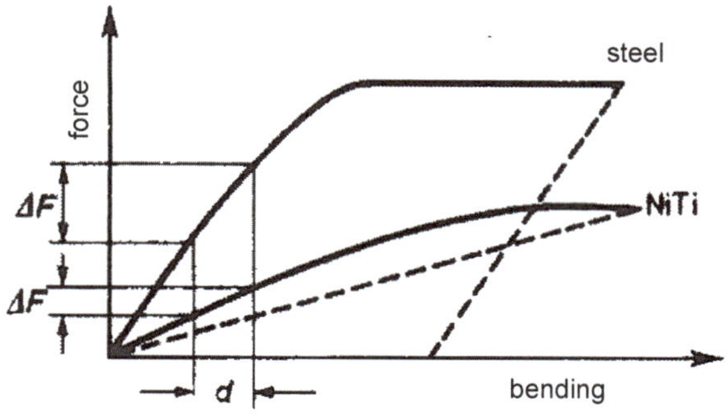

Fig. 15. Deformation forces for wires from NiTi and of austenitic stainless steel [1, 3]

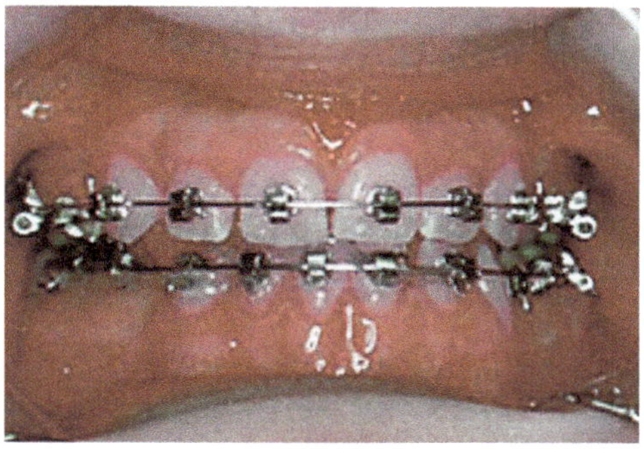

Fig. 16. Dental braces of nitinol

CONCLUSIONS

Increasingly high competition and higher expectations from consumers limit the use of classical construction materials. Broader use of modern, smart materials is expected. Their use improves efficiency and reliability of devices and allows designing new equipment that had been impossible to construct before. The use of these materials is more and more evident in such industries as aviation and medicine.

With their property of dimming mechanical vibrations, smart materials are used to construct rudder or carrying surfaces, but also to produce elements subjected to variable loads in medicine. Due to their simplicity of use and low cost they should also be used in other branches of industry.

REFERENCES

1. Ćwikła A.: Medyczne zastosowania materiałów inteligentnych. VII Konferencja informatyki stosowanej, Chełm 2008.
2. Giergiel J., Ćwikła A.: Materiały piezoceramiczne w aktywnej redukcji drgań i hałasu. Seminarium: Stan obecny i perspektywy rozwoju mechaniki i budowy maszyn, Chełm 2007, Scientific Bulletin of Chełm, Section of Technical Sciences 1, 2007: 51-70,
3. Bojarski Z., Morawiec H.: Metale z pamięcią kształtu. Wydawnictwo Naukowe PWN, Warszawa 1989.
4. www.aviationnow.com
5. www.matint.pl
6. http://www.continuum-dynamics.com/solution-me-sma.html
7. http://www.mitre.org/news/events/tech03/briefings/modeling_simulation/odonnell.pdf
8. http://www3.ntu.edu.sg/mae/Admin/Divisions/aerospace/aero_lect_abstract_liuyong.pdf
9. http://www.continuum-dynamics.com/solution-ae-sma.html

SELECTED PHYSICAL PROPERTIES OF EXTRUDED COMPOSITES TYPE OF POROUS PVC-METAL

Aneta Tor-Świątek[1]

[1] Department of Polymer Processing, Lublin University of Technology, Nadbystrzycka 36, 20-618 Lublin, Poland, e-mail: a.tor@pollub.pl

ABSTRACT

The article presents studies of selected physical and mechanical properties of hybrid materials type of polymer-metal. In the frame of this work modification of PVC with the iron and copper powder in amount of 0, 1.5 and 3% and blowing agent in amount of 0, 0.5, 1% was done. Extrudates in a form of pipe were tested to determine density, porosity, maximum tensile stress, stress at break, modulus of elasticity and elongation with break. The samples were also observed in a microscope. The studies have shown significant influence of the added components on the properties tested.

Keywords: poly(vinyl chloride), extrusion, fillers, metallic powders.

INTRODUCTION

Polymer composites are now widely used in construction. They form a group of materials whose properties depend on the characteristics of individual components, the degree of miscibility, volume fraction, technology and processing conditions. A commonly used method of physical modification is the addition to polymeric materials auxiliary agents in the form of powder of fibrous fillers (e.g. calcium carbonate, magnesium oxide, beryllium oxide, talc, chalk and metal oxides such as aluminum, brass, nickel, copper, iron) and also blowing agents (porophors). Fillers in the form of powders are added to polymers to improve mechanical, dielectric, thermal, chemical and processing (better flow, decreased shrinkage) properties.

A new, interesting solution is hybrid materials type of polymer-metal. They are mostly produced by injection molding [1] and metal components usually take form of mould insert. Hybrid elements combine the advantageous properties of metals, such as high rigidity and mechanical strength with the preferred features of polymeric materials, such as easy processing, low specific gravity and corrosion resistance [1, 2]. Another method of preparing polymer –metal composites is adding a metal powder in the extrusion process.

Adding metallic powders to the polymer causes significance changes in properties and structure of the composition. Adding aluminum and brass increases acoustic conductivity, nickel powder improves impact resistance, whereas copper improves stability. Adding metallic powders also causes different specialist properties, such as electric, magnetic and tribological properties. They also increase barrier properties and strength properties such as hardness and tensile strength [3]. Conversely, changing the physical structure of the polymer materials form solid to porous results in a significant reduction of gravity of the product [6], the change of individual properties [4] and physical and geometric structure [5, 6], what have an influence on lower manufacturing costs.

Literature analyses have shown that research of hybrid materials containing polymer – metal compositions concern polymer in solid form [4, 8, 9]. In all analyzed cases the metal had a form of fibers [10, 3, 11] and was usually made in injection moulding process or plating [4, 12, 13, 14].

The aim of the study was to analyse the influence of adding a blowing agent and the filler in the form of a powder iron and copper on selected physical properties of the obtained composite.

MATERIALS AND METHODS

A modification of plastified PVC Alfavinyl GFM/4-31-TR was done. The plastic has density 1210 kg/m^3 (in 23 °C) and tensile strength 21 MPa. The plastic was modified with iron and copper powder, which was added in the amount of 0% and 3% and mixture of Fe i Cu in the amount of 1.5%. The characteristic of metallic powders used was shown in Table 1. In addition, the PVC was modified with a blowing agent (BA) in a form of Hydrocerol 530 granulate (prod. Clariant). Porophor was dosaged in the amounts of 0%, 0.5% i 1%. The prepared mixtures were extruded in a technological line of a profile extrusion (by free cellular method) fitted with a single screw extrusion T32-25, extrusion head to pipe extrusion, cooling device and receiving device. The process was done with a constant screw rotational speed of 70 rpm. The other process conditions were: temperature of selected plasticizing system zones amounting to 110, 120, 130, 150 °C respectively, the temperature of extruder head equal was 165 °C and the temperature of cooling factor amount 18 ±2 °C. In the frame of extrusion the pipe with 19 mm external diameter and 3 mm thickness wall was obtained.

In the framework of the research program of physical properties the measurements of density, porosity, tensile strength and elongation at break was done. The visualization of the structure of the surface and cross section of the obtained extrudates was also done. The research of density and porosity was done in accordance to PN-EN ISO 845:2000. The research of mechanical properties was made on tensile machine Zwick/Roel Z010 according to PN-EN ISO 527-1:1998 and PN-EN ISO 1798:2001. The structure observations of the extrudates was made in a passing light on metallographic microscope Nikon ECLIPSE LV100ND with DS-U3 Digital Camera Control Unit.

RESULTS AND DISCUSSION

Results of selected research was show in Table 2 and 3. The studies have shown the effect of the mechanical properties of metal powders added to the measured factors. The addition of metal powder resulted in an increased density by about 3% for the solid extrudates with an iron content of 3% and 5% for the solid extrudates with 3% of copper. Further adding a blowing agent resulted in a significant decrease in the density of the

Table 1. Characteristics of powder fillers used (producer data)

Type of filler	Chemical composition, %	Bulk density, kg/m^3	Granulation	Geometrical form
Copper powder (Cu) Lt 16 (Stanchem Sp. j.)	Cu – 99.7 O_2 – 0.3	1660	>63 µm – 2% >45 µm – 10% <45 µm – 88%	Irregular
Iron powder (Fe) (Stanchem Sp. j.)	O_2 – 0.1 C – 0.007 S – 0.01 Fe – 99.8	3200	+0,150 mm – 19% +0,045 mm – 98%	Irregular

Table 2. Mean values (from 10 measurements) selected physical properties of obtained composite PVC porous-metal

Dosage of porophor BA, %	Dosage of iron Fe, %	Dosage of copper Cu, %	Density, kg/m^3	Porosity, %
0	3	0	1247.10	0
0.5	3	0	1017.37	18
1.0	3	0	1013.25	24
0	0	3	1265.77	0
0.5	0	3	1019.37	19
1.0	0	3	1017.82	26
0	1.5	1.5	1260.32	0
0.5	1.5	1.5	1018.13	19
1.0	1.5	1.5	1016.63	25

Table 3. Mean values (from 10 measurements) of mechanical properties at tension of composite PVC porous-metal

Dosage of porophor BA, %	Dosage of iron Fe, %	Dosage of copper Cu, %	Maximum tensile stress, MPa	Stress at break, MPa	Modulus of elasticity, MPa	Elongation with break, %
0	3	0	14.9	14.8	8.01	170
0.5	3	0	12.2	13.7	12.6	140
1,0	3	0	8.55	8.34	6.75	120
0	0	3	19.9	19.8	13.6	180
0.5	0	3	9.85	9.57	8.76	130
1.0	0	3	7.34	6.96	6.54	120
0	1.5	1.5	17.4	17.4	11.4	150
0.5	1.5	1.5	14.7	14.2	10.9	140
1.0	1.5	1.5	7.60	7.09	6.22	120

extrudates with Fe from an average of 16% for dispensing 0.5% BA, and 17% for 1% BA. The values obtained for the density and porosity of the extrudates containing the blowing agent and Cu are greater average of 2%, which may be due to the larger size of the copper particles. In contrast, a slight decrease of the density of the extrudates with 1% BA in relation to the extrudates of 0.5% BA may result of insufficient expansion of the pores, which prevented the added metal powder.

The study of the mechanical properties showed a significant decrease in the tested factors by the addition of auxiliary agents. Tensile stress and stress at break decreased by 30% after the addition of Fe and 5% after the addition of Cu. Adding a blowing agent resulted in a further reduction of the mechanical strength of the extrudates by 42% for samples with 0.5% BA and 3% Fe and about 53% for samples with 1% BA and 3% Cu. The lowest values of mechanical examined is observed with the addition of 1% of blowing agent and 3% of copper.

Modification of PVC with a mixture of iron and copper powder allowed to obtain intermediate values in relation to the results of the extrudates and with iron and the extrudates with copper.

In the result of microscopic observations images of surface and cross sections morphology of polymer composite type of porous PVC-metal was obtained. Sample images of selected extrudates are shown in Figures 1 and 2. The observations showed the homogeneity of the physical structure of the resulting product. Visible is a relatively uniform distribution of added blowing agent in the entire cross-section of the pipe. Distribution of added metal powders is not so uniform. The largest amount of fillers is located directly on the surface of the section. It is also unsatisfactory that the surface of the extrudate, which is due to the addition of the foaming agent significantly increased its roughness. Therefore, it becomes necessary to use more intensive cooling of the extrudates, e.g. through the use of a calibration device.

CONCLUSIONS

When combining and processing components with known properties it is difficult to predict all the properties of the obtained composite porous polymer-metal. The study indicates the possibility of simultaneous modification of polymeric recipients in a form of blowing agent and powder metals. The studies have shown that the addition of the aforementioned ingredients significantly change the characteristics and structure of the resulting polymer composition. Adding a blowing agent significantly reduces the molecular weight of the material, which is reduced by the addition of copper and iron powder. Noteworthy is the fact that the highest distribution of the metal powder takes place mainly on the outer surface of the composite, while the porous structure exists in the whole cross-section of extrudates. The analyzed results show that properly selected ingredients and processing technologies allow to obtain composites with modified physical and mechanical properties.

REFERENCES

1. Kaczmar J., Wróblewski R., Nakonieczny L., Iwko I.: Wytwarzanie i właściwości elementów hybrydowych typu metal-tworzywo polimerowe. Polimery 53, 2008, 7–8, 519–525.

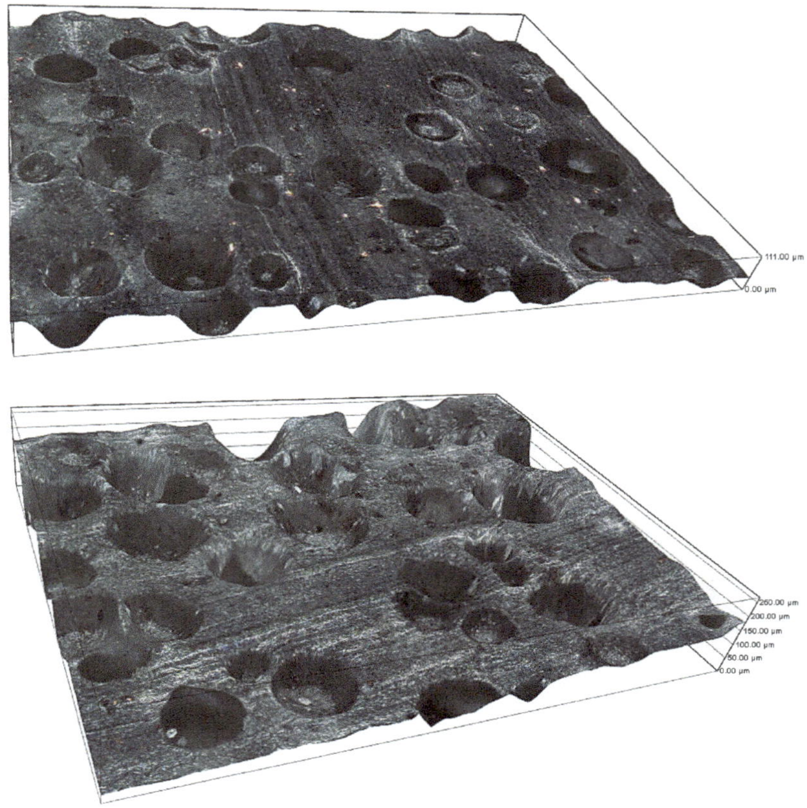

Fig. 2. Topography of cross section of composite PVC porous-metal: a) composite with 0.5% BA and 3%FE, b) composite with 1% BA and 1.5%Cu and 1.5 %Fe

Fig. 1. Image of surface morphology and shape of powder iron and copper: a) composite with 0% BA and 1.5%Fe and 1.5%Cu, b) composite with 0.5% BA and 3%Fe c) composite with 1% BA and 1.5%Cu and 1.5%Fe

2. Gnatowski A.: Wpływ napełniacza na właściwości wybranych mieszanin polimerowych. Kompozyty (Composites) 5, 2005, 2, 63–68.
3. Panwar V., Cha K., Park J., Park S.: High actuation response of PVDF/PVP/PSSA based ionic polymer-metal composite actuactor. Sensors and Actuators B: Chemical 2012, 161, 460–470.
4. Tor-Świątek A. Samujlo B.: Use of thermovision research to analyze the thermal stability of microcellular extrusion process of poly(winyl chloride). Eksploatacja i Niezawodnosc – Maintenance and Reliability 15, 2013, 1, 58–61.
5. Tor-Świątek A., Sikora J.W.: Analiza wpływu modyfikacji poli(chlorku winylu) mokrosferami na strukturę geometryczną wytłoczyny. Przemysł Chemiczny 2013, 4, 538–541.
6. Garbacz T.: Właściwości trójwarstwowych powłok z PVC wytworzonych w procesie współwytłaczania mikroporującego. Polimery 56, 2011, 2, 129–134.
7. Gościański M., Maciejewski H., Guliński J., Leda H.: Badanie wpływu modyfikacji mineralnych napełniaczy proszkowych silanami na wybrane właściwości mechaniczne liniowego polietylenu małej gęstości. Polimery 49, 2004, 1, 15–22.
8. Panwar V., et al.: Dynamic mechanical, electric al and actuation properties of ionic polimer metal composites using PVDF/PVP/PSSA blend membranes. Materials Chemistry and Physics 135, 2012, 2–3, 928–937.
9. Jo C. et al.: Recent advances in ionic polymer-metal composite actuators and their modeling and applications. Progress in Polymer Science 38, 2013, 7, 1037–1066.
10. Kim S. J. et al.: A bio-inspired multi degree of freedom actuator based on novel cylindrical ionic polymer-metal composite materials. Robotics and Autonomous Systems 63, 2014, 1, 53–60.
11. Królikowski W., Rosłaniec Z.: Nanokompozyty polimerowe. Kompozyty (Composites) 4, 2004, 9, 3–16.
12. Dobrzański L.A., Drak M.: Materiały kompozytowe o osnowie polimerowej zbrojone cząstkami stopu ze sproszkowanej szybko chłodzonej taśmy Nd-Fe-B. Journal of Alloys and Compounds 449, 2008, 1, 88–92.
13. Tasdemir M., Gulsoy H.O.: Mechanical properties of polymers filled with iron powder. International Journal of Polymeric Materials and Polymeric Biomaterials 57, 2008, 3, 258–265.
14. Steiner G., Gerndorf R.: Hart-Weich-Verbunde mit Metallinsert: Kunststoff-Metal-Hybride der neuen Generation. Kunststoffe 2004, 7, 83–87.

Permissions

All chapters in this book were first published in ASTRJ, by Society of Polish Mechanical Engineers and Technicians; hereby published with permission under the Creative Commons Attribution License or equivalent. Every chapter published in this book has been scrutinized by our experts. Their significance has been extensively debated. The topics covered herein carry significant findings which will fuel the growth of the discipline. They may even be implemented as practical applications or may be referred to as a beginning point for another development.

The contributors of this book come from diverse backgrounds, making this book a truly international effort. This book will bring forth new frontiers with its revolutionizing research information and detailed analysis of the nascent developments around the world.

We would like to thank all the contributing authors for lending their expertise to make the book truly unique. They have played a crucial role in the development of this book. Without their invaluable contributions this book wouldn't have been possible. They have made vital efforts to compile up to date information on the varied aspects of this subject to make this book a valuable addition to the collection of many professionals and students.

This book was conceptualized with the vision of imparting up-to-date information and advanced data in this field. To ensure the same, a matchless editorial board was set up. Every individual on the board went through rigorous rounds of assessment to prove their worth. After which they invested a large part of their time researching and compiling the most relevant data for our readers.

The editorial board has been involved in producing this book since its inception. They have spent rigorous hours researching and exploring the diverse topics which have resulted in the successful publishing of this book. They have passed on their knowledge of decades through this book. To expedite this challenging task, the publisher supported the team at every step. A small team of assistant editors was also appointed to further simplify the editing procedure and attain best results for the readers.

Apart from the editorial board, the designing team has also invested a significant amount of their time in understanding the subject and creating the most relevant covers. They scrutinized every image to scout for the most suitable representation of the subject and create an appropriate cover for the book.

The publishing team has been an ardent support to the editorial, designing and production team. Their endless efforts to recruit the best for this project, has resulted in the accomplishment of this book. They are a veteran in the field of academics and their pool of knowledge is as vast as their experience in printing. Their expertise and guidance has proved useful at every step. Their uncompromising quality standards have made this book an exceptional effort. Their encouragement from time to time has been an inspiration for everyone.

The publisher and the editorial board hope that this book will prove to be a valuable piece of knowledge for researchers, students, practitioners and scholars across the globe.

List of Contributors

Myron Czerniec
Institute of Technological Systems of Information, Lublin University of Technology, Nadbystrzycka 36, 20-612 Lublin, Poland

Jan Varga
Department of Technologies and Materials, Faculty of Engineering, Technical University in Košice, Mäsiarska 74, 040 01 Košice, Slovak Republic

Jozef Stahovec
Department of Technologies and Materials, Faculty of Engineering, Technical University in Košice, Mäsiarska 74, 040 01 Košice, Slovak Republic

Jozef Beno
Department of Technologies and Materials, Faculty of Engineering, Technical University in Košice, Mäsiarska 74, 040 01 Košice, Slovak Republic

Marek Vrabeľ
Department of Technologies and Materials, Faculty of Engineering, Technical University in Košice, Mäsiarska 74, 040 01 Košice, Slovak Republic

Jakub Matuszak
Department of Production Engineering, Mechanical Engineering Faculty, Lublin University of Technology 36 Nadbystrzycka Str., 20-618 Lublin, Poland

Kazimierz Zaleski
Department of Production Engineering, Mechanical Engineering Faculty, Lublin University of Technology 36 Nadbystrzycka Str., 20-618 Lublin, Poland

Martin Novák
Faculty of Production Technology and Management, Jan Evangelista Purkyně University in Ústí nad Labem, Pasteurova 1, 400 96 Ústí nad Labem, Czech Republic

Nataša Náprstková
Faculty of Production Technology and Management, Jan Evangelista Purkyně University in Ústí nad Labem, Pasteurova 1, 400 96 Ústí nad Labem, Czech Republic

Jerzy Józwik
Faculty of Mechanical Engineering, Lublin University of Technology, 36 Nadbystrzycka Str., 20-618 Lublin, Poland

Agnieszka Benduch
Department of Industrial Furnaces and Environmental Protection, Technical University of Czestochowa, Al. Armii Krajowej 19, 42-200 Częstochowa, Poland

Rafał Wyczółkowski
Department of Industrial Furnaces and Environmental Protection, Technical University of Czestochowa, Al. Armii Krajowej 19, 42-200 Częstochowa, Poland

Jerzy Józwik
Department of Production Engineering, Mechanical Engineering Faculty, Lublin University of Technology 36 Nadbystrzycka Str., 20-816 Lublin, Poland

Paweł Pietras
Graduate student of Mechanical Engineering Faculty, Lublin University of Technology, Poland

Umit Huner
Faculty of Engineering, Department of Mechanical Engineering, Trakya University, Edirne 22030, Turkey

Kazimierz Zaleski
Lublin University of Technology, 36 Nadbystrzycka Str., 20-618 Lublin, Poland

Agnieszka Skoczylas
Lublin University of Technology, 36 Nadbystrzycka Str., 20-618 Lublin, Poland

V. I. Marchuk
Lutsk National Technical University, 75 Lvivska Str., Lutsk 43018, Ukraine

L. M. Ravenets
Lutsk National Technical University, 75 Lvivska Str., Lutsk 43018, Ukraine

S. V. Marchuk
Lutsk National Technical University, 75 Lvivska Str., Lutsk 43018, Ukraine

Antoni Świć
Institute of Technological Information Systems, Lublin University of Technology, Nadbystrzycka 36, 20-618 Lublin, Poland

Tomasz Jachowicz
Department of Polymer Processing, Mechanical Engineering Faculty, Lublin University of Technology, 36 Nadbystrzycka Str., 20-618 Lublin, Poland

Janusz W. Sikora
Department of Polymer Processing, Mechanical Engineering Faculty, Lublin University of Technology, 36 Nadbystrzycka Str., 20-618 Lublin, Poland

Jerzy Józwik
Department of Production Engineering, Mechanical Engineering Faculty, Lublin University of Technology, Nadbystrzycka 36, 20-816 Lublin, Poland

Dariusz Mika
Center for the Study of Engineering, Institute of Technical Sciences and Aviation, The State School of Higher Education in Chelm, Depułtycze Królewskie 55, 22-100 Chelm, Poland

Piotr Penkała
Institute of Technical Sciences and Aviation, The State School of Higher Education in Chełm, Pocztowa 54, 22-100 Chełm, Poland

Beata Pławaś
Institute of Technical Sciences and Aviation, The State School of Higher Education in Chełm, Pocztowa 54, 22-100 Chełm, Poland

Kazimierz Zaleski
Department of Mechanical Engineering, Lublin University of Technology, 36 Nadbystrzycka Str., 20-618 Lublin, Poland

Agnieszka Skoczylas
Institute of Technical Sciences and Aviation, The State School of Higher Education in Chełm, Pocztowa 54, 22-100 Chełm, Poland

Fatih Karaçam
Department of Mechanical Engineering, Faculty of Engineering, Trakya University, Edirne 22030, Turkey

Taner Timarci
Department of Mechanical Engineering, Faculty of Engineering, Trakya University, Edirne 22030, Turkey

Paweł Mazur
Institute of Materials Science and Engineering, Cracow University of Technology, Aleja Jana Pawła II 37, 31-155 Kraków, Poland

Janusz Mikuła
Institute of Materials Science and Engineering, Cracow University of Technology, Aleja Jana Pawła II 37, 31-155 Kraków, Poland

Jerzy S. Kowalski
Institute of Materials Science and Engineering, Cracow University of Technology, Aleja Jana Pawła II 37, 31-155 Kraków, Poland

Mariusz Krakowski
Department of Computer Modelling and Metal Forming Technologies, Mechanical Engineering Faculty, Lublin University of Technology, 36 Nadbystrzycka Str., 20-618 Lublin, Poland
The Communication Equipment Manufacturing „PZL-Świdnik" S.A., Lotników Polskich 1, 21-045 Świdnik, Poland

Jarosław Bartnicki
The Communication Equipment Manufacturing „PZL-Świdnik" S.A., Lotników Polskich 1, 21-045 Świdnik, Poland

Tomasz Urbański
Department of Ship Structure, Mechanics and Fabrication, Faculty of Maritime Engineering and Transport, West Pomeranian University of Technology in Szczecin, Poland

Katarzyna Kozłowicz
Department of Refrigeration and Food Industry Energetics, University of Life Sciences in Lublin, Doświad-czalna 44, 20-280 Lublin, Poland

Zbigniew Pater
Mechanical Engineering Faculty, Lublin University of Technology, 36 Nadbystrzycka Str., 20-618 Lublin, Poland

Arkadiusz Tofil
Mechanical Engineering Faculty, Lublin University of Technology, 36 Nadbystrzycka Str., 20-618 Lublin, Poland

Mirosław Szala
Department of Materials Engineering, Faculty of Mechanical Engineering, Lublin University of Technology, Nadbystrzycka 36, 20-618 Lublin

Tadeusz Hejwowski
Department of Materials Engineering, Faculty of Mechanical Engineering, Lublin University of Technology, Nadbystrzycka 36, 20-618 Lublin, Poland

Iwona Lenart
Student of Mechanical Engineering Faculty, Lublin University of Technology, Nadbystrzycka 36, 20-618 Lublin, Poland

Rafał Wyczółkowski
Department of Industrial Furnaces and Environmental Protection, Częstochowa University of Technology, Al. Armii Krajowej 19, 42-200 Częstochowa, Poland

Agnieszka Benduch
Department of Industrial Furnaces and Environmental Protection, Częstochowa University of Technology, Al. Armii Krajowej 19, 42-200 Częstochowa, Poland

José Miguel Mendivil-Escalante
Universidad Autónoma de Sinaloa, Facultad de Ingeniería Mochis, Los Mochis, Sinaloa, Mexico

Jorge Luis Almaral-Sánchez
Universidad Autónoma de Sinaloa, Facultad de Ingeniería Mochis, Los Mochis, Sinaloa, Mexico

José Manuel Gómez-Soberón
Universidad Politécnica de Cataluña, Departamento de Construcciones II-EPSEB, Barcelona, Spain

Susana Paola Arredondo-Rea
Universidad Autónoma de Sinaloa, Facultad de Ingeniería Mochis, Los Mochis, Sinaloa, Mexico

Ramón Corral-Higuera
Universidad Autónoma de Sinaloa, Facultad de Ingeniería Mochis, Los Mochis, Sinaloa, Mexico

Andrés Castro-Beltrán
Universidad Autónoma de Sinaloa, Facultad de Ingeniería Mochis, Los Mochis, Sinaloa, Mexico

Francisca Guadalupe Cabrera-Covarrubias
Universidad Politécnica de Cataluña, Escuela de Ingenieros Caminos Canales y Puertos, Barcelona, Spain

Anna Dziubińska
Faculty of Mechanical Engineering, Lublin University of Technology, 36 Nadbystrzycka Str., 20-618 Lublin, Poland

Andrzej Gontarz
Faculty of Mechanical Engineering, Lublin University of Technology, 36 Nadbystrzycka Str., 20-618 Lublin, Poland

Grzegorz Winiarski
Faculty of Mechanical Engineering, Lublin University of Technology, 36 Nadbystrzycka Str., 20-618 Lublin, Poland

Rafał Grzejda
Faculty of Mechanical Engineering and Mechatronics, West Pomeranian University of Technology, Szczecin, 19 Piastów Ave., 70-310 Szczecin, Poland

Anna Rudawska
Department of Production Engineering, Faculty of Mechanical Engineering, Lublin University of Technology, Nadbystrzycka 36, 20-618 Lublin, Poland

Oladipupo Olaosebikan Ogunleye
Department of Chemical Engineering, Faculty of Engineering and Technology, Ladoke Akintola University of Technology, P.M.B. 4000, Ogbomoso, Nigeria

Taofeeq Olalekan Salawudeen
Department of Chemical Engineering, Faculty of Engineering and Technology, Ladoke Akintola University of Technology, P.M.B. 4000, Ogbomoso, Nigeria

Muyibi A. Suleyman
Department of Biotechnology Engineering, Kulliyyah of Engineering, International Islamic University Malaysia, Jalan Gombak, Kuala Lumpur, Malaysia

Yusof Faridah
Department of Biotechnology Engineering, Kulliyyah of Engineering, International Islamic University Malaysia, Jalan Gombak, Kuala Lumpur, Malaysia

Kazimierz Zaleski
Faculty of Mechanical Engineering, Lublin University of Technology, 36 Nadbystrzycka Str., 20-618 Lublin, Poland

Stanisław Bławucki
Faculty of Mechanical Engineering, Lublin University of Technology, 36 Nadbystrzycka Str., 20-618 Lublin, Poland

Agnieszka Benduch
Department of Industrial Furnaces and Environmental Protection, Faculty of Process & Material Engineering and Applied Physics, Technical University of Częstochowa, Al. Armii Krajowej 19, 42-200 Częstochowa, Poland

Jarosław Boryca
Department of Industrial Furnaces and Environmental Protection, Faculty of Process & Material Engineering and Applied Physics, Technical University of Częstochowa, Al. Armii Krajowej 19, 42-200 Częstochowa, Poland

Waldemar Witkowski
Rzeszow University of Technology, Powstańców Warszawy 12, 35-959 Rzeszów, Poland

Andrzej Kubit
Department of Manufacturing Processes and Production Engineering, The Faculty of Mechanical Engineering and Aeronautics, Rzeszow University of Technology, Powstanców Warszawy 12, 35-959 Rzeszow, Poland

Barbara Cecińska
Department of Manufacturing Processes and Production Engineering, The Faculty of Mechanical Engineering and Aeronautics, Rzeszow University of Technology, Powstanców Warszawy 12, 35-959 Rzeszow, Poland

Kamil Drozd
Department of Manufacturing Processes and Production Engineering, The Faculty of Mechanical Engineering and Aeronautics, Rzeszow University of Technology, Powstanców Warszawy 12, 35-959 Rzeszow, Poland

Ireneusz Zagórski
Department of Production Engineering, Mechanical Engineering Faculty, Lublin University of Technology, 36 Nadbystrzycka Str., 20-618 Lublin, Poland

Józef Kuczmaszewski
Department of Production Engineering, Mechanical Engineering Faculty, Lublin University of Technology, 36 Nadbystrzycka Str., 20-618 Lublin, Poland

Andrzej Rysak
Faculty of Mechanical Engineering, Lublin University of Technology, 20-618 Lublin, Poland

Michael Müller
Technische Universität Chemnitz, Institut für Werkzeugmaschinen und Produktionsprozesse Reichenhainer, 09126 Chemnitz, Germany

Marek Borowiec
Faculty of Mechanical Engineering, Lublin University of Technology, 20-618 Lublin, Poland

Jarosław Zubrzycki
Faculty of Mechanical Engineering, Lublin University of Technology, 20-618 Lublin, Poland

Grzegorz Litak
Faculty of Mechanical Engineering, Lublin University of Technology, 20-618 Lublin, Poland

Anna Godlewska-Lach
Faculty of Mechanical Engineering, Lublin University of Technology, 20-618 Lublin, Poland

Volker Wittstock
Technische Universität Chemnitz, Institut für Werkzeugmaschinen und Produktionsprozesse Reichenhainer, 09126 Chemnitz, Germany

Krzysztof Ciecieląg
Department of Mechanical Engineering, Lublin University of Technology, ul. Nadbystrzycka 36, 20-618 Lublin, Poland

Kazimierz Zaleski
Department of Mechanical Engineering, Lublin University of Technology, ul. Nadbystrzycka 36, 20-618 Lublin, Poland

Wiktor Harmatys
Laboratory for Coordinate Metrology (M-10), Faculty of Mechanical Engineering, Cracow University of Technology, Jana Pawła II 37, 31-864 Kraków, Poland

Piotr Gąska
Laboratory for Coordinate Metrology (M-10), Faculty of Mechanical Engineering, Cracow University of Technology, Jana Pawła II 37, 31-864 Kraków, Poland

Adam Gąska
Laboratory for Coordinate Metrology (M-10), Faculty of Mechanical Engineering, Cracow University of Technology, Jana Pawła II 37, 31-864 Kraków, Poland

Maciej Gruza
Laboratory for Coordinate Metrology (M-10), Faculty of Mechanical Engineering, Cracow University of Technology, Jana Pawła II 37, 31-864 Kraków, Poland

Jerzy Sładek
Laboratory for Coordinate Metrology (M-10), Faculty of Mechanical Engineering, Cracow University of Technology, Jana Pawła II 37, 31-864 Kraków, Poland

Olga Gziut
Student of Mechanical Engineering Faculty, Lublin University of Technology, 36 Nadbystrzycka Str., 20-618 Lublin, Poland

Józef Kuczmaszewski
Department of Production Engineering, Mechanical Engineering Faculty, Lublin University of Technology, 36 Nadbystrzycka Str., 20-618 Lublin, Poland

Ireneusz Zagórski
Department of Production Engineering, Mechanical Engineering Faculty, Lublin University of Technology, 36 Nadbystrzycka Str., 20-618 Lublin, Poland

Paweł Fudali
The Faculty of Mechanical Engineering and Aeronautics, Rzeszow University of Technology, al. Powstańców Warszawy 8, 35-959 Rzeszów

Waldemar Witkowski
The Faculty of Mechanical Engineering and Aeronautics, Rzeszow University of Technology, al. Powstańców Warszawy 8, 35-959 Rzeszów

Dawid Wydrzyński
The Faculty of Mechanical Engineering and Aeronautics, Rzeszow University of Technology, al. Powstańców Warszawy 8, 35-959 Rzeszów

Józef Stokłosa
Faculty of Transport and Computer Science, University of Economics and Innovations, Projektowa 4, 20-209 Lublin, Poland

Marek Jaśkiewicz
Faculty of Mechatronics and Machine Building, Kielce University of Technology, 1000-lecia Państwa Polskiego 7, 25-314 Kielce, Poland

Tomasz Jachowicz
Lublin University of Technology, Faculty of Mechanical Engineering, Department of Polymer Processing; Nadbystrzycka 36, 20-618 Lublin, Poland

Ivan Gajdoš
Technical University of Kosice, Faculty of Mechanical Engineering, Department of Technologies and Materials, Masiarska 74, 04001 Kosice, Slovakia

Volodymyr Krasinskyi
Lviv Polytechnic National University, Department of Chemical Technology of Plastics, S. Bandera str. 12, 79013 Lviv, Ukraine

Filip Górski
Chair of Production Engineering and Management, Faculty of Mechanical Engineering and Management, Poznan University of Technology, 3 Piotrowo Str., 60-965 Poznań, Poland

Wiesław Kuczko
Chair of Production Engineering and Management, Faculty of Mechanical Engineering and Management, Poznan University of Technology, 3 Piotrowo Str., 60-965 Poznań, Poland

Radosław Wichniarek
Chair of Production Engineering and Management, Faculty of Mechanical Engineering and Management, Poznan University of Technology, 3 Piotrowo Str., 60-965 Poznań, Poland

Kazimierz Zaleski
Mechanical Engineering Faculty, Lublin University of Technology, 36 Nadbystrzycka Str., 20-618 Lublin, Poland

Andrzej Zyśko
Mechanical Engineering Faculty, Lublin University of Technology, 36 Nadbystrzycka Str., 20-618 Lublin, Poland

Adam Ćwikła
Institute of Technical Sciences and Aviation, The State School of Higher Education in Chełm, Pocztowa 54, 22-100 Chełm, Poland

Aneta Tor-Świątek
Department of Polymer Processing, Lublin University of Technology, Nadbystrzycka 36, 20-618 Lublin, Poland